# ANNUAL REVIEW OF NUCLEAR AND PARTICLE SCIENCE

## EDITORIAL COMMITTEE (1987)

# ANNUAL REVIEW OF NUCLEAR AND PARTICLE SCIENCE

## VOLUME 37, 1987

J. D. JACKSON, *Editor*
University of California, Berkeley


HARRY E. GOVE, *Associate Editor*
University of Rochester


ROY F. SCHWITTERS, *Associate Editor*
Harvard University

ANNUAL REVIEWS INC.   4139 EL CAMINO WAY   P.O. BOX 10139   PALO ALTO, CALIFORNIA 94303-0897

A̶R̶ ANNUAL REVIEWS INC.
Palo Alto, California, USA

*International Standard Serial Number: 0163-8998*
*International Standard Book Number: 0-8243-1537-5*
*Library of Congress Catalog Card Number: 53-995*

TYPESET BY AUP TYPESETTERS (GLASGOW) LTD., SCOTLAND
PRINTED AND BOUND IN THE UNITED STATES OF AMERICA

# PREFACE

The present volume is the thirty-seventh in this series and the tenth to appear under our editorship. It is fitting therefore to pause, take stock of the past, acknowledge our debts, and speak of the future. Our tenure began with the addition of the words "and Particle" after "Nuclear" in the name of the series. The new title was not an innovation totally of our making, but we welcomed it as an appropriate recognition of what our field had become and a title that more clearly represented our vision of what the series would attempt to review. In our first preface we voiced our expectation that there would be some narrowing of the subject matter, with mainline nuclear physics and particle physics providing the lion's share of the reviews. A survey of the 140 reviews in the 10 volumes bears this out. Very close to two thirds of the chapters are equally divided between nuclear physics and particle physics (with a small amount of subjectivity in the allocation of topics in relativistic heavy ions and the quark-gluon plasma to one or the other of the two fields). Roughly 10% of the chapters are on astrophysics and space science, with slightly more on instrumentation or the physics of accelerators. The remaining 10–12% of the reviews are divided among applications in other sciences, as well as applied nuclear science itself. Whether the regular reader has noticed it or not, there has been a slight narrowing of subject matter.

In that first preface there was another intention expressed—to have every review present a winning face through a generally accessible introduction and summary. With some success and some failure, we state our determination at the 10-year mark to continue with this goal. Having thus expressed our bona fides on striving for excellence, it is appropriate to acknowledge our debt to those who really count in making the *Annual Review of Nuclear and Particle Science* as good as it has been for the past 10 years. The largest and most important group is the 241 authors of the 140 reviews. We are indeed fortunate to have expert and responsible colleagues who share our view of the value of serious reviews. Smaller groups of great importance are the successive Editorial Committees who plan the volumes, suggesting topics and prospective authors as well as advising on a diversity of issues. Their dedication and endurance at our marathon annual editorial meeting, at which $2\frac{1}{2}$ volumes are being juggled, are wondrous to see. The importance and responsibility of the members of the Editorial Committee are evidenced by the appearance of their names

(*continued*)   v

opposite the title page in each volume. The Editors are also in the debt of various staff members at Annual Reviews, Inc., but most of all, Ms. Margot E. Platt, Production Editor. In our ten years, we have known, and wish to know, no other. In a somewhat archaic and mixed metaphor, Margot is both main spring and balance wheel of our stately time piece; Editors, authors, and Committee members are the cogwheels. Without her ticking away cheerfully and most effectively, none of us could go smoothly on our appointed rounds.

This tenth volume is by chance weighted on the nuclear physics side. Some of it is not by chance, but rather from the broadening of nuclear physics to include pion and kaon interactions with nuclei and also relativistic collisions between nuclei. The nuclear physics purist might object, just as the particle physics purist would disown the topics, but they surely are part of nuclear and particle science. The blending of the fields (or a reassertion of the basic unity) is evidenced by chapters on pion production in heavy-ion collisions and pion-nucleus interactions by Braun-Munzinger & Stachel and Gibbs & Gibson, respectively, and also by the review of nuclear effects in deep inelastic lepton scattering by Berger & Coester. Experimental reviews in nuclear physics cover magnetic dipole excitations (Berg & Kneissl), recoil mass spectrometers (Cormier), electron scattering and nuclear structure (Frois & Papanicolas), and nuclear fragmentation in both proton- and heavy-ion-induced reactions (Lynch). Theoretical topics in nuclear physics are computer simulations of dynamics (Boal) and a relativistic model for nucleon-nucleus scattering (Wallace). The interesting topic of signatures of the putative quark-gluon plasma produced in high-energy nuclear collisions is addressed by Kajante & McLerran, while theorems of perturbative quantum chromodynamics are summarized by Collins & Soper. Theory and experiment on heavy-quark systems (mostly $c\bar{c}$ and $b\bar{b}$) are reviewed extensively by Kwong, Rosner & Quigg.

There are no astrophysics reviews this time, but very high energy interactions produced by cosmic rays are reviewed by Jones, Takahashi, Wosiek & Miyamura (the JACEE collaboration). The subjects of instrumentation and accelerator physics are represented by a review of new computer technologies for particle physics (Gaines & Nash) and a discussion of advanced accelerator concepts, particularly for electron-positron colliders (Siemann). Interesting applications of nuclear physics to subsurface geology are surveyed by Ellis, Schweitzer & Ullo.

The results of a recent informal survey on the need for reviews in general and the usefulness of the *Annual Review of Nuclear and Particle Science* in particular provide opinion on past performance and food for thought about future directions. Most respondents to our questionnaire believed the need for good reviews was steadily increasing, although the definition of a review varied. Some stressed conference and summer school proceedings as their primary sources beyond the telephone and preprints. The *Annual Review of Nuclear and Particle Science* received high marks for providing the reader with reliable introductions to new fields. The unification of nuclear and particle physics under one roof was praised by many, but not all. The more focused and specialized researchers found the broad coverage detrimental, preferring reviews (in their field) of greater length, detail, and immediacy. Mild criticism of the composition of the Editorial Committee, of the balance of coverage, and of the lack of focus of some of the theoretical reviews seems more to reflect the personal biases of the respondents than to be an indicator of serious deficiencies in the series. Not surprisingly, those who work in interface areas or related fields would like to see greater coverage of such topics.

The Editors appreciate the thoughtfulness and frankness of those who responded to the questionnaire. From this admittedly small sample of old friends and "strangers" has come evidence that the *Annual Review of Nuclear and Particle Science* is a valuable if not unique enterprise and that its Editors and Editorial Committee are generally pointed in the right direction. A review cannot be all things to all people. The *Annual Review of Nuclear and Particle Science* appears to have a niche that pleases many, but complacency is unwarranted. We can do better. More awareness of burgeoning new areas, especially between established fields, perhaps greater flexibility on length limitations, and greater breadth of interests on the Editorial Committee are among issues to be addressed in the future. As always, a review series depends for its success on its authors. In the flush of solicitations of future articles, the Editors often wish for more willingness on the part of those approached. In the flurry of manuscript deadlines, they long for fewer defaults, but in the glow of appearance of successive volumes, they marvel at the overall quality and diversity of the reviews and the dedication and conscientiousness of the authors. Those readers who have contributed to this or other review series must know that their efforts are appreciated by their colleagues, present and future. We hope those people to be invited in the coming years will share their willingness and sense of community, and agree to write, then actually write (and deliver on time) the concise, scholarly, and impeccable reviews of which we know they are capable. With such help, we can look to the future with confidence.

H. E. GOVE
J. D. JACKSON
R. F. SCHWITTERS

# SOME RELATED ARTICLES IN OTHER *ANNUAL REVIEWS*

From the *Annual Review of Astromony and Astrophysics*, Volume 25 (1987):

*Clustering of Astronomers*, W. H. McCrea

*Existence and Nature of Dark Matter in the Universe*, Virginia Trimble

From the *Annual Review of Computer Science*, Volume 2 (1987):

*Techniques and Architecture for Fault-Tolerant Computing*, Roy A. Maxion, Daniel P. Siewiorek, and Steven A. Elkind

*Linear Programming (1986)*, Nimrod Megiddo

From the *Annual Review of Fluid Mechanics*, Volume 19 (1987):

*Ludwig Prandtl and His Kaiser-Wilhelm-Institut*, K. Oswatitsch and K. Wieghardt

*Tsunamis*, S. S. Voit

From the *Annual Review of Physical Chemistry*, Volume 38 (1987):

*Theoretical Studies of Silicon Chemistry*, Kim K. Baldridge, Jerry A. Boatz, Shiro Koseki, and Mark S. Gordon

*Physical and Chemical Properties of Alkalides and Electrides*, James L. Dye and Marc G. DeBacker

Annual Review of Nuclear and Particle Science
Volume 37, 1987

# CONTENTS

*Ann. Rev. Nucl. Part. Sci. 1987. 37: 1–31*

# COMPUTER SIMULATIONS OF NUCLEAR DYNAMICS

## David H. Boal

Department of Physics, Simon Fraser University, Burnaby, British Columbia V5A 1S6, Canada

CONTENTS

## 1. INTRODUCTION

The problem of describing a nuclear reaction is obviously a very complicated one that does not lend itself easily to analytical treatment. There are not a very large number of particles in the system, yet the system is not always so dilute that the interactions are strictly two-body in nature. The energy and distance scales are such that quantum mechanical effects are certain to be critically important to some stages of the reaction. While statistics may dominate certain experimental observables, on average the systems begin the reaction with a phase space configuration very different

1

0163–8998/87/1201–0001$02.00

from that of the reaction products, and only part of the system may reach any form of equilibrium.

About a quarter of a century ago (107) it was recognized that computers could be used to simulate some aspects of nuclear reactions. Even though the machines were of limited speed and memory judged by today's standards, the experimental observables that were the subject of the simulations were also relatively simple. As the experimental measurements have become both more detailed and more accurate, so too has the need for more sophisticated simulations.

As an illustration of the physics ingredients necessary for a successful simulation of the time evolution of a nuclear reaction, we show in Figure 1 a simulation (14) of a Ca + Ca collision at a bombarding energy of 200 MeV/nucleon in the lab frame. The figure shows the positions of the nucleons in a central (zero-impact parameter) collision shortly after the period of maximum overlap of the projectile and target. One can see that there are significant density fluctuations present, rendering a simple description of the process in terms of homogeneous nuclear matter inappropriate. Similarly, the nuclear material has not totally vaporized to produce a dilute gas, and hence quantum effects such as the Pauli exclusion principle remain important. Several regions of high density will emerge at the end of the reaction as fragments.

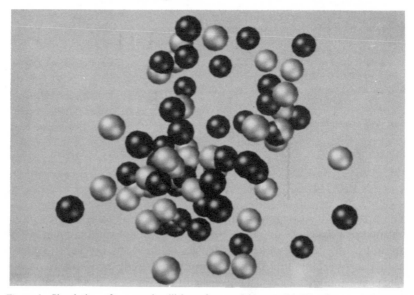

*Figure 1* Simulation of a central collision of two calcium nuclei. The time evolution was stopped shortly after the time of maximum overlap (14).

Figure 1 provides an example of how complicated the reaction process can be; nevertheless, there *are* processes that lend themselves to treatment analytically. Simple knockout or transfer reactions are unlikely to need the computational effort required to describe a nearly central heavy ion collision. Further, some characteristics of the reaction products may find their description in statistical mechanics because of ensemble averaging or other causes.

The purpose of a computer simulation is not simply to describe reactions; it allows us to test the sensitivity of the reaction to the nuclear physics incorporated in the simulation. Thermal model analyses (27, 54, 82a) of intermediate energy nuclear reactions indicate that the reaction trajectories traverse regions characterized by temperatures ranging from a few to many tens of MeV. In this temperature range, infinite nuclear matter shows several interesting properties. The pressure vs density diagram predicted (26) for a gas of neutral nucleons interacting via a zero-range Skyrme potential is shown in Figure 2. The isotherms are labeled by their respective temperatures (measured in MeV). For densities much above normal nuclear matter density, the pressure rises fairly rapidly. This region may be accessible in the early stages of a high energy, heavy ion reaction (125).

The form of the isotherms is reminiscent of that of a van der Waals fluid: the usual Maxwell construction yields a phase diagram with distinct liquid and vapor phases and with a critical temperature of 15.3 MeV for the parametrization used in the figure. Of particular interest is the mechanical instability region (21) $dP/d\rho < 0$, in which density fluctuations grow with time. As mentioned in the discussion of Figure 1, similar density fluctuations may be important for understanding the emission of complex fragments. One of the challenges for simulations is to make the connection between experimental measurements and the physics, such as the equation of state, that underlies the reaction mechanism.

Because the field of computer simulations is advancing so rapidly, several of today's burning issues will be resolved and replaced by new ones between the writing of this review and its publication. While reference is made to current controversial questions, they do not form the bulk of the material in this review. Rather, we examine several different approaches to numerical simulations and highlight the similarities and differences between them. Results from several techniques are illustrated and contrasted. In choosing examples for the application of the techniques, emphasis is placed on examining general issues in reaction dynamics and the properties of nuclear matter, rather than dealing with specific experimental comparisons.

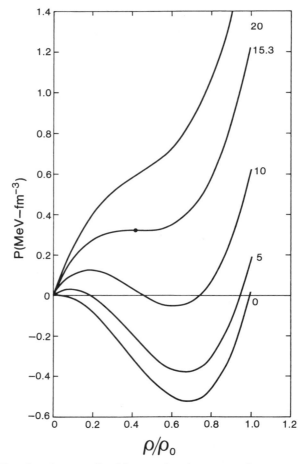

*Figure 2*  Equation of state predicted for neutral nuclear matter using a zero-range Skyrme-type interaction. The temperatures of the isotherms are given in MeV (26).

## 2.  HIGHLY EXCITED NUCLEAR MATTER

### 2.1  *The Nuclear Equation of State*

The nucleon-nucleon interaction is repulsive at short distances and attractive at large distances. This behavior is qualitatively similar to the molecular interaction of substances that exist in liquid and gaseous phases. Based on such considerations, nuclear matter also may be expected to exist in liquid and gaseous phases (52, 53, 100, 103, 118, 120).

The properties of nuclear matter at high densities have been investigated for some time because of their importance in astrophysics, for example, in the study of the formation and characteristics of neutron stars (see 100 and references therein). The problem of accessing the equation of state experimentally has only been addressed in more recent years. Research has focussed on two main areas: the determination of the nuclear equation of state at high densities and the search for evidence of a liquid-vapor phase transition.

To extract the essential features of the equation of state, we will adopt a simple model (26, 67, 104) in which the nucleon-nucleon potential is approximated by a zero-range Skyrme-type interaction:

$$V_{12} = \left[ -t_0 + \frac{t_3}{6}\rho \right]\delta(\mathbf{r}_1 - \mathbf{r}_2), \qquad\qquad 1.$$

where $\rho$ denotes the density at $(\mathbf{r}_1 + \mathbf{r}_2)/2$ for nucleons at positions $\mathbf{r}_1$ and $\mathbf{r}_2$. The parameters $t_0$ and $t_3$ in Equation 1 are chosen to fit the properties of cold nuclear matter. The potential energy associated with this interaction has the form

$$U(\rho) = -\frac{3}{4}t_0\rho + \frac{3}{16}t_3\rho^2. \qquad\qquad 2.$$

This is commonly referred to as a "stiff" equation of state. There are many other popular parametrizations of the equation of state that also use a power law form (for a more detailed discussion, see 45, 109, 110, 125, 126):

$$U(\rho) = -a\rho + b\rho^\sigma. \qquad\qquad 3.$$

The choice of $\sigma = 7/6$ is referred to as a soft equation of state. The pressure corresponding to Equation 2 can be expressed as the sum of two terms: a kinetic term,

$$P_{kin} = -\frac{T}{V}\ln z, \qquad\qquad 4.$$

where $z$ is the grand partition function for independent fermions, and an interaction term,

$$P_{int} = -\frac{3}{8}t_0\rho + \frac{1}{8}t_3\rho^2. \qquad\qquad 5.$$

A particular choice of $t_0$ and $t_3$ yields (26) the pressure vs density curves shown in Figure 2. The phase diagram associated with Figure 2 is shown in Figure 3. Maxwell's construction is used to find the liquid-gas coexistence curve (LGC); the isothermal (ITS) and isentropic (IES) spinodal curves are defined by the conditions $(\partial P/\partial \rho)_{T=0}$ and $(\partial P/\partial \rho)_{S=0}$, respectively. Also shown in the figure are dotted curves of constant entropy $(S/A)$. A more detailed discussion of the properties of the phase diagram can be found in (67).

The boundaries of the phase diagram are not particularly sensitive to the specific choice of parametrization for the nucleon-nucleon interaction; rather similar results were obtained by two groups (26, 104) using different parametrizations. However, the inclusion of the Coulomb interaction and the finite size of the nucleus lower the predicted value of the critical temperature considerably (32, 83, 131).

The trajectory that a reaction will take through the phase diagram varies according to projectile and energy. As discussed below, proton-induced reactions are unlikely to compress the nucleus very much, and the reaction trajectory of the target region is likely to involve excursions through the coexistence region and perhaps into the mechanical instability region (26).

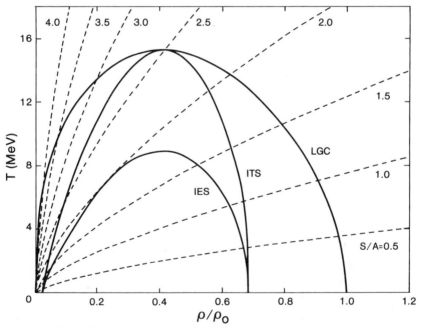

*Figure 3*  Phase diagram of neutral nuclear matter. Isentropes of given values of $S/A$ are shown by the dashed curves. The liquid-gas coexistence curve is labeled as LGC; the isothermal and isentropic spinodals are labeled as ITS and IES, respectively (26).

Central collisions of heavy ions, on the other hand, produce considerable compression and heating. Such systems, once thermalized, should expand approximately isentropically (19), perhaps passing near the critical point (81).

## 2.2    Hydrodynamics

The reaction trajectory can be determined by solving the equations governing the time evolution of quantities such as the energy density, number density, etc. An example (133) of an early attempt to do this analytically is the use of the diffusion equation to follow the dissipation of heat deposited in nuclear matter. However, the complex geometry of a reaction generally requires that the equations be solved numerically.

If the mean free path of a nucleon in nuclear matter is sufficiently short, then a local equilibrium may be established within the reaction zone and the region may evolve according to hydrodynamics. Estimates of the mean free path (86) suggest that it is of the order of the internucleon spacing. Although this does not put one well into the hydrodynamic regime, neither is it so far away from that regime that hydrodynamics is irrelevant. The attraction of hydrodynamics is, of course, the simplicity of its ingredients: conservation laws and an equation of state. The sensitivity of experimental observables to the equation of state can then be found in a straightforward manner.

In relativistic form, the ideal single-fluid hydrodynamic equations read (45, 102, 110, 126):

$$\frac{\partial N}{\partial t} + \mathbf{V} \cdot (\mathbf{v}N) = 0, \qquad\qquad 6.$$

$$\frac{\partial \mathbf{M}}{\partial t} + \mathbf{V} \cdot (\mathbf{v}\mathbf{M}) = -\mathbf{V}p, \qquad\qquad 7.$$

$$\frac{\partial E}{\partial t} + \mathbf{V} \cdot (\mathbf{v}E) = -\mathbf{V} \cdot (\mathbf{v}p), \qquad\qquad 8.$$

where $N$, $\mathbf{M}$, and $E$ are the nucleon number density, momentum density, and energy (including rest mass energy) density, respectively, in the laboratory (or computational) frame of reference. The velocity $\mathbf{v}$ of the fluid frame can be defined in more than one way. The common choice in nuclear problems is to demand that the baryon number flux vanish in the fluid frame (see 56, 102 for further discussion). The number density, energy density, and pressure ($n$, $e$, $p$ respectively) in the rest frame are then related to their computational frame counterparts by:

$$N = \gamma n \qquad\qquad 9.$$

$$\mathbf{M} = \gamma^2(\varepsilon+p)\mathbf{v} \qquad\qquad\qquad 10.$$

$$E = \gamma^2(\varepsilon+p)-p, \qquad\qquad\qquad 11.$$

where $\gamma$ is the usual Lorentz factor. The numerical solution of these equations does not require a particularly large amount of CPU (central processing unit) time, but does involve a large amount of computer memory unless some simplifying geometry such as cylindrical symmetry is available. The numerical methods involve the use of a spatial mesh of fluid cells; the quantities of interest are averaged across the cell and hence the cells should be sufficiently small to insure numerical accuracy. Both space-fixed (Eulerian) and moving-boundary (Lagrangian) cells have been used. Several finite difference methods of performing the numerical integrations are in use; their accuracy and limitations are compared in (45, 114). Since the system must go out of equilibrium at some point, the calculations are truncated once a quantity such as the number density has fallen below a specified value.

Several numerical solutions of the hydrodynamic equations for heavy ion collisions have been performed (for a sample, see 7–9, 38, 111, 121, 126, 127; for other properties of nuclear fluid dynamics, see 136). Early simulations of heavy ion collisions used only a single fluid for the nuclei: the projectile and target nuclear matter became indistinguishable on impact. A way of allowing the interpenetration of the nuclei expected from a finite mean free path is to use separate fluids for each nucleus and introduce a drag term to couple them. A comparison (45) of the time evolution of the spatial density predicted by the one- and two-fluid models is shown in Figure 4. Clearly, there is less momentum transfer in the two-fluid calculations since the nuclei are partially transparent to each other.

The single-particle momentum distributions predicted by the simulations make a good test of their normalization, although such inclusive quantities are not likely to provide a detailed probe of the validity of the theoretical description. An example (45) of the comparison of the predictions with data is shown in Figure 5. The predictions of both one- and two-fluid hydrodynamics are shown for the reaction Ne+U at 250 MeV/nucleon bombarding energy. Certainly the trend of the energy spectra to steeper slopes at wider emission angles is qualitatively supported by the calculations.

## 2.3   Collective Flow

It is clear from Figure 4 that sideways collective motion or flow of the nuclear material has developed by the end of the time sequence studied. This is intuitively expected from hydrodynamics and was one of its first novel predictions. Such collective motion has been observed in heavy ion

$$^{238}U + {}^{238}U$$

$$E_{LAB} = 2.1 \text{ GeV/nucleon}$$

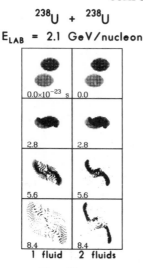

1 fluid    2 fluids

*Figure 4*  Matter distributions for a $^{238}U + {}^{238}U$ collision at a bombarding energy of 2.1 GeV/nucleon calculated with the one-fluid model (*left*) and the two-fluid model (*right*) as observed from the center-of-mass frame (45).

collisions (see 57 references therein). The magnitude of the effect is shown in Figure 6. The middle left frame shows the sideways peaking in the experimental data. Shown for comparison are the fluid dynamical predictions as well as those of two cascade models (described in the following section) in which the equation of state is neglected and only random nucleon-nucleon collisions affect the otherwise straightline motion of an individual nucleon. While the more sophisticated one-body theories, such as the Vlasov-Uehling-Uhlenbeck approach treated in the following section, also predict sideways peaked cross sections (108; see, however, 87), the measurements do indicate compressional effects in nuclear reactions. We return to another possible measure of the equation of state, meson production, in Section 3.4.

## 3.  SIMULATIONS OF SINGLE-PARTICLE DISTRIBUTIONS

Hydrodynamics assumes that individual N-N collisions are frequent enough to maintain a local equilibrium during the course of a reaction. Although there are conditions under which this scenario may be applicable, many reactions (for example proton-induced reaction or intermediate energy, heavy ion collisions) are not subject to such simplifications. We begin our discussion of the more sophisticated methods required for these

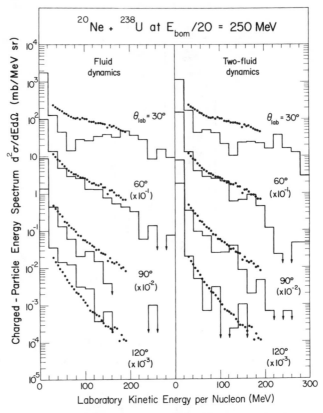

*Figure 5* Charged-particle energy spectra for $^{20}$Ne + $^{238}$U at a bombarding energy of 250 MeV/nucleon calculated using one-fluid (*left*) and two-fluid (*right*) hydrodynamics (45). (Data are from 68.)

reactions by examining the properties of single-particle distributions, to which many simulation techniques have been applied.

## 3.1   *The Intranuclear Cascade Model*

The earliest simulations of nuclear collisions used a fairly simple model (1, 17, 107): the nucleons were propagated in space by means of classical mechanics and were allowed to scatter from one another at their distance of closest approach so long as that distance was less than the classical scattering radius determined from measured cross sections. This model goes under the name of the intranuclear cascade model (we shorten the title here to "cascade model" because there have been many variants of it in the last two decades: for example 35, 36, 42, 47–51, 73, 87, 92, 124, 138).

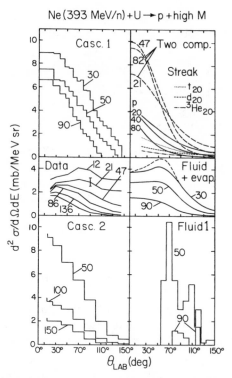

Ne (393 MeV/n) + U → p + high M

*Figure 6* Comparison of data (*middle left*) with calculations for high-multiplicity-selected inclusive proton differential cross section from Ne + U at 393 MeV/nucleon. Strong forward peaking is observed in both the cascade calculations (*top* and *bottom left*) and the thermal model (*top right*). Fluid dynamical models show sideways peaking (126).

The model is computationally both fast and simple, an advantage when comparisons with experiment are required.

In the cascade model, the positions and momenta are initialized in a Monte Carlo sense, i.e. they are chosen randomly according to a prescribed distribution. For example, the momentum distribution is often chosen to be that of a zero-temperature Fermi gas, with the Fermi momentum determined by the local density. Ignored are effects arising from Pauli blocking of the N-N collisions, nuclear binding, etc, so the model cannot realistically describe fragment formation. So long as the outcome of the collision event is not sensitive to the nature of the correlations in the initialization, which are random in this model, then the model can be applied to aspects of reactions for which a classical $A$-body description is appropriate (see Section 4.4).

Even if one does not trust the correlations, the model is still useful. The single-particle phase space density $f(r, p; t)$ of the classical nucleons of the cascade model should obey the Boltzmann equation with collisions but no mean field (10):

$$\frac{\partial}{\partial t} f + \mathbf{v} \cdot \frac{\partial f}{\partial \mathbf{r}} = - \int \frac{d^3 p_2 d^3 p_1' d^3 p_2'}{(2\pi)^6}$$

$$\times v_{12} \sigma(\mathbf{p} + \mathbf{p}_2 \rightarrow \mathbf{p}_1' + \mathbf{p}_2') (ff_2 - f_1'f_2') \delta^3(\mathbf{p} + \mathbf{p}_2 - \mathbf{p}_1' - \mathbf{p}_2'), \quad 12.$$

where $v_{12}$ is the relative velocity of the pair of particles colliding with cross section $\sigma(\mathbf{p} + \mathbf{p}_2 \rightarrow \mathbf{p}_1' + \mathbf{p}_2')$. Hence the cascade model could be regarded as a means of solving for the time evolution of $f(r, p; t)$ by taking a Monte Carlo sample and propagating it according to the microscopic physics underlying the Boltzmann equation.

Although there have been many applications of the cascade model to reactions, in particular comparisons with proton and neutron energy spectra, here we wish only to mention its use in elucidating reaction mechanisms (see 48 for a more extensive compilation of applications). The cascade model has been used to address a number of questions about the internal dynamics of a reaction, such as the approach to thermal and chemical equilibrium (49, 89). What we wish to discuss here is the time evolution of the entropy, defined in terms of the single-particle density by means of

$$S = - \frac{g}{(2\pi)^3} \int d^3 r \ d^3 p [f \ln f - (1-f) \ln (1-f)], \qquad 13.$$

where $g$ is the spin degeneracy factor. While the numerical integration over phase space can be very difficult, the two systems discussed here allow a substantial simplification.

The first example comes from heavy ion physics. Using a cascade model, Bertsch & Cugnon (19) followed the entropy produced in the central collision of two calcium nuclei at a bombarding energy of 800 MeV/ nucleon in the lab. The time dependence of the calculated entropy is shown in Figure 7 (19). The entropy changes fairly rapidly during the equilibration stages of the reaction, reaching its asymptotic value not long after 10 fm/$c$ elapsed time. The density reaches a maximum at 8 fm/$c$. Shown in the figure as well is the number of particles that have undergone a collision. Again, this number rises rapidly during the thermalization stage. Although the actual predicted value of the entropy may be changed by the inclusion of a nuclear binding interaction, nevertheless the simulation shows that the system expands isentropically and will pass near the critical point of the liquid-gas phase diagram. The fact that the entropy is fixed relatively early in the reaction has led to attempts at developing a

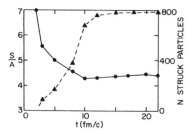

*Figure 7*   Time dependence of the entropy and number of nucleons that have undergone scatterings calculated for central $^{40}Ca + ^{40}Ca$ collisions at $E = 800$ MeV/nucleon in a cascade model (19).

reliable method of determining the entropy from the measured products of the reactions (46).

The second example is the entropy generated in a nuclear target by the passage of a proton through it. In this calculation (26) the nuclear interaction is approximated by placing the target nucleons in a spherically symmetric step function potential. Since few nucleons are scattered out of this well during the course of the reaction, and since the momentum transferred from the projectile to the target nucleus is small, the well is spatially fixed. The reaction trajectory in excitation energy and entropy is shown in Figure 8 for those nucleons inside the target region. The system remains in a state of relatively low entropy and excitation energy throughout the reaction. Whether the system breaks up depends in part upon whether it has enough energy to carry it into the mechanical instability region (21, 34), and one can see from the figure that this occurs for the two bombarding energies shown. The simulation shows that fragmentation in intermediate energy, proton-induced reactions resembles more the breakup of cool matter near the mechanical instability region (4, 33) than it does the condensation of clusters from a hot nuclear vapor (67, 74, 81; see also 60).

## 3.2   The Vlasov Equation and TDHF

At low excitation energies, so many of the individual N-N collisions in a reaction are Pauli-blocked that the nuclear mean field plays an increasingly important role. The time-dependent Hartree-Fock (TDHF) equation represents the limit in which two-body collisions are neglected and a particle interactions only with the nuclear mean field. TDHF calculations have been successfully applied at low excitation energies to describe such phenomena as fusion and compound nucleus formation (see 55, 96, 109 for examples). The classical analogue of TDHF is the Vlasov equation, in which the time evolution of $f(r, p; t)$ is given by (10, 132)

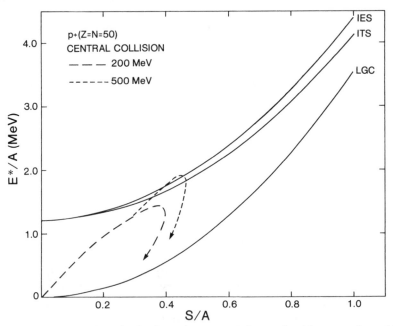

*Figure 8* Reaction trajectories for the nuclear target region predicted by a cascade model for 200 and 500 MeV p+$(Z = N = 50)$ central collisions (*dashed lines*). Shown for comparison are the isothermal (ITS) and isentropic (IES) spinodals and the liquid-gas coexistence curve (LGC) (26).

$$\frac{\partial}{\partial t}f + \mathbf{v}\cdot\frac{\partial}{\partial \mathbf{r}}f - \nabla U\cdot\frac{\partial}{\partial \mathbf{p}}f = 0. \qquad\qquad 14.$$

The predictions of the Vlasov equation should be similar to those of TDHF, and this is demonstrated in Figure 9 (3). The figure shows the density projected on the reaction plane for five different impact parameters in the $^{12}C + {}^{16}O$ reaction at 25 MeV/nucleon bombarding energy in the lab. The density is sampled at more than 100 fm/$c$ after the nuclei touch. While the distributions are not identical, one can see that they are fairly similar. The Vlasov equation has also been applied with qualitative success to low energy reactions; whether it is appropriate for reactions at higher energies is addressed in the next subsection.

### 3.3 A Boltzmann Equation for Fermions

For many problems of interest in nuclear reaction studies, the simplifications made in the cascade model or the Vlasov equation are not justifiable. The Boltzmann equation for $f(r, p; t)$ includes both a mean field

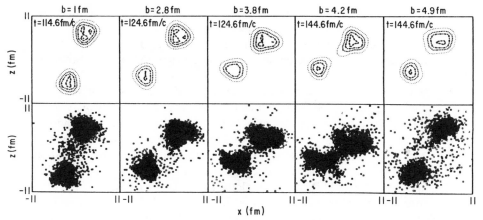

*Figure 9* Density projection on the reaction plane calculated for $^{12}C + ^{16}O$ collisions at $E = 25$ MeV/nucleon by means of the time-dependent Hartree-Fock (*upper part*) and Vlasov (*lower part*) equations (3).

interaction and a collision term but is not appropriate for a low temperature gas of fermions because effects such as Pauli blocking are excluded. An equation resembling the classical Boltzmann equation has been developed for fermion distributions [Nordheim (112), Landau (101), Uehling & Uhlenbeck (129); see 10, 18, 85, 116 for more detailed discussion]. Referred to as the Vlasov-Uehling-Uhlenbeck (VUU) or Landau-Vlasov equation in its application to intermediate energy, heavy ion reactions (2, 3, 5, 20, 70, 98, 99, 108, 128a; see also 119), the equation reads

$$\frac{\partial}{\partial t}f + \mathbf{v} \cdot \frac{\partial}{\partial \mathbf{r}}f - \nabla U \cdot \frac{\partial}{\partial \mathbf{p}}f = -\int \frac{d^3 p_2 \, d^3 p_1' \, d^3 p_2'}{(2\pi)^6}$$
$$\times \, v_{12}\sigma(\mathbf{p} + \mathbf{p}_2 \rightarrow \mathbf{p}_1' + \mathbf{p}_2')\,[ff_2(1 - f_1')(1 - f_2') - f_1'f_2'(1 - f)(1 - f_2)]$$
$$\times \, \delta^3(p_1 + p_2 - p_1' - p_2'). \qquad 15.$$

Here the $f$'s are the single-particle Wigner functions from which the single-particle density can be obtained via

$$\rho(\mathbf{r}) = (2\pi)^{-3} \int d^3 p \, f(\mathbf{r}, \mathbf{p}). \qquad 16.$$

The effects of the Pauli principle are partly included in the collision term: the $(1 - f)$ terms inhibit scattering to regions of high density in phase space. To the extent that the collisions are fast, this term may be all that is required to avoid saturating phase space. However, there may be circumstances where the phase space density builds up faster than the collision term can deplete it, and one must find a better way to handle the consequences of Fermi-Dirac statistics. We return to this point in Sections 3.4 and 4.1.

Solutions of this equation are usually found using the Monte Carlo sampling technique. Similar problems have been solved by evaluating $f$ and its derivatives on a spatial grid, but in general the computer memory requirements necessary to store the distribution are immense, particularly in regions of phase space where $f$ is varying rapidly. However, progress can be made using a mesh in circumstances where the geometry can be simplified (93, 94, 115), for example in the collision of slabs of nuclear matter. In the Monte Carlo approach, $f$ is sampled by means of test particles, $N$ particles per nucleon, by running $N$ events with one particle per nucleon simultaneously. The value of $f$, when it is needed to evaluate the Pauli blocking term for example, is calculated by summing over all $N$ events at the same time step. The test particles are propagated by classical mechanics, with the force generated by the derivative of the nuclear mean field. Collisions are allowed to occur between test particles within each event. The simulation technique resembles closely the technique used in the cascade model, except that nuclear binding and some aspects of the Pauli principle now have been incorporated.

The extension of the Vlasov equation to include collisions changes its predictions for intermediate-energy heavy-ion reactions substantially since nuclei are largely transparent to each other in the Vlasov equation. A comparison (98) between the Vlasov and VUU predictions is shown in Figure 10. The momentum space distributions are shown after an elapsed time of 60 fm/$c$ for a central Ar + Ca collision at a bombarding energy of 137 MeV/nucleon. In the Vlasov approach, the momentum distribution is still fairly close to that of the initial projectile and target. In contrast, the VUU approach shows far more thermalization, as is observed experimentally.

For a test of the absolute normalization of the predictions, Figure 11 shows a VUU calculation (3) of the proton spectra observed in the $^{12}$C + $^{16}$O reaction at 25 MeV/nucleon. The VUU predictions have been averaged over impact parameter; the agreement is impressive. Such agreement is not always expected from the current level of sophistication of the simulations, since many codes simplify the initialization of $f$ (e.g. uniform spherically symmetric density distributions in coordinate space) or its propagation (e.g. energy-independent N-N cross sections). Perhaps statistics plays such an important role (54, 59, 88) in the outcome of the reactions that these simplifications are irrelevant, or perhaps the effects of such simplifications

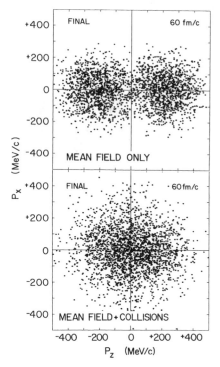

*Figure 10*   Final momentum-space distribution predicted for a central Ar + Ca collision at $E = 137$ MeV/nucleon by solutions of the Vlasov (*upper part*) and Vlasov-Uehling-Uhlenbeck (*lower part*) equations (98).

cancel out. In any event, the predicted normalizations for the single-particle distributions are generally within a factor of two of experiment.

## 3.4   *Meson Production*

So far in this review we have concentrated on the time evolution of the baryons in the nucleus. Of course, intermediate and high energy nuclear reactions can result in emission of other particles as well, such as photons, pions, and kaons. Several analytical models (see 37 and 106 for further references) have been developed to describe particle production; what we wish to mention here is the use of simulations in understanding the production mechanism and the role therein of the equation of state.

The cascade model has been used to predict photon, pion, and kaon production (for examples, see 91, 47, and 11 respectively). In this approach, meson production is envisaged to occur through the production of unstable

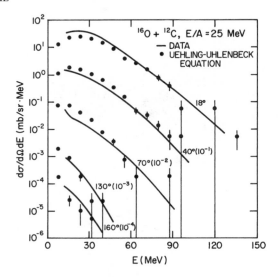

*Figure 11*    Proton spectra for the reaction $^{12}C+^{16}O$ at $E = 25$ MeV/nucleon. The solid line represents the data (44) while the points are the predictions of the VUU equation (3).

resonances, such as $NN \rightarrow \Delta(1230)N$, which subsequently decay. These calculations are in agreement with some of the meson production data, but not all: pion yields are typically overpredicted by a factor of two. Models that assume the establishment of thermal and chemical equilibrium also overpredict the pion yields; an example (69) is shown in Figure 12 (78). The negative pion yield per participant nucleon is shown in the figure as a function of bombarding energy for two systems: La + La (78) and Ar + KCl (77). [The term "participant nucleon" is an experimentally defined quantity that attempts to distinguish between the participant nucleons in the geometrically overlapping region of the projectile and target and the "spectator nucleons" whose momenta are not radically changed in the reaction (see 78).]

It has been suggested that the suppression of the pion yield results from nucleons losing some of their kinetic energy because of compressional effects during the early high density stages of the reaction (125). Taken at face value, Figure 12 argues that a considerable amont of energy must be removed from the system in the form of compressional potential energy before the thermal model prediction would come close to experiment. Calculations based on the VUU equation have been performed to measure the importance of the equation of state. Not all codes agree with each other on the magnitude of the effect (20, 126), but those that find a sensitivity to the equation of state tend to favor the stiff form. Currently,

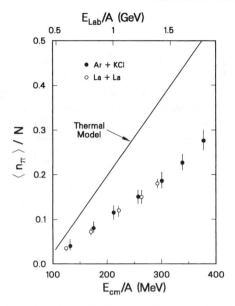

*Figure 12*   Ratio of the mean pion multiplicity to the number of participant nucleons as a function of incident center-of-mass energy (*bottom scale*) and laboratory energy (*top scale*). The data points are for La + La collisions (78) and Ar + KCl collisions (77). Also displayed is a thermal model prediction (based on 69) that does not include a compression term.

there remain difficulties in dealing with pion absorption and anti-symmetrization effects other than those included through Pauli blocking in the collision term.

## 4.   FLUCTUATIONS AND CORRELATIONS

The microscopic description of correlation functions and nuclear fragmentation process requires one to move beyond the one-body kinetic equations of the previous section. On the quantum mechanical level, this task is exceedingly difficult. Progress has been made by introducing fluctuations into TDHF calculations (90, 128), within which multiplicity distributions and mass spectra have been calculated for isolated nuclei evolving from given initial temperatures and densities. The calculations confirm the qualitative expectation of increasing particle multiplicities and the disappearance of large compound nuclear residues at higher temperatures. At the time of writing, these calculations have been limited to two spatial dimensions.

## 4.1    *Classical Equation of Motion Approach*

The development of full three-dimensional calculations began at the classical level, in which nucleon-nucleon interactions can be incorporated by means of a two-body potential. However, many of the applications to which this approach has been applied have been to single-particle distributions, particularly proton energy spectra (29–31). The difficulty in extending the calculations to fragmentation processes lies in the problem of incorporating the Pauli principle. A resolution to the problem was proposed by Wilets and coworkers (40, 134, 135; see also 56a, 62), who introduced a momentum-dependent term in the nucleon-nucleon potential between identical particles

$$V_{\text{pauli}} \propto r_{ij}^{-2} \exp\left\{ -\alpha[(\beta \mathbf{p}_{ij} \cdot \mathbf{r}_{ij})^4 - 1] \right\}, \qquad\qquad 17.$$

where $\mathbf{r}_{ij}$ and $\mathbf{p}_{ij}$ are the relative positions and momenta of the two-particle pair. The parameters of the potential are adjusted to leave intact the asymptotic properties of N-N scattering. The potential is clearly repulsive in regions of high phase space density. Not only does this idea allow one to calculate $A$-body quantities, it also begins to address the problem of how to incorporate the Pauli principle in a way other than through N-N scattering. Even if the usual N-N scattering terms were turned off, regions of high phase space density would be avoided in this model. Unfortunately, the model has yet to be applied to fragmentation.

## 4.2    *Including the Pauli Principle at the Semiclassical Level*

An alternative to the use of classical equations of motion with two-body potentials is to extend the mean field methods given in Section 3. The first such calculation for nucleus-nucleus collisions was a hybrid model (63). The approach and interpenetration phase of the reaction was treated in terms of the cascade model in which binding effects were neglected. At the end of this phase, the collision term was turned off. The phase space fluctuations generated during the first stage were then propagated on an event-by-event basis by replacing each nucleon with a number of test particles whose positions and momenta were spread out with Gaussian distributions centered at the classical position and momentum of the corresponding nucleon. The resulting distribution of test particles was then propagated by solving the Vlasov equation with a density-dependent mean field such as Equation 2.

The effect of introducing the mean field is dramatic (63). Figure 13 shows the projection of the coordinate space density onto the reaction plane for a single collision of two $A = 20$ nuclei at a bombarding energy of 400 MeV/nucleon. The simulation was stopped after the main collision

(a)

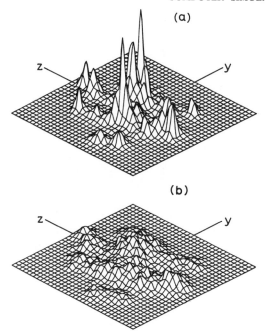

(b)

*Figure 13*   Contour plots of coordinate space density projected onto the reaction plane for a single collision between two $A = 20$ nuclei at a bombarding energy of 400 MeV/nucleon. The mean field interaction is included in part (*a*) and omitted in part (*b*) (63).

epoch, while the system was still expanding. The upper part of the figure shows the result of a calculation for which the mean field was included; the lower part shows the result when it was omitted. Clearly, the mean field is essential to preserve density fluctuations and produce bound clusters.

Several groups have gone further in this direction by including both a mean field and collision terms simultaneously to avoid breaking up the calculation into two distinct parts (6, 12, 14, 15, 71). Unlike the VUU calculations, these semiclassical models propagate the phase space fluctuations on an event-by-event basis. At relatively low excitation energies the fluctuations must be known with good accuracy. To ensure this accuracy, Gregoire and coworkers (71) have adopted the prescription of representing each nucleon by a number of test particles whose phase space distribution is spread out by means of Gaussian distributions. This method requires large amounts of computing time for the simulation of a single event; hence comparisons with data are not available at the time of writing. The calculations have, however, reproduced several qualitative features of nucleus-nucleus collisions at incident energies of several tens of MeV per nucleon, such as the onset of incomplete fusion and the energy dependence of linear momentum transfer in fusion-like rections.

At higher energies, nucleon-nucleon collisions populate larger volumes of phase space, and so the numerical accuracy required for the calculation of fluctuations is reduced. One method (15), which has been used to investigate several different aspects of correlations in intermediate-energy heavy-ion collisions, uses a single test particle per nucleon, spread out in phase space with a Gaussian distribution. The savings in computation time are considerable and this allows a more thorough investigation of the reaction mechanism and comparisons with experiment. As an example, we discuss results (15) obtained for Ca + Ca collisions at a bombarding energy of 100 MeV/nucleon.

The collisions were followed for a duration of 150 fm/$c$, by which time most fragments had separated in coordinate space. Bound clusters were defined by linking groups of nucleons that were separated from their nearest neighbors by less than 3 fm. The momentum distributions of the reaction products predicted by the simulations showed the characteristic features expected from the participant-spectator picture: light mass systems, particularly nucleons, were widely dispersed in phase space, while heavier systems were more tightly clustered near the beam and target rapidity. The simulation allows one to check whether this observation arises mainly from geometry. To answer this question, a large number of collision events were generated, each with an impact parameter randomly chosen according to a geometrical distribution that weights peripheral collisions more heavily than central ones. At the end of each event, the fragment masses were determined. Then an average over the entire sample was made of the collision impact parameter associated with each fragment mass. This quantity is shown in Figure 14. Clearly, light fragments arise from less peripheral collisions than do heavy ones. This figure also serves as a warning against analyzing impact-parameter-averaged data with models assuming a common origin for all products.

In the simulations described, both the coulomb interaction and a phenomenological isospin-dependent mean field were included. This makes it possible to calculate isotopic distributions, such as are shown in Figure 15. The calculated distributions are peaked along the $Z = N$ curve but show a higher abundance on the neutron-rich side than the proton-rich side. This is in qualitative agreement with experiment, as are the predicted cross sections.

## 4.3    Reaction Trajectories

Numerical simulations may be particularly useful in elucidating the relation between the observed reaction products and the temporal evol-

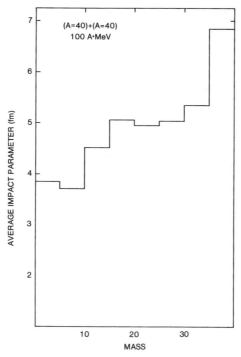

*Figure 14*  Average impact parameter associated with a given range of fragment mass predicted by a semiclassical calculation for Ca + Ca collisions at $E = 100$ MeV/nucleon. The bin size corresponds to 5 mass units (15).

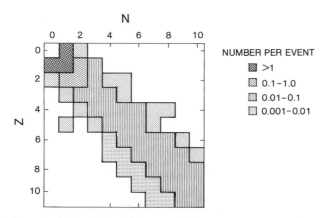

*Figure 15*  Isotopic yields of light fragments predicted by a semiclassical calculation for Ca + Ca collisions at $E = 100$ MeV/nucleon. The calculations were averaged over impact parameter (15).

ution of the reaction. There are several scenarios for fragment emission, each of which may have its domain of applicability. Among the possibilities are long time-scale emission such as is observed in compound systems (22, 23, 79), statistical breakup (4, 33, 54, 72, 88), and coalescence of light fragments from nucleons late in the reaction (39, 74, 75, 117).

One question of interest is whether and how the reaction is influenced by the liquid-gas coexistence region. On the atomic scale, this problem has been investigated by means of simulations for the expansion of hot argon droplets (130). The fermionic nature of their constituents makes nuclei much more difficult to handle than classical particles, so progress here has not been so rapid.

Because of the uncertainty in calculating the excitation energy of droplets in the semiclassical simulations, one method used (15) is to follow the evolution of the local coordinate and phase space densities in the vicinity of individual nucleons during a collision. At the end of the collision, it can be determined whether the nucleon has emerged as a free particle or as part of a bound cluster. The results of such an analysis for central Ca + Ca reactions at 100 MeV/nucleon bombarding energy are shown in Figures 16 and 17.

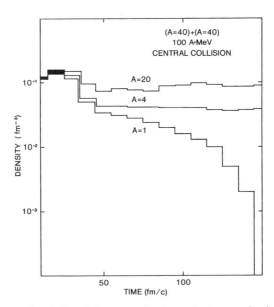

*Figure 16*  Temporal evolution of the average local coordinate space density in the vicinity of nucleons finally emitted as free nucleons ($A = 1$) or in bound clusters ($A = 4, 20$). The simulation is a semiclassical calculation of central Ca + Ca collisions at 100 MeV/nucleon (15).

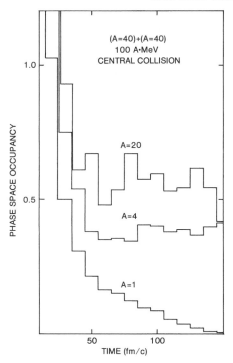

*Figure 17*  Temporal evolution of the average phase space density in the vicinity of nucleons finally emitted as free nucleons ($A = 1$) or in bound clusters ($A = 4, 20$). Conditions as in Figure 16 (15).

The figures show the temporal evolution of the average local coordinate space and phase space densities for free nucleons and nucleons that finally emerge in clusters of mass 4 and 20. As shown in Figure 16, after the initial phase of the reaction ($t < 30$ fm/c) the spatial density begins to decrease as the system expands. The density drops to about half of normal nuclear matter density and then the system breaks up into free nucleons and bound clusters. Although there is some oscillation in the density of the clusters, nucleons bound in clusters do not come from regions of very low density that contract into regions of higher density as the clusters are formed.

This interpretation is substantiated by the temporal evolution of the phase space densities shown in Figure 17. While the local coordinate space densities are fairly similar for all nucleons during the early violent stages of the reaction, the phase space densities are not. Particles that ultimately emerge as free nucleons are scattered into regions of low phase space density at a very early stage of the reaction. Particles that ultimately emerge in the form of clusters remain in regions of relatively high phase space density throughout the reaction. This behavior bears qualitative resemblance to the phenomenon of percolation (13, 41; for a review, see 123).

## 4.4  *Two-Particle Correlation Functions*

As a final example of the application of computer simulations to nuclear reactions, we wish to examine two-particle correlation functions. The measurement of two-photon correlations at small relative momentum was advanced some time ago as a means of estimating stellar sizes (76). It was subsequently suggested that such measurements could also be used in nuclear and particle physics to measure reaction volumes (66, 95, 97, 137). Measurements of correlations between pairs of pions (61, 139), protons (140) and even nuclear fragments (43) showed the feasibility of the technique, and it has now been used extensively in nuclear physics (for a recent summary, see 65). The measured reaction volumes agree qualitatively with what is expected both intuitively and from simple numerical simulations (28, 82).

However, there are ambiguities in the measurements, and simulations may be of use in their resolution (16, 25, 82). One such observation is the variation in the measured reaction volume found as a function of the summed energy of the two particles (65). The correlations become weaker—that is, the correlation function $R(\mathbf{p}_1, \mathbf{p}_2)$ moves closer to zero— as the summed energy of the particles decreases (the relative momentum being held constant). In a simple thermal model approach to these measurements (84), the correlation function is proportional to the reciprocal of the volume, if one assumes that all particles leave the reaction region simultaneously. Thus, if the emission time scale is unimportant, the measurements can be interpreted to mean that energetic particles come from small spatial regions while slow particles come from large regions.

However, the interpretation of the energy dependence can be complicated by at least two factors, the hydrodynamic expansion of the reaction zone expected at high temperatures (113) and the long evaporative time scale expected at low temperatures. A cascade model simulation has been used to assess the importance of the finite emission time effect (25). The two-proton correlation function was simulated for emission from a single nucleus at finite temperature. Nuclear binding was included by putting the nucleons in a fixed step function potential well, as was described in the entropy calculation used to generate Figure 8. The correlation function obtained from the analysis of 20,000 events is shown in Figure 18. Three energy cuts have been made in generating the computer "data" for the correlation function shown in the figure.

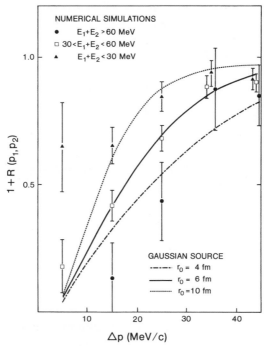

*Figure 18* A cascade model simulation of two-proton correlation functions predicted for an isolated nucleus of fixed size at an initial temperature of 15 MeV. The smooth curves correspond to predictions for Gaussian sources of zero lifetime and radius parameters of $r_0 = 4$, 6, and 10 fm (25).

The smooth curves in the figure are the results from an analytical calculation of the correlation function assuming zero-lifetime source of Gaussian spatial extent (95). The Gaussian is characterized by a parameter $r_0$, which has the values 4, 6, and 10 fm in the figure. The apparent change in the size of the source region predicted by the simulation as obtained from the analysis is very similar to what is observed experimentally. However, in the simulation the spatial dimension of the source is fixed. The simulations show that the variation in apparent size arises from the longer emission times associated with the lower energy ejectiles.

## 5. SUMMARY AND OUTLOOK

The speed and storage capabilities of modern computers have allowed nuclear reaction studies to move beyond specific reactions that lend themselves to analytical treatment and toward an accurate description of the complex reactions found in heavy ion physics. The success achieved in describing reactions at the low (TDHF, Vlasov equation) and high (cascade) extremes of bombarding energy have encourged the development of more sophisticated simulations that include the nuclear mean field, individual N-N collisions, and antisymmetrization effects. These calculations have been successful at the quantitative level in comparison with single-particle inclusive measurements. The incorporation of fluctuations in the simulations at the semiclassical level has opened up the study of fragment formation and two-particle correlation functions. Because of the large CPU-time requirements of such simulations, only qualitative comparisons with data are currently warranted.

The simulations allow the investigation of such questions as the nature of the nuclear equation of state at high densities and the effects of the nuclear liquid-gas phase transition on fragmentation processes. Evidence is present within the simulations for phase transition effects, although the extraction of an experimental signature of the effects remains elusive. At high densities, the equation of state may be making itself felt through collective flow and reduced meson production. The inclusion of antisymmetrization effects through more than the Uehling-Uhlenbeck collision term will allow one to probe more deeply into the high density phase of energetic reactions.

At very high energy densities, one expects (105) that the parton constituents of the nucleons will become deconfined and form a plasma state. Simulations based on parton degrees of freedom have already been used to estimate the conditions achievable in ultrarelativistic heavy ion collisions (24) and undoubtedly will play an important role in the search for experimental signatures of this state. Progress in the development of a tractable transport theory is expected to be rapid in the coming years (64, 58, 80).

As is always the case with computer-based research, progress is at least in part subject to machine limitations. However, the near-term future of this field of study looks bright for at least two reasons: the easy accessibility of supercomputers to an increasing number of researchers and the availability of modestly priced, yet powerful smaller scale processors that lend themselves to dedicated use for simulation purposes. Further, the application of computer animation techniques as exemplified by Figure 1 will allow us to see inside a collision in a manner never before available.

ACKNOWLEDGMENT

The author wishes to thank the Natural Sciences and Engineering Research Council of Canada for financial support.

*Literature Cited*

1. Abate, E., Bellini, G., Fiorini, E., Ratti, S. *Nuovo Cimento* 22: 1206 (1961)
2. Aichelin, J. *Phys. Lett.* 164B: 261–64 (1985)
3. Aichelin, J., Bertsch, G. *Phys. Rev.* C31: 1730–38 (1985)
4. Aichelin, J., Hufner, J., Ibarra, R. *Phys. Rev.* C30: 107–18 (1984)
5. Aichelin, J., Stocker, H. *Phys. Lett.* 163B: 59–65 (1985)
6. Aichelin, J., Stocker, H. *Phys. Lett.* 176: 14 (1986)
7. Amsden, A. A., Bertsch, G. F., Harlow, F. H., Nix, J. R. *Phys. Rev. Lett.* 35: 905–8 (1975)
8. Amsden, A. A., Ginocchio, J. N., Harlow, F. H., Nix, J. R., Danos, M., et al. *Phys. Rev. Lett.* 38: 1055–58 (1977)
9. Amsden, A. A., Harlow, F. H., Nix, J. R. *Phys. Rev.* C15: 2059–71 (1977)
10. Balescu, R. *Equilibrium and Nonequilibrium Statistical Mechanics.* New York: Wiley (1975)
11. Barz, H. W., Iwe, H. *Phys. Lett.* 143B: 55–59 (1984)
12. Bauer, W., Bertsch, G. F., Das Gupta, S. *Mich. State Univ. Rep.* (1986)
13. Bauer, W., Dean, D. R., Mosel, U., Post, U. *Phys. Lett.* 150B: 53–56 (1985)
14. Beauvais, G. E., Boal, D. H., Glosli, J. *Nucl. Phys. A.* In press (1987)
15. Beauvais, G. E., Boal, D. H., Wong, J. C. K. *Phys. Rev.* C35: 545–55 (1987)
16. Bernstein, M. A., Friedman, W. A., Lynch, W. G. *Phys. Rev.* C29: 132–38; C30: 412 (E) (1984)
17. Bertini, H. W. *Phys. Rev.* 131: 1801–21 (1963); 138: AB2 (E) (1965)
18. Bertsch, G. *Prog. Part. Nucl. Phys.* 4: 483 (1980)
19. Bertsch, G., Cugnon, J. *Phys. Rev.* C24: 2514–20 (1981)
20. Bertsch, G., Kruse, H., Das Gupta, S. *Phys. Rev.* C29: 673–75 (1984)
21. Bertsch, G., Siemens, P. J. *Phys. Lett.* 126B: 9–12 (1983)
22. Blann, M. *Ann. Rev. Nucl. Sci.* 25: 123–66 (1975)
23. Blatt, J. M., Weisskopf, V. F. *Theoretical Nuclear Physics.* New York: Wiley (1952)
24. Boal, D. H. *Phys. Rev.* C33: 2206–8 (1986)
25. Boal, D. H., DeGuise, H. *Phys. Rev. Lett.* 57: 2901–4 (1986)
26. Boal, D. H., Goodman, A. L. *Phys. Rev.* C33: 1690–98 (1986)
27. Boal, D. H., Reid, J. H. *Phys. Rev.* C29: 973–84 (1984)
28. Boal, D. H., Shillcock, J. C. *Phys. Rev.*

29. C33: 549–56 (1986)
29. Bodmer, A. R., Panos, C. N. *Phys. Rev.* C15: 1342–58 (1977)
30. Bodmer, A. R., Panos, C. N. *Nucl. Phys.* A356: 517–22 (1981)
31. Bodmer, A. R., Panos, C. N., Mackellar, A. D. *Phys. Rev.* C22: 1025–54 (1980)
32. Bonche, P., Levit, S., Vautherin, D. *Nucl. Phys.* A436: 265–93 (1985)
33. Bondorf, J. *Nucl. Phys.* A387: 25c–36c (1982)
34. Bondorf, J. P., Donangelo, R., Shulz, H., Sneppen, K. *Phys. Lett.* 162B: 30–34 (1985)
35. Bondorf, J. P., Feldmeier, H. T., Garpman, S., Halbert, E. C. *Phys. Lett.* 65B: 217–20 (1976)
36. Bondorf, J. P., Siemens, P. J., Garpman, S., Halbert, E. *Z. Phys.* 279: 385–94 (1976)
37. Braun-Munzinger, P., Stachel, J. *Ann. Rev. Nucl. Part. Sci.* 37: 97 (1987)
38. Buchwald, G., Graebner, G., Theis, J., Maruhn, J. A., Greiner, W., Stocker, H. *Phys. Rev.* C28: 1119–22 (1983)
39. Butler, S. T., Pearson, C. A. *Phys. Rev. Lett.* 7: 69–71 (1961)
40. Callaway, D. J. E., Wilets, L., Yariv, Y. *Nucl. Phys.* A327: 250–68 (1979)
41. Campi, X., Desbois, J., Lipparini, E. *Phys. Lett.* 142B: 8–13 (1984)
42. Chen, K., Fraenkel, Z., Friedlander, G., Grover, J. R., Miller, J. M., Shimamoto, Y. *Phys. Rev.* 166: 949–67 (1968)
43. Chitwood, C. B., Aichelin, J., Boal, D. H., Bertsch, G., Fields, D. J., et al. *Phys. Rev. Lett.* 54: 302–5 (1985)
44. Chitwood, C. B., Fields, D. J., Gelbke, C. K., Klesch, D. R., Lynch, W. G., et al. *Phys. Rev.* C34: 858–71 (1986)
45. Clare, R. B., Strottman, D. *Phys. Rep.* 141: 177–280 (1985)
46. Csernai, L. P., Kapusta, J. *Phys. Rep.* 131: 223–318 (1986)
47. Cugnon, J. *Phys. Rev.* C22: 1885–96 (1980)
48. Cugnon, J. Lectures given at the Cargese Summer School (1984)
49. Cugnon, J., Jaminon, M. *Phys. Lett.* 123B: 155–59 (1983)
50. Cugnon, J., Knoll, J., Randrup, J. *Nucl. Phys.* A360: 444–58 (1981)
51. Cugnon, J., Mizutani, T., Vandermeulen, J. *Nucl. Phys.* A352: 505–34 (1981)
52. Curtin, M. W., Toki, H., Scott, D. K. *Phys. Lett.* 123B: 289–92 (1983)
53. Danielewicz, P. *Nucl. Phys.* A314: 465–84 (1979)

54. Das Gupta, S., Mekjian, A. Z. *Phys. Rep.* 72: 131–83 (1981)
55. Davies, K. T. R., Devi, K. R. S., Koonin, S. E., Strayer, M. R. In *Heavy Ion Science*, ed. D. A. Bromley. New York: Plenum (1984)
56. de Groot, S. R., van Leeuwen, W. A., van Weert, Ch. G. *Relativistic Kinetic Theory*. Amsterdam: North Holland (1980)
56a. Dorso, C., Duarte, S., Randrup, J. *Phys. Lett. B*. In press (1987)
57. Doss, K. G. R., Gustafsson, H. A., Gutbrod, H. H., Kampert, K. H., Kolb, B., et al. *Phys. Rev. Lett.* 57: 302–5 (1986)
58. Elze, H.-Th., Gyulassy, M., Vasak, D. *Lawrence Berkeley Lab. Rep. 21137* (1986)
59. Fai, G., Randrup, J. *Nucl. Phys.* A381: 557–76 (1982)
60. Fischer, M. E. *Physics* 3: 255 (1967)
61. Fung, S. Y., Gorn, W., Kiernan, G. P., Lu, J. J., Oh, Y. T., Poe, R. T. *Phys. Rev. Lett.* 41: 1592–94 (1978)
62. Gale, C., Bertsch, G., Das Gupta, S. *Univ. Minn. Rep.* (1986)
63. Gale, C., Das Gupta, S. *Phys. Lett.* 162B: 35–38 (1985)
64. Gavin, S. *Nucl. Phys.* A435: 826–43 (1985)
65. Gelbke, C. K., Boal, D. H. *Prog. Nucl. Part. Phys.* In press (1987)
66. Goldhaber, G., Goldhaber, S., Lee, W., Pais, A. *Phys. Rev.* 20: 300–12 (1960)
67. Goodman, A. L., Kapusta, J. I., Mekjian, A., *Phys. Rev.* C30: 851–65 (1984)
68. Gosset, J., Gutbrod, H. H., Meyer, W. G., Poskanzer, A. M., Sandoval, A., et al. *Phys. Rev.* C16: 629–57 (1977)
69. Gosset, J., Kapusta, J. I., Westfall, G. D. *Phys. Rev.* C18: 844–55 (1978)
70. Gregoire, C., Remaud, B., Scheuter, F., Sebille, F. *Nucl. Phys.* A436: 365–96 (1985)
71. Gregoire, C., Remaud, B., Sebille, F., Vinet, L., Raffray, Y. GANIL preprint (1986)
72. Gross, D. H. E., Satpathy, L., Meng, T.-C., Satpathy, M. *Z. Phys.* A309: 41–48 (1982)
73. Gudima, K. K., Toneev, V. D. *Sov. J. Nucl. Phys.* 27: 351 (1978)
74. Gutbrod, H. H., Sandoval, A., Johansen, P. J., Poskanzer, A. M., Gosset, J., et al. *Phys. Rev. Lett.* 37: 667–70 (1976)
75. Gyulassy, M., Frankel, K., Remler, E. A. *Nucl. Phys.* A402: 596–611 (1983)
76. Hanbury-Brown, R., Twiss, R. Q. *Nature* 178: 1046 (1956)
77. Harris, J. W., Bock, R., Brockman, R., Sandoval, A., Stock, R., et al. *Phys. Lett.* 153B: 377–81 (1985)
78. Harris, J. W., Odyniec, G., Pugh, H. G., Schroeder, L. S., Tinckhell, M. L., et al. *Lawrence Berkeley Lab. Rep. LBL-22003* (1986)
79. Hauser, W., Feshbach, H. *Phys. Rev.* 87: 366–73 (1952)
80. Heinz, U. *Phys. Rev. Lett.* 51: 351–54 (1983)
81. Hirsch, A. S., Bujak, A., Finn, J. E., Gutay, L. J., Minich, R. W., et al. *Phys. Rev.* C29: 508–25 (1984)
82. Humanic, T. J. *Phys. Rev.* C34: 191–95 (1986)
82a. Jacak, B. V., Westfall, G. D., Gelbke, C. K., Harwood, L. H., Lynch, W. G., et al. *Phys. Rev. Lett.* 51: 1846–49 (1983)
83. Jaqaman, H. R., Mekjian, A. Z., Zamick, L. *Phys. Rev.* C27: 2782–91 (1983)
84. Jennings, B. K., Boal, D. H., Shillcock, J. C. *Phys. Rev.* C33: 1303–6 (1986)
85. Kadanoff, L. P., Baym, G. *Quantum Statistical Mechanics*. New York: Benjamin (1962)
86. Kikuchi, K., Kawai, M. *Nuclear Reactions at High Energy*. Amsterdam: North Holland (1968)
87. Kitazoe, Y., Sano, M., Yamamura, Y., Furutani, H., Yamamoto, K. *Phys. Rev.* C29: 828–36 (1984)
88. Knoll, J. *Phys. Rev.* C20: 773–80 (1979)
89. Knoll, J., Randrup, J. *Nucl. Phys.* A324: 445–63 (1979)
90. Knoll, J., Strack, B. *Phys. Lett.* 149B: 45–49 (1984)
91. Ko, C. M., Bertsch, G., Aichelin, J. *Phys. Rev.* C31: 2324–26 (1985)
92. Kodama, T., Duarte, S. B., Chung, K. C., Nazareth, R. A. M. S. *Phys. Rev. Lett.* 49: 536–39 (1982)
93. Kohler, H. S. *Nucl. Phys.* A378: 159–80 (1982)
94. Kohler, H. S., Neilsson, B. S. *Nucl. Phys.* A417: 541–63 (1984)
95. Koonin, S. E. *Phys. Lett.* 70B: 43–47 (1977)
96. Koonin, S. E., Davies, K. T. R., Maruhn-Rezwani, V., Feldmeier, H., Kreiger, S. J., Negele, J. W. *Phys. Rev.* C15: 1359–74 (1977)
97. Kopylov, G. I. *Phys. Lett.* 50B: 472–74 (1974)
98. Kruse, H., Jacak, B. V., Molitoris, J. J., Westfall, G. D., Stocker, H. *Phys. Rev.* C31: 1770–74 (1985)
99. Kruse, H., Jacak, B. V., Stocker, H. *Phys. Rev. Lett.* 54: 289–92 (1985)
100. Lamb, D. Q., Lattimer, J. M., Pethick, C. J., Ravenhall, D. G. *Phys. Rev. Lett.* 41: 1623–26 (1978)

101. Landau, L. D., *Phys. Z. Sov. Union* 10: 154 (1936)
102. Landau, L. D., Lifshitz, E. M. *Fluid Mechanics*. Oxford: Pergamon (1959)
103. Lattimer, J. M., Pethick, C. J., Ravenhall, D. G., Lamb, D. Q. *Nucl. Phys.* A432: 646–742 (1985)
104. Lopez, J. A., Siemens, P. J. *Nucl. Phys.* A431: 728–44 (1984)
105. Kajantie, K., McLerran, L. *Ann. Rev. Nucl. Part. Sci.* 37: 293 (1987)
106. Mekjian, A. Z. *Nucl. Phys.* A384: 492–536 (1982)
107. Metropolis, N., Bivins, R., Storm, M., Miller, J. M., Friedlander, G., Turkevich, A. *Phys. Rev.* 110: 185–203 (1958)
108. Molitoris, J. J., Stocker, H. *Phys. Lett.* 162B: 47–54 (1985)
109. Negele, J. *Rev. Mod. Phys.* 54: 913–1015 (1982)
110. Nix, J. R. *Prog. Part. Nucl. Phys.* 2: 237–84 (1979)
111. Nix, J. R., Strottman, D., Yariv, Y., Fraenkel, Z. *Phys. Rev.* C25: 2491–97 (1982)
112. Nordheim, L. W. *Proc. R. Soc. Ser. A* 119: 689 (1928)
113. Pratt, S. *Phys. Rev. Lett.* 53: 1219–21 (1984)
114. Potter, D. *Computational Physics*. New York: Wiley (1973)
115. Randrup, J. *Nucl. Phys.* A314: 429–53 (1979)
116. Remler, E. A. *Ann. Phys.* 95: 455–95 (1975)
117. Remler, E. A. *Ann. Phys.* 136: 293–316 (1981)
118. Sauer, G., Chandra, H., Mosel, U. *Nucl. Phys.* A264: 221–43 (1976)
119. Sebille, F., Remaud, B. *Nucl. Phys.* A420: 141–61 (1984)
120. Shulz, H., Munchow, L., Ropke, G., Schmidt, M. *Phys. Lett.* 119B: 2–16 (1982)
121. Sierk, A. J., Nix, J. R. *Phys. Rev.* C22: 1920–26 (1980)
122. Deleted in proof
123. Stauffer, D. *Phys. Rep.* 54: 1–74 (1979)
124. Stevenson, J. D. *Phys. Rev. Lett.* 41: 1702–5 (1978)
125. Stock, R. *Phys. Rep.* 135: 259–315 (1986)
126. Stocker, H., Greiner, W. *Phys. Rep.* 137: 277–392 (1986)
127. Stocker, H., Maruhn, J. A., Greiner, W. *Phys. Rev. Lett.* 44: 725–28 (1980)
128. Strack, B., Knoll, J. *Z. Phys.* A315: 249–50 (1984)
128a. Tsang, M. B., Ronningen, R. M., Bertsch, G., Chen, Z., Chitwood, C. B., et al. *Phys. Rev. Lett.* 57: 559–62 (1986)
129. Uehling, E. A., Uhlenbeck, G. E. *Phys. Rev.* 43: 552 (1933)
130. Vincentini, A., Jacucci, G., Pandharipande, V. R. *Phys. Rev.* C31: 1783–93 (1985)
131. Vinet, L., Sebille, F., Gregoire, C., Remaud, B., Shuck, P. *Phys. Lett.* 172B: 17–22 (1986)
132. Vlasov, A. *Zh. Eksp. Teor. Fiz.* 8: 291 (1938)
133. Weiner, R., Weststrom, M. *Nucl. Phys.* A286: 282–96 (1977)
134. Wilets, L., Henley, E. M., Kraft, M., Mackellar, A. D. *Nucl. Phys.* A282: 341 (1977)
135. Wilets, L., Yariv, Y., Chestnut, R. *Nucl. Phys.* A301: 359–64 (1978)
136. Wong, C. Y., Welton, T. A., Maruhn, J. A. *Phys. Rev.* C15: 1558–70 (1977)
137. Yano, F. B., Koonin, S. E. *Phys. Lett.* 78B: 556–59 (1978)
138. Yariv, Y., Fraenkel, Z. *Phys. Rev.* C20: 2227–43 (1979)
139. Zajc, W. A., Bistirlich, J. A., Bossingham, R. J., Bowman, H. R., Clawson, C. W., et al. *Phys. Rev.* C29: 2173–87 (1984)
140. Zarbakhsh, F., Sagle, A. L., Brochard, F., Mulera, T. A., Perrez-Mendez, V., et al. *Phys. Rev. Lett.* 46: 1268–70 (1981)

Ann. Rev. Nucl. Part. Sci. 1987. 37: 33–69

# RECENT PROGRESS ON NUCLEAR MAGNETIC DIPOLE EXCITATIONS

## Ulrich E. P. Berg[1] and Ulrich Kneissl

Institut für Kernphysik, Strahlenzentrum der Justus-Liebig-Universität Giessen, D-6300 Giessen, Federal Republic of Germany

CONTENTS

## 1. INTRODUCTION AND OUTLINE

Spin vibrations of the atomic nucleus have been studied intensively during recent years with the help of diverse reactions. The study of spin forces, which are based largely on the pion fields of the nucleons, gives us an insight into the magnetic properties of a nucleus. Spin vibrations are observed in charge-exchange reactions, inelastic proton and electron scattering, and nuclear resonance fluorescence experiments. Much new experimental data concerning spin-flip isospin-flip transitions have been obtained during the past decade at modern accelerator laboratories, for example at the Indiana

---

[1] Present address: Festo Didactic, Ruiter Str. 82, D-7300 Esslingen 1, FRG.

0163–8998/87/1201–0033$02.00

University Cyclotron Facility (IUCF) in Bloomington, Institute de Physique Nucléaire (IPN) in Orsay, the high-resolution electron scattering
facility of the Technische Hochschule in Darmstadt, the high-duty-cycle
electron accelerator of the University of Illinois in Urbana, and the polarized bremsstrahlung beam of the Universität Giessen.

The giant Gamow-Teller resonance was discovered in (p,n) experiments
ten years ago at the Michigan State University cyclotron (1) and has
since been systematically studied at IUCF (2–4) with great success. A
breakthrough for the investigation of spin-flip transitions with proton
scattering was the experiments at very small scattering angles with 200-
MeV protons at Orsay (5). These experiments established the existence of
the magnetic dipole resonance in heavy nuclei.

Since the nucleon spin is associated with the magnetic moment of the
nucleus, it can interact through the electromagnetic field. Therefore, photons and electrons are two interesting probes for studying spin properties
of nuclei.

Even before the existence of the giant Gamow-Teller resonance was
known, magnetic dipole transitions were being vigorously investigated via
inelastic electron scattering, particularly 180° electron scattering. Backward electron scattering is very selective for magnetic multipole excitations.
This field of nuclear physics was essentially initiated by Barber and collaborators (6), followed by Fagg and collaborators, who opened up a new
line of M1 transition studies (7). Currently the backward-angle inelastic
electron scattering experiments with superb energy resolution at Darmstadt and Amsterdam provide much important data for the investigation
of spin vibrations in nuclei (8–10).

A new method to probe magnetic dipole excitations with real photons
has been developed during the past five years at the University of Giessen
linear electron accelerator (11). With the use of linearly polarized
bremsstrahlung in nuclear resonance fluorescence experiments it became
possible to determine transition probabilities, multipole orders, and parities of dipole transitions to bound states in a completely model-independent manner and with the high accuracy of $\gamma$-ray spectroscopy.

Novel observations have been made as a result of studying spin-flip
isospin-flip transitions, as discussed below. A completely surprising
phenomenon, namely a collective M1 mode, was discovered by Bohle et
al (12). A powerful tool to investigate this mode, which does not involve
spin-flip transitions but is purely orbital, is the combination of photon (13)
and electron scattering data (14).

This review article deals mainly with the investigation of the spin-flip
and orbital M1 resonance, with the help of the electromagnetic interaction.
Experiments with real photons are emphasized. In the following section the

two different categories of M1 transitions are summarized and connections between different reactions used to study M1 transitions are pointed out. That section is followed by a short description of excitations induced by virtual and real photons at modern electron accelerators. Then the method of investigating M1 excitations with nuclear resonance fluorescence is discussed in some detail and experimental results from that work are presented. Finally first conclusions from the current studies of the collective orbital M1 mode are drawn and new prospects of experiments with real photons are given.

# 2. CATEGORIES OF M1 EXCITATIONS

The nuclear reactions treated here involve a simultaneous flipping of nucleon spins and isospins. This can be effected in the experiment by the hadronic or the electromagnetic interaction. Theoretical relations between cross sections of such different probes as electrons, nucleons, and pions were recently pointed out by Petrovich et al (15). The Hamilton operator for M1 transitions contains a spin and an orbital part:

$$\tfrac{1}{2}(g_s^p - g_s^n)\vec{\sigma}\vec{\tau} + (g_l^p - g_l^n)\vec{l}\vec{\tau}, \qquad\qquad 1.$$

while the Gamow-Teller operator

$$(g_s^p - g_s^n)\vec{\sigma}\vec{\tau} \qquad\qquad 2.$$

has no orbital contribution. Spin-flip transitions will be excited by the $\vec{\sigma}\vec{\tau}$ term and orbital excitations by the $\vec{l}\vec{\tau}$ term. In nature, excitations are usually not of pure spin-flip or orbital type but are dominated by one or the other. Under the kinematical conditions of the (p,n) and (p,p') investigations (2, 16) of the Gamow-Teller and the M1 resonance, respectively, the $V_{\sigma\tau}$ term of the effective nucleon-nucleus interaction is dominant (17).

## 2.1  Spin-Flip M1 Excitations

The relationship between M1, Gamow-Teller (p,n), and $\beta^+$ transitions is depicted in Figure 1. This example shows the different $1^+$ states, which can be reached via $(\gamma,\gamma')$, (p,p'), (e,e'), (p,n), or $(\pi^+,\gamma)$. The initial ground-state spin and parity of $^{26}$Mg are $J_0^\pi = 0^+$; its ground-state isospin is $T_0 = 1$ and the third component in isospin space is $T_0^{(3)} = 1$. Absorption of magnetic dipole radiation leads to $1^+$ states with isospin $T = 1$ and 2. The strength distribution between the two isospin components, calculated on simple geometrical grounds, is indicated in Figure 1. The M1 resonance is located at about 10 MeV. The analog resonance can be reached by a (p,n) reaction. These states should be located at nearly the same excitation

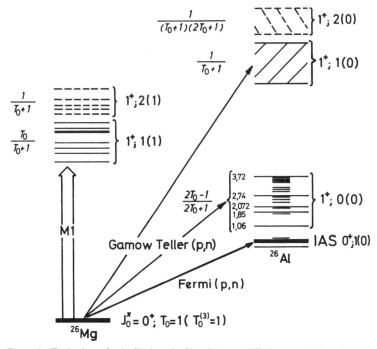

*Figure 1*    Excitation of spin-flip isospin-flip $1^+$ states, differing only in isospin space.

energies above the $J^\pi = 0^+$, $T = 1$ $[T^{(3)} = 0]$ isobaric analog state of the $^{26}$Mg ground state in $^{26}$Al. The two $T = 1$ and 2 excitation regions of $1^+$ states in $^{26}$Al are shown with cross-hatching in Figure 1. The antianalog states of the M1 resonance in $^{26}$Mg, differing only in the isospin quantum numbers $[T = 0, T^{(3)} = 0]$ are expected to be in the low-energy range of the $^{26}$Al level scheme.

If neutron excess is high, as in heavy nuclei, it becomes obvious from the estimate of Gamow-Teller strength distribution among the analog and antianalog states that the antianalog resonance will mainly be populated in a (p,n) experiment. These antianalog $1^+$ states of the M1 resonance were the giant Gamow-Teller resonance discovered in (p,n) experiments at the Michigan State University cyclotron (1). Only light nuclei offer the chance to study the analog states of the M1 resonance, and by comparing $B$(M1) values with (p,n) cross sections, it becomes possible to investigate the orbital part of the M1 transition operator (18–20).

In Figure 2 is shown how the M1 resonance is revealed in a nuclear resonance fluorescence experiment. Peaks due to excitation of $1^+$ states

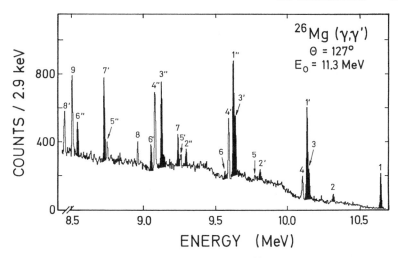

*Figure 2*  Nuclear resonance fluorescence spectrum from ²⁶Mg. (The peaks are numbered in order.) The solid peaks stem from excitations of 1⁺ states. Because of the response of a Ge(Li) γ-ray detector to monoenergetic high-energy γ-rays, single- and double-escape peaks, labeled by one and two primes respectively, occur in addition to full-energy peaks.

are marked. The $(\gamma, \gamma')$ spectrum was recorded with bremsstrahlung of 11.3-MeV endpoint energy at a scattering angle of 127°. The parities of the dipole transitions observed were determined with polarized photons, as discussed below.

Roughly the same energy region, but in ²⁶Al as measured in a ²⁶Mg(p,n)²⁶Al experiment, is depicted in Figure 3. Peaks whose angular distributions demonstrate that they belong to excitation of 1⁺ states (18) in ²⁶Al are marked too. In addition, $B(M1)$ values from the photon scattering experiment are plotted in Figure 3. The location of 1⁺ states in ²⁶Mg and ²⁶Al as well as the distribution of transition strengths are in nice agreement.

## 2.2  Orbital Collective M1 Excitations

Besides the spin-flip excitations discussed in the previous section, a new class of low-lying, collective M1 excitations has been predicted for deformed nuclei in different nuclear models. This collective mode is closely related to the orbital motion of neutrons with respect to protons. The famous electric giant dipole resonance (GDR) (21) represents a familiar example of another kind of an isovector motion. As well known, the main properties of the GDR can be described by hydrodynamical models (22, 23) as an oscillation of neutrons against protons. The mean excitation energies of the GDR are given by

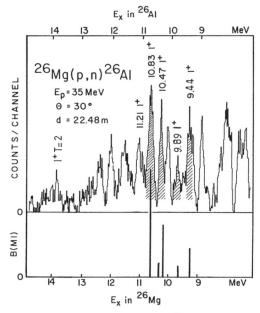

*Figure 3* Comparison of M1 transition strengths in $^{26}$Mg from nuclear resonance fluorescence (*lower part*) with peaks in a $^{26}$Mg(p,n) spectrum due to excitations of the analog $1^+$ levels in $^{26}$Al.

$$E(\text{GDR}) = 77A^{-1/3} \text{ (MeV)} \qquad (23) \qquad\qquad 3.$$

$$E(\text{GDR}) = 34A^{-1/6} \text{ (MeV)} \qquad (22) \qquad\qquad 4.$$

depending on the choice of the restoring force and the boundary conditions, respectively. For deformed nuclei the GDR splits into two components corresponding to vibrations along the short and long symmetry axes of the deformed nuclei (24).

The macroscopic picture of the hydrodynamical models, describing the GDR, has been adopted by Lo Iudice & Palumbo (25) to construct a collective isovector magnetic mode. In their two-rotor model (TRM) the neutrons and protons are assumed to act as rigid deformed bodies that may rotate against each other around a common axis. An appropriate restoring force leads to a scissor-like oscillation. This geometrical picture of the so-called Scissor Mode makes obvious the predominant orbital character of this excitation. The excitation energy is given in the framework of the TRM by

$$E(\text{M1}) = 42\delta A^{-1/6} \text{ (MeV)} \qquad (26), \qquad\qquad 5.$$

which shows the same $A$ dependence as the GDR resonance energy in the

Goldhaber-Teller model (22). However, the absolute scale is reduced by the deformation parameter $\delta$, which is about 0.25 for strongly deformed nuclei. Therefore, the energy of the M1 mode is lowered to 3–4 MeV whereas the GDR is concentrated at 12–15 MeV in heavy nuclei. This low excitation energy of the M1 mode implies that this mode should be observed in isolated, low-energy, bound states, in contrast to the GDR located in the continuum of highly excited nuclear states.

A large number of theoretical papers have been published on this subject, most of them after the discovery of the new magnetic mode by Richter and coworkers (12). Therefore, this article restricts itself to the rather typical descriptions outlined in the following.

In a sum rule approach (SRA), Lipparini & Stringari (27) showed that the isovector M1 mode can occur by the coupling of a rigid rotation to the isovector giant quadrupole resonance. For the excitation energy, an $A$ dependence has been derived:

$$E(\text{M1}) = 56\delta A^{-1/3} \text{ (MeV)}. \qquad\qquad 6.$$

Microscopic shell-model calculations using the random phase approximation (RPA) also succeeded in describing the new magnetic model to some extent. The excitation energy shows a $\delta A^{-1/3}$ dependence. The scaling factor is about 66 MeV (28). It changes slightly in different calculations, which differ mainly in the choice of the residual interaction. Furthermore, the RPA calculations suggest a fragmentation of the strength into several states (28–31).

Hilton (32) treated the rotational oscillation in his RPA calculation analogously to the linear oscillations of the GDR. In this so-called giant-angle dipole (GAD) model, the excitation energy of the M1 mode is proportional to the GDR energy

$$E(\text{GAD}) = \delta E(\text{GDR}), \qquad\qquad 7.$$

in fair agreement with the TRM predictions.

Very recently, parameter-free microscopic calculations of $B(\text{M1})$ strength distributions and decay branchings of the corresponding $1^+$ states have been performed by Hammaren et al (33) for some deformed even-even nuclei in the mass region $A = 130$. The total strength of 6–9 $\mu_0^2$ has been predicted to split mainly into two regions at excitation energies of 4 and 5 MeV, respectively.

The interacting boson model can be considered to be between the macroscopic and the shell-model description. In its second version (IBA-2) (34–37), neutron and proton degrees of freedom are treated separately. Neutrons and protons are assumed to couple to pairs (bosons) with angular momentum $L = 0, 2$, s and d bosons, respectively (in the simplest version). The Hamiltonian, in a simplified form, can be written as

$$H \approx \kappa Q_v Q_\pi + \lambda M.$$ 8.

The first term takes into account the quadrupole-quadrupole interaction between neutron and proton bosons and describes the well-known low-lying collective bands in deformed nuclei (see Figure 4). The second so-called Majorana term represents the analogy to the symmetry energy term in hydrodynamical models and determines the separation of states of complete and mixed symmetry with respect to proton and neutron degrees of freedom. The Majorana force parameter $\lambda$ has to be determined experimentally (see Section 5.3). In the framework of the IBA-2, the orbital M1 mode corresponds to the excitation of a $1^+$ state of mixed symmetry, which represents the head of a $K = 1$ band (see Figure 4). The geometrical interpretation of this M1 excitation is a small-angle oscillation of the protons against the neutrons *outside* an inert core. This motion of the "nuclear wobble" (38) likewise suggests that the new M1 mode corresponds to an orbital excitation. Furthermore, a lower strength as compared to the TRM prediction seems to be plausible since only nucleons outside the core participate in the collective motion. The transition strength (39) in the SU(3) limit for axially symmetric rotators amounts to

$$\frac{3}{4\pi} \frac{8 N_v N_\pi}{2(N_v + N_\pi) - 1} (g_v - g_\pi)^2,$$ 9.

(where $N_v$, $N_\pi$ and $g_v$, $g_\pi$ are the neutron and proton boson numbers and $g$-factors, respectively).

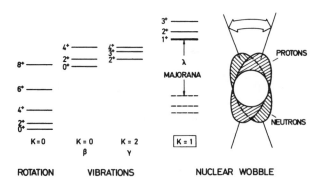

*Figure 4* Collective bands of a deformed nucleus and geometrical interpretation of the "nuclear wobble."

In Table 1 the excitation energies and transition strengths as obtained by different model calculations are summarized together with the numerical values for $^{156}$Gd, which is the most extensively investigated nucleus so far (14). The theoretical results are compared with the experimental data in Section 5.3.

An extension of the two-rotor model to triaxially deformed nuclei predicts an energetic splitting of the orbital M1 mode (41):

$$\Delta E = E_x \frac{1}{\sqrt{3}} \tan \gamma. \qquad 10.$$

The usual deformation parameter $\gamma$ describes the deviation from axial symmetry. The strength should be shared within the two states according to

$$B_2(M1)/B_1(M1) = \left(1 - \frac{1}{\sqrt{3}} \tan \gamma\right) \Big/ \left(1 + \frac{1}{\sqrt{3}} \tan \gamma\right), \qquad 11.$$

which is on the order of unity for small deformation parameters.

In conclusion, the signatures of the new orbital M1 mode in deformed nuclei are

1. an excitation energy of about 3 MeV (in the rare earth region),
2. a transition strength on the order of 3 $\mu_0^2$, and
3. a predominantly orbital character corresponding to convection current type excitations; no or very weak excitation in (p,p′) reactions.

Adequate models must describe the electron scattering form factors as well as the photon scattering data (at the photon point).

## 3.  ELECTROMAGNETIC PROBES

### 3.1  *Electromagnetic Excitations by Virtual and Real Photons*

The application of electromagnetic probes (electrons or photons) in nuclear structure studies offers some fundamental advantages. First of all, the excitation is due to the well-known electromagnetic interaction of the incident electrons or photons with the nuclear charge, current, and magnetization densities. Furthermore, the interaction is weak; therefore, the processes can be treated in perturbation theory, and precise, detailed, and rather model-independent information on the nuclear structure can be extracted. In the following, some basics of electron and photon scattering are summarized that are relevant for understanding the experiments discussed later in this article. For a deeper insight the reader is referred to

**Table 1** Model predictions for the excitation energy and $B(M1)\uparrow$ values of the orbital M1 mode

| Model | Ref. | Excitation energy (MeV) | Total strength $B(M1)\uparrow$ ($\mu_0^2$) | Numerical values for $^{156}$Gd | |
|---|---|---|---|---|---|
| | | | | $E_x$ (MeV) | $B(M1)\uparrow$ ($\mu_0^2$) |
| TRM | (25, 26) | $42\,\delta A^{-1/6}$ | $0.035\,\delta A^{3/2}$ | 4.54 | 17.12 |
| SRA | (27) | $56\,\delta A^{-1/3}$ | $0.043\,\delta A^{4/3}$ | 2.61 | 9.06 |
| RPA | (28) | $66\,\delta A^{-1/3}$ | $0.027\,\delta A^{4/3}$ | 3.08 | 5.69 |
| | (29) | | $0.044\,\delta A^{4/3}$ | | 9.27 |
| | (30) | | $0.043\,\delta A^{4/3}$ | 2.5–4.1 | 9.06 |
| GAD | (32) | $\delta E(\mathrm{GDR})$ | $0.189\,\delta\,\dfrac{(N\cdot Z)^{4/3}}{N^{4/3}+Z^{4/3}}\,(g_\mathrm{p}-g_\mathrm{n})^2$ | 3.59 | 7.53 |
| IBA-2 | (39) | | $\dfrac{3}{4\pi}\,\dfrac{8\cdot N_\pi\cdot N_\nu}{2(N_\pi+N_\nu)-1}\,(g_\pi-g_\nu)^2$ | | 2.91[a] |
| | | | | | 3.84[b] |

[a] Bare $g$-factors.
[b] Effective $g$-factors (40).

the textbooks of Überall (42) or one of the excellent review articles on this subject (43, 44).

Inelastic scattering of an electron on a nucleus can be understood as an exchange of a virtual photon of energy $\omega$ and momentum $q$ (see Figure 5). The energy of the virtual photon corresponds to the excitation energy transferred to the nucleus and is determined by the energy loss of the scattered electron. The transfer of momentum $q$ can be varied for a fixed excitation energy by changing the scattering angle and/or the energy of the incident electrons according to

$$q = (k_1^2 + k_2^2 - 2k_1 k_2 \cos\theta)^{1/2}$$

$$\omega = k_1 - k_2$$

12.

($k_{1,2}$ are the momenta of the incident and scattered electrons, respectively; in units of $\hbar = c = 1$). This possibility of the $q$ variation in electron scattering enables one to map out the spatial distributions of the change, current, and magnetization transition densities, whereas in experiments with real photons the transfer of momentum is fixed and equal to the excitation energy. Therefore only transition probabilities can be extracted from real photon experiments.

Although electron scattering experiments, in particular on heavy nuclei, have to be analyzed in a Distorted Wave Born Approximation (DWBA), in the following the relevant features of electron scattering are summarized in a Plane Wave Born Approximation (PWBA) description since in this treatment the underlying physical processes are more transparent.

In PWBA the inelastic electron scattering cross section can be written as a sum of cross sections for excitation of different electric and magnetic multipolarities:

$$\frac{d\sigma}{d\Omega} = \sum_{\lambda=0}^{\infty} \frac{d\sigma^\lambda}{d\Omega_{el}} + \sum_{\lambda=1}^{\infty} \frac{d\sigma^\lambda}{d\Omega_{magn}}.$$

13.

The cross section for the excitation of a transition with multipolarity $\lambda$

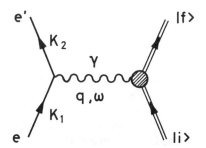

*Figure 5*  Kinematics of inelastic electron scattering in the one-photon-exchange approximation

can be factorized (neglecting recoil effects) into the Mott cross section, describing the electron scattering on the Coulomb potential of a point-like proton, and a sum of form factors taking into account the spatial extension of the nucleus and being responsible for the diffraction pattern of the scattering cross section:

$$\frac{d\sigma^\lambda}{d\Omega} = 4\pi\sigma_{\text{Mott}}\left\{|F_L^\lambda(q)|^2 + \left[\frac{1}{2} + \tan^2(\theta/2)\right] \cdot (|F_E^\lambda(q)|^2 + |F_M^\lambda(q)|^2)\right\}. \qquad 14.$$

The form factors correspond to matrix elements of the multipole operators. The longitudinal (with respect to the $q$ axis) Coulomb form factor $F_L$ is connected to the charge transition density; the transverse electric and magnetic form factors $F_E$ and $F_M$ are related to the convection current and magnetization transition densities. A measurement of the scattering cross section at a fixed $q$ and at different scattering angles allows one to disentangle longitudinal and transverse contributions (Rosenbluth plot; see Equation 14).

For the discussion of low-$q$ electron scattering and its relation to photon scattering it is useful to express the form factors by the corresponding transition probabilities:

$$\begin{aligned} B(C\lambda, q) &= [(2\lambda+1)!!/q^\lambda]^2 &\cdot |F_L^\lambda(q)|^2 \\ B(E\lambda, q) &= [\lambda/(\lambda+1)] \cdot [(2\lambda+1)!!/q^\lambda]^2 \cdot |F_E^\lambda(q)|^2 \\ B(M\lambda, q) &= [\lambda/(\lambda+1)] \cdot [(2\lambda+1)!!/q^\lambda]^2 \cdot |F_M^\lambda(q)|^2. \end{aligned} \qquad 15.$$

The scattering cross sections are then

$$\frac{d\sigma^\lambda}{d\Omega_{\text{el}}} \sim q^{2\lambda} \cdot [V_L(\theta) \cdot B(C\lambda, q) + V_T(\theta) \cdot B(E\lambda, q)]$$

$$\frac{d\sigma^\lambda}{d\Omega_{\text{magn}}} \sim q^{2\lambda} \cdot [V_T(\theta) \cdot B(M\lambda, q)]. \qquad 16.$$

In the context of magnetic excitations it should be noted that the kinematical function $V_L(\theta)$ vanishes at $\theta = 180°$ whereas $V_T(180°)$ remains finite. Therefore magnetic transitions should be studied at backward angles.

The transition probabilities $B(\lambda, q)$ as measured in (e, e′) experiments depend on the transfer of momentum $q$; on the other hand, the same transition probabilities at the photon point ($q = \omega$) describe the photoabsorption and the ground-state $\gamma$ decay of excited levels. The photoabsorption cross section integrated over the level width equals

$$\int \sigma(\omega) \, d\omega = (2\pi)^3 \alpha \sum_{\lambda=1}^{\infty} \frac{(\lambda+1)\omega^{2\lambda-1}}{\lambda[(2\lambda+1)!!]^2} \cdot [B(E\lambda, \omega) + B(M\lambda, \omega)] \qquad 17.$$

and the ground-state decay width of an excited level with spin $J$ amounts to

$$\Gamma_0 = 8\pi\alpha \sum_{\lambda=1}^{\infty} \frac{(\lambda+1)\omega^{2\lambda+1}}{\lambda[(2\lambda+1)!!]^2} \frac{2J_0+1}{2J+1} \cdot [B(E\lambda, \omega) + B(M\lambda, \omega)]. \qquad 18.$$

The transition probabilities $B(\lambda, q)$ as extracted from $(e, e')$ experiments can be related to the photon scattering data by an extrapolation of the form factors to the photon point. The photon point is not accessible in electron scattering since it corresponds to a scattering angle of $0°$. Furthermore the increased background of the radiation tail of the elastic peak makes $(e, e')$ experiments extremely difficult at forward scattering angles. In addition one deals with rather low cross sections near the photon point since the form factors scale proportionally to $q^\lambda$. The model-independent PWBA analysis of electron scattering data is limited to experiments on light nuclei and to the use of high electron energies. In a realistic analysis, in particular when dealing with heavy nuclei, the necessary DWBA treatment leads to a certain model dependence of the results. Therefore, the comparison of $(e, e')$ and $(\gamma, \gamma')$ data is of crucial importance in many cases. The information from virtual and real photon work complement one another and supply detailed insights into nuclear structure. As an example of the power of this tool, form factors of the 2.974-MeV $1^+$ state in $^{156}$Gd (see Section 5.3), predicted by IBA-2 and two-quasi-particle $(2\nu f_{7/2} \rightarrow 2\nu f_{5/2})$ calculations (14, 12), are depicted in Figure 6. The limited,

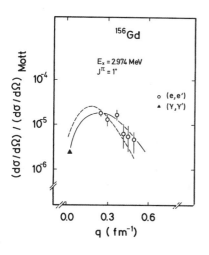

*Figure 6* Form factor of the 2.974-MeV state in $^{156}$Gd as calculated in IBA-2 (*solid line*) (14) and a two-quasiparticle $(2\nu f_{7/2} \rightarrow 2\nu f_{5/2})$ calculation (*dashed line*) (12) in comparison with $(e, e')$ and $(\gamma, \gamma')$ data (14).

accessible $q$ range makes it impossible to distinguish between the two model calculations on the basis of the $(e, e')$ data only. However, both calculations differ considerably at the photon point. Therefore, the photon scattering data (13, 14) enable one to reach a clear conclusion as to which model is appropriate.

## 3.2   Experimental Progress

During the past years considerable progress was made in experimental techniques for investigating nuclear magnetic excitations with electromagnetic probes. In present day inelastic electron scattering experiments, measurements can be performed with resolutions comparable to those obtained in hadron-induced reaction studies. An important innovation has been the development of energy loss spectrometer arrangements. By matching the dispersion of the beam transport system to that of the spectrometer, electrons with a certain fixed energy loss were focussed into a single spot of the focal plane, independent of the momentum uncertainty of the incoming beam. Typical relative energy resolutions $dE/E$ of some $10^{-4}$ can be achieved at modern electron spectrometers working in the energy loss mode. Furthermore, these installations make available beam intensities higher than conventional spectrometer arrangements because a rather broad momentum interval can be used.

An example of this technique's success is the energy loss arrangement at the 70-MeV Darmstadt electron accelerator (DALINAC) (45), where systematic studies of magnetic excitations have been performed by low-energy, but high-resolution inelastic electron scattering experiments (8, 9). The excellent experimental performance of this facility made possible, for example, the recent discovery of low-lying collective M1 excitations (2–3 MeV) in heavy deformed nuclei (12) (see Section 5.3).

In photon scattering work essential experimental developments can be stated too. The availability of modern electron accelerators led to an intensified use of bremsstrahlung photon beams in nuclear resonance fluorescence (NRF) experiments. This technique has the advantage that all states with considerable ground-state widths can be excited simultaneously as a result of the continuous energy spectrum of bremsstrahlung. There are no limitations as in experiments using monenergetic photons from $(n, \gamma)$ capture reactions, where a random overlap of the incident $\gamma$-ray energy and the energy of the state to be excited is necessary (46, 47). Since the pioneering work in photon scattering of bremsstrahlung on low-lying nuclear states carried out by Metzger (48), high-current electron linear accelerators (linacs) in a wide energy range are in operation; furthermore high-resolution $\gamma$-ray spectrometers of good efficiency are now available [large volume Ge(Li) or Ge(HP) detectors].

The high beam intensities of present electron accelerators make it possible for the first time to use off-axis bremsstrahlung with reasonable intensity as a source of linearly polarized photons for nuclear fluorescence experiments. Figure 7 shows the setup of the bremsstrahlung facility as installed at the Giessen linac (11, 49). The electron beam from the accelerator (mean currents up to 300 $\mu$A) is bent, its energy analyzed by a magnet system, and then focused on a bremsstrahlung radiator target. A 1 mm thick water-cooled tungsten target is used for the production of an intense unpolarized photon beam (about $10^9$ photons/s $\cdot$ MeV). Behind the bremsstrahlung target a dumping magnet cleans the photon beam from electrons. A sophisticated system of collimators and beam hardeners within a 3-m concrete wall delivers a well-collimated photon beam in the experimental area. Two sets of steering coils in front of the radiator target make it possible to change the angle of incidence of the electron beam on the target in order to produce linearly polarized off-axis bremsstrahlung. In this operation mode thin aluminum foils (12–50 $\mu$m) are used as a radiator. In the experimental area four germanium $\gamma$-ray detectors are installed. The degree of polarization can be measured online by a polarimeter based on the photodisintegration of the deuteron (see Section 4.2). The good collimation of the beam, the excellent shielding of the detectors (which reduces the background level), and the high beam intensity enabled systematic NRF studies at the Giessen facility (11, 50) even on enriched isotopes available only in quantities of a few grams.

Conventional electron linacs such as the Giessen machine suffer from

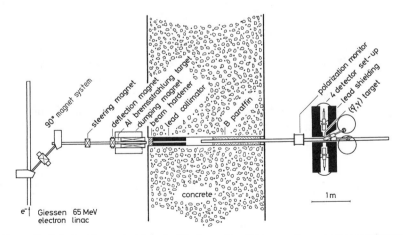

*Figure 7*  Polarized bremsstrahlung facility for nuclear resonance fluorescence experiments as installed at the 65-MeV Giessen electron linear accelerator (11).

their low duty cycle (on the order of 0.1%). Typical repetition rates are a few hundred hertz. Hence the possible counting rates of the $\gamma$-ray spectrometers are limited to less than 10% of the repetition rate in order to avoid pile-up effects. This drawback has been overcome by the construction of the first continuous wave (CW) electron accelerators such as the Microtron Using a Superconducting Linac (MUSL 2) at the University of Illinois (51) and the first stage (14 MeV) of the Mainz Microtron (MAMI) (52). At the continuous beams of such facilities the maximum counting rates in NRF experiments are limited only by the performances of the $\gamma$-ray spectrometers and amount to about 5 kHz as compared to about 100 Hz at conventional linacs.

A high duty cycle of the beam also lets one use coincidence techniques, e.g. an efficient production of monochromatic "tagged photons" (bremsstrahlung monochromator) (53). The use of monochromatic photons is a big advantage for the investigation of highly excited continuum states by photon scattering (54). Photon intensities of some $10^5 \gamma/s$ (in an energy bin of 50–100 keV) could be used in photon scattering experiments (55) at the tagged photon facility of the University of Illinois. Because of their low intensity, the scattered photons were detected by large-volume NaI/Tl $\gamma$-ray spectrometers in these experiments. The $\gamma$-ray energy resolution of the tagging electron spectrometer (50–100 keV), which is not sufficient to resolve individual bound nuclear states at increased level densities. However, interesting information about mean multipole strength distributions can be obtained from photon scattering with monochromatic tagged photons, in particular when using linearly polarized photons with an enhanced degree of polarization (55, 56) (see Section 4.2).

Besides the new CW electron accelerators, based on high-frequency electron linacs and particular recirculation systems, electrostatic accelerators represent very useful, unique machines for low-energy NRF experiments (48). At the Dynamitron accelerator of the University of Stuttgart (57) a continuous electron beam of up to 4 mA at a maximum energy of 4.3 MeV is available. A high-intensity bremsstrahlung facility has been set up by a Giessen/Köln/Stuttgart collaboration (58) and successfully used for the investigation of low-lying collective M1 excitations in heavy deformed nuclei (13, 14, 59).

## 4.   NUCLEAR RESONANCE FLUORESCENCE WITH LINEARLY POLARIZED PHOTONS

Resonant scattering of bremsstrahlung on an atomic nucleus has become an important spectroscopic method to determine spin vibrations and collective M1 modes in nuclei. The measurement of transition probabilities, spins, and parities in an NRF experiment is completely model independent. On the other hand, exciting a nucleus with photons is very selective; mainly dipole transitions will be induced and to a much lesser degree electric quadrupole transitions.

A breakthrough for NRF experiments with bremsstrahlung was the development of the linearly polarized bremsstrahlung beam facility at the University of Giessen 65-MeV electron linear accelerator (11, 49). With this experimental arrangement it was possible for the first time to determine parities of highly excited dipole states close to the particle threshold in an NRF measurement and to study systematically the spin-flip M1 resonance, which is located around 10 MeV.

## 4.1   *Formalism of Photon Scattering*

If an atomic nucleus is irradiated by a continuous photon spectrum, it can be excited by multipole radiation of order $L_1$ or $L_{1'}$ with an energy corresponding to the excitation energy of an excited state. This process and the quantities that influence the photon scattering cross section are shown in Figure 8. The probability of being excited depends on the ground-state transition width $\Gamma_0$. The excited level then decays back to the ground state of the nucleus or to a low-lying excited state. The multipole order of the radiation in the exit channel is $L_2$ or $L_{2'}$ (see Figure 8). The photon scattering intensity for scattering on a nuclear state is directly proportional to the ground-state decay width $\Gamma_0$ of a level and the branching ratio to the ground state $\Gamma_0/\Gamma$, where $\Gamma$ is the total decay width of a state including the decay widths to excited states $\Gamma_f$ and the particle decay widths, if the level is above particle threshold. The dependence of the photon scattering cross section, integrated over a single resonance, on the decay widths is

$$\frac{d\sigma(\hat{\gamma}, \gamma')}{d\Omega} = \frac{2J+1}{2J_0+1}(\pi\lambda)^2\Gamma_0\frac{\Gamma_f}{\Gamma}\frac{W(\theta, \phi)}{4\pi}, \qquad 19.$$

where $\lambda$ is the reduced wavelength of the absorbed photon.

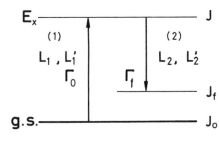

*Figure 8*  Definition of multipole orders $L$, decay widths $\Gamma$, and spins $J$; (1) and (2) denote entrance and exit channels.

If thin photon scattering targets are used, self-absorption within the target becomes negligible and the ground-state decay widths $\Gamma_0$ can be calculated directly from the photon scattering intensities by using Equation 19. An example of an NRF spectrum obtained in a photon scattering experiment on $^{26}$Mg was shown in Figure 2.

The multipole order of a $\gamma$-ray transition can be determined by measuring the angular intensity distribution $W(\theta, \phi)$ of the scattered photons. In a $(\gamma, \gamma')$ experiment, the axis of quantization is the direction of the well-collimated bremsstrahlung beam. The scattering $\theta$ angle is measured between bremsstrahlung beam and emitted deexcitation $\gamma$ ray; $\phi$ is the angle between scattering and polarization plane defined by the electrical field vector and the beam direction. The angles relevant for photon scattering are shown in Figure 9.

The angular distributions for dipole and quadrupole radiation from a nucleus with ground-state spin zero are plotted in Figure 10. It is sufficient in this case to measure the intensities of scattered photons at two angles (e.g. at 90° and 127°) in order to determine the multipolarity of a $\gamma$-ray transition.

If the bremsstrahlung beam is linearly polarized, an azimuthal asymmetry of the scattered photons will be observed, which in turn can be used to determine the parity of an excitation. The angular distribution for scattering of linearly polarized photons will be

$$W(\theta, \phi) = W(\theta) + (\pm)_{L_1} \frac{\cos 2\phi}{(1+\delta_1^2)(1+\delta_2^2)} \sum_\nu K_\nu(1) A_\nu(2) P_\nu^{(2)} (\cos \theta), \quad 20.$$

where $(\pm)_{L_1}$ equals $+1$ and $-1$ for electric and magnetic transitions, respectively. $W(\theta)$ is the angular distribution function for scattering of unpolarized photons. Mixing of the radiations of different multipole orders in the entrance channel (1) and the exit channel (2) is determined by the mixing ratio $\delta$. Numerical values for the angular correlation coefficients

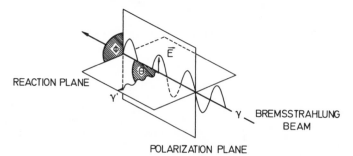

REACTION PLANE

$\gamma'$

$\bar{E}$

$\gamma$  BREMSSTRAHLUNG BEAM

POLARIZATION PLANE

*Figure 9*  Definition of photon scattering angle $\theta$ and azimuthal angle $\phi$ being relevant in NRF experiments with linearly polarized bremsstrahlung.

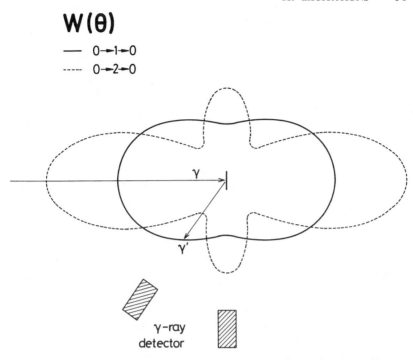

*Figure 10*  Angular distribution for pure dipole and quadrupole scattering on a ground-state spin $J_0 = 0$ nucleus.

$A_\nu(1)$, $A_\nu(2)$, and $K_\nu(1)$ as well as for the unnormalized associated Legendre functions $P_\nu^{(2)}$ (cos $\theta$) can be found in the review article by Fagg & Hanna (60), "Polarization Measurements on Nuclear Gamma Rays."

Usually, the polarization of the bremsstrahlung beam is $< 100\%$. Therefore, the terms with cos $2\phi$ in Equation 20 have to be multiplied by the degree of photon polarization. Figure 11 shows the azimuthal elastic photon scattering intensities at $\theta = 90°$ for pure E1 and M1 excitations of a nucleus with ground-state $J = 0$. Polarization of the incoming photons is assumed to be 100%. In addition, the locations at which the $\gamma$-ray detectors are placed to measure scattering asymmetries are shown.

It is evident from Figure 11 that the number of photons scattered within the polarization plane $N_\parallel$ differ from the number of photons scattered perpendicular to it, $N_\perp$. The measured asymmetry

$$\varepsilon = \frac{N_\perp - N_\parallel}{N_\perp + N_\parallel} = P_\gamma \cdot \Sigma(\theta) \qquad\qquad 21.$$

depends on the degree of bremsstrahlung polarization $P_\gamma$ and the analyzing

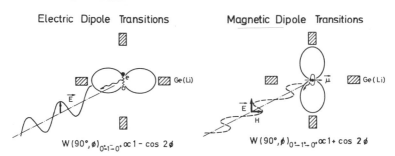

*Figure 11* Azimuthal angular distribution of photon scattering cross sections for nuclear resonance fluorescence experiments (electric and magnetic dipole excitations). In both illustrations the direction of the electric field vector $E$ of the incoming $\gamma$-ray beam was assumed to be the same. The radiation pattern of oscillating charges $e$ is perpendicular to the photon intensity distribution emitted by a magnetic dipole.

power $\Sigma(\theta)$ of the $(\vec{\gamma}, \gamma')$ reaction. Since the absolute value of $\Sigma(\theta = 90°)$ equals 1 for a spin 0-1-0 cascade, the magnitude of the measured asymmetry is determined by the degree of bremsstrahlung polarization $P_\gamma(E_x)$.

For electric dipole transitions, one will measure asymmetries of $1 > \varepsilon > 0$; for magnetic dipole transitions as well as for electric quadrupole transitions, the range will be $0 > \varepsilon > -1$.

These asymmetries, which are shown later in an asymmetry plot obtained from a measurement with polarized bremsstrahlung on $^{206}$Pb, are the basis for parity determination.

## 4.2    Production of Linearly Polarized Photons

As mentioned before, an urgent demand exists for sources of linearly polarized photons with a high degree of polarization, with a high spectral intensity (photons/s · eV), and with variable energy. Unfortunately, these requirements for an ideal source of polarized photons cannot be fulfilled by present techniques. Up to now the application of off-axis bremsstrahlung has been the most successful method for NRF experiments (11).

Since the 1950s, it has been known that off-axis electron bremsstrahlung is partially linearly polarized (61, 62). As schematically shown in Figure 12, the electric field vector $E$ of the bremsstrahlung photons is preferentially perpendicular to the emission plane of the photons and is aligned tangentially to a circle around the incident beam direction. The optimal off-axis angle is about $mc^2/E_0$ ($mc^2$ = electron rest energy, $E_0$ = electron bombarding energy). In order to select off-axis angles that correspond to different directions of the plane polarization, the reaction target can be moved around the central beam (54). Another possibility is to change, by steering coils, the angle of incidence of the electron beam on the

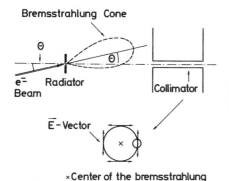

Bremsstrahlung Cone

Θ

e⁻    Radiator
Beam

Collimator

Ē-Vector

×Center of the bremsstrahlung

*Figure 12* The production of linearly polarized off-axis bremsstrahlung (schematically; see text).

bremsstrahlung radiator target and to select the off-axis angle by a subsequent fixed collimator. The electric field vector then can easily be switched in four directions, as indicated in Figure 12. Furthermore such an arrangement, as installed at the Giessen polarized bremsstrahlung facility (11, 49), offers the advantage of a fixed reaction target position, which allows a fully symmetric setup of four detectors (up, down, left, right). Therefore, all systematic asymmetries of the apparatus can be cancelled in first order.

The degree of polarization can be measured via the photodisintegration of the deuteron. The analyzing power of this fundamental photonuclear reaction has been studied extensively both theoretically (63) and experimentally (64). The analyzing power amounts to nearly unity for moderate energies ($< 30$ MeV) and emission angles of $90°$ (63, 64). This means the protons and neutrons are preferentially emitted in the direction of the electric field vector of the polarized bremsstrahlung radiation (for predominant E1 absorption). Because of the two-body disintegration of the deuteron, the energy of the photon inducing the reaction can be determined, even if one is using continuous bremsstrahlung, by measuring the proton or neutron energy. A typical polarimeter consists of four detectors arranged azimuthally around the photon beam. Two types have been realized, based on proton or neutron detection, respectively. When using proton detectors (Si surface barrier detectors or telescopes), the energy dependence of the degree of polarization can be determined more easily (49). However, the neutron detection technique (liquid scintillation counters with pulse shape discrimination) offers the advantage of high counting rates, since thick targets can be used (11).

In Figure 13 a typical dependence of the degree of polarization $P_\gamma(E_\gamma)$ on the photon energy is plotted. $P_\gamma$ is zero at the bremsstrahlung endpoint energy $E_0$ (energy of the incident electrons) and amounts to about 30% at

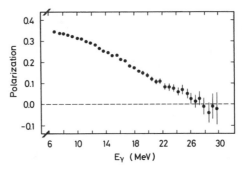

*Figure 13*    Degree of polarization of off-axis bremsstrahlung as a function of the photon energy ($E_0 = 30$ MeV, $\theta = 1.4°$) (49).

one third of $E_0$. The photon flux for moderate energies ($E_0 < 30$ MeV) is reduced by a factor of 20 to 30 as compared to that of unpolarized bremsstrahlung beams because of the lower bremsstrahlung yields at off-axis angles and the requirement of a thin radiator target (in order to avoid multiple scattering of the electrons). Therefore, high-current electron accelerators (mean currents of $\sim 100$ $\mu$A) are needed to obtain reasonable polarized photon intensities.

The degree of polarization can be considerably increased by collimation of both the bremsstrahlung quanta and the post-bremsstrahlung electrons. This technique as proposed by Laszweski et al (55) can be applied at tagged photon facilities (bremsstrahlung monochromator). Degrees of polarization on the order of 50% can be achieved at photon energies of about 10 MeV and bombarding energies of 20 MeV. Recently, extensive calculations, in particular of the polarization enhancement, were performed by Ahrens (65) and Sherman et al (66).

The application of polarized off-axis bremsstrahlung in photon scattering work is nicely demonstrated in Figure 14, where results are depicted from a $^{30}$Si$(\vec{\gamma}, \gamma')$ experiment (50). Four high-resolution Ge(Li) detectors were installed at azimuthal angles of 0, 90, 180, and 270° with respect to the polarization plane defined by the photon beam and the electric field vector. The upper part of Figure 14 shows the pulse height spectra recorded by detectors perpendicular to the polarization plane; in the lower part the corresponding spectrum taken by detectors parallel to the polarization plane is plotted. In $^{30}$Si two dipole transitions of 9.357 and 9.792 MeV have been observed in an NRF-experiment using unpolarized bremsstrahlung (67). The corresponding full-energy, single-escape, and double-escape peaks are marked in the spectra as M, M′, M″ and E, E′, E″ respectively. Enhancements of relative intensities are clearly evident: of the "E-peaks"

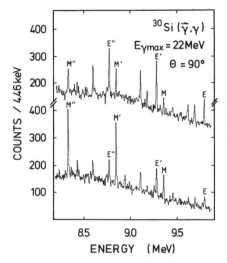

*Figure 14* The $(\vec{\gamma}, \gamma')$ spectra in an NRF experiment using polarized photons as recorded perpendicular (*upper part*) and parallel (*lower part*) to the polarization plane (see text).

in the upper spectrum and of the "M-peaks" in the lower spectrum. This measured azimuthal asymmetry now allows unique parity assignments (9.357 MeV: magnetic dipole, $1^+$; and 9.792 MeV: electric dipole, $1^-$). For nuclei with zero ground-state spin, such as $^{30}$Si, the analyzing power in a $(\vec{\gamma}, \gamma')$ reaction is maximal ($|\Sigma(90°)| = 1$) and changes its sign ($+1$ for electric and $-1$ for magnetic excitations, respectively). The magnitude of the observed asymmetry then equals the degree of polarization of the photon beam used.

## 5.  RESULTS AND DISCUSSION

Investigation of M1 resonances in nuclei has always been an interesting and challenging puzzle for nuclear physicists. LS closed shell nuclei, such as $^{16}$O or $^{40}$Ca where no M1 strength was presumed in a single shell model (68), showed considerable spin-flip strength; on the other hand, in jj closed nuclei, such as $^{90}$Zr or $^{208}$Pb where at least one M1 giant resonance is expected (69), it was difficult to identify the total spin vibration strength (70). Another surprise in the field of nuclear spectroscopy was the discovery of the rotational vibrations of protons against neutrons in deformed nuclei. The status of studies of M1 transitions has been reviewed by Richter (8, 9); here we discuss contributions from the investigation of M1 excitations with polarized and unpolarized bremsstrahlung.

### 5.1  *Closed Shell Nuclei*

5.1.1  $^{16}$O    Snover et al (71) found three $1^+$ states in $^{16}$O with a surprisingly large ground-state M1 strength totalling $B(M1)\uparrow = 0.72 \, \mu_0^2$. Some high-

lying M1 strength in addition was found in inelastic electron scattering
(72). This M1 strength has been explained by the mixing of core-excited
states into the ground state (68) and in second-order perturbation theory
(73).

The 16.2-MeV $1^+$ state in $^{16}$O could also be observed in an experiment
with polarized bremsstrahlung (49), and it was shown that the $(\vec{\gamma}, \text{particle})$
reaction provides new information about decay amplitudes of overlapping
unbound states of different parity and/or different multipolarity, information not obtainable from unpolarized measurements. The measured
$^{16}$O$(\gamma, p)$ spectrum is depicted in Figure 15 together with the observed
analyzing power of this reaction using polarized photons. A clear deviation
at 16.2 MeV of the measured from a calculated analyzing power, assuming
pure E1 absorption, gives direct evidence that a $1^+$ state was excited
overlapping with a broad E1 resonance. The $^{16}$O$(\vec{\gamma}, p)$ experiment yielded
decay amplitudes of the 16.2-MeV resonance for the proton decay to the
ground state of $^{15}$N of $|^1P|^2 = 0.28 \pm 0.010$ and $|^3P|^2 = 0.07 \pm 0.13$ for the
singlet and triplet amplitudes.

5.1.2  $^{40}$Ca   An unexpectedly strong isovector M1 transition was observed
in the spin-saturated nucleus $^{40}$Ca too (75, 76), which was explained by
intense ground-state correlations (75) and by second-order perturbation
theory (73). This excitation has also been investigated in an NRF experiment with unpolarized bremsstrahlung (77). The reduced transition probability of the 10.318-MeV level $[B(M1)\uparrow = 1.30 \pm 0.19 \ \mu_0^2]$ as obtained in
the photon scattering experiment agrees within the error bars with the

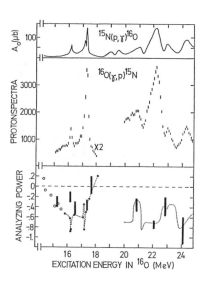

*Figure 15*  Nuclear photoeffect with polarized bremsstrahlung. The upper panel
shows the cross section from the inverse
reaction (74). The middle part gives the
measured $^{16}$O$(\gamma, p)$ cross section. In the
lower part the observed analyzing power
at 90° is presented as error bars together
with calculated values taking $a_2$ coefficients
from literature and assuming pure E1
absorption. The data are as follows: open
circles from (74); solid circles and dashed
line from (71); and dotted line from (71a).

results of the inelastic electron scattering experiment (75) ($1.12 \pm 0.07 \ \mu_0^2$), but in addition to the ground-state decay width, the partial decay widths to other states could be measured in NRF (77).

5.1.3 $^{90}$Zr    Inelastic scattering of 200-MeV protons reveals a pronounced bump at an excitation energy between 8 and 10 MeV (16); this bump has been assigned to the spin vibration mode in this nucleus. More recent $^{90}$Zr(p, p′) measurements with polarized protons were able to resolve fine structure in the M1 resonance bump and to settle the spin-flip character of that excitation (78).

High-resolution NRF experiments with unpolarized photons revealed that a bump observed at 9 MeV in a photon scattering measurement with tagged photons consisted of a large number of individual transitions (11). First results from a measurement with polarized bremsstrahlung, however, showed that none of the strongest dipole transitions observed in NRF are due to an M1 excitation (11). In this case the M1 excitation region is covered by very strong E1 transitions. On the other hand, the measurements with polarized photons show that the M1 resonance must be strongly fragmented between many states, because the strong transitions, for which parities could be determined, were E1 excitations. Here measurements with tagged polarized photons, which are currently carried out at Urbana, can help to determine total M1 and E1 strengths (56, 79).

5.1.4 $^{206}$Pb AND $^{208}$Pb    The NRF experiments with polarized bremsstrahlung led to the discovery of the isoscalar M1 transition in $^{208}$Pb (80). The isoscalar character of this state, which has also been excited in a $^{209}$Bi(d, $^3$He) reaction (81), was established later by inelastic electron electron and proton scattering experiments (82, 83). This M1 transition is explained in a two-state model as a destructive interference of neutron and proton spin-flip excitations:

$$|1^+\rangle = \alpha|\pi h_{11/2}^{-1}h_{9/2}\rangle - \beta|\nu i_{13/2}^{-1}i_{11/2}\rangle, \qquad\qquad 22.$$

with $\beta < 0$ and $\alpha^2 + \beta^2 = 1$. The discovery of the isoscalar magnetic dipole transition in $^{208}$Pb at $E_x = 5846$ keV solved a long-standing problem concerning the isoscalar M1 strength in this nucleus. The present state of knowledge concerning M1 strength in $^{208}$Pb has been summarized by Laszewski & Wambach (84). The existence of the isoscalar M1 state in $^{208}$Pb prompted the question of whether this $J^\pi = 1^+$ state survives when the two $3p_{1/2}$ neutrons close to the Fermi surface are removed.

Figure 16 shows an asymmetry plot from an NRF experiment with polarized bremsstrahlung on $^{206}$Pb (85). Only two states exhibited a negative asymmetry (positive parity): the one at 4116 keV, which was a previously known $2^+$ state, and a state at 5800 keV. An angular distribution

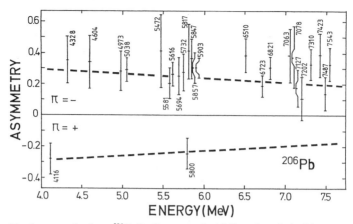

*Figure 16*   Asymmetries from $^{206}$Pb$(\vec{\gamma}, \gamma')$ measured with linearly polarized bremsstrahlung (85).

measurement gave evidence that a dipole transition had been detected. It is believed that the 5800-keV transition in $^{206}$Pb is an analog transition to the isoscalar M1 transition in $^{208}$Pb. A comparison of both M1 transitions in $^{206}$Pb and $^{208}$Pb from NRF is given in Table 2.

In addition, considerable isovector M1 strength of $19 \pm 2 \ \mu_0^2$ between 6.7 and 8 MeV has been detected in $^{206}$Pb by Laszewski et al using polarized tagged photons (56).

## 5.2   sd-Shell Nuclei

Two principal mechanisms are considered as likely sources of quenching M1 and Gamow-Teller excitation strength, namely (*a*) "core polarization," or mixing between the valence shell-model space and many highly excited orbits, and (*b*) subnucleonic effects, mesonic exchange currents, and, in particular, $\Delta$-hole excitations (86–88). Quantitative analysis of the degree of quenching must also take into account the effects of configuration mixing within the valence shell-model space.

In the following, a survey of dipole transition strengths to bound states in sd-shell nuclei is given. It was not possible to determine parities of excitations observed in earlier NRF experiments. With the polarized

**Table 2**   Comparison of the isoscalar M1 states in $^{208}$Pb and $^{206}$Pb

| Isotope | Excitation energy (keV) | $B$(M1) values ($\mu_0^2$) |
|---------|-------------------------|----------------------------|
| $^{208}$Pb | $5846 \pm 1$ | $1.6 \pm 0.5$ |
| $^{206}$Pb | $5800 \pm 1$ | $1.5 \pm 0.4$ |

bremsstrahlung facility at Giessen, parity determinations became feasible and most of the parities of transitions detected in NRF experiments (50) on even-even nuclei in the sd shell could be determined. In cases where spins were unknown, they were obtained by angular distribution measurements.

The $(\vec{\gamma}, \gamma')$ experiments were performed on $^{22}Ne$, $^{26}Mg$, $^{28}Si$, $^{30}Si$, $^{32}S$, and $^{34}S$. The results of these measurements and a comparison to shell-model calculations are shown in Figure 17. The observed M1 strength in the $T = 0$ isotopes of sd-shell nuclei shows a centroid energy of 11.0 MeV, if the two strong unbound M1 transitions to levels at 11.14 and 11.62 MeV in $^{32}S$ are included, as reported by Fagg and coworkers (89). Deviations are within 0.5 MeV. It has been pointed out by Kurath (90) that M1 strength should be concentrated in a few levels at the low-energy end of the $1^+$ level spectrum.

If two neutrons are added to the $T = 0$ nuclei, the picture changes drastically. A spreading of the M1 strength distribution is observed and strong E1 excitations appear in the realm of the $1^+$ states. The $B(M1)$ values of single transitions in $4N+2$ nuclei are, on the average, only 20 and 50% of those in the $4N$ nuclei. The experimental centroid energies of the $T_<$ components of the M1 resonance in $T = 1$ sd-shell nuclei are at 9.2 MeV, with deviations falling within a 1-MeV spread. These results were compared with shell-model calculations, and it was found that the reduction and spreading of M1 strength compared to the extreme jj limit of the shell model could be explained very well by taking into account intra-sd-shell configuration mixing (50). Only about 10% of the configuration mixed shell-model predictions of M1 strengths remain missing, to be accounted for in terms of subnucleonic effects and higher-order core polarization.

## 5.3  Deformed Nuclei

Strong, low-energy M1 excitations have been predicted for heavy, deformed nuclei by several nuclear models, as outlined in Section 2.2. Very recently this new magnetic dipole mode was discovered by Richter and coworkers (12) in rare earth nuclei by high-resolution inelastic electron scattering experiments at the Darmstadt linac. The new mode is of a rather pure orbital character since it could not be excited by inelastic proton scattering (91, 92) where the spin part of the M1 transition operator dominates. Detailed information about the distribution of the orbital, magnetic dipole strength can be extracted from a combined analysis of electron scattering and resonance fluorescence experiments. The reduced transition probabilities obtained by extrapolating the electron scattering form factors to the photon point can be compared with the $B(M1)\uparrow$ values measured directly at the photon point in photon scattering experiments.

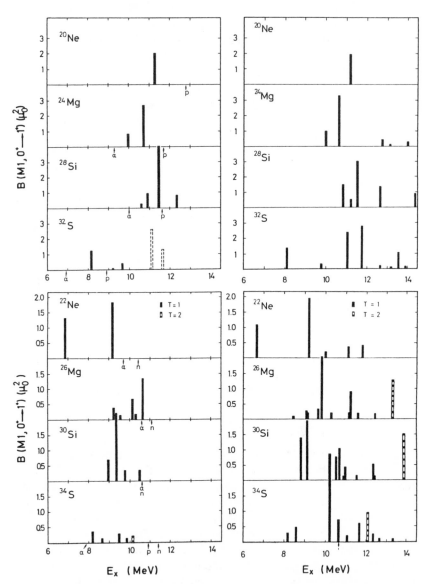

*Figure 17*   Reduced transition probabilities of bound $1^+$ states in $T = 0$ and 1 sd-shell nuclei from nuclear resonance fluorescence measurements performed at the Giessen electron linear accelerator and results from shell model calculations (50). The two $B$(M1) values plotted as dashed bars were obtained in $^{32}$S (e, e′) experiments (89). Particle emission thresholds are indicated by p, n, and $\alpha$.

As an example, the results for $^{156}$Gd are discussed in more detail (14). Figure 18 shows experimental spectra for $^{156}$Gd: in the upper part a $(\gamma, \gamma')$ spectrum as measured by a Ge(Li) detector at a bremsstrahlung endpoint energy of 3.5 MeV. The marked peaks indicate ground-state transitions. Satellite peaks shifted by 89 keV to lower energies correspond to transitions to the first excited $2^+$ states, respectively. The observed branching ratios $R = \Gamma_0/\Gamma_{2^+}$ amount to about 2, as is expected for $\Delta K = 1$ transitions within the validity of the Alaga rules (93). The angular distributions of the marked transitions show a clear dipole pattern. In the lower part of Figure 18 the sum of all background-subtracted $^{156}$Gd (e, e') spectra is plotted. Besides the most prominent peak at 3.070 MeV, five weaker magnetic transitions could be detected by the comparison with the $(\gamma, \gamma')$ data. The corresponding lines have been cross-hatched in the line decomposition of the (e, e') spectrum.

In the case of the strong 3.070-MeV transition, the behavior of the electron scattering form factor alone identified the excitation as being due

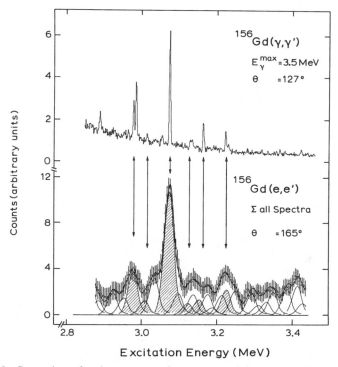

*Figure 18*  Comparison of nuclear resonance fluorescence and electron scattering spectra in the region of the new M1 mode (14).

to the new mode. However, for the weaker transitions, only a combined analysis of both the (e, e') and (γ, γ') experiments succeeded in distinguishing between orbital and spin excitations. This is nicely demonstrated in Figure 19, where theoretical M1 transition form factors are compared to the (e, e') and (γ, γ') data represented by circles and triangles, respectively. The dashed lines show form factors assuming a spin-flip excitation mechanism (12). The full lines correspond to microscopic IBA-2 calculations (14) (orbital excitation). For the weaker transitions, contributions of electric multipoles have been indicated in forward-angle electron scattering data. Therefore, appropriate Tassie model form factors have been added in these cases (14). Figure 19 clearly shows the crucial importance of the photon scattering data enabling an unambiguous distinction between the different form factor calculations. The form factors of all discussed transitions (except the weak 3.158-MeV transition) are well described by the IBA-2 calculations and therefore can be ascribed to the new orbital magnetic mode. Independently, the positive parity of the 3.070-MeV state has been confirmed by a $(\vec{\gamma}, \gamma')$ experiment at the Giessen linac using linearly polarized bremsstrahlung (94).

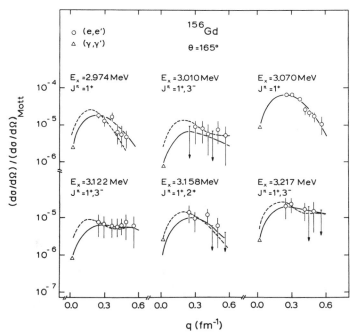

*Figure 19*  M1 transition form factors in comparison with data from electron scattering and nuclear resonance fluorescence experiments (14). The downward-pointing arrows represent upper limits.

The experimental results of the $(\gamma, \gamma')$ and $(e, e')$ experiments are in good agreement. The observed orbital M1 strength is concentrated in five transitions near 3.1 MeV and amounts to $2.1 + 0.3$ $\mu_0^2$ in the $(\gamma, \gamma')$ work compared to $2.3 + 0.5$ $\mu_0^2$ in the $(e, e')$ experiment. This strength lies below the value of about 3.8 $\mu_0^2$ as predicted by the IBA-2 model in the SU(3) limit (39).

Evidence for the collective magnetic dipole mode in further deformed nuclei in the rare earth region came from systematic $(e, e')$ experiments performed by the Darmstadt group (95). In the energy range 3.1–3.5 MeV, strong M1 transitions $(1-1.5$ $\mu_0^2)$ of orbital character could be detected in the isotopes $^{154}$Sm, $^{158}$Gd, $^{164}$Dy, $^{168}$Er, and $^{174}$Yb. In $^{164}$Dy a splitting of the strength (about 50 keV) into two main components has been observed and may be connected with a triaxial nuclear deformation (41).

These electron scattering experiments have been complemented by recent NRF experiments of a Giessen/Köln/Stuttgart collaboration. The excellent energy resolution of modern $\gamma$-ray spectroscopy (about 3 keV at 3 MeV) and the optimized arrangement at the Stuttgart bremsstrahlung facility (58) led to an increased detection sensitivity of about 0.1 $\mu_0^2$ for M1 transitions in the excitation energy range of about 3 MeV. These experiments are powerful tools for investigating the fine structure and fragmentation of the orbital magnetic dipole strength. Final results have been obtained for the even Gd isotopes $(A = 156, 158, 160)$ and the even Dy isotopes $(A = 160, 162, 164)$ (13, 14, 59, 96). As an example of the observed fragmentation of the strength, the results for the Gd isotopes are plotted in Figure 20. The fragmentation increases with higher mass numbers. The shift of the centroid excitation energy roughly follows the expected 66 $\delta A^{-1/3}$ MeV dependence (28). The total strength, assuming all excitations can be ascribed to the orbital M1 mode, is about 2.4 and 2.7 $\mu_0^2$ for $^{158}$Gd and $^{160}$Gd, respectively. In the investigated Dy isotopes, the strength is concentrated mainly in two or three states of comparable transition strength (59, 96). The center of excitation energies is shifted from about 2.8 MeV in $^{160}$Dy to 3.15 MeV in $^{164}$Dy. The total strength observed in this energy range is about 2.5–3.6 $\mu_0^2$.

The new M1 mode is not restricted to the rare earth region. It can also be observed in the other well-known island of deformed nuclei, the actinide region. In spite of the increased experimental difficulties in both electron and photon scattering on these heavy nuclei, a joint Darmstadt/Giessen/Köln/Stuttgart collaboration succeeded in detecting the orbital M1 mode in $^{238}$U and $^{232}$Th (59, 97, 98). In $^{238}$U four prominent transitions near 2.2 MeV could be observed in the $(\gamma, \gamma')$ experiments (see Figure 21) whereas in $^{232}$Th the strength seems to be mainly concentrated in one transition at 2.043 MeV. The corresponding electron scattering

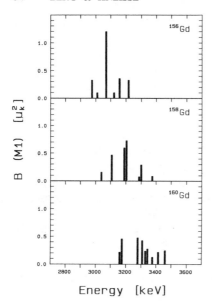

*Figure 20* M1 strength distribution in
$^{156}$Gd and strength distribution of $\Delta K = 1$
dipole transitions in $^{158,160}$Gd in the energy
range of the orbital M1 mode.

form factors show the M1 character of these excitations. Furthermore, in both isotopes weaker M1 transitions were detected at higher excitation energies: near 2.5 and 2.3 MeV respectively.

Similar collective M1 excitations occur even in light nuclei, as predicted first by Zamick for nuclei in the fp shell (99). The Darmstadt group detected a corresponding $1^+$ state in $^{46}$Ti at 4.3 MeV (93). In $^{48}$Ti, a triaxially deformed nucleus (100) in contrast to the symmetric rotator $^{46}$Ti, a pronounced splitting of the M1 strength (3.7 and 5.7 MeV) has been observed (101).

Comparing the experimental results with the theoretical values as summarized in Table 1, it is obvious that the excitation energies can be reproduced well by SRA, GAD, and RPA calculations whereas the TRM prediction overestimates the excitation energy. However, it should be mentioned that a microscopic calculation of the restoring force reduced the excitation energy as calculated in macroscopic models to fairly exactly the experimental values (102).

The experimental excitation energies can be used to adjust the Majorana parameter $\lambda$ of the IBA-2 Hamiltonian (see Equation 8). The ratio of the Majorana parameter to the deformation parameter $\delta$ turned out to be the same for nuclei of different deformations but of the same mass number $A$. The ratio follows the empirical relation

$$\lambda/\delta = 4.3(N_v N_\pi)^{-1/2},$$

where $N_v$, $N_\pi$ are the neutron and proton boson numbers respectively (98).

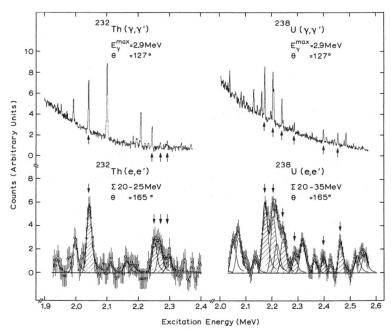

*Figure 21* Comparison of NRF and electron scattering spectra for $^{232}$Th and $^{238}$U (50, 97, 98). The marked peaks correspond to orbital M1 excitations. The strong peaks in the $^{232}$Th $(\gamma, \gamma')$ spectrum at 2103 keV ($^{208}$Pb, single-escape peak) and at 2212 keV ($^{27}$Al) are due to the radioactivity of the target and a calibration reaction, respectively.

The total strength as calculated in macroscopic and RPA models (see Table 1) seems to be too high in comparison to the experimental values. Furthermore, the $A$ dependence given by the RPA calculations (increasing strength proportional to $A^{4/3}$) cannot be confirmed by the experimental findings so far. The IBA-2 results (39) qualitatively agree with the trend of the observed $A$ dependence of the M1 strengths. In particular the predicted strength maximum at $A = 164$ has been confirmed by the (e, e') data (95). These findings could be corroborated by recent photon scattering experiments (59, 96) in which the highest $B$(M1) values have been found in $^{164}$Dy, too. The absolute values of the M1 strengths observed in the nuclei investigated so far are lower than the model predictions cited in Section 2.3 (TRM, SRA; RPA, GAD, and IBA-2). Closest to the experimental results are the IBA-2 predictions. As an upper limit an exhaustion of 70–90% of the IBA-2 sum rule values (in the rotational limit, bare boson $g$-factors) can be given for the Gd and Dy isotopes by summing up the strengths of all transitions detected in the high-sensitivity photon scattering experiments in the energy range of interest.

The current knowledge on the orbital M1 dipole mode was reviewed in more detail recently by Richter (97, 101). A survey of the most recent experimental and theoretical progress can be found in the proceedings of the International Conference of Nuclear Structure, Reactions, and Symmetries (Dubrovnik 1986) (103).

# 6.  CONCLUSION AND OUTLOOK

The investigation of M1 excitations is a very lively field in nuclear physics. It was not possible here to cover the whole field completely. Therefore this article focussed mainly on experiments with real photons. Some interesting contributions, both experimental and theoretical, were not discussed: the successful studies of the fp shell with protons (104), electrons (105), and polarized photons; the many approaches to explain the experimentally observed reduction of M1 strength compared to sum rules; or a large number of theoretical articles dealing with the orbital M1 mode.

Some highlights were pointed out, such as the discovery of the isoscalar M1 transitions in Pb nuclei or the discovery of spin vibrations in closed LS-shell nuclei. The importance of ground-state correlations was demonstrated in the systematic investigation of even-even sd-shell nuclei. That nuclear spectroscopy is still exciting was shown by the discovery of a so far unknown phenomenon, namely the orbital rotational oscillations of protons against neutrons in deformed nuclei.

Nuclear resonance fluorescence techniques with linearly polarized photons have progressed considerably. The development of new high-duty-cycle electron accelerators and the construction of new powerful sources of polarized synchrotron radiation (106) give confidence that this method of investigating nuclei has not been exhausted but has a very promising future.

ACKNOWLEDGMENTS

The authors are very grateful to the members of the Giessen photonuclear group for an engaging collaboration during many years: in particular to K. Wienhard, who initiated the Giessen polarized bremsstrahlung facility; to our longtime coworkers K. Ackermann, K. Bangert, C. Bläsing, R. D. Heil, and R. Stock; and to W. Arnold and his linac team. It is a pleasure to thank C. Wesselborg, P. von Brentano (Köln); D. Bohle and A. Richter (Darmstadt), and B. Fischer, H. Hollick, and D. Kollewe (Stuttgart) for a pleasant and stimulating collaboration in the investigations of the new orbital M1 mode. These experiments have been performed at the Stuttgart Dynamitron, where we enjoyed the kind hospitality of the Institut für

Strahlenphysik, under the guidance of Prof. K. W. Hoffmann. A fruitful collaboration with the photonuclear group at Urbana (Illinois) is much appreciated.
The financial support of the Deutsche Forschungsgemeinschaft is gratefully acknowledged.

*Literature Cited*

1. Doering, R. R., Galonsky, A., Patterson, D. M., Bertsch, G. F. *Phys. Rev. Lett.* 35: 1691–93 (1975)
2. Horen, D. J., Goodman, C. D., Bainum, D. E., Foster, C. C., Goulding, C. A., et al. *Phys. Lett.* 99B: 383–86 (1981)
3. Gaarde, C., Larsen, J. S., Rapaport, J. See Ref. 4, pp. 65–89
4. Petrovitch, F., Brown, G. E., Garvey, G. T., Goodman, C. D., Lindgren, R. A., Love, W. G., eds. *Spin Excitations in Nuclei.* New York: Plenum (1984)
5. Djalali, C. *J. Phys. C* 4: 375–87 (1984)
6. Barber, W. C. *Ann. Rev. Nucl. Sci.* 12: 1–42 (1962)
7. Fagg, L. W. *Rev. Mod. Phys.* 47: 683–711 (1975)
8. Richter, A. *Proc. Int. Conf. on Nucl. Struct.,* Florence, 1983, ed. P. Blasi, R. A. Ricci, pp. 189–217. Bologna: Tipografica Compositori (1983)
9. Richter, A. *Prog. Part. Nucl. Phys.* 13: 1–62 (1985)
10. van der Bijl, L. T., Blok, H., Blok, H. P., Ent, R., Heisenberg, J., et al. *J. Phys. C* 4: 465–69 (1984)
11. Berg, U. E. P. *J. Phys. C* 4: 359–73 (1984)
12. Bohle, D., Richter, A., Steffen, W., Dieperink, A. E. L., Lo Iudice, N., et al. *Phys. Lett.* 137B: 27–31 (1984)
13. Berg, U. E. P., Bläsing, C., Drexler, J., Heil, R. D., Kneissl, U., et al. *Phys. Lett.* 149B: 59–63 (1984)
14. Bohle, D., Richter, A., Berg, U. E. P., Drexler, J., Heil, R. D., et al. *Nucl. Phys. A* 458: 205–16 (1986)
15. Petrovitch, F., Carr, J. A., McManus, H. *Ann. Rev. Nucl. Part. Sci.* 36: 29–81 (1986)
16. Crawley, G. M., Anantaraman, N., Galonsky, A., Djalali, C., Marty, N., et al. See Ref. 4, pp. 91–109
17. Love, W. G., Franey, M. A. *Phys. Rev. C* 24: 1073–94 (1975)
18. Berg, U. E. P. See Ref. 19, pp. 387–99
19. Goodman, C. D., Austin, S. M., Bloom, S. D., Rapaport, J., Satchler, G. R., eds. *The (p,n) Reaction and the Nucleon-Nucleon Force.* New York: Plenum (1980)
20. Anderson, B. D., McCarthy, R. J., Ahmad, M., Fazely, A., Kalenda, A. M., et al. *Phys. Rev. C* 26: 8–13 (1982)
21. Baldwin, G. C., Klaiber, G. S. *Phys. Rev.* 71: 3–10 (1947)
22. Goldhaber, M., Teller, E. *Phys. Rev.* 74: 1046–49 (1948)
23. Steinwedel, H., Jensen, J. H. D. *Z. Naturforsch.* 5a: 413–70 (1950)
24. Danos, M. *Nucl. Phys.* 5: 23–32 (1958)
25. Lo Iudice, N., Palumbo, F. *Phys. Rev. Lett.* 41: 1532–34 (1978); *Nucl. Phys. A* 236: 193–208 (1979)
26. De Franceschi, G., Palumbo, F., Lo Iudice, N. *Phys. Rev. C* 29: 1496–1509 (1984)
27. Lipparini, E., Stringari, S. *Phys. Lett.* 130B: 139–43 (1983)
28. Bes, D. R., Broglia, R. A. *Phys. Lett.* 137B: 141–44 (1984)
29. Kurasawa, H., Suzuki, T. *Phys. Lett.* 144B: 151–54 (1984)
30. Hamamoto, I., Åberg, S. *Phys. Lett.* 145B: 163–66 (1984); Internal Rep. Lund-MPh 86/07 Univ. Lund (Sweden)
31. Iwasaki, S., Hara, K. *Phys. Lett.* 144B: 9–12 (1984)
32. Hilton, R. R. *Z. Phys. A* 316: 121–22 (1984); *J. Phys. C* 6: 255–64 (1984)
33. Hammaren, E., Schmid, K. W., Faessler, A., Grümmer, F. *Phys. Lett.* 171B: 347–52 (1986)
34. Iachello, F., ed. *Interacting Bosons in Nuclear Physics.* New York: Plenum (1979)
35. Iachello, F. *Nucl. Phys. A* 358: 89c–112c (1981)
36. Dieperink, A. E. L. *Prog. Part. Nucl. Phys.* 9: 121–46 (1983)
37. Dieperink, A. E. L., Wenes, G. *Ann. Rev. Nucl. Part. Sci.* 35: 77–105 (1985)
38. Iachello, F. *Phys. Today* 38: 40–41 (1985)
39. van Isacker, P., Heyde, K., Jolie, J., Waroquier, M., Moreau, J., Scholten, O. *Phys. Lett.* 144B: 1–4 (1984)
40. Sambataro, M., Scholten, O., Dieperink, A. E. L., Piccito, G. *Nucl. Phys. A* 423: 333–49 (1984)

41. Palumbo, F., Richter, A. *Phys. Lett.* 156B: 101–2 (1985); Lo Iudice, N., Lipparini, E., Stringari, S., Palumbo, F., Richter, A. *Phys. Lett.* 161B: 18–20 (1985)
42. Überall, H. *Electron Scattering from Complex Nuclei*, Parts A and B. New York: Academic (1971); *Springer Tracts in Modern Physics* 49: 1–89. Berlin/Heidelberg/New York: Springer (1969)
43. Donnelly, T. W., Walecka, J. D. *Ann. Rev. Nucl. Sci.* 25: 329–405 (1975)
44. Heisenberg, J., Blok, H. P. *Ann. Rev. Nucl. Part. Sci.* 33: 569–609 (1983)
45. Walcher, Th., Frey, R., Gräf, H. D., Spamer, E., Theissen, H. *Nucl. Instrum. Methods* 153: 17–28 (1978)
46. Freund, A., ed. *Proc. Int. Workshop on application of intense capture gamma-ray sources*, *Nucl. Instrum. Methods*, Vol. 166 (1979)
47. Moreh, R. *Nucl. Instrum. Methods* 163: 275–76 (1979)
48. Metzger, F. R. *Prog. Nucl. Phys.* 7: 54–88 (1959); *Phys. Rev.* 187: 1700–4 (1969)
49. Wienhard, K., Schneider, R. K. M., Ackermann, K., Bangert, K., Berg, U. E. P. *Phys. Rev. C* 14: 1363–66 (1981)
50. Berg, U. E. P., Ackermann, K., Bangert, K., Bläsing, C., Naatz, W., et al. *Phys. Lett.* 140: 191–96 (1984)
51. Axel, P., Cardman, L. S., Hanson, A. O., Harlan, J. R., Hoffswell, R. A., et al. *IEEE Trans.* NS 24(3): 1133–35 (1977)
52. Herminghaus, H., Feder, A., Kaiser, K. H., Manz, W., von der Schmitt, H. *Nucl. Instrum. Methods* 138: 1–12 (1976)
53. O'Connell, J. S., Tipler, P. A., Axel, P. *Phys. Rev.* 126: 228–39 (1962)
54. Nathan, A. M., Starr, R., Laszewski, R. M., Axel, P. *Phys. Rev. Lett.* 42: 221–23 (1979)
55. Laszewski, R. M., Rullhusen, P., Hoblit, S. D., Le Brun, S. F. *Nucl. Instrum. Methods* 228: 334–42 (1985)
56. Laszewski, R. M., Rullhusen, P., Hoblit, S. D., Le Brun, S. F. *Phys. Rev. Lett.* 54: 530–33 (1985)
57. Hammer, J. W., Fischer, B., Hollick, H., Trauvetter, H. P., Kettner, K. U., et al. *Nucl. Instrum. Methods* 161: 189–98 (1979)
58. Heil, R. D. *Proc. Int. Symp. on Symmetries and Nucl. Struct. (SANS)*, Dubrovnik, 5–14 June (1986), ed. R. A. Meyer, F. Iachello, V. Paar, P. von Brentano. Singapore: World Sci. (1986)
59. Kneissl, U. See Ref. 103, 1: 362–67
60. Fagg, L. W., Hanna, S. S. *Rev. Mod. Phys.* 31: 711–58 (1959)
61. May, M., Wick, G. C. *Phys. Rev.* 81: 628–29 (1951)
62. Olsen, H., Maximon, L. C. *Phys. Rev.* 114: 887–904 (1959)
63. Partovi, F. *Ann. Phys.* 27: 79–113 (1964)
64. De Pascale, M. P., Giordani, G., Matone, G., Babusi, D., Bernabei, R., et al. *Phys. Rev. C* 32: 1830–41 (1985)
65. Ahrens, J. Internal rep. Max Planck Inst. Chemie, Mainz. Unpublished (1982)
66. Sherman, N. K., Aniel, T., de Miniac, A. *Natl. Res. Council—Rep. PXNR-2634.* Ottawa, Canada (1982)
67. Ackermann, K. PhD thesis. Giessen (1982)
68. Arima, A., Strottman, D. *Phys. Lett.* 96B: 23–25 (1980)
69. Bohr, A., Mottelson, B. R. *Nuclear Structure*, 2: 636–41. Reading, Mass: Benjamin (1975)
70. Brown, G. E., Raman, S. *Comments Nucl. Part. Phys.* 9: 79–88 (1980)
71. Snover, K. A., Ikossi, P. G., Trainor, T. A. *Phys. Rev. Lett.* 43: 117–20 (1979); Snover, K. A., Adelberger, E. G., Ikossi, P. G., Brown, B. A. *Phys. Rev. C* 27: 1837–65 (1983)
71a. O'Connell, W. J., Hanna, S. S. *Phys. Rev. C* 17: 892–902 (1978)
72. Küchler, G., Richter, A., Spamer, E., Steffen, W., Knüpfer, W. *Nucl. Phys. A* 406: 473–92 (1983)
73. Adachi, S., Lipparini, E., van Giai, N. *Nucl. Phys. A* 438: 1–14 (1985)
74. Earle, E. D., Tanner, N. W. *Nucl. Phys. A* 95: 241–70 (1979)
75. Gross, W., Meuer, D., Richter, A., Spamer, E., Titze, O., Knüpfer, W. *Phys. Lett.* 84B: 296–300 (1979); Steffen, W., Graf, H. D., Gross, W., Meuer, D., Richter, A., et al. *Phys. Lett.* 95B: 23–26 (1980); Steffen, W. PhD thesis. Technische Hochschule, Darmstadt (1984)
76. Crawley, G. M., Anantaraman, N., Djalali, C., Galonsky, A., Jourdain, J. C., et al. *Phys. Rev. C* 26: 87–90 (1982)
77. Moreh, R., Sandefur, W. M., Sellyey, W. C., Sutton, D. C., Vodhanel, R. *Phys. Rev. C* 25: 1824–29 (1982)
78. Nanda, S. K., Glasshausser, C., Jones, K. W., McGill, J. A., Carey, T. A., et al. *Phys. Rev. Lett.* 51: 1526–29 (1983)
79. Laszewski, R. M., Rullhusen, P., Hoblit, S. D., Le Brun, S. F. *Phys. Rev. C* 34: 2013–15 (1986)
80. Wienhard, K., Ackermann, K., Bangert, K., Berg, U. E. P., Bläsing, C., et al. *Phys. Rev. Lett.* 49: 18–21 (1982)
81. Hayakawa, S. I., Fujiwara, M., Imanishi, S., Fujita, Y., Katayama, I., et al. *Phys. Rev. Lett.* 49: 1624–27 (1982)

82. Müller, S., Richter, A., Spamer, E., Knüpfer, W., Metsch, B. C. *Phys. Lett.* 120B: 305–8 (1983); Müller, S., Küchler, G., Richter, A., Blok, H. P., Blok, H., et al. *Phys. Rev. Lett.* 54: 293–96 (1985)

83. Fujiwara, M., Fujita, Y., Katayama, I., Morinobu, S., Yamazaki, T., et al. *J. Phys. C* 4: 453–57 (1984)

84. Laszewski, R. M., Wambach, J. *Comments Nucl. Part. Phys.* 14: 321–40 (1985)

85. Ratzek, R., Berg, U. E. P., Bläsing, C., Jung, A., Schennach, S., et al. *Phys. Rev. Lett.* 56: 568–71 (1986)

86. Bertsch, G. F. *Nucl. Phys. A* 354: 157c–71c (1981)

87. Knüpfer, W., Dillig, M., Richter, A. *Phys. Lett.* 122B: 7–10 (1983); Knüpfer, W., Müller, W., Metsch, B. C., Richter, A. *Nucl. Phys. A* 457: 292–300 (1986)

88. Lawson, R. D. *Phys. Lett.* 125B: 255–59 (1983)

89. Burt, P. E., Fagg, L. W., Crannell, H., Sober, D. I., Stapor, W., et al. *Phys. Rev. C* 29: 713–21 (1984)

90. Kurath, D. *Phys. Rev.* 130: 1525–29 (1963)

91. Djalali, C., Marty, N., Morlet, N., Willis, A., Jourdain, J. C., et al. *Phys. Lett.* 164B: 269–73 (1985)

92. Wesselborg, C., Schiffer, K., Zell, K. O., von Brentano, P., Bohle, D., et al. *Z. Phys. A* 323: 485–86 (1986)

93. Alaga, G., Alder, K., Bohr, A., Mottelson, B. R. *Dan. Mat. Fys. Medd.* 29: 1–22 (1955)

94. Berg, U. E. P. Habilitation thesis. Giessen. Unpublished (1985)

95. Bohle, D., Küchler, G., Richter, A., Steffen, W. *Phys. Lett.* 148B: 260–64 (1984)

96. Wesselborg, C., Berg, U. E. P., von Brentano, P., Fischer, B., Heil, R. D., et al. *Proc. Int. Nucl. Phys. Conf., Harrogate, UK*, p. 88. Inst. Phys. (1986)

97. Richter, A. In *Nuclear Structure 1985*, ed. R. Broglia, G. B. Hagemann, B. Herskind, pp. 469–88. Amsterdam: Elsevier Sci. (1985)

98. Bohle, D., Guhr, Th., Hartmann, U., Hummel, K. D., Kilgus, G., et al. *Proc. Int. Symp. on Weak and Electromagnetic Interactions in Nuclei, Heidelberg*, July 1–5, 1986, ed. H. V. Klapdor, pp. 311–20. Heidelberg: Springer Verlag (1986)

99. Zamick, L. *Phys. Rev. C* 31: 1955–56 (1985)

100. Rebel, H., Hauser, G., Schweimer, G. W., Nowicki, G., Wiesener, W., Hartmann, D. *Nucl. Phys. A* 218: 13–42 (1974)

101. Richter, A. *Phys. Blätter* 49: 313–21 (1986)

102. Nojarov, R., Bochnacki, Z., Faessler, A. *Z. Phys. A* 324: 289–98 (1986)

103. Meyer, R. A., Paar, V., eds. *Proc. Int. Conf. on Nuclear Structure, Reactions and Symmetries (NSRS)*, Dubrovnik, 5–14 June 1986, Vols. 1, 2. Singapore: World Sci. (1986)

104. Marty, N., Djalali, C., Morlet, M., Willis, A., Jourdain, J. C., et al. *Nucl. Phys. A* 296: 145c–52c (1983); Djalali, C., Marty, N., Morlet, M., Willis, A., Jourdain, J. C., et al. *Nucl. Phys. A* 388: 1–18 (1982)

105. Eulenberg, G., Sober, D. I., Steffen, W., Gräf, H. D., Küchler, G., et al. *Phys. Lett.* 116B: 113–17 (1982); Sober, D. I., Metsch, B. C., Knüpfer, W., Eulenberg, G., Küchler, G., et al. *Phys. Rev. C* 31: 2054–70 (1985)

106. Chrien, R., Hofmann, A., Molinari, A. *Phys. Rep.* 64: 249–389 (1980)

*Ann. Rev. Nucl. Part. Sci. 1987. 37: 71–95*
*Copyright © 1987 by Annual Reviews Inc. All rights reserved*

# A COSMIC-RAY EXPERIMENT ON VERY HIGH ENERGY NUCLEAR COLLISIONS[1]

*W. V. Jones,[2] Y. Takahashi[3] and B. Wosiek[4]*

Department of Physics and Astronomy, Louisiana State University, Baton Rouge, Louisiana 70803, USA

*O. Miyamura*

Department of Applied Mathematics, Osaka University, Osaka 560, Japan

CONTENTS

[1] The JACEE Collaboration: T. H. Burnett, S. Dake, J. H. Derrickson, W. F. Fountain, M. Fuki, J. C. Gregory, T. Hayashi, To. Hayashi, R. Holynski, J. Iwai, A. Jurak, J. J. Lord, C. A. Meegan, H. Oda, A. Olszewski, T. Ogata, T. A. Parnell, E. Roberts, T. Saito, S. Strausz, T. Tabuki, T. Tominaga, J. W. Watts, J. P. Wefel, B. Wilczynska, H. Wilczynski, R. J. Wilkes, W. Wolter, and B. Wosiek.
[2] On temporary leave at NASA Headquarters, Astrophysics Division, Code EZ, Washington, DC, 20546, USA (Visiting Senior Scientist Program, California Institute of Technology, Jet Propulsion Laboratory).
[3] Now at the Department of Physics, University of Alabama in Huntsville, Huntsville, AL 35899, USA.
[4] On temporary leave from the Institute of Nuclear Physics, 30-055 Krakow, Poland.

71

## 1.  INTRODUCTION

The Japanese-American Cooperative Emulsion Experiment (JACEE) was conceived in October of 1978 for the dual purpose of studying the physics of high energy nucleus-nucleus collisions and for investigating the composition and energy spectra of primary cosmic rays at energies extending up to about $10^{15}$ eV. The collaboration involves four institutions in the USA (Louisiana State University, NASA Marshall Space Flight Center, University of Alabama in Huntsville, and the University of Washington), four institutions in Japan (Institute for Cosmic Ray Research of the University of Tokyo, Kobe University, Osaka University, and Waseda University) and one institution in Poland (Institute of Nuclear Physics, Krakow). This collaboration was formed to test the premise that balloon-borne exposures of large-area emulsion chambers could break the artificial barrier that had limited the exposure factor (product of area, solid angle, and exposure time) of conventional, electronic counter, cosmic-ray experiments to a few tens $m^2$-sr-hr. It was planned to expose a $0.8 \times 1$ m emulsion chamber, with an acceptance aperture exceeding $\pi$ steradians, at a nominal altitude of 4 $g/cm^2$ for approximately 40 hours during the high altitude wind-turnaround seasons, nominally May and September. A single flight could achieve approximately 50–100 $m^2$-sr-hr exposure, and it was hoped that annual flights in an evolving experiment would eventually accumulate about 1000 $m^2$-sr-hr total exposure.

The specific nuclear astrophysics objectives, which are not discussed in detail in this paper, include (a) measuring the relative abundances of the major charge groups (C+O, Ne+Mg+Si, and the Fe peak) for several energy intervals and (b) determining which of the currently popular models of cosmic-ray confinement (1–3) is most nearly correct by measuring the ratio of light (Li+Be+B) secondary nuclei to medium (C+O) source nuclei. The experiments were also designed to provide a bridge between the abundant low energy cosmic-ray data and the rare extensive air shower observations by measuring directly the charge composition and energy spectra from a few tens of GeV/amu up to about $10^{15}$ eV total energy.

The scientific objectives relating to the interactions of heavy nuclei included, among other things, searches for new channels of particle production at high nuclear energy densities and temperatures by (a) measuring transverse momenta, (b) studying particle correlations in the pionization region, and (c) looking for a signature of the postulated quark-gluon plasma (4–8). As a minimum, it was expected that the experiments would determine whether or not nucleus-nucleus interactions exhibit coherent particle production or other collective mechanisms that cannot be reduced to the superposition of independent nucleon-nucleon collisions. The experimental results would also check whether or not the apparent threshold around 100 TeV (1 TeV = $10^{12}$ eV) for "anomalous" cosmic-ray events (9) is real.

Table 1 gives some relevant statistics about the flights carried out to date. With the exception of the hybrid counter-emulsion experiment, all the flights employed similar payloads, even though some missions had unique emulsion chamber configurations. Seven separate flights (JACEE-0 through JACEE-6) have so far yielded a total exposure of about 350 $m^2$-sr-hr.

Since many of the research goals are still limited by statistics, long-duration exposures remain the highest priority. With the approval of an around-the-world balloon flight for January/February 1987, the original goal of accumulating about 1000 $m^2$-sr-hr exposures seems imminent. Measurements from that transglobal flight, if successful, will concentrate on the highest energy events, in order to accumulate data up to much higher energies than any previous direct observations of primary cosmic rays. Almost in parallel with this move to higher energies, a modification of the chamber design for the recent experiments has lowered the nominal energy threshold ($\sim 1$ TeV/amu for the earlier experiments) to about 200 GeV/amu.

Including the 20–65-GeV/amu energy range of the hybrid counter-emulsion experiment (JACEE-3), our present exposures collectively span more than four decades in energy. Although the coverage is not complete, there is sufficient overlap with the 15-GeV/amu beams from the Alternating Gradient Synchrotron (AGS) at the Brookhaven National Laboratory (BNL) and the 60–200-GeV/amu beams from the Super-Proton Synchrotron (SPS) at the European Center for Nuclear Research (CERN) to ensure accurate normalization of the entire data set. Note that our data do not extend down to the lower energies available at Dubna (4 GeV/amu) and the Bevalac (2 GeV/amu) at the Lawrence Berkeley Laboratory (LBL).

**Table 1**   JACEE balloon flights

| Flight | Date | Launch site | Residual altitude ($g/cm^2$) | Time (hr) | Area ($m^2$) |
|--------|------|-------------|------------------------------|-----------|--------------|
| JACEE-0 | 5/79 | Sanriku (Japan) | 8.0 | 29.0 | 0.20 |
| JACEE-1 | 9/79 | Palestine (USA) | 3.7 | 26.5 | 0.80 |
| JACEE-2 | 10/80 | Palestine (USA) | 4.0 | 29.6 | 0.80 |
| JACEE-3 | 6/82 | Greenville (USA) | 5.0 | 39.0 | 0.25 |
| JACEE-4 | 9/83 | Palestine (USA) | 4.5 | 59.5 | 0.80 |
| JACEE-5 | 10/84 | Palestine (USA) | 5.0 | 15.0 | 0.80 |
| JACEE-6 | 5/86 | Palestine (USA) | 5.0 | 30.0 | 0.80 |

Figure 1 illustrates the energy and rapidity ranges covered by present accelerator and balloon experiments, as well as by potential experiments at the planned Relativistic Heavy Ion Collider (RHIC). Note that, until RHIC is in operation, balloon experiments such as JACEE remain the only source of data on ultrarelativistic nuclear interactions.

## 2.  HIGH ENERGY COSMIC-RAY NUCLEI

It should be noted that cosmic rays, which appear to contain nuclei of all the elements found on earth, are observed to have a steep, differential power law, energy spectrum ($\propto E^{-2.7}$). With energies extending beyond $10^{20}$ eV, they have long been used as natural particle beams for unique studies of nuclear interactions.

Measurements of cosmic-ray abundances in the region 1–10 GeV/amu show that the cosmic-ray composition generally follows the solar system abundances, although there are some striking differences (10). For example, secondary nuclei, which result from fragmentations of the primary (source) nuclei during traversal of the interstellar medium, are considerably more abundant in cosmic rays than in solar system matter. In addition, it has been found experimentally that these secondary nuclei are relatively less abundant at higher energies (11–15), which implies that the amount of matter traversed by cosmic rays in the galaxy probably depends on the particle energy.

Prior to JACEE the only direct measurements available on the spectra

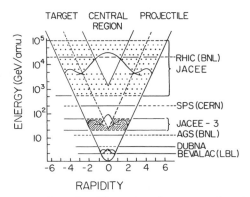

*Figure 1*  Laboratory energies and rapidity regions (target, central, and projectile) covered by current and proposed high energy, heavy ion accelerators and by the JACEE cosmic-ray experiments. The scale growth of the central rapidity range and particle density are illustrated for three representative energies. The rapidity, $y$, is defined as $y = \frac{1}{2} \ln [(E' + P_L)/E' - P_L)]$, where $E'$ is the energy of the secondary particle produced in the collision and $P_L$ is the longitudinal component (parallel to the beam direction) of its momentum, both in the center-of-mass frame.

of cosmic-ray nuclei with energies above a few TeV were those carried out by the PROTON satellites (16–18), in which an ionization calorimeter was used to make observations above 40 GeV. Those measurements extended up to about 2 TeV/amu for helium, 20 TeV for protons, and $10^{16}$ eV total energy for "all-particles." The helium and all-particle components maintained the same spectral power law index over the energy range covered, but it was reported that the proton integral spectral index changed from $-1.7$ to $-2.3$ at approximately 2 TeV. The proton and helium spectra measured by JACEE, which are presented in Figure 2 along with the results for heavy nuclei, show no major change in the spectral slopes over the energy interval 1–100 TeV/amu (19, 20). Although the JACEE results agree with extrapolations of the absolute fluxes and spectral indices of lower energy balloon data (21–23), an independent check of this apparent discrepancy is still needed.

For the purpose of this paper, the particle fluxes in Figure 2 illustrate the cosmic-ray beams available for nuclear interaction studies.

## 3. APPARATUS AND TECHNIQUES

Figure 3 shows a schematic diagram of a typical event profile in the emulsion chamber, which consists of interleaved layers of double-sided

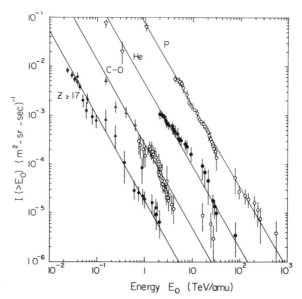

*Figure 2*  Integral energy spectra for protons (P), helium nuclei (He), the CNO-group nuclei (C-O), and heavy nuclei ($Z \geq 17$). Legend: $\bigcirc$ and $\bullet$, JACEE high energy data; $\blacksquare$, low energy JACEE-3 DATA; $\triangledown$, data from Ref. (21); $\triangle$ and $\blacktriangle$, data from Ref. (23).

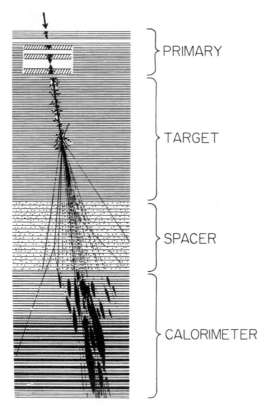

*Figure 3*  Schematic diagram of the JACEE emulsion chamber showing, from the top, the primary module, the target section, the spacer, and the calorimeter.

nuclear emulsion plates, x-ray films, etchable plastic (CR-39) detectors, low density spacers, and lead plates. The identifiable chamber segments are (from the top): (*a*) a primary charge module; (*b*) a target section; (*c*) a spacer section; and (*d*) a lead/emulsion/x-ray film calorimeter. In order to exploit the submicrometer spatial resolution of nuclear emulsions, all components of the chamber were assembled in $40 \times 50 \text{ cm}^2$ precision-machined, light-tight, plastic boxes. The plates were aligned with an accuracy better than 150 $\mu$m.

The JACEE-3 apparatus, shown schematically in Figure 4, was a hybrid instrument that combined an emulsion chamber with a gas Cherenkov counter, two solid Cherenkov counters, wire proportional counters, and a plastic scintillator shower counter. The counter system was used to determine the energy, charge, and trajectory of incident nuclei, thereby providing in situ, supplementary information for checking and calibrating

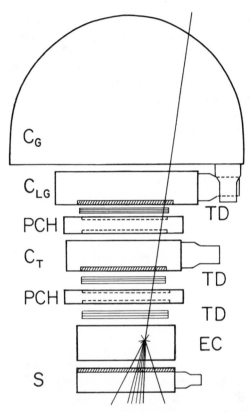

*Figure 4*  Schematic diagram of the hybrid counter-emulsion chamber instrument used for JACEE-3. Legend: $C_G$ = 1-atmosphere freon-12 gas Cherenkov counter; $C_{LG}$ = lead glass Cherenkov counter; PCH = proportional counter hodoscope; $C_T$ = Teflon Cherenkov counter; EC = emulsion chamber; TD = passive tracking detectors; S = plastic scintillator. A hypothetical track through the counters, with an interaction in the emulsion chamber, is illustrated.

the emulsion measurements. Because of the restricted geometry factor and the gas counter threshold, this was a "low energy" experiment. In addition to providing independent calibrations of the emulsion measurement methods, it yielded a systematic study of 20–65-GeV/amu iron-group nuclei interacting in plastic, emulsion, and lead targets.

## 3.1   *Emulsion Chambers*

The emulsion chambers satisfy several requirements for the experiment: (*a*) large geometrical factor for collection of events at high energies with efficient rejection of low energy background; (*b*) excellent resolution for angular measurements of charged secondary particles; (*c*) accurate charge measurements on the primary particles; and (*d*) reliable energy measurements (24). The emulsion plates consist of Fuji nuclear emulsion types 7B (electron-sensitive), 6B (alpha particle–sensitive), and/or 2F (heavy particle–sensitive) deposited on both surfaces of an 800 $\mu$m thick acrylic sheet. The dimensional stability of the acrylic base plates avoids problems of shrinkage and distortion usually associated with emulsion pellicles, thereby providing the accuracy needed for following tracks through the stack of plates.

In addition to the emulsion layers, the chambers include CR-39 plastic track detectors and x-ray films. The CR-39 sheets aid the scanning for heavy incident nuclei and provide an independent medium for a refined measurement of the particle charge. The x-ray films provide the scanning medium and, therefore, the "trigger" for large cascades generated by nuclear interactions in the emulsion chamber. The high energy cascades reveal themselves in the x-ray films as dark spots that can be seen during naked-eye scanning. The detection threshold, which depends on the background level, is typically about 1000 cascade electrons per mm$^2$, which corresponds, in turn, to an electromagnetic cascade of energy around 300 GeV.

## 3.2   *Event Measurements*

Events with x-ray film spots exceeding the selected threshold are located in an adjacent emulsion plate and then followed "plate-to-plate" with a microscope in the upstream direction to the first interaction vertex. The primary particle track is identified in the first emulsion layer above the vertex and followed further upstream to its entry into the chamber. Subsequently, the primary particle's charge is measured, its energy is determined, and the secondary particles emerging from the interaction site are studied (24).

PRIMARY CHARGE   Charge determinations are based on ionization measurements by the traditional methods of grain, gap, blob, and/or delta-ray counting (25) in the thick (200–400 $\mu$m) emulsions of the primary charge module. The emulsion layers in the target are also used to identify the $Z > 1$ projectile fragments. The charge resolution, which is limited mainly by the observable track length, has varied from $0.2e$ for protons and helium nuclei to $4e$ for iron nuclei measured in a single plate. The resolution for the highest charge events can, of course, be improved considerably by performing multiple measurements in several consecutive emulsion plates and using independent etch-pit mesurements in several CR-39 layers. For iron-group nuclei the charge resolution is typically better than one charge unit.

SECONDARY PARTICLES    The multiplicities and angular distributions of both the charged secondaries and the gamma rays from neutral particles produced in the interaction constitute the basic data avilable for analyzing the events. The emulsion chamber technique requires the incident particle to undergo a nuclear interaction inside the chamber. Interactions occur in the target section, which is comprised of thin (50–75 $\mu$m) nuclear emulsions alternated with either plastic sheets (in order to have a substantial mass of low-$Z$ material) or with thin plates of iron or lead (in order to have interactions in heavier material). The target section has a vertical thickness of approximately 0.25 proton interaction lengths, which corresponds to about 1.0 interaction mean free path for iron nuclei. Particles interacting elsewhere in the chamber, e.g. the calorimeter, can also be analyzed.

The track resolution of nuclear emulsion permits the hundreds of particles emerging from an interaction vertex to be unambiguously detected, and the charged-particle multiplicities $N_{ch}$ to be determined. The emission angles of the charged secondaries can be measured in emulsion plates downstream of the vertex with a typical error of 0.02–0.1 units over the pseudorapidity range $\eta = 0$–10, where $\eta$ is defined in terms of the polar emission angle $\theta$ as $\eta = -\ln \tan \theta/2$.

PRIMARY ENERGY    The energy measurements depend upon the development of high energy cascades in the calorimeter, which typically contains about seven radiation lengths of lead for the development of vertical cascades. Neutral pions emitted from an interaction site decay into gamma rays, which, together with any directly produced gamma rays, separate laterally in the spacer module before initiating cascades in the calorimeter. The energy in each cascade is determined by counting the number of associated electron tracks in several emulsion layers of the calorimeter and then comparing those numbers with the number of cascade electrons predicted by Monte Carlo simulations of the cascade development. Calibrations of this method by direct checks with cascades generated by electron beams (26) and by analyzing two-photon invariant mass distributions (24) indicate an accuracy of about 22% for the energy, $E_\gamma$, assigned to photon-induced cascades. Figure 5 shows typical cascade data plotted over calculated curves for different gamma-ray energies. The threshold energy is approximately 30–50 GeV for essentially 100% probability of detecting individual photon-initiated cascades.

From the total energy in gamma-ray-initiated cascades, $\Sigma E_\gamma$, the incident hadron energy $E_0$ is determined via the relation $\Sigma E_\gamma = k_\gamma E_0$, where $k_\gamma$, the average gamma-ray inelasticity, is typically about 0.1–0.2. The shape of the $k_\gamma$ distribution is relatively independent of energy, although it exhibits large fluctuations and depends on the mass of the primary particle (27).

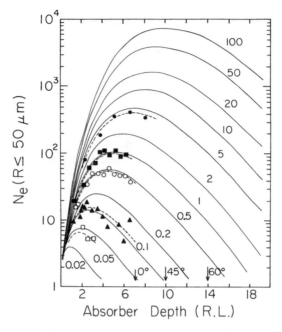

*Figure 5* Longitudinal development for cascades initiated by gamma rays, expressed as the number of electron tracks $N_e$ located within a circle radius $R \le 50$ μm about the cascade axis, given as a function of the absorber depth in radiation lengths (R.L.). The curves are labeled with the primary photon energy $E_\gamma$ in TeV. The dashed lines indicate fits to the data points for typical cascades. The arrows near the bottom show the maximum thickness of the calorimeter for photons incident at 0°, 45°, and 60° zenith angles.

The primary energy determinations are accurate to 40–50% (standard deviation) for individual proton interactions and improve somewhat for heavier nuclei because of the superposition of multiple cascades. Although a resolution of 25% is possible for some events, the average for the ensemble of interactions discussed here is about 40%, which is adequate for our experimental objectives.

## 3.3    *Experimental Material*

The data from JACEE-3 represent an unbiased sample of 20–65 GeV/amu iron-group ($Z > 22$) interactions in plastic (CHO), emulsion (Em), and lead (Pb) targets. The multiplicities and angular distributions of charged secondaries have been measured for about 130 of these low energy interactions (28). The other flights listed in Table 1 provide data collected under a $\Sigma E_\gamma$ trigger, which is the source of our data set on "high energy," above 1 TeV/amu, interactions of nuclei heavier than carbon in

CHO, Em, Pb, and iron targets (29). In addition to measurements of charged particles, the emission angles and energies of photons have been determined for about 120 of these high energy events. The measurements in the calorimeter, which are limited to the partially central and forward pseudorapidity regions, allow determination of individual photon transverse momenta $P_{T_\gamma}$ with an accuracy $\Delta P_{T_\gamma}/P_{T_\gamma} \approx 0.25$. The average transverse momentum $\langle P_{T_\gamma} \rangle$ for an event is estimated (30) from an exponential fit to either the differential or integral distribution of $P_{T_\gamma}$ values measured for all of the individual showers from an interaction. The average gamma-ray transverse momentum is related to the average transverse momentum of parent $\pi^0$ mesons through the relationship (31)

$$\frac{dN}{dP_{T_{\pi^0}}} = q_T \frac{d^2N}{dq_{T_\gamma}^2}, \tag{1.}$$

where

$$q_T = [P_{T_{\pi^0}} + (P_{T_{\pi^0}}^2 + m_\pi^2)^{1/2}]/2 \tag{2.}$$

and

$$q_{T_\gamma} = [P_{T_\gamma} + (P_{T_\gamma}^2 + m_\pi^2)^{1/2}]/2. \tag{3.}$$

For events with overlapping individual photon showers (e.g. interactions occurring in the calorimeter and/or extremely high energy interactions in the target), $\langle P_{T_\pi} \rangle$ is obtained by comparing the three-dimensional cascade development with Monte Carlo simulations. The simulations are based on the measured pseudorapidity distribution of the charged particles, assuming isospin symmetry for pions and an invariant $P_T$ distribution.

The energy densities $\varepsilon$ (GeV/fm$^3$) of the high energy interactions are evaluated at the time $t = 1$ fm/$c$ after the collision by using the formula proposed by Bjorken (32):

$$\varepsilon = \frac{3}{2} (\langle P_{T_\pi} \rangle^2 + m_\pi^2)^{1/2} \left. \frac{dN}{d\eta_c} \right/ (2\pi A_{\text{Min}}^{2/3}), \tag{4.}$$

where $A_{\text{Min}}$ is the atomic mass number of the smaller nucleus in the collision, $\langle P_{T_\pi} \rangle$ is the mean $\pi^0$ transverse momentum, $dN/d\eta_c$ is the measured density of charged particles in the central pseudorapidity region ($|\eta_{\text{CMS}}| \leq 1$), and $\eta_{\text{CMS}}$ denotes the laboratory pseudorapidity transformed to the proton-proton center-of-mass system (CMS).

# 4.  NUCLEUS-NUCLEUS COLLISIONS

Our data on nucleus-nucleus collisions are discussed both in terms of the conservative idea that a nucleus can be approximated as a cluster of free nucleons and in view of the exciting possibility of creating a new state of matter.

The conservative picture has evolved from studies of high energy hadron-nucleus collisions, in which the incident hadron and the produced secondaries were expected to undergo multiple scatterings within the target nucleus, thereby generating an intranuclear, hadronic shower. However, the accumulated data (33–35) showed that the number of particles produced in nuclear targets is only moderately greater than the number produced in a hydrogen target. This "nuclear transparency" was initially surprising because a hadron must penetrate several mean free paths of nuclear matter during collision with a nucleus. The absence of significant intranuclear cascading implies that secondary particle production is not an instantaneous process, i.e. a finite time ($\sim 1$ fm/$c$) is required in the particle's rest frame for its creation. Alternatively one may think that bare particles are produced instantaneously but a finite time is needed for them to dress in the gluon field and become observable, real hadrons. Consequently, because of time dilatation in the laboratory frame, the fast particles are hadronized outside the nucleus. Only the incident hadron and slow secondaries can re-interact inside the target nucleus (36, 37).

The so-called superposition models (38–43) are based on nuclear transparency, with two additional assumptions: ($a$) the incident hadron (or its constituents) undergoes independent collisions inside the nucleus and ($b$) slow particles modify only slightly the observed final state. Different degrees of nuclear transparency (nuclear stopping power) are incorporated into different models, which can be extended in a straightforward way from hadron-nucleus to nucleus-nucleus interactions (44–47).

The recent surge of interest in nucleus-nucleus collisions is related to the possibility of observing a new state of matter (4–8). The currently accepted theory of strong interactions (quantum chromodynamics) predicts that, under extreme conditions of high temperature and high density, ordinary nuclear matter may transit into a phase of deconfined quarks and gluons, i.e. a quark-gluon plasma. Different theories and models agree that the phase transition is likely to occur at energy densities exceeding $\sim 2$ GeV/fm$^3$ (4–8, 48–52).

A major difficulty in current quark-gluon plasma searches is that data interpretation must rely on qualitative features only; in particular on differences between expected manifestations of the plasma formation and predictions of more conventional particle production mechanisms. Among the hadronic signals, one expects high multiplicities of produced particles, an enhanced ratio of strange to nonstrange particles, high transverse momenta, and unusual event structure. Leptonic signals, such as direct photons emitted as plasma electromagnetic radiation or direct dileptons produced in quark-antiquark annihilation, should provide information about the early stage of plasma formation, particularly its temperature (5–7).

## 4.1  Inclusive Data

The inclusive characteristics of the JACEE data samples for average multiplicities ($\bar{N}$), multiplicity dispersions ($D$), and pseudorapidity distributions $dN_{ch}/d\eta$ for nucleus-nucleus (A + B) collisions indicate significant differences from the existing data on proton-nucleus interactions. For example, the ratio of the dispersion to the average multiplicity, $(D/\bar{N})_{AB}$, is approximately twice as large for nucleus-nucleus collisions as for proton-nucleus collisions (53). Their multiplicity distributions are compared in Figure 6. Although the nucleus-nucleus data are distinctly different from the proton-nucleus data, the former can be explained as a simple superposition of elementary collisions provided the spatial structure of nuclear density is taken into account. The observed differences, e.g. Figure 6, reflect simple, anticipated features, such as large impact parameter variations.

Figure 7 shows the inclusive pseudorapidity distributions for particles produced in 20–65-GeV/amu iron interactions in CHO, Em, and Pb targets. The dependence of the central pseudorapidity densities on target

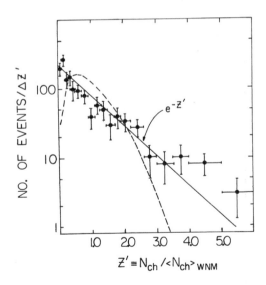

*Figure 6*  Normalized charged-particle multiplicity distribution for nucleus-nucleus interactions. The dashed curve represents a fit to proton-nucleus data (54); $\langle N_{ch} \rangle_{WNM}$ is the average multiplicity predicted by the wounded nucleon model (44).

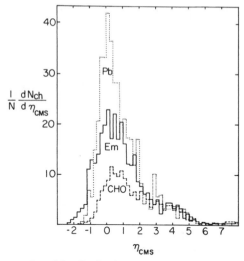

*Figure 7* Inclusive pseudorapidity distributions of the particles produced in 20–65-GeV/amu Fe-group interactions in three targets: lead (*dotted histogram*), emulsion (*solid histogram*), and CHO (*dashed histogram*).

mass B is observed to be approximately $B^{0.45 \pm 0.05}$, which is consistent with the $B^{0.4-0.5}$ dependence expected from superposition models (44, 46).

On the basis of the above observations, one may conclude that nuclei are "transparent" in interactions with other high energy nuclei in much the same way as they are transparent in collisions with single nucleons. On the other hand, inclusive secondary particle multiplicities and angular distributions are rather insensitive to the dynamics of the collision, so it would be premature to conclude on the basis of these data that nucleus-nucleus collisions are simply superpositions of nucleon-nucleon inter-actions. Superposition of independent, elementary collisions cannot fully explain the data presented below.

Figure 8 shows the discrepancy between the distribution for the number of wounded nucleons (nucleons that undergo at least one inelastic collision) calculated from the Glauber model (55) and the distribution for $N_{part}$, the number of nucleons that participated in an interaction as defined by

$$N_{part} = (Z_p - \Sigma Z_f - N_{sp}) A / Z_p,$$
                                                                                    5.

where $Z_p$ is the projectile charge, $Z_f$ is the heavy fragment charge, and $N_{sp}$ is the number of spectator nucleons (28). [Experimentally, $N_{sp}$ is deter-mined by consecutively adding all relativistic singly charged tracks, starting with the most forward one, until the root-mean-square of the emission angles satisfies Goldhaber's (56) evaporation formula, $\langle \theta_{lab}^2 \rangle^{1/2} = 0.12/E_0.$]

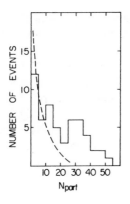

*Figure 8* Distribution of the number of participant nucleons $N_{part}$ in 55 "Fe"+ CHO collisions. The dashed curve indicates the Glauber model expectation for the number of wounded nucleons.

The larger spread (by about a factor of two) in the distribution of the participant nucleons can be understood in terms of rescattering of non-pion-producing nucleons inside the target nucleus, analogous to the (approximately factor of two) greater number of recoil nucleons relative to the number of proton-nucleon collisions in the proton-nucleus interactions (57, 58). However, it is not certain that every wounded nucleon in a heavy projectile nucleus can produce recoil nucleons in the same manner as in proton-nucleus collisions.

Another example of the inconsistency with a simple superposition picture is given in Figure 9, which illustrates the combined transverse momentum distribution from 23 C+CHO interactions at energies above 1 TeV/amu. This figure shows a flat high $P_T$ tail with exponential slope of $2\langle P_{T\gamma}\rangle \geq 1$ GeV/$c$ (59, 60). Accelerator data on proton-nucleon collisions also show high $P_T$ enhancements, but at much higher $P_T$ values (61, 62). No significant change in the slope of the $P_T$ distribution is expected from

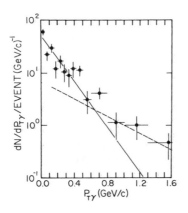

*Figure 9* Inclusive $P_{T\gamma}$ distribution in high energy ($E \geq 1$ TeV/amu) C+CHO collisions.

simple superposition models, even if multiple scattering inside the nucleus is taken into account (63). Other possibilities, such as the contribution of minijets (64) to the C+CHO data, have not yet been tested.

Frequent close pairings of secondaries were observed while measuring the pseudorapidities and azimuthal angles. Therefore, we have analyzed short-range two-particle correlations, which may be more sensitive to various interaction models than single-particle spectra (65, 66). The distributions of the azimuthal separation $dN/d(\Delta\phi)$ of two particles emitted close to one another in the pseudorapidity variable are shown in Figures 10 and 11, respectively, for our low energy and high energy data. The excess of close pairs with $\Delta\eta \le 0.1$ and $\Delta\phi < 30°$ relative to the uniform level is about 8–9%. Trivial backgrounds, e.g. Dalitz pairs, photon conversions in the vicinity of the vertex, and/or random pairs, account for less than one fifth of the observed excess. More interesting explanations, such as interference of identical Bose-Einstein particles (67–69) or decays of resonances with masses shifted to lower values (65, 70) (due to a partial restoration of the chiral symmetry that preceded the hadronization process) require better particle identification. Nevertheless, the observed data suggest significant contributions from such effects.

Production of electron pairs downstream of the vertex has been investigated in both high and low multiplicity interactions. The ratios of photons to charged hadrons in high multiplicity ($N_{ch} = 30–400$) events do not differ

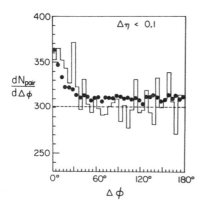

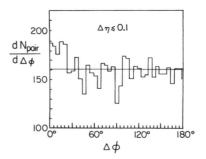

*Figure 10* Inclusive $\Delta\phi$ distribution for pairs with $\Delta\eta \le 0.1$ in 20–65-GeV/amu "Fe" collisions. Monte Carlo results for the Hanbury Brown-Twiss effect (65, 69), using the individual track data and a source radius of 3 fm, are given by the solid circles. The dashed line denotes a uniform background for $\Delta\phi > 30°$.

*Figure 11* Inclusive $\Delta\phi$ distribution for pairs with $\Delta\eta \le 0.1$ in high energy ($E \ge 1$ TeV/amu) interactions of nuclei heavier than carbon. The solid line indicates a uniform background for $\Delta\phi > 30°$.

much from the expectations based on charge symmetry, although for individual events this ratio varies from 0.3 to 2.5 (71, 72). For low multiplicity events (less than 15 tracks), numbers of parent photons moderately larger than predicted from isospin symmetry have been observed (73, 74).

## 4.2   Central Collisions

Inclusive data are dominated by peripheral collisions, whereas high density nuclear matter is expected to be created only in central collisions. Since the latter contribute only 5–10% of the total inelastic cross section, they are not readily studied with the statistics available from most cosmic-ray experiments. From the JACEE data, we have selected the highest multiplicity interactions, i.e. events with multiplicity several times the average multiplicity in the low energy data and events with more than 400 produced particles in the high energy data, as candidates for central collisions. Such events often have enough produced particles for statistically significant analysis from an individual interaction.

Characteristics of the highest multiplicity events for iron-group primaries from the low energy experiment are listed in Table 2. The number of produced pions in the forward direction $N_{\pi f}$ is several times the average multiplicity $\langle N_{\pi f} \rangle$ for each target nucleus. The observed central pseudo-rapidity densities given in Column 5 significantly exceed the calculated values (Column 6), which represent impact parameter $b = 0$ expectations from a superposition model (46).

The central collision events at high energies, which are mainly interactions in lead targets, are listed in Table 3. Figures 12 and 13 show the pseudorapidity distributions for the two highest multiplicity ($N_{ch} > 1000$) events, whose central rapidity densities, $dN/d\eta_c \approx 200$, are almost 100 times greater than those of proton-proton collisions at CERN Intersecting Storage Rings (ISR) energies ($<1.8$ TeV). The number of particles predicted, respectively, by the multichain model (46) and the wounded nucleon model (44) for the first two events in Table 3 are Ca + C event, 860 and 431; Si + Em event, 960 and 564. The multiplicities predicted by the multichain model, which takes into account the deceleration of

**Table 2**   Highest multiplicity events from JACEE-3

| A + B | $E_0$ (GeV/amu) | $N_{\pi f}$ | $\langle N_{\pi f} \rangle$ | $dN/d\eta_c$ (measured) | $dN/d\eta_c$ (calculated) |
|---|---|---|---|---|---|
| Fe + C | 54.0 | 101 | $19.7 \pm 2.7$ | 45 | 22 |
| Ti + Em | 28.0 | 124 | $27.3 \pm 7.3$ | 91 | 58 |
| Ti + Pb | 42.0 | 133 | $41.6 \pm 11.1$ | 97 | 60 |

**Table 3**   High energy central collision candidates ($N_{ch} > 400$)

| A + B | $E_0$ (TeV/amu) | $N_{ch}$ | $dN/d\eta_c$ | $\langle P_{T\pi} \rangle$ (GeV/c) | $\varepsilon$ (GeV/fm$^3$) |
|---|---|---|---|---|---|
| Ca + C | 100.0 | $760 \pm 30$ | $81 \pm 8$ | $0.53 \pm 0.04$ | 2.0 |
| Si + Em | 4.1 | $1010 \pm 30$ | $183 \pm 10$ | $0.55^{+0.12}_{-0.05}$ | 2.7 |
| V + Pb | 1.5 | $1050^{+300}_{-50}$ | $258 \pm 12$ | $0.55 \pm 0.05$ | 2.5 |
| Ti + Pb | 1.0 | 416 | $139 \pm 8$ | $1.00^{+0.2}_{-0.15}$ | 2.4 |
| Ca + Pb | 1.8 | 452 | $100 \pm 16$ | $1.1^{+1.0}_{-0.35}$ | 2.2 |
| Ca + Pb | 0.5 | $670 \pm 40$ | $142 \pm 8$ | $1.1 \pm 0.5$ | 3.3 |
| Si + Pb | 4.0 | $790^{+40}_{-25}$ | $147 \pm 8$ | $1.0^{+0.2}_{-0.15}$ | 3.8 |
| C + Pb | 71.5 | $400^{+15}_{-30}$ | $81 \pm 7$ | $\geq 1.2$ | 4.5 |

nucleons inside the nucleus, are consistent with our observations; however, the wounded nucleon model, which incorporates full nuclear transparency, is obviously incompatible with the data (75, 76).

The average transverse momenta given in Table 3 exceed the asymptotic values established by the CERN ISR and CERN Proton-Antiproton Collider experiments (61, 62). It should also be noted that the calculated energy densities for all the events listed in Table 3 exceed 2 GeV/fm$^3$, which is believed to be the critical density for quark-gluon plasma formation.

One strong signature of plasma formation would be a positive correlation between the average transverse momentum, which is related to the temperature, and the particle density, which measures the entropy. Qualitatively, one would expect the dependence shown in Figure 14 as the equation of state for a specific A + B combination (77). The JACEE nucleus-nucleus results displayed in Figure 15 (30, 78–81), show a gradual rise of average $P_T$ and an increasing spread of data points for high pseudorapidity densities. The large overall spread in the data may reflect the dispersion caused by mixing a variety of A + B combinations.

The correlation between $\langle P_{T\pi} \rangle$ and the energy density $\varepsilon$ for individual nucleus-nucleus interactions is displayed in Figure 16. The growth of $\langle P_T \rangle$ with increasing energy density is obvious, and the change is rather striking above 2 GeV/fm$^3$. Potential contributions to the observed increase may include multiple scattering (45), impact parameter effects (82), and contributions from quantum-chromodynamics minijets (64). While these effects have not yet been studied quantitatively, they could explain some

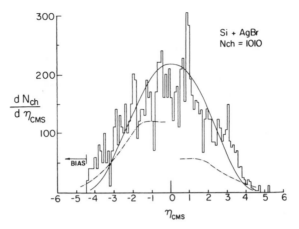

*Figure 12*  Pseudorapidity distribution for the high multiplicity ($N_{ch} > 1000$) Si+AgBr interaction at 4.1 TeV/amu. The solid and dashed curves denote expectations from the multichain (46) and wounded nucleon (44) models, respectively.

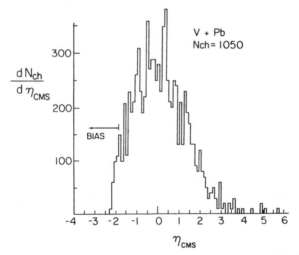

*Figure 13*  Pseudorapidity distribution for the high multiplicity ($N_{ch} > 1000$) V+Pb interaction at 1.5 TeV/amu.

of the increase of $\langle P_T \rangle$ with both increasing incident energy and mass number. On the other hand, our statistics for the events of greatest interest ($\varepsilon > 2$ GeV/fm$^3$) are still low, and any interpretation of the observed growth of $\langle P_T \rangle$ as the formation of new states of matter must await further tests with larger statistical samples.

Fluctuations in the angular distributions of final-state particles provide

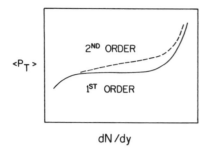

dN/dy

*Figure 14* Schematic representation of the expected dependence of average transverse momentum on rapidity density for first- and second-order phase transitions.

yet another signal for quark-gluon plasma formation. The violent flow resulting from release of the latent heat in the hadronization process (83, 84) is expected (85, 86) to produce either isolated maxima with a width of one rapidity unit or broader, bumpy regions with widths of a few rapidity units. Various structural fluctuations in the pseudorapidity and azimuthal distributions have been observed in the data. The events shown in Figures 12 and 13 are representative of several events exhibiting fluctuations with width $\Delta\eta \approx 0.1$–$0.2$ pseudorapidity units. Analysis of those fluctuations (87–89) indicated a nonstatistical origin, but their very short-range nature

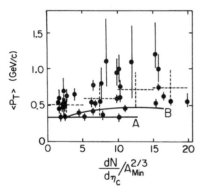

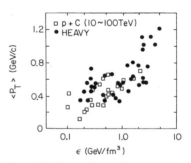

*Figure 15* Dependence of the average transverse momentum on the central rapidity density per unit colliding volume. The average dependence is indicated by dashed crosses. Curve $A$ denotes proton-proton data (61) up to 1.7 TeV with $\langle P_T \rangle = 0.34$ GeV/$c$. Curve $B$ represents the CERN proton-antiproton data (62) at 150-TeV laboratory energy.

*Figure 16* Dependence of average transverse momentum on energy density. The solid circles are for individual events ($2 \le Z_p \le 26$) and the open squares are for proton + CHO events.

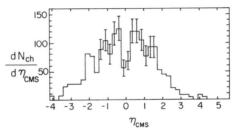

*Figure 17*  Example of an event with a "central-dip" in the $\eta$ distribution: $N_{ch} = 452$ for this 1.8-TeV/amu Ca + Pb interaction.

makes it difficult to relate them in a straightforward way to the violent process of plasma expansion.

Figure 17 is representative of a few events exhibiting a central dip in the $\eta$ distribution. It is interesting to note that this 1.8 TeV/amu, Ca + Pb interaction has the highest transverse momentum of all events recorded in the JACEE data, although it cannot be concluded from the present statistics that high $\langle P_T \rangle$ is correlated with the central dip.

Similarly, prominent structure exists in the azimuthal angle distribution of several high multiplicity events. Figure 18 shows both the $\eta$-$\phi$ scatter plot and the azimuthal angle distribution for the highest multiplicity Fe + CHO event in the low energy data, while Figure 19 shows the azimuthal distribution for the highest multiplicity event in the high energy data. Fourier analysis of the latter event indicates the dominant bimodal structure (88),

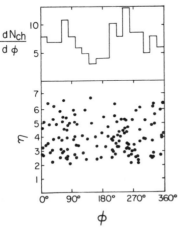

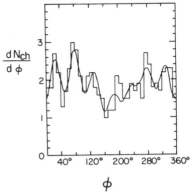

*Figure 18*  Azimuthal angle distribution (*top*) and $\eta$-$\phi$ scatter plot (*bottom*) for the 54-GeV/amu "Fe" + CHO interaction.

*Figure 19*  Azimuthal angle distribution for the 1.5-TeV/amu V + Pb interaction. The solid curve represents the results from Fourier analysis of the data.

as well as prominent higher order harmonics. The dipole nature may suggest collective behavior, but theoretical models are not yet quantitative enough for determining its significance.

## 5.  CONCLUDING REMARKS

In spite of the relatively low statistics, cosmic-ray data on very high energy nuclear collisions have shown that nucleus-nucleus interactions cannot be fully explained as a superposition of nucleon-nucleon interactions. Some of the data even suggest the onset of new, perhaps drastic, phenomena, although the current statistical and theoretical justifications are insufficient for drawing conclusions about their relationships to the postulated quark-gluon plasma and/or chiral-symmetry phase transitions.

Inclusive studies based on parameters such as $dN_{ch}/d\eta$, $N_{ch}/\langle N_{ch}\rangle$, and $D$ generally favor simple superposition models. However, the large excess of participant nucleons relative to Glauber model predictions requires other considerations, as do (a) the large $P_T$ tail observed in the inclusive transverse momentum distribution and (b) the short-range, two-particle correlations. The observed high multiplicities can be explained by multi-chain-type models, but not by wounded nucleon models that assume full nuclear transparency. Central collisions with very high multiplicities also exhibit high average transverse momenta, and the observed gradual rise of $\langle P_T\rangle$ with particle and/or energy density becomes increasingly dramatic above $2\ \text{GeV}/\text{fm}^3$. Some of the events exhibit structure in the distributions of both pseudorapidity and azimuthal angles.

Several of the questions raised by our cosmic-ray studies, as well as other phenomena predicted by the latest theories, should be clarified by the detailed investigations that have been initiated with the 15–200-GeV/amu, heavy ion beams at the BNL-AGS and the CERN-SPS during the past year. Explanations of other fundamental features of very high energy, heavy ion collisions may have to await higher energy, heavy ion accelerators, such as the Relativistic Heavy Ion Collider (RHIC).

Further progress in cosmic-ray studies of very high energy collisions requires an increase in collecting power by at least a factor of five over that already achieved by the JACEE exposures. This increase is possible with existing technology and balloon flight operations being used in long-duration flights that circumnavigate the globe in 14–18 days, with recovery near the original launch site. Unique opportunities for long exposures may become available in conjunction with the Space Station, currently scheduled for permanently manned operation in late 1994. Pioneering investigations with cosmic rays, followed by detailed, statistically reliable studies with ultrarelativistic accelerated heavy ion beams, should ultimately provide concrete information on the high density states of matter.

ACKNOWLEDGMENTS

This work has been supported in the USA by the National Science Foundation, which supports the efforts at Louisiana State University and the University of Alabama in Huntsville; by the Department of Energy, which supports the efforts at the University of Washington; and by the National Aeronautics and Space Administration, which supports the efforts at the Marshall Space Flight Center and, in addition, provides balloons and flight support at the National Scientific Balloon Facility (NSBF) in Palestine, Texas. We also acknowledge invaluable financial support from the Louisiana Center for Energy Studies.

In Japan the work has been supported by the Japan Society for the Promotion of Science, the University of Tokyo Institute for Cosmic Ray Research, and the Kashima Foundation. In Poland the efforts have been supported by the Institute of Nuclear Physics.

We wish to give special thanks to the NSBF staff for the excellent services in conducting the balloon flights and to the scanning and technical staffs of all the institutions participating in the project for their efforts.

*Literature Cited*

1. Peters, B., Westergaard, N. J. *Astrophys. Space Sci.* 48: 21–46 (1977)
2. Cowsik, R., Wilson, L. W. *13th ICRC Conf. Pap.* 1: 500–5. Denver, Colo: Univ. Denver (1973)
3. Ormes, J., Freier, P. *Astrophys. J.* 222: 471–83 (1978)
4. Chin, S. A. *Phys. Lett.* 78B: 552–55 (1978)
5. Shuryak, E. V. *Phys. Rep.* 115C: 151–314 (1984)
6. Satz, H. *Ann. Rev. Nucl. Part. Sci.* 35: 245–70 (1985)
7. Cleymans, J., Gavai, R. V., Suhonen, E. *Phys. Rep.* 130C: 217–92 (1986)
8. Gustafsson, H. A., Jakobsson, B., Otterlund, I., Aleklett, K. (eds., Proc. 2nd Int. Conf. on Nucleus-Nucleus Coll., Visby, Sweden). *Nucl. Phys.* A447: 1–700c (1985)
9. Halzen, F., Liu, H. C. *Phys. Rev. Lett.* 48: 771–73 (1982)
10. Simpson, J. A. *Ann. Rev. Nucl. Part. Sci.* 33: 323–81 (1983)
11. Smith, L. H., Buffington, A., Smoot, G. F., Alvarez, L. W., Wahlig, M. A. *Astrophys. J.* 180: 987–1010 (1973)
12. Webber, W. R., Lezniak, J. A., Kish, J. See Ref. 2, 6: 248–53
13. Juliusson, E. *Astrophys. J.* 191: 331–48 (1974)
14. Orth, C. D., Buffington, A., Smoot, G. F., Mast, T. S. *Astrophys. J.* 226: 1147–61 (1978)
15. Balasubrahmanian, V. K., Ormes, J. F. *Astrophys. J.* 186: 109–22 (1973)
16. Grigorov, N. L., Gubin, Yu. V., Rapoport, I. D., Savenko, I. A., Yakovlev, et al. *12th ICRC Conf. Pap.* 5: 1746–51. Hobart, Tasmania: Univ. Tasmania (1971)
17. Grigorov, N. L., Mamontova, N. A., Rapoport, I. D., Savenko, I. A., Akimov, V. V., et al. See Ref. 16, pp. 1752–59
18. Grigorov, N. L., Rapoport, I. D., Savenko, I. A., Nesterov, V. E., Prokhin, V. L. See Ref. 16, pp. 1760–68
19. Burnett, T. H., Dake, S., Fuki, M., Gregory, J. C., Hayashi, T., et al (The JACEE Collaboration). *Phys. Rev. Lett.* 51: 1010–13 (1983)
20. Burnett, T. H., Dake, S., Fuki, M., Gregory, J. C., Hayashi, T., et al. *19th*

*ICRC Conf. Pap.* 2: 32–35, 48–51. La Jolla, Calif: Univ. Calif. (1985)

21. Ryan, M., Ormes, J. F. and Balasubrahmanyan, V. K. *Phys. Rev. Lett.* 28: 985–88 (1972) and erratum *op cit.*, 1497 (1972)

22. Schmidt, W. K. H., Pinkau, K., Pollvogt, U., Huggett, R. W. *Phys. Rev.* 184: 1279–82 (1969)

23. Simon, M., Spiegelhauer, H., Schmidt, W. K. H., Siohan, F., Ormes, J. F., et al. *Astrophys. J.* 239: 712–24 (1980)

24. Burnett, T. H., Dake, S., Fuki, M., Gregory, J. C., Hayashi, T., et al. *Nucl. Instrum. Methods* A251: 583–95 (1986)

25. Powell, C. F., Fowler, P. H., Perkins, D. H. *The Study of Elementary Particles by the Photographic Method.* New York: Pergamon 669 pp. (1959)

26. Hotta, N., Munakata, H., Sakata, M., Yamamoto, Y., Dake, S., et al. *Phys. Rev. D* 22: 1–12 (1980)

27. Gaisser, T. K., Stanev, T. *Phys. Rev.* D28: 464–67 (1983)

28. Burnett, T. H., Dake, S., Fuki, M., Gregory, J. C., Hayashi, T., et al. *Phys. Rev.* D35: 824–32 (1987)

29. Burnett, T. H., Dake, S., Fuki, M., Gregory, J. C., Hayashi, T., et al. See Ref. 20, 6: 152–64

30. Takahashi, Y., Dake, S. (The JACEE Collaboration). *Nucl. Phys.* A461: 263–78 (1987)

31. Kopylov, G. I. *Phys. Lett.* 41B: 371–74 (1972)

32. Bjorken, J. D. *Phys. Rev.* D27: 140–51 (1983)

33. Ellias, J. E., Busza, W., Halliwell, C., Luckey, D., Swartz, P., et al. *Phys. Rev.* D22: 13–35 (1980)

34. Babecki, J., Czachowska, Z., Furmanska, B., Gierula, J., Holynski, R., et al. *Phys. Lett.* 52B: 247–8 (1974)

35. Otterlund, I. *Nucl. Phys.* A418: 87–116c (1984)

36. Feinberg, E. L. *Sov. Phys. JETP* 23: 132–40 (1966)

37. Gottfried, K. *Phys. Rev. Lett.* 32: 957–61 (1974)

38. Anisovich, V. V., Shabelsky, Yu. M., Shekhter, V. M. *Nucl. Phys.* B133: 477–89 (1978)

39. Bialas, A., Czyz, W., Furmanski, W. *Acta Phys. Pol.* B8: 585–89 (1977)

40. Nikolaev, N. N. *Phys. Lett.* B70: 95–8 (1977)

41. Capella, A., Tran Thanh Van, J. *Phys. Lett.* B93: 146–50 (1980)

42. Chao, W. Q., Chiu, C., He, Z., Tow, D. M. *Phys. Rev. Lett.* 44: 518–21 (1980)

43. Kinoshita, K., Minaka, A., Sumiyoshi, H. *Prog. Theor. Phys.* 61: 165–75 (1979)

44. Bialas, A., Bleszynski, M., Czyz, W.

*Nucl. Phys.* B111: 461–76 (1976)

45. Capella, A., Pajares, C., Ramallo, A. V. *Nucl. Phys.* B241: 75–98 (1985)

46. Kinoshita, K., Minaka, A., Sumiyoshi, H. *Z. Phys.* C8: 205–13 (1981)

47. Sumiyoshi, H. *Phys. Lett.* B131: 241–46 (1983)

48. Fucito, F., Solomon, S. *Phys. Rev. Lett.* 55: 2641–44 (1985)

49. Kogut, J. *Phys. Rev. Lett.* 56: 2557–60 (1986)

50. Fukugita, F., Ukawa, A. *Phys. Rev. Lett.* 57: 503–6 (1986)

51. Redlich, K., Satz, H. *Phys. Rev.* D33: 3747–52 (1986)

52. Nakamura, A. *Phys. Lett.* 149B: 391–95 (1984)

53. Wosiek, B. See Ref. 20, 9: 509–18

54. Abdurazakova, Y. A., Abduzhamilov, A., Abduzhamilov, Sh., Azimov, S. A., Barbier, L., et al. *Acta Phys. Pol.* B18: 249–53 (1987)

55. Glauber, R. J., Matthiae, G. *Nucl. Phys.* B21: 135–57 (1970)

56. Goldhaber, A. S. *Phys. Lett.* 53B: 306–8 (1974)

57. Hegab, M. K., Hufner, J. *Nucl. Phys.* A384: 353–70 (1982)

58. Suzuki, N. *Nucl. Phys.* A403: 553–71 (1983)

59. Burnett, T. H., Dake, S., Fuki, M., Gregory, J. C., Hayashi, T., et al. *AIP Conf. Proc.* 85: 552–61 (1982)

60. Jones, W. V. *Nucl. Phys.* A418: 139–60c (1984)

61. Breakstone, A., Campanini, R., Crawley, H. B., Dallawalle, G. M., Deninno, M. M., et al. *Phys. Lett.* 132B: 463–66 (1983)

62. Arnison, G., Astbury, A., Aubert, B., Bacci, C., Bernabei, R., et al. *Phys. Lett.* 118B: 167–72 (1982)

63. Pajares, C. *Nucl. Phys.* A418: 613–24 (1984)

64. Jacob, M. *CERN Preprint TH-3515* (1983)

65. Takahashi, Y., Parnell, T. A., Eby, P. B. See Ref. 20, 6: 133–37

66. Burnett, T. H., Dake, S., Fuki, M., Gregory, J. C., Hayashi, T., et al. See Ref. 20, 6: 168–71

67. Cocconi, G. *Phys. Lett.* 49B: 459–64 (1974)

68. Kopylov, G. I., Podgoretsky, M. *Sov. J. Nucl. Phys.* 18: 336–41 (1974)

69. Hanbury Brown, R., Twiss, R. O. *Nature* 177: 27–29 (1956)

70. Pisarski, R. D. *Phys. Lett.* 110B: 155–58 (1982)

71. Burnett, T. H., Dake, S., Fuki, M., Gregory, J. C., Hayashi, T., et al. *Phys. Rev. Lett.* 50: 2062–65 (1983)

72. Burnett, T. H., Dake, S., Fuki, M.,

Gregory, J. C., Hayashi, T., et al. *Proc. Int. Symp. on Cosmic Ray Super-High Energy Interactions*, pp. 7-17–7-22. Beijing, China: Acad. Sinica, Inst. High Energy Phys.

73. Burnett, T. H., Dake, S., Fuki, M., Gregory, J. C., Hayashi, T., et al. *Nucl. Phys.* A447: 197–202 (1985)

74. Burnett, T. H., Dake, S., Fuki, M., Gregory, J. C., Hayashi, T., et al. See Ref. 20, 6: 172–75

75. Burnett, T. H., Dake, S., Fuki, M., Gregory, J. C., Hayashi, T., et al. *AIP Conf. Proc.* 123: 723–34 (1984)

76. Burnett, T. H., Dake, S., Derrickson, J. M., Fountain, W., Fuki, M., et al. *Physics in Collisions* 5: 443–62. Yvette, France: Editions Frontiere (1985)

77. Van Hove, L. *Z. Phys.* C27: 135–44 (1985)

78. Burnett, T. H., Dake, S., Fuki, M., Gregory, J. C., Hayashi, T., et al. *Lect. Notes Phys.* 221: 187–95 (1985)

79. Burnett, T. H., Dake, S., Fuki, M.,

Gregory, J. C., Hayashi, T., et al. *Nucl. Phys.* A449: 189–96 (1985)

80. Burnett, T. H., Dake, S., Fuki, M., Gregory, J. C., Hayashi, T., et al. See Ref. 20, 6: 164–67

81. Burnett, T. H., Dake, S., Fuki, M., Gregory, J. C., Hayashi, T., et al. *Phys. Rev. Lett.* 57: 3249–52 (1986)

82. Barshay, S. *Phys. Rev.* D29: 1010–2 (1984)

83. Shuryak, E. V., Zhirov, O. V. *Phys. Lett.* 89B: 253–5 (1980)

84. Shuryak, E. V., Zhirov, O. V. *Phys. Lett.* 171B: 99–102 (1986)

85. Van Hove, L. *Z. Phys.* C21: 93–8 (1983)

86. Gyulassy, M., Kajantie, K., Kurki-Suonio, H., McLerran, L. *Nucl. Phys.* B237: 477–501 (1984)

87. Takagi, F. *Phys. Rev. Lett.* 53: 427–30 (1984)

88. Miyamura, O., Tabuki, T. *Z. Phys.* C31: 71–6 (1986)

89. Bialas, A., Peschanski, R. *CEN-Saclay Preprint SPhT/85/101* (1985)

*Ann. Rev. Nucl. Part. Sci. 1987. 37 : 97–131*

# PION PRODUCTION IN HEAVY-ION COLLISIONS

## Peter Braun-Munzinger and Johanna Stachel

Physics Department, State University of New York, Stony Brook, New York 11794

CONTENTS

## 1. INTRODUCTION

In collisions between two nucleons, pions can be produced if the center-of-mass energy of the nucleon-nucleon system exceeds the pion mass $m_\pi$. Since $m_\pi c^2$ is approximately 140 MeV (135 MeV for neutral pions), this translates into a laboratory energy of about 290 MeV (280 MeV for neutral pions) as the threshold energy for an incoming nucleon hitting a nucleon at rest and producing a pion.

In collisions between complex nuclei, pion production is governed by the same energy conservation principle: the center-of-mass energy has to

97

0163–8998/87/1201–0097$02.00

exceed $m_\pi c^2$. Since this energy is now shared among many nucleons, this implies that in collisions between two heavy nuclei pions can be produced at beam energies per nucleon that are more than an order of magnitude below the 290 MeV discussed above.

To illustrate this we show in Figure 1 the absolute threshold for pion production in energy per nucleon in the laboratory system for collisions between two nuclei of equal mass number $A$. For nuclei with mass number larger than oxygen, the threshold is below 20 MeV/nucleon and in heavier systems it comes close to the Coulomb barrier (dashed-dotted line in Figure 1), which would eventually provide a natural cutoff for the process. At these low beam energies the center-of-mass energy in any given collision between a nucleon from the target and a nucleon from the projectile is in general not sufficient to produce a pion.

In fact, pion production very close to the absolute threshold requires that more and more nucleons in the projectile and target be completely stopped in the overall center-of-mass system and their relative kinetic energy (except for a small $Q$-value effect) be converted into the total energy of the pion. This is illustrated in Figure 2, where for a projectile near $^{14}N$ we show the minimum number of target nucleons required to interact with the projectile of a given beam energy per nucleon to produce a pion of kinetic energy $0 < T_\pi < 250$ MeV in the center of mass. As the beam energy decreases, the number of "active" target nucleons has to increase

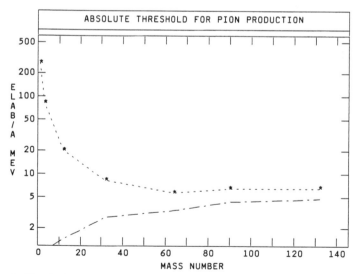

*Figure 1*  Threshold energy for pion production in collisions between two equal nuclei as a function of their mass number. The dashed-dotted line is the laboratory Coulomb energy.

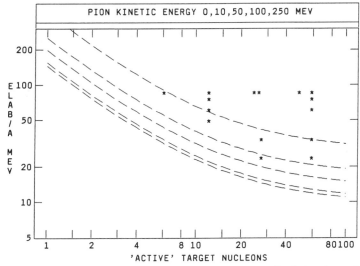

*Figure 2* Minimum number of target nucleons required in a $^{14}$N-induced reaction to produce a pion of kinetic energy 0, 10, 50, 100, and 250 MeV (*dashed lines* from bottom to top) in the laboratory. The stars indicate target masses and beam energies for which pion production has been observed experimentally.

to provide the required center-of-mass energy for pion production. In this sense, pion production from nucleus-nucleus collisions requires more and more "cooperative" interaction among the nucleons in target and projectile the closer one gets to the absolute threshold.

Studying the probability of pion production near the absolute threshold should thus provide a unique means of discovering whether or not cooperative mechanisms play a role in nuclear reactions. It is this prospect, along with the availability of particle beams in the appropriate energy range of a few tens of MeV/nucleon, that has made pion production in nucleus-nucleus collisions a focus of much recent research.

One should note that the possibility of creating pions at beam energies well below 290 MeV in nucleus-nucleus collisions was recognized very early by McMillan & Teller (1). Indeed, the first "artificially" produced pions resulted from interactions of 95-MeV/nucleon alpha particles (generated in the Berkeley 184″ cyclotron) with carbon nuclei (2). The theoretical interpretation of these results was that, because of the Fermi motion of the nucleons in the alpha particle and the carbon nucleus, some of the nucleon-nucleon collisions would exceed the threshold value of 290 MeV and lead to pion production even at 95-MeV/nucleon beam energy.

This Fermi gas model was taken up again and analyzed in detail in 1977 by Bertsch (3). He showed that under reasonable assumptions for the momentum distributions of nucleons in nuclei and taking into account the Pauli principle one expects pion production from this mechanism to cease somewhere near 50 MeV/nucleon, i.e. well above the absolute threshold displayed in Figure 1. Following these investigations a first exploratory study of pion production in the range of a few hundred MeV/nucleon was undertaken by Benenson et al (4). Since 1982, a number of groups have studied pion production in nucleus-nucleus collisions in the energy range $20 < E_{Lab}/A < 90$ MeV. To give an impression of where these data typically fall with respect to the threshold, we also show in Figure 2 some of the target masses and projectile energies for which pion production has been observed (assuming that all target nucleons are "active").

In this review we discuss the currently available data, with special emphasis on whether or not collective effects are at work in these reactions. In Section 2, experimental techniques are presented for measuring charged and neutral pions, along with a discussion of the pertinent resolutions of the various setups. Section 3 contains an overview of the available experimental results and comparisons of results from various groups for neutral and charged pions. In Section 4 we discuss the data in the light of various models based on the assumption of independent nucleon-nucleon collisions. Section 5 presents a discussion of thermal models in which pions are created by the statistical decay of a hot zone created during the collision. In Section 6 conclusions concerning possible collective effects are discussed in the context of models assuming more coherent production processes.

## 2.    EXPERIMENTAL METHODS

In the following we review experimental methods used for the detection of charged and neutral pions in heavy-ion reactions. The main emphasis is on pions produced close to the threshold. Most of the pions are produced with relatively low kinetic energies ($T_\pi < 250$ MeV). Consequently, the instrumental requirements to detect such particles are somewhat different from a typical high energy experiment.

### 2.1  Charged-Pion Detection

Two methods have been used to detect charged pions from intermediate-energy heavy-ion collisions: magnetic spectrometers and scintillator range telescopes. They differ in the attainable energy resolution and in their variations in the angular and kinetic energy acceptance and background suppression capability.

2.1.1 MAGNETIC SPECTROMETERS Various setups have been used (4–7) combining either a fixed magnetic field with a relatively large momentum acceptance (4, 6), or alternatively, a variable magnetic field with a smaller momentum and solid-angle acceptance for each field setting (5). Depending on background suppression and resolution requirements, the pion detection methods range from a simple four-element plastic scintillator telescope (4) to sets of several drift chambers before and after the magnet combined with scintillator time-of-flight hodoscopes and Čerenkov threshold counters (5, 7). The pion flight paths can vary over a wide range; distances from 1.6 to 30 m have been used. This flight path combined with the charged-pion lifetime of 26 ns imposes a threshold on the lowest pion kinetic energies that can be measured with such spectrometers. It ranges from low values of $\sim 25$ MeV and small corrections for pion decay to 66 MeV for the largest distances combined with large corrections for all measured pion momenta. Starting with pions of 100-MeV kinetic energy only 7% of these have not decayed after a 30-m flight path. This severely limits the measurement of cross sections for pions with low kinetic energies. On the other hand, these instruments provide excellent energy resolution and good background suppression. Cross sections of the order of pb/(sr MeV/$c$) have been reported by Chiavassa et al (5).

The main advantage of magnetic spectrometers for pion detection lies in their good pion kinetic energy resolution; values of $\Delta p/p = 1.7\%$ (FWHM) at $p = 160$ MeV/$c$ (6) or $\Delta p/p < 1\%$ at $p = 150$–300 MeV/$c$ (5) have been obtained with momentum acceptances of 50 and 11%, respectively. Spectrometers with the best resolutions usually cover only one angle, in general 0°, and a small solid angle of a few msr. From other spectrometers, data at larger angles are obtainable; Sullivan et al (6) covered $\theta_\pi = 0$–20°, Nagamiya et al (7) covered $\theta_\pi = 20$–90°.

2.1.2 RANGE TELESCOPES An alternative approach to detecting charged pions is the use of segmented scintillator range telescopes, as described by Johansson et al (8) and Bernard et al (9). A relatively large solid angle, as compared to magnetic spectrometers, can be covered. Telescope setups covering $\sim 50$ msr have been used and, in principle, much larger solid angles can be employed.

In Figure 3 the setup used recently by Bernard et al (9) is shown. The range telescopes typically consist of about ten scintillators, of which the deposited energy $\Delta E$ and their time correlation are measured. The integrated depth stops about 80-MeV pions, and the thickness of the individual scintillators is chosen to correspond to range differences of about 8–10 MeV. As such scintillator represents one pion kinetic energy bin, this represents also the energy resolution of these devices. Since no magnetic

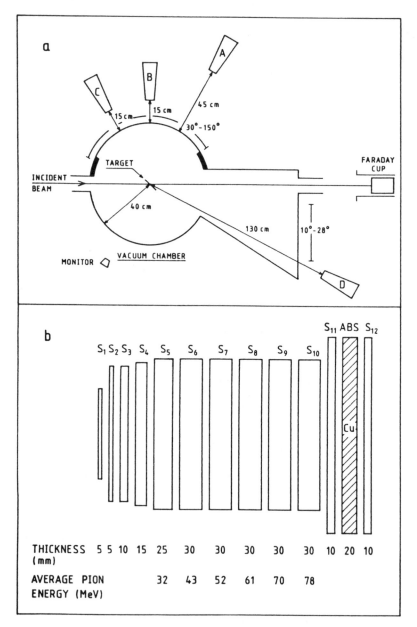

*Figure 3*  Experimental setup (*a*) and scintillator range telescope (*b*) used by Bernard et al (9) to measure charged pions.

fields are applied, these telescopes see a high rate of background events, predominantly protons, which outnumber pions at $E_{Lab}/A = 100$ MeV by four orders of magnitude. Usually the first 10–20 mm of plastic are used as an active or passive absorber to discriminate against this background. This imposes a lower threshold of 20–30 MeV on detectable pion kinetic energies. Further background suppression is achieved by use of $\Delta E_i$-$\Delta E_j$ and $\Sigma \Delta E_i$-$\Delta E_j$ correlations; a sensitivity of $\sim 0.1$ $\mu$b/sr for pion production has been achieved at laboratory angles greater than 27°.

The pion angular resolution is defined by the opening angles of the individual telescopes and is therefore of the order of a few degrees [in Ref. (9) it is $\pm (3$–$4)°$]. Since the flight paths for these setups are short (order of 1 m), only 6–20% of the pions decay before detection, and the corresponding corrections can easily be made. Discrimination between $\pi^+$ and $\pi^-$ is achieved by detecting the 4.2-MeV $\mu^+$ in delayed coincidence with an initial stop signal from a $\pi^+$ over 100–200 ns.

In summary, with this technique larger solid angles and relatively simple detection devices are traded against a reduced energy resolution and a limited range in detectable pion kinetic energies, typically 25–80 MeV. The upper limit is actually determined by the increasing number of secondary reactions of pions in the detector material before stopping.

## 2.2 Neutral Pion Detection

In recent years much attention has been focussed on neutral pion production in heavy-ion reactions at $E_{Lab}/A < 100$ MeV. There, the bulk of the cross section is at pion kinetic energies below the lower thresholds obtainable for charged pions (see Sections 2.1.1 and 2.1.2). Furthermore, the cross sections are in the nb/MeV regime. Under these circumstances the detection of neutral pions has several advantages over charged-pion detection. Neutral pions decay with 98.8% probability and on a fast time scale of $0.87 \times 10^{-16}$ s, i.e. still in the target, into two $\gamma$-rays, which can then be detected in any large solid-angle $\gamma$-detector.

Several setups incorporating lead-glass telescopes have been used recently (10, 11). Because in intermediate-energy heavy-ion reactions one must detect $\gamma$-rays from $\pi^0$ decay in the presence of a seven to eight orders of magnitude greater background of low energy statistical $\gamma$-rays from nuclear deexcitation, the lead-glass telescopes are segmented longitudinally, and prompt (within a few ns) coincidences between two "complete" telescopes are required.

The apparatus we used (10), consisting of 20 such telescopes, is shown schematically in Figure 4. The incoming photon is converted into an electromagnetic shower in a section of F2 glass 1.6 radiation lengths deep. This is backed by an absorber section (SF5 glass) 15 radiation lengths deep

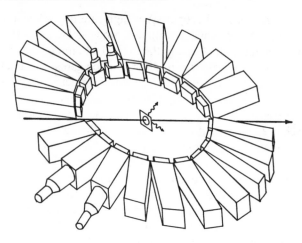

*Figure 4*    Schematic drawing of Pb-glass telescope array used to detect neutral pions (10).

where showers of up to 300 MeV are still practically fully contained. The lateral dimensions are $9.5 \times 9.5$ cm$^2$ and $14.5 \times 14.5$ cm$^2$ for converters and absorbers, respectively. The design of the setup described by Michel et al (11) is similar but somewhat less deep ($1.8 + 6$ radiation lengths), which implies some leakage for $T_\pi \geq 70$ MeV.

The $\gamma$-energy resolution for the setup described in (10) has been measured and follows a $0.1/[E_\gamma(\text{GeV})]^{1/2}$ behavior determined basically by photo-electron statistics. Because of the pion decay kinematics, the pion kinetic energy resolution there is mainly determined by the detector opening angles of $\Delta\theta_\gamma = \pm 8°$. The response function is nearly Gaussian and its variance $\sigma$ can be parametrized as $\sigma(T_\pi) = 2.5 + 0.233 T_\pi$ (MeV). The use of glass with lower lead content leads to a better energy resolution (the best value of $0.076/[E_\gamma(\text{GeV})]^{1/2}$ was achieved by Baer et al (12) with LF5 glass) but requires substantially more material.

The $\gamma$-angular resolution can be improved substantially by the use of multiwire proportional chambers between converters and absorbers. In this case resolutions $\sigma(T_\pi)$ about a factor 25 smaller than quoted above have been obtained (12). This reduces, however, the detection efficiency for a pion by several orders of magnitude and has thus not been applied to measure nb cross sections. Lead-glass arrays with typical total solid angles of about 1 sr have been used (10, 11). This leads to a net pion detection efficiency of 0.5% (0.1%) at $T_\pi = 100$ MeV and $\theta_\pi = 0-20°$ (80–100°) as determined from Monte Carlo simulations and the measured photon conversion probability (10).

Pions are unambiguously identified in these setups by reconstructing their invariant mass. An example is shown in Figure 5 for the reaction 35-MeV/nucleon $^{14}$N + Ni. Pion production cross sections of a few pb/MeV at $T_\pi = 100$ MeV, corresponding to integrated cross sections as low as 1 nb, have been measured and this, probably, is close to a lower limit in sensitivity for such experiments.

It can be concluded that, by detecting neutral pions in lead-glass telescopes, large solid angles can be covered with relatively simple devices and with no cutoff in pion kinetic energy. This is obtained at the expense of energy and angular resolution, as discussed. As can be seen in the next section this is, however, not such a serious drawback for the reactions discussed here, in which the pion spectra are more or less structureless.

## 3.  EXPERIMENTAL RESULTS

Inclusive data for neutral pion production in heavy-ion reactions are available from two $\pi^0$ spectrometers (10, 13) using heavy-ion beams of 25–84 MeV/nucleon. Data using 22-MeV/nucleon $^{32}$S beams are now being analyzed (by the Oak Ridge/Stony Brook collaboration). For all beam energies, data exist using various targets between $^6$Li to $^{238}$U. The data

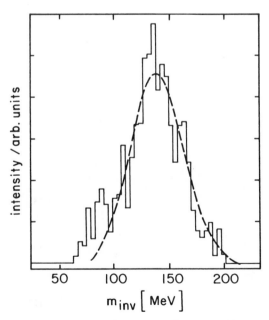

*Figure 5*   Invariant mass spectrum measured for 35-MeV/nucleon $^{14}$N + Ni (10). The dashed line is the result of a Monte Carlo simulation including the detector response and geometry.

cover all pion emission angles $\theta_\pi$ from 0 to 180° and kinetic energies from 0 to 100–200 MeV (depending on the cross section). At the higher energies, i.e. 84 or 74 MeV/nucleon, charged-pion data for various pion kinetic energy and angular ranges are available for comparison.

Table 1 gives a synopsis of the experiments performed so far. At higher beam energies, i.e. between 100 MeV/nucleon and the threshold for pion production in free nucleon-nucleon collisions, data exist for charged pions and they are also included in Table 1.

In the following we present the characteristics and systematic trends of pion production data in heavy-ion collisions below $E_{Lab}/A = 100$ MeV. Data at higher beam energies are not directly comparable, in a model-independent way, among themselves nor to data at lower beam energies because of the different pion angles and kinetic energies covered. We refer, however, to the systematic trends observed in these reactions whenever needed.

## 3.1    Pion Kinetic Energy Distributions

All kinetic energy spectra exhibit a broad maximum at low pion kinetic energies and a more or less exponential decay toward higher pion kinetic

**Table 1**    Inclusive pion production experiments in heavy-ion reactions

| Accelerator | $E_{Lab}/A$ (MeV) | Projectile | Pion | Detection method[a] | $T_\pi$ (MeV) | $\theta_\pi$ (degrees) | Ref. |
|---|---|---|---|---|---|---|---|
| BEVALAC | 125–400 | $^{20}$Ne | $\pi^+, \pi^-$ | MS | 35–160 | 0–30 | 5 |
| BEVALAC | $\geq 280$ | $^{20}$Ne, $^{40}$Ar | $\pi^+, \pi^-$ | MS | 25–60 | 0–20 | 6 |
| BEVALAC | 183 | $^{20}$Ne, $^{40}$Ar | $\pi^-$ | MS | 50–285 | 20–90 | 7 |
| CERN SC | 85 | $^{12}$C | $\pi^+$ | RT | 25–72 | 55–145 | 8 |
| CERN SC | 85 | $^{12}$C | $\pi^+, \pi^-$ | RT | 32–75 | 27–150 | 9 |
| CERN SC | 75 | $^{12}$C | $\pi^+$ | RT | 32–75 | 27–150 | 9 |
| CERN SC | 86 | $^{12}$C | $\pi^+\pi^-$ | MS | 65–190 | 0 | 5 |
| CERN SC | 84 | $^{12}$C, $^{18}$O | $\pi^0$ | LG | 0–200 | 0–180 | 13 |
| CERN SC | 74 | $^{12}$C | $\pi^0$ | LG | 0–200 | 0–180 | 13 |
| CERN SC | 60 | $^{12}$C | $\pi^0$ | LG | 0–150 | 0–180 | 13 |
| CERN SC | 48 | $^{12}$C | $\pi^0$ | LG | 0–100 | 0–180 | 13 |
| GANIL | 44 | $^{40}$Ar, $^{82}$Kr | $\pi^0$ | LG | 0–100 | 0–180 | 14 |
| MSU K = 500 | 35 | $^{14}$N | $\pi^0$ | LG | 0–150 | 0–180 | 10 |
| ORNL HHIRF | 25 | $^{16}$O | $\pi^0$ | LG | 0–110 | 0–180 | 15 |
| ORNL HHIRF | 22 | $^{32}$S | $\pi^0$ | LG | 0–110 | 0–180 | See text |

[a] MS = magnetic spectrometer, RT = range telescope, LG = lead glass.

energies. The maximum shifts down with decreasing projectile energy from $\sim 60$ MeV at $E_{Lab}/A = 183$ MeV (7) to $\sim 10$ MeV at $E_{Lab}/A = 25$ MeV. Figure 6 shows a $\pi^+$ and $\pi^-$ momentum spectrum as measured by Chiavassa et al (5) at $0°$ and using 86-MeV/nucleon $^{12}$C as projectile. Figure 7 displays angle-integrated $\pi^0$ kinetic energy spectra using the same projectile at 60, 74, and 84 MeV/nucleon to bombard $^{12}$C and $^{238}$U targets.

To compare the two sets of data for charged and neutral pions one has to transform the momentum spectra of Ref. (5) into $d^2\sigma/dT_\pi\,d\Omega$ and then fold them with the experimental resolution with which the neutral pions were measured. It should be remembered that the resolution for the charged-pion energy spectra is negligibly small as compared to the $\pi^0$ data (see previous section). We prefer here to use the much simpler procedure of folding charged-pion data with a resolution function $\sigma(T_\pi)$. Unfolding the $\pi^0$ data would imply that data are known over a very large kinetic energy range with very good statistical precision if one wants to avoid substantial errors.

From the measured $\gamma$-energy resolution of the spectrometer used by Noll et al (13) and the $\gamma$-angular resolution as shown by Michel et al (11) we have determined, by a Monte Carlo simulation, the resolution function $\sigma(T_\pi) = 5.86 + 0.24T_\pi$ (MeV) and folded the $\pi^+$ and $\pi^-$ spectra (5)

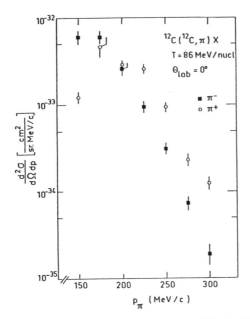

*Figure 6*   Momentum spectrum for charged pions produced at $0°$ in the laboratory in C+C interactions (5).

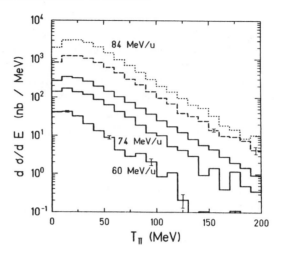

*Figure 7* Neutral pion kinetic energy spectra for C+C collisions measured at 60, 74, and 84 MeV/nucleon (*solid lines*) and for C+Ni (*dashed line*) and C+U (*dotted line*) at 84 MeV/nucleon (13).

assuming a Gaussian distribution. In Figure 8 they are compared to the corresponding $\pi^0$ spectrum at $0°$ (13). As can be seen, the slope of the folded $\pi^-$ spectrum agrees reasonably well with the one measured for the $\pi^0$ distribution. When fitted between 120 and 160 MeV with an exponential distribution of $d\sigma/dT_\pi \propto \exp(-T_\pi/E_0)$, values of $E_0 = 25.9$ and 29.8 MeV are obtained for $\pi^-$ and $\pi^0$, respectively. Folding of the $\pi^+$ distribution yields a harder spectrum. However, this is mostly due to the scatter of the unfolded experimental points between 90 and 130 MeV. After folding of the charged-pion distributions there remains, however, a difference in the overall cross section; the reported $\pi^0$ cross section between 100 and 140 MeV is about 2.7 times larger than that for either $\pi^+$ or $\pi^-$. The quoted uncertainties are $\pm 30\%$ for $\pi^+$, $\pi^-$ and $\pm 15\%$ for $\pi^0$.

At lower pion kinetic energies a comparison can be made to the data of Bernard et al (9) and Johansson et al (8). Figure 9 shows the unfolded $\pi^+$, $\pi^-$, and $\pi^0$ distributions measured at $\theta_\pi = 90°$. In addition, the charged-pion distributions folded again with the resolution function for the neutral pions are displayed. It can be seen that there is good agreement between the two sets of $\pi^+$ data (8, 9). The slopes of the folded $\pi^+$ and $\pi^-$ spectra again agree well with the $\pi^0$ result.

It should be noted, however, that too few data points in the charged-pion distributions are available to make the folding with the relatively poor resolution $\sigma(T_\pi)$ very meaningful; in particular, a continuation of the

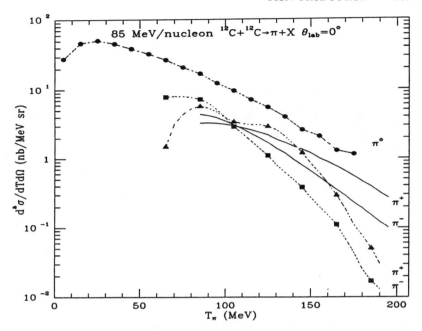

*Figure 8*  Comparison of charged-pion spectra (5) and neutral pion spectra (13) at $\theta_{\text{Lab}} = 0°$ (*dashed lines*). The solid lines are obtained by folding the $\pi^+$ and $\pi^-$ data with the resolution function for the $\pi^0$ detection system (11).

charged-pion spectrum to lower energies would make the folded spectrum somewhat steeper. Again the $\pi^0$ cross sections are larger than the values for $\pi^+$ and $\pi^-$ production. For the folded distributions this difference is, however, comparable to the quoted uncertainties for both sets of data.

As far as the pion production is concerned there should be no differences between $\pi^+$, $\pi^-$, and $\pi^0$ for the $N = Z$ system $^{12}\text{C} + {}^{12}\text{C}$ and only a small difference for systems with $N \neq Z$. Therefore only Coulomb final-state interaction effects remain to be considered. At larger projectile velocities a strong Coulomb focussing effect at $0°$ has been established (4, 6). It was attributed to a Coulomb distortion of the pion wave function in the vicinity of the projectile charge. Since at $E_{\text{Lab}}/A = 84$ MeV a 50-MeV pion is much faster than the projectile, no such effects should occur and only a shift in the spectra due to Coulomb interaction with the target and projectile remnants is expected. Bernard et al (9) found the shift between $\pi^+$ and $\pi^-$ spectra to be about 10 MeV, which corresponds approximately to the Coulomb potential of the compound nucleus $^{24}\text{Mg}$. One would then expect the $\pi^0$ data to be bracketed by the charged-pion data. The behavior of the $\pi^+$ data points of Chiavassa et al (5) below $p_\pi = 225$ MeV/$c$ is also not

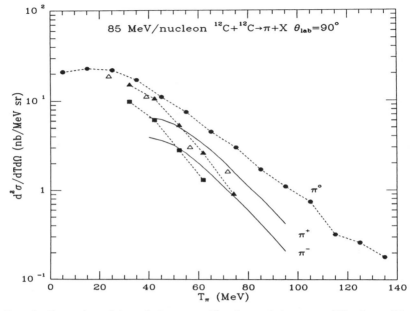

*Figure 9* Comparison of charged-pion spectra (9) and neutral pion spectra (13) at $\theta_{Lab} = 90°$ (*dashed lines*). The solid lines are obtained by folding the $\pi^+$ and $\pi^-$ data with the resolution function for the $\pi^0$ detection system (11).

explained and is, presumably, an instrumental effect in the vicinity of the detection threshold.

Figure 10 shows a $\pi^0$ kinetic energy spectrum obtained (10) for 35-MeV/nucleon $^{14}$N + Ni collisions integrated over all pion emission angles (*top*) and for three different angular regions. Each spectrum exhibits the same overall features. A possible structure around 60 MeV is not significant. These spectra have been corrected for various cuts needed for background suppression (see 10) with correction factors from Monte Carlo simulations. The appropriate uncertainties are contained in the error bars shown. A recent higher statistics simulation shows that the region between 40 and 60 MeV has been slightly undercorrected and the region between 60 and 100 MeV slightly (maximally 20%) overcorrected. The properly corrected histogram is very close to the solid line in the top histogram for $T_\pi = 40$–100 MeV.

It should be noted that in this reaction pions of up to 150 MeV are observed; this corresponds to 75% of the total energy available in the center-of-mass frame. In the reaction 25-MeV/nucleon $^{16}$O + Al, the highest observed pion kinetic energy corresponds to the situation in which all the energy available in the center of mass is transferred to an energetic

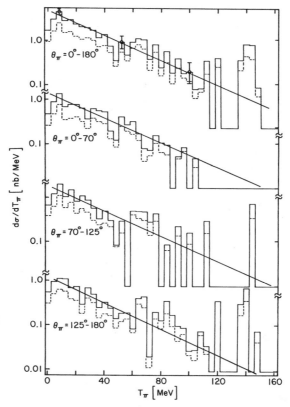

*Figure 10* Pion kinetic energy spectra (*solid histograms*) in the laboratory for 35-MeV/ nucleon $^{14}$N + Ni (10) integrated over all pion emission angles and for various angle bins. The solid line shows an exponential with an inverse slope constant of 23 MeV. Dashed lines are evaluated assuming energy-independent conversion probabilities (for details, see Reference 10).

pion. It should be noted, however, that at $T_\pi = 120$ MeV the resolution is $\sigma(T_\pi) = 30$ MeV so that the statement just made refers to an excitation energy range of $\sim 70$ MeV in the compound nucleus populated at a cross-section level of $4 \pm 3.5$ pb/MeV.

Figure 11 shows exponential slope constants of angle-integrated $\pi^0$ spectra in the vicinity of $T_\pi = 50$ MeV that have been extracted from published data. The error bars shown reflect the arbitrariness in determining a slope constant rather than propagation of statistical errors. It is remarkable that these slope constants change little with decreasing beam energy. They drop from values around 29 MeV at $E_{\text{Lab}}/A = 84$ MeV to 22 MeV at $E_{\text{Lab}}/A = 60$ MeV but are essentially constant for lower beam

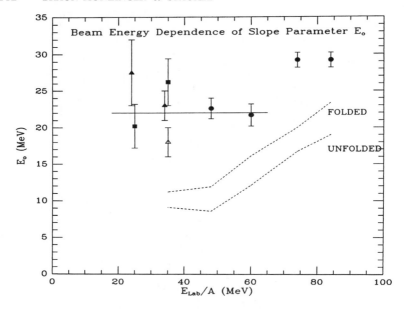

*Figure 11*  Beam energy dependence of slope parameters deduced from pion kinetic energy distributions. Full circles: C+C (13); full triangles: $^{16}O(^{14}N)+Ni$ (10); full squares: $^{16}O(^{14}N)+Al$ (10); open triangle: $^{14}N+W$ (10). The dashed lines are predictions of Prakash et al (30) with and without taking into account the experimental resolution.

energies, as indicated by the solid line in Figure 11. Again, it should be remembered that these slope constants contain effects of the finite pion energy resolution. Calculations should be folded with the appropriate resolution function before comparing to the data. The effect is displayed by showing the primary slope constants from a thermal model discussed below (Section 5) together with values obtained after folding the respective theoretical spectra with the experimental resolution. Although there is substantial change, the overall trend is little affected.

## 3.2   Pion Rapidity and Transverse Momentum Distributions

In order to understand the pion production mechanism, another interesting quantity to study is the invariant cross section:

$$\sigma_{\text{inv}} = E_\pi \, d^3\sigma/dp^3 = 1/p_\pi \, d^2\sigma/(d\Omega \, dT_\pi) = 1/(2\pi p_\perp) \, d^2\sigma/(dp_\perp \, dy). \qquad 1.$$

Experimental data covering large enough angular and energy ranges contain this information and may be analyzed accordingly. If this invariant cross section is plotted as a function of the rapidity $y$ and the pion momentum perpendicular to the beam axis $p_\perp$, all the information about the velocity of the pion-emitting source is contained in one variable ($y$) while the other variable is free of kinematic effects and is determined only by internal characteristics of the system emitting the pions. In a thermal model one might, for example, use a temperature to describe properties of the source (see Section 5).

If the statistics are high enough, contour plots of equal invariant cross section versus $y$ and $p_\perp$ are obtained [see for example Figure 5 in (9) and Figure 1 in (13)]. If there is a single moving source emitting pions, these contour plots have to be symmetric about the source rapidity. For symmetric target-projectile combinations this source rapidity has been found to be half the projectile rapidity both for $\pi^0$ (13) and $\pi^+$ (9). For asymmetric systems this information is available for C+Ni,U at beam energies of 84, 74, and 60 MeV/nucleon. In all cases a source rapidity intermediate between half the beam rapidity and the rapidity of the center-of-mass frame is found. One interpretation of this is that more target than projectile nucleons (but not the entire compound nucleus) take part in pion production. Similar findings have been reported (8, 9) for charged pions produced in reactions of 85-MeV/nucleon $^{12}$C on heavy target nuclei (Au, Pb).

One should, however, keep in mind that pions, once created, will re-interact in the nuclear medium. For kinetic energies below $T_\pi = 100$ MeV the main contributing process is pion absorption with a mean free path of 2–3 fm, which is small as compared to the size of a U nucleus. For a more detailed discussion of this problem, see chapter IV of (10) and references cited there.

For low beam energies indications are that the projectile will stop in the center-of-mass system before traversing the whole target nucleus; this therefore leads to a shadowing effect in the forward direction resulting in a downward shift in rapidity of the invariant cross section. The data of (10) have been interpreted in this way. There the apparent source rapidity is below the value for the center-of-mass frame and even slightly negative [see Figure 12 (top)]. This and the clear asymmetry of the rapidity distribution have been interpreted in a semiquantitative way as being due to stopping of the projectile in the center-of-mass system over a few femtometers combined with strong pion reabsorption.

One should therefore consider measured pion production cross sections from nucleus-nucleus collisions at low pion kinetic energies always in connection with pion reabsorption. The primary production cross sections can be factors of two to five larger, depending on the target size, pion mean free path, impact parameter, and projectile range inside the target. At larger pion kinetic energies, where rescattering is important, these effects are expected to be more complicated.

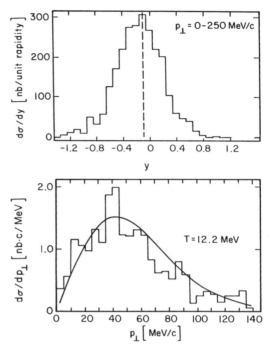

*Figure 12*  Pion spectra from $^{14}N + Ni$ at 35 MeV/nucleon (10) vs rapidity (*top*) and perpendicular momentum (*bottom*). The dashed line marks the centroid of the rapidity spectrum. The solid line (*bottom*) is a thermal model prediction (see 10) with temperature $T = 12$ MeV.

Figure 12 (*bottom*) shows a transverse momentum spectrum $d\sigma/dp_\perp$ typically obtained in many pion production experiments (13, 10, 15). With decreasing beam energy the maximum moves toward lower values of $p_\perp$ and the spectrum decreases more steeply toward larger transverse momenta. This is, of course, directly related to the similar trend observed in the pion kinetic energy spectra (see Section 3.1) but the interpretation here is independent of the (in general) unknown source velocity. It is frequently interpreted in terms of a pion source temperature (see Section 5).

## 3.3  *Angular Distributions*

Both neutral and charged-pion angular distributions at $E_{Lab}/A = 85$ MeV have been found to be strongly forward peaked in the laboratory frame because of kinematic effects. Figure 13 shows the $\pi^0$ angular distributions obtained by Noll et al (13), transformed to the nucleon-nucleon center-of-

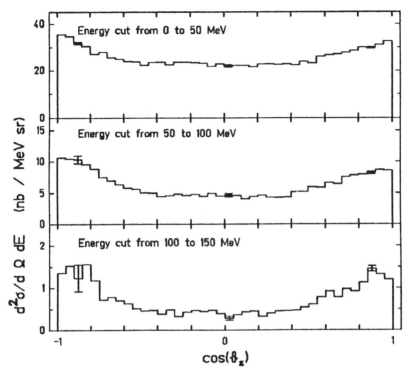

*Figure 13*   Angular distributions for $\pi^0$ production in C+C collisions at 84 MeV/nucleon (13).

mass system. The distributions are forward-backward symmetric in this frame, with a minimum at 90° that is getting more pronounced with increasing pion kinetic energy. The $\pi^+$ angular distribution for the same system (9), measured from 30 to 150°, shows quantitatively the same anisotropy. Again the same anisotropy is observed by Johansson et al (8), who also studied $\pi^+$ production in the same reaction. They also find the pronounced increase of this anisotropy with increasing pion kinetic energy.

For 35- and 25-MeV/nucleon beams, the shape of $\pi^0$ angular distributions depends strongly on the target mass (see Figure 14). It is forward peaked in the laboratory frame, as one would expect, for the Al target. For the Ni target, however, the distribution is forward-backward symmetric with a minimum in the vicinity of 90°. As with the rapidity distributions, this has been attributed to a combined effect of stopping of the projectile and pion reabsorption. Schematic Monte Carlo simulations (10) indicated that the observed distributions for different targets are consistent with the same primary angular distribution, which has to be forward-

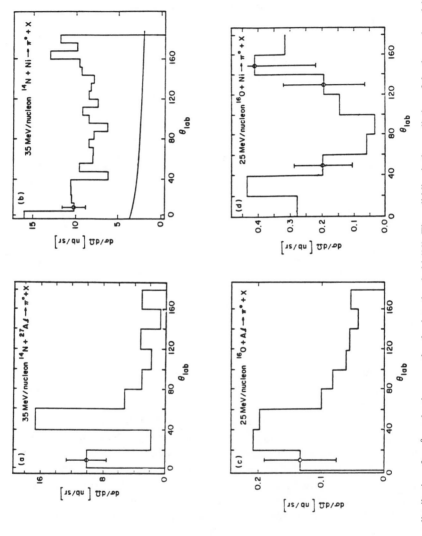

*Figure 14*  Angular distributions for $\pi^0$ production near the absolute threshold (10). The solid line is the prediction of the thermal model discussed in Section 5.

backward peaked (in the rest frame of the source). For the reaction 35-MeV/nucleon $^{14}$N+W $\rightarrow \pi^0$+X (10) the same group again finds a distribution similar to that shown in Figure 14 for $^{14}$N+Ni, with a pronounced minimum [the ratio of $d\sigma/d\Omega(0°)$ to $d\sigma/d\Omega(90°)$ equals 2:1].

In conclusion, existing data on pion production in heavy-ion reactions between 25 and 85 MeV/nucleon consistently show pion angular distributions that are not isotropic; this indicates that besides the expected dominant s-wave character there must be at least a significant p-wave contribution.

## 3.4  Mass Dependence

Because of the different kinematic situation for different measured systems, comparisons of the target and/or projectile mass dependence are meaningful only if data over a large fraction of the total angular and kinetic energy distributions are available. To visualize a simple power law behavior, the integrated cross sections may be plotted logarithmically as a function of $(A_T A_P)^{1/2}$. This is done in Figure 15 for the integrated inclusive $\pi^0$ cross sections (10, 13–15). Data for many different target nuclei are available at 84 MeV/nucleon (13), and one can see that the target mass dependence

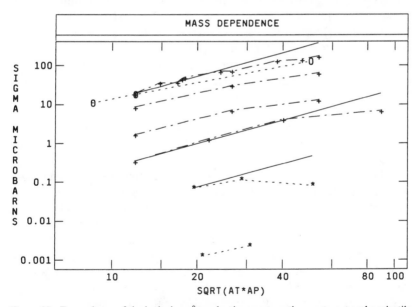

*Figure 15*  Dependence of the inclusive $\pi^0$ production cross section on target and projectile mass number. Data at 25 and 35 MeV/nucleon (∗) are from (10); the data between 48 and 84 MeV/nucleon (+) are from (13). Solid lines show a $(A_T * A_P)^{2/3}$ dependence for comparison. Open circles are data for $\pi^+$ production (9) normalized to the $\pi^0$ data.

does not follow a simple power law over the whole mass range. It flattens for heavier targets, while at lower masses it would be consistent with an $A^{2/3}$ dependence as indicated by the solid line.

For comparison this dependence is also shown for the $\pi^+$ data of Bernard et al (9). Since only part of the angular and kinetic energy distributions have been measured (see Table 1), the curve has been normalized to the $\pi^0$ curve at $(A_T A_P)^{1/2} = 12$ to allow a better comparison. Apparently the overall trend is in agreement with the $\pi^0$ data. Qualitatively the same behavior is observed at lower beam energies but the discrepancies with $\sigma \propto A^{2/3}$ get more pronounced, and finally, for 35-MeV/nucleon $^{14}$N + W there is even a reduction in cross section as compared to the Ni target.

Again this trend probably indicates strong pion reabsorption effects. There is, however, no obvious argument concerning the beam energy dependence of this effect. This would require a detailed dynamical calculation that keeps track of all target and projectile remnants and allows the pions to reinteract with them.

### 3.5    Pionic Fusion

The process in which two colliding nuclei amalgamate and cool by pion emission, leaving the compound nucleus close to or at its ground state, is termed pionic fusion. It was first observed in $^3$He + $^3$He $\to \pi^+ + ^6$Li by the Orsay group (16), with cross sections in the few tens of nb/sr range. Studying this process for heavier systems proved much more difficult because of the very low cross sections encountered. For systems such as $^3$He + $^6$Li $\to \pi^+ + ^9$Be, for example, in the bombarding energy range of 235–283 MeV, the ground-state transition is observed (16) with a cross section reduced by approximately three orders of magnitude compared to that for $^3$He + $^3$He. For $^9$Be and $^{13}$C targets the measured cross sections are still in the 50–100-pb/sr range, which indicates no further drop with mass number.

All these experiments were performed by detecting the scattered pion in a magnetic spectrometer. Although this allows good energy resolution, the achievable solid angles are only in the few msr range. This makes experiments with heavier targets and projectiles unfeasible because of the necessary restriction in target thickness. An interesting technique that might alleviate some of these restrictions was recently developed by Schott et al (17). This group studied the reaction $^3$He + $^{12}$C $\to \pi^+ + ^{15}$N at a bombarding energy of 181 MeV by detecting the $^{15}$N recoil nucleus in a magnetic spectrometer. Because of kinematic focussing, a detection efficiency of 84% could be achieved. However, very thin targets must be used to obtain good energy resolution and background suppression. At 181-MeV beam energy this reaction is only 19 MeV above threshold for pionic fusion. Angle-integrated cross sections of $1.3 \pm 0.3$ nb were obtained for population of states in $^{15}$N with $6.5 < E_x < 10.4$ MeV. The limit for ground-state population is 70 pb. This increase of the cross section with excitation energy may indicate preferential population of states with higher spin and/or isospin.

# 4. PION PRODUCTION FROM INDEPENDENT NUCLEON-NUCLEON COLLISIONS

As a first step, it is natural to consider pion production from nucleus-nucleus collisions in terms of nucleon-nucleon collisions leading to a free pion. The process can, however, be much more complicated than collisions between free nucleons leading to a pion. In the approaching nuclei, the nucleons are initially bound in their (shell model) orbits because of the nuclear mean field. Collisions between the nucleons of projectile and target will then take place within the nuclear medium, with some of the final states blocked by other nucleons (the Pauli exclusion principle). Furthermore, when the nuclear densities start to overlap the initial nucleon distributions will rapidly be altered by the interactions. In the following we discuss the predictions of various models for pion production calculated from the sequence of nucleon-nucleon collisions taking place in a typical interaction.

## 4.1  Simple Considerations

Pion production is thought to occur in the very initial phase of the reaction where high relative momenta between projectile and target nucleons are possible if one takes into account the coupling of the relative momentum between projectile and target and the Fermi momenta $p_F$ of the nucleons. In the simplest picture (e.g. 3) the maximum relative energy between a nucleon from the projectile and one from the target that can be obtained from this coupling is

$$E_{\text{max}} = (p_F + P/2)^2/m_N.  \qquad 2.$$

Here, $P$ is the projectile momentum per nucleon and $m_N$ is the nucleon mass. In Equation 2 we used nonrelativistic kinematics as appropriate for the energy range below 90 MeV/nucleon. If $E_{\text{max}}$ exceeds $m_\pi c^2$ a pion can be produced in such a collision.

To illustrate the effect of coupling beam and Fermi momentum we show in Figure 16 $E_{\text{max}}$ and the corresponding maximum kinetic energy, in the laboratory, of the created pion under the assumptions that (a) only nucleons with antiparallel momenta and (b) nucleons with their momenta

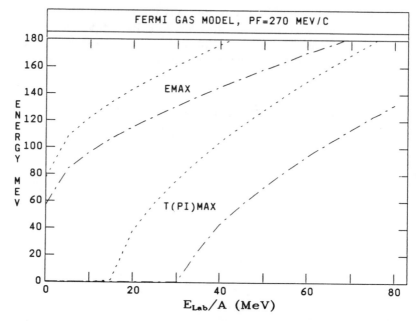

*Figure 16*  Maximum relative energy for two colliding nucleons in the Fermi gas approach and corresponding maximum pion laboratory kinetic energies. The two sets of curves are calculated under the assumption of (*a*) antiparallel momenta (*short dashes*) and (*b*) momenta antiparallel within 30° (*dashed-dotted lines*).

antiparallel within a cone angle of 30° contribute to the production. From this naive consideration one expects pion production to start somewhere between 20- and 30-MeV/nucleon beam energy, in qualitative agreement with experimental observations (see previous section).

One should note, however, that overall energy conservation for this process implies that near threshold the two colliding nucleons carry a large fraction of the total center-of-mass energy. Consider, for example, two colliding carbon nuclei at $E_{\text{Lab}}/A = 30$ MeV. Inspection of Figure 16 reveals that $E_{\text{max}} = 160$ MeV while the overall center-of-mass energy is only 180 MeV. Nearly 90% of the center-of-mass energy is then carried by the two colliding nucleons and the other 22 nucleons have to "conspire" coherently for this process to happen!

This problem of energy conservation in pion production was studied in a schematic model by Shyam & Knoll (18). They noted that, in a shell-model approach, the relative energy between nucleons from target and projectile cannot simply be calculated from their relative momentum since nucleons in shell-model orbits have a momentum distribution but fixed energy. Estimating available energies from harmonic oscillator wave functions, they find much lower values than obtained from the naive Fermi gas model. Consequently, their calculated pion production cross sections are greatly (by orders of magnitude) reduced and underestimate the experimental data even at $E_{Lab}/A = 84$ MeV.

Through overall energy conservation some degree of coherence is implicitly introduced into the calculations, although none of the models takes this into account explicitly. These difficulties notwithstanding, it is interesting to study the consequences of models based on independent nucleon-nucleon collisions, especially at higher beam energies, where overall energy conservation is of less concern.

## 4.2   Models Incorporating Some Reaction Dynamics

After the pioneering work of Bertsch (3), the Fermi gas model was examined in greater detail by several authors (19–21). While these investigations differ in how they treated the reaction dynamics, they all obtained the inclusive cross section for pion production by folding momentum distributions of nucleons in projectile and target with the basic cross section $\sigma(NN \rightarrow NN\pi)$. Although this cross section could possibly be strongly modified by the surrounding nuclear medium, it is, in all investigations, taken to be equal to $\sigma^{free}(NN \rightarrow NN\pi)$ describing pion production between free nucleons.

In an effort to avoid the energy conservation problem, Guet & Prakash (19) consider only processes in which the colliding nucleons are scattered into the continuum along with the produced pion. They find that this mechanism leads to considerable cross sections at beam energies above 100 MeV/nucleon, but even with inclusion of a strong mean field contribution it underpredicts, by orders of magnitude, the data in the energy range below 50 MeV/nucleon. In Figure 17 the results of this calculation are compared to the experimental excitation function (long-dashed line).

To improve on the Fermi gas model of Bertsch (3) one must take into account the time dependence of the nucleon momentum distributions during the collision process. This has been accomplished by Cassing (20), who incorporated distortions introduced through the relative motion of the colliding nuclei and through the nucleon-nucleon collisions, and by Blann (21) in an approach based on the Boltzmann master equation. Both authors use the simple relative energy prescription discussed above. Blann (21) in addition restricts the relative energy by the overall excitation energy of the composite system. As shown in Figure 17 (dashed-dotted line) these models qualitatively reproduce the observed bombarding energy dependence of the inclusive pion production cross section.

Aichelin (22) includes the nuclear mean field potential by employing the

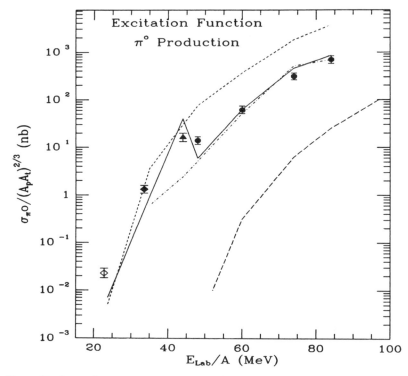

*Figure 17* Comparison of the measured energy dependence of the $\pi^0$ production cross section with various model predictions. Long dashes: (19); dashed-dotted line: (21); solid line: (30); short dashes: (31, 32).

Boltzmann-Uehling-Uhlenbeck theory. Because of numerical difficulties this method is restricted to energies above 60 MeV/nucleon. At these higher energies, calculations agree rather well with the available inclusive cross sections. We note that both Blann and Aichelin take into account, in a schematic way via the mean free path, the possible reabsorption of the pions after their production. Since the mean free path for absorption of pions in nuclear matter is relatively short ($\lambda_\pi = 2$–3 fm in the energy range $0 < T_\pi < 60$ MeV), this effect reduces the observed cross sections by a factor of two to five, depending on the size of the system.

While the beam energy dependence of the data is reasonably well predicted by these models, only Blann (21) reports kinetic energy spectra. The somewhat surprising result of his calculations is that these kinetic energy spectra generally are much harder than is experimentally observed. In the range $20 < T_\pi < 80$ MeV his calculations at $E_{Lab}/A = 35$ MeV, for example, are essentially flat while the data drop by approximately an order of magnitude (see Figure 6 of Ref. 21). This apparent availability of too much energy for the pion may again be related to the energy prescription used in these calculations.

Summarizing the status of models based on independent nucleon-nucleon collisions we note a rather surprising success of the more sophisticated models in reproducing the measured beam energy dependence of the inclusive pion production cross section. At the higher beam energies ($E_{lab}/A > 60$ MeV) where the details of the energy prescription may not be so essential, this suggests that taking proper account of the two-body collision dynamics and the influence of the mean field leads to a rather good description of the data and no further collective enhancement need be invoked. More information, especially on energy and angular distributions, is, however, urgently needed to substantiate this conclusion. When extended to energies close to threshold, these calculations still predict cross sections not too different from those observed experimentally. In view of the discussion at the beginning of this section, we feel that this success, in a kinematic regime where the models should not work, is related to the treatment of overall energy conservation with its implicitly built-in coherence and, in particular, to the specific prescription used in these models to calculate relative energies.

## 5. THERMAL MODELS

There is by now considerable evidence (23–25) that highly excited nuclear systems can be formed in intermediate-energy nucleus-nucleus collisions. The underlying picture for this scenario is that collisions between nucleons from target and projectile lead to a thermalization of the initial center-of-mass energy. This thermalization is thought to take place within a time scale short compared to typical expansion times of the system, so that temporarily a local, hot, compressed zone can be formed. If the total excitation energy in this "hot spot" exceeds $m_\pi c^2$ then a pion can in principle be created as one of the decay products. Moreover, if one further assumes that the decay of this hot zone follows statistical laws then one can straightforwardly calculate the decay probabilities for all decay products, including pions, by properly counting the available phase space. Once thermalization is assumed, this method also avoids problems with overall energy conservation. Considering the difficulties discussed above in the context of the nucleon-nucleon scattering model, it was therefore tempting to consider what such a thermal model would predict.

In the first such study, Aichelin & Bertsch (26) assumed that target and projectile form a compound nucleus. The compound nucleus excitation energy $E^*$ is then simply obtained from the center-of-mass energy and mass excesses of the colliding nuclei. The decay of this highly excited system was then described by applying Weisskopf theory (27) of compound nucleus decay. In this approach, the decay probabilities are determined by nuclear level densities and the inverse (absorption) cross section for the decay particle under consideration. For pions and nucleons (the latter presumed to constitute the dominant decay mode) the inverse cross sections are reasonably well known. Furthermore, at the high excitation energies encountered the nuclear level densities should be well described by the Fermi gas model.

Under these circumstances the pion production cross section can be immediately specified once one fixes the cross section $\sigma_0$ for compound nucleus formation. Since $\sigma_0$ is, at the energies discussed here, mostly determined by the geometrical overlap of the colliding nuclei and, therefore, only weakly dependent on beam energy, this approach allows a straightforward evaluation of the beam energy dependence of the inclusive pion production cross section. Aichelin & Bertsch (26) applied this model first to carbon-carbon collisions and found agreement between data and calculations for $\sigma_0 = 160$ mb. It was soon realized, however, that this compound nucleus model predicts cross sections varying approximately as $\exp(-m_\pi/T)$, where $T$ is the nuclear temperature. For asymmetric systems at a constant bombarding energy per nucleon the temperature decreases with increasing size of the system (since roughly the same energy is shared among many more nucleons). The calculated pion production cross sections are then predicted to fall with increasing target mass number, at variance with the experimental observations (see Section 3).

This difficulty can be circumvented if one notes that, at the relatively high beam energies considered here, full compound nucleus formation is rather unlikely. The data discussed in (23–25), for example, indicate that only a subset of projectile and target nucleons contributes to the formation of the hot, compressed zone. The temperature in this zone then depends only little on the mass number of target and projectile. Aichelin (28) and Gale & Das Gupta (29) found that one can understand the trend of the data, including asymmetric systems, if one makes assumptions about the number of nucleons $N$ in the hot spot and about the dependence of $\sigma_0$ on $N$.

This thermal model was subsequently put on a firmer footing by Prakash et al (30) by combining the geometrical aspects of the fireball model (23) with the compound nucleus model of Aichelin & Bertsch (26). Note that the pion decay probability as calculated in the original fireball model cannot be used in the present context because it assumes chemical equilibrium for the pions and, consequently, would strongly overestimate their yield.

A similar approach was independently developed by Shyam & Knoll (31). Although they use slightly different language, their formulation is actually quite similar to that of Prakash et al (30). Assuming equilibration over a local hot zone and decay governed by the rules of many-particle phase space, the invariant cross section, in both models, can be factorized into a formation cross section and a decay probability. Shyam & Knoll (31) evaluate the formation cross section for each impact parameter by extrapolating the results of three-dimensional cascade calculations, while Prakash et al (30) take the geometrical prescription of the original fireball model. The most significant difference between the two approaches is in the evaluation of the decay probabilities. They are calculated by Shyam & Knoll (31) by counting with equal probability all possible phase-space cells including multibody final states and those comprising clusters of nucleons. In the compound nucleus approach of Prakash et al (30) only two-body final states are evaluated using Weisskopf theory.

The results calculated with both models are compared in Figure 17 to the experimental data. With the recently revised version of the Shyam & Knoll model (32) the data are generally overpredicted by about a factor of five for the higher beam energies, and the slope of the calculated excitation function is significantly steeper than the data. In contrast, the model of Prakash et al (30) reproduces the measured data surprisingly well. Note that in this plot, data for different systems have been scaled according to $(A_P * A_T)^{2/3}$. The "kink" near 44 MeV/nucleon indicates the deviations of the data from this simple scaling law (the system is $^{40}Ar + {}^{40}Ca$) and is well reproduced by the thermal model of Prakash et al (30).

At present, the differences between the two thermal models discussed above are not fully understood. One notable difference is the use by Prakash et al (30) of experimental information on pion absorption in the calculation of the decay probabilities whereby some coherence is implicitly built into the model.

While the agreement between data and the model of Prakash et al (30) is good for the energy dependence of the total inclusive cross section, there is a serious discrepancy in the calculated and measured pion kinetic energy spectra. This is illustrated in Figures 11 and 18 for various systems and beam energies. Note that folding the theoretical calculations with the response functions of the different $\pi^0$-detection systems does not appreciably alter this conclusion (see dashed lines in Figure 18). The folding procedure increases the theoretically determined slope constants but not sufficiently to reproduce the experimentally observed values.

We conclude that at least the pions produced at relatively low kinetic energies are probably emitted from a locally heated zone formed in the

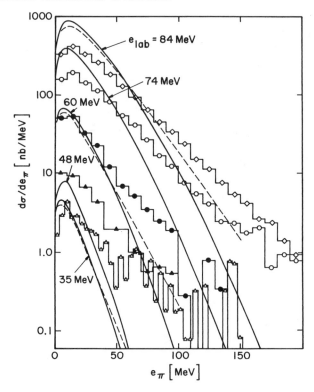

*Figure 18* Comparison of calculated (30) and experimental pion spectra. The data are from (13) for C + C (*upper four curves*) and from (10) for $^{14}N + Ni$. The dashed lines are obtained by folding the calculations (*solid lines*) with the experimental resolutions.

course of the reaction. Since pions are not initially present in this system and since they cannot be created after the system has expanded, the pion signal is easily distinguished from that of spectators in the reaction and should provide information about the formation and decay of such a fireball that is complementary to that obtained by measuring spectra of nucleons. The mere presence of a pion signal at energies as low as 25 MeV/nucleon indicates a large concentration of energy in the fireball. More quantitatively, the analysis of transverse momentum spectra indicates temperatures of about 12 MeV at 35 MeV/nucleon (see Section 3 and Figure 11), which is in very good agreement with the values deduced from the model of Prakash et al (30).

The experimentally observed surplus of pions at high kinetic energies cannot easily be explained in such thermal models and might be taken as evidence for a nonstatistical component in the production mechanism. In the next section we discuss the status and predictions of models incorporating some degree of collectivity into the reaction dynamics.

# 6. MODELS INVOLVING COLLECTIVE PROCESSES

Probably the most interesting aspect of pion production at energies close to the absolute threshold is its possible sensitivity to collective or coherent mechanisms in the production process. A precise definition of coherence or collectivity is useful only with a specific model in mind. We adopt here the operational definition that coherent or collective production is a mechanism whereby energy is extracted from the relative motion by slowing down the projectile and/or target as a whole and converted into one degree of freedom, the pion. Within this definition, a large fraction of the nucleons have to interact with a common "phase." Thermal production, although it involves pooling of the kinetic energy of many nucleons, is, following this definition, not a collective process. Note that even at the absolute threshold one expects a small, but nonzero, noncollective contribution from thermal production much like the Hauser-Feshbach formalism for nucleon decay, which describes two-body reactions to definite final states with complete loss of information on the phases of the interacting nucleons.

With this preamble in mind we discuss the available evidence for coherent effects in pion production. We first focus on the collective model for pion production proposed by the Frankfurt group (33). There it is assumed that the rapid deceleration of projectile and target during the collision gives rise to pionic bremsstrahlung much like a rapidly decelerated nucleus of charge number $Z$ radiates photons with intensity proportional to $Z^2$. In electromagnetic bremsstrahlung, this collectivity arises because the nuclear current is proportional to the charge number $Z$. Pionic bremsstrahlung is slightly more complicated since the source term or current has to involve (33) the spin and isospin of the colliding nuclei. For most nuclei in their ground state, i.e. before the collision, the expectation values for spin and isospin, $\langle S \rangle$ and $\langle T \rangle$, are small or zero so that one expects little cross section from such a mechanism.

However, Vasak et al (33) have pointed out that spin and isospin might be coherently generated in the initial phase of the reaction. Simple, although statistical, estimates (34) indicate that such a buildup of spin and isospin may be possible. In the absence of a microscopic description one has to assume, however, that the spin-isospin current depends linearly on the mass number of projectile or target to get the same coherence as in the photon case. Once this coherence is assumed, the model of Vasak et al (33) can be used to predict pion production cross sections for many different systems and energies with only one common model parameter related to the deceleration of the colliding nuclei. In the most recent version a classical friction model is used (35) to describe the reaction dynamics for arbitrary projectile-target combinations.

The results from this model for the energy dependence of the inclusive pion production cross section are compared in Figure 19 to the experimental observations. The two agree over the full energy range. The model also reproduces some of the observed pion angular distributions (35) but generally predicts too steep slopes for the pion kinetic energy spectra. In addition, this model was recently extended to predict high energy photon production in these collisions, with some success. Given this impressive agreement with data, microscopic calculations are called for to justify the coherence assumption on the spin-isospin current. Further elucidation of

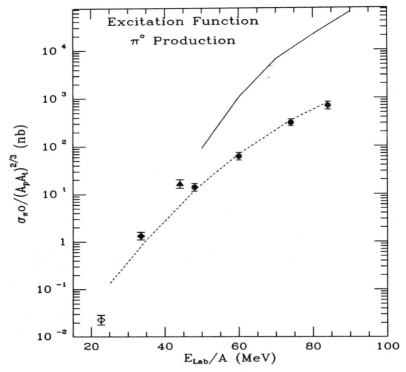

*Figure 19* Comparison of the measured energy dependence of the $\pi^0$ production cross section with the coherent models of (35) (*dashed line*) and (40) (*solid line*).

the effects of overall energy conservation on the bremsstrahlung mechanism would also be important to provide a firm basis for this model.

Coherent production of pions has also been investigated in an approach first suggested by Brown & Deutchman (36). In their picture the projectile, passing by the target, excites the latter into a $\Delta$-nucleon hole state that subsequently decays into a free pion. Projectile and target remain essentially intact during the collision, and the loss of relative energy is directly converted into the pion. Energy $\omega$ and momentum component $q_z$ transferred to the target follow a simple dispersion relation $\omega = q_z v_P$, where $v_P$ is the velocity of the projectile. As was pointed out by Feshbach & Zabek (37) and elaborated on by Pirner et al (38), this can be understood in a Weizsaecker-Williams-like approach whereby the nuclear field of the projectile creates virtual pions that are then brought onto the mass shell by interaction with the target. The predictions of this coherent model have been worked out mainly by two groups (39, 40). As a result of the fully quantum mechanical approach, realistic calculations are quite difficult and, moreover, should be compared to data in which pions are measured in coincidence with peripherally scattered projectile nuclei. One can estimate, however, an upper limit for the production cross section if one neglects energy loss of the projectile as well as pion rescattering and absorption in the nuclear medium. The results of such a calculation (40) are shown as a solid line in Figure 19. The comparatively large cross sections obtained suggest the importance of these collective contributions from peripheral collisions.

In another, ambitious approach, Tohyama et al (41) and very recently D. Ernst and M. Strayer (private communication) have begun to incorporate time-dependent mean field theory into a model for pion production. This approach is potentially the most powerful as it provides a fully quantum mechanical framework for the study of pion production. The calculations are rather schematic still but indicate the importance of coupling the pion production to the mean field of the colliding nuclei.

Finally, we note that direct (coherent) mechanisms have been identified (16) for pion production in very light systems such as $^3\text{He} + {}^3\text{He} \rightarrow \pi^+ + {}^6\text{Li}$ (see Section 3) and successfully interpreted in terms of both a direct reaction model (42) and a model based on coherent propagation of $\Delta$-nucleon hole excitations (43). The extension to heavier systems of these exclusive models is, however, not straightforward.

# 7. CONCLUSIONS AND OUTLOOK

The production of pions in nucleus-nucleus collisions has been observed at energies close to the absolute threshold. In some cases, more than half the initial center-of-mass energy is converted during the reaction into just one degree of freedom, the pion. The observed cross sections, especially at high pion kinetic energies, cannot be fully understood in terms of thermal models or models invoking independent nucleon-nucleon collisions. This suggests the presence of a coherent mechanism in the production process.

In the near future one foresees experiments with better resolution and sufficient solid-angle coverage to allow studies of pion production even closer to the kinematic limit, as well as simultaneous detection of heavy fragments. Theoretical descriptions of these processes will ultimately require a fully quantum mechanical approach, but the first steps in this direction are now being taken.

ACKNOWLEDGMENTS

It is our pleasure to thank Dr. M. Prakash for many discussions and a careful reading of the manuscript. This work was supported in part by the National Science Foundation.

*Literature Cited*

1. McMillan, W. G., Teller, E. *Phys. Rev.* 72: 1 (1947); Horning, W., Weinstein, M. *Phys. Rev.* 72: 251 (1947)
2. Gardner, E., Lattes, C. M. G. *Science* 107: 270 (1948)
3. Bertsch, G. *Phys. Rev.* C15: 713 (1977)
4. Benenson, W., Bertsch, G., Crawley, G. M., Kashy, E., Nolen, J. A., et al. *Phys. Rev. Lett.* 43: 683 (1979)
5. Chiavassa, E., Costa, S., Dellacasa, G., De Marco, N., Gallio, M., et al. *Nucl. Phys.* A422: 621 (1984)
6. Sullivan, J. P., Bistirlich, J. A., Bowman, H. R., Bossingham, R., Buttke, T., et al. *Phys. Rev.* C25: 1499 (1982)
7. Nagamiya, S., Hamagaki, H., Hecking, P., Kadota, S., Lombard, R., et al. *Phys. Rev. Lett.* 48: 1780 (1982)
8. Johansson, T., Gustafsson, H. A., Jakobsson, B., Kristiansson, P., Noren, B., et al. *Phys. Rev. Lett.* 48: 732 (1982)
9. Bernard, V., Girard, J., Julien, J., Legrain, R., Poitou, J., et al. *Nucl. Phys.* A423: 511 (1984)
10. Braun-Munzinger, P., Paul, P., Ricken, L., Stachel, J., Zhang, P. H., et al. *Phys. Rev. Lett.* 52: 255 (1984); Stachel, J., Braun-Munzinger, P., Freifelder, R. H., Paul, P., Sen, S., et al. *Phys. Rev.* 33: 1420 (1986); The SUNY Stony Brook-Oak Ridge collaboration, unpublished
11. Michel, C., Grosse, E., Noll, H., Dabrowski, H., Heckwolf, H., et al. *Nucl. Instrum. Methods* A243: (1986)
12. Baer, H. W., Bolton, R. D., Bowman, J. D., Cooper, M. D., Cverna, F. H., et al. *Nucl. Instrum. Methods* 180: 445 (1981)
13. Noll, H., Grosse, E., Braun-Munzinger, P., Dabrowski, H., Heckwolf, H., et al. *Phys. Rev. Lett.* 52: 1284 (1984); Grosse, E. *Nucl. Phys.* A447: 611c (1985)
14. Heckwolf, H., Grosse, E., Dabrowski, H., Klepper, O., Michel, C., et al. *Z. Phys.* A315: 243 (1984)
15. Young, G. R., Obenshain, F. E., Plasil, F., Braun-Munzinger, P., Freifelder, R., et al. *Phys. Rev.* C33: 742 (1986)
16. LeBornec, Y., Bimbot, L., Koori, N., Reide, F., Willis, A., et al. *Phys. Rev. Lett.* 47: 1870 (1981); Willis, N., Bimbot, L., Hennino, T., Jourdain, J. C., LeBornec, Y., Reide, F. *Phys. Lett.* 136B: 334 and references quoted there (1984)
17. Schott, W., Wagner, W., Kienle, P., Pollock, R., Bent, R., et al. *Phys. Rev.* C34: 1406 (1986)
18. Shyam, R., Knoll, J. *Phys. Lett.* 136B: 221 (1984)
19. Guet, C., Prakash, M. *Nucl. Phys.* A428: 119c (1984)

20. Cassing, W. *Proc. Int. Workshop Gross Prop. Nuclei Nucl. Excitations, Hirsch-egg*, ed. H. Feldmeier. GSI Rep. (1985)
21. Blann, M. *Phys. Rev.* C32: 1231 (1985)
22. Aichelin, J. *Phys. Lett.* 164B: 261 (1985)
23. Westfall, G. D., Gosset, J., Johannsen, P. J., Poskanzer, A. M., Meyer, W. G., et al. *Phys. Rev. Lett.* 37: 1202 (1976)
24. Fields, D. J., Lynch, W. G., Chitwood, C. B., Gelbke, C. K., Tsang, M. B., et al. *Phys. Rev.* C30: 1912 (1984)
25. Westfall, G. D., Jacak, B. V., Anantaraman, N., Curtin, M. W., Crawley, G. M., et al. *Phys. Lett.* 116B: 118 (1982)
26. Aichelin, J., Bertsch, G. *Phys. Lett.* 138B: 350 (1984)
27. Weisskopf, V. *Phys. Rev.* 52: 295 (1937)
28. Aichelin, J. *Phys. Rev. Lett.* 52: 2340 (1984)
29. Gale, C., Das Gupta, S. *Phys. Rev.* C30: 414 (1984)
30. Prakash, M., Braun-Munzinger, P., Stachel, J. *Phys. Rev.* C33: 937 (1986)
31. Shyam, R., Knoll, J. *Nucl. Phys.* A426: 606 (1984)
32. Shyam, R., Knoll, J. *Nucl. Phys.* A459: 732 (1986)
33. Vasak, D., Stoecker, H., Mueller, B., Greiner, W. *Phys. Lett.* 93B: 243 (1980); Vasak, D., Mueller, B., Greiner, W., J. *Phys.* G11: 1309 (1985)
34. Uhlig, M., Schaefer, A., Vasak, D. Z. *Phys.* A319: 97 (1984)
35. Stahl, T., Uhlig, M., Mueller, B., Greiner, W., Vasak, D. *GSI-Preprint 86-46* (1986)
36. Brown, G. E., Deutchman, P. A. *Proc. Workshop High Resolution Heavy Ion Phys.*, pp. 212. Saclay, France (1978)
37. Feshbach, H., Zabek, M. *Ann. Phys.* 107: 110 (1977)
38. Pirner, H. J. *Phys. Rev.* C22: 1962 (1980); Hiller, B., Pirner, H. J. *Phys. Lett.* 109B: 338 (1982)
39. Norbury, J. W., Deutchman, P. A., Townsend, L. W. *Nucl. Phys.* A433: 691 (1985)
40. Prakash, M., Guet, C., Brown, G. E. *Nucl. Phys.* A447: 626c (1986)
41. Tohyama, M., Kaps, R., Masak, D., Mosel, U. *Phys. Lett.* 136B: 226 (1984)
42. Germond, J. F., Wilkin, C. *Phys. Lett.* 106B: 311 (1981)
43. Klingenbeck, K., Dillig, M., Huber, M. *Phys. Rev. Lett.* 47: 1655 (1982)

*Ann. Rev. Nucl. Part. Sci. 1987. 37: 133–76*

# ELECTRON SCATTERING AND NUCLEAR STRUCTURE

## Bernard Frois

Service de Physique Nucleaire–Haute Energie, CEN Saclay,
91191 Gif-sur-Yvette Cedex, France

## Costas N. Papanicolas

Department of Physics and Nuclear Physics Laboratory,
University of Illinois, Urbana, Illinois 61801, USA

CONTENTS

133

0163–8998/87/1201–0133$02.00

## 1. INTRODUCTION

Traditional nuclear models attempt to explain nuclear structure in the framework of nucleonic degrees of freedom only. The interaction between nucleons is described in terms of an effective potential whose microscopic origin would be found in the suppressed degrees of freedom. The properties of the nucleon are assumed to be the same in the nuclear medium as in vacuo. Until quite recently, this simple description was found satisfactory for most nuclear observables.

During the last five years the limits of this framework have been carefully mapped by electron scattering experiments. For heavy nuclei, the mean field theory based on phenomenological effective interactions provides a reliable and consistent description. However, its shortcomings have been clearly seen. In few-body systems, the discrepancies between NN potential theory and experiment are dramatic. The explicit introduction of non-nucleonic degrees of freedom has been demonstrated to be essential.

The search for the appropriate degrees of freedom to describe nuclei is the central focus of nuclear physics today. Therefore we explore in this review our current understanding of nuclear structure as defined by electromagnetic data. The precision of the electromagnetic probe allows us to define accurately the limits of present theoretical descriptions. We review here a broad range of subjects that have been addressed by recent experiments, from the study of meson exchange currents and single-particle distributions to collective excitations in heavy nuclei. However, we do not discuss elastic magnetic scattering, inelastic excitation of discrete states, or single-nucleon knockout reactions since these reactions were recently reviewed (1–3). The principal aim of this review is to offer a fresh perspective on nuclear structure, based on the new generation of electron scattering data presented here and in the above-mentioned articles.

## 2. ELECTRON-NUCLEUS SCATTERING

### 2.1  *The Role of Electron Scattering*

Electron scattering offers unique and widely appreciated advantages for the study of nuclear structure—the weakness of the interaction and the knowledge of the reaction mechanism being the most significant. Multiple scattering effects are negligible and perturbation of the initial state of the nucleus is minimal. The ability to vary independently the momentum and the energy transferred to the nucleus allows the mapping of spatial distributions of the constituent particles. Because electrons are point particles, they offer superb spatial resolution that can be adjusted to the scale of processes that need to be studied. This scale is related to the momentum transfer $q(\lambda \approx 1/q)$. Incident electron energies of 500 MeV result in resolution of the order 0.5 fm, which is ideally suited for the study of nucleon distributions in nuclei. At much higher energies, $E \geq 10$ GeV, electrons have a spatial resolution sufficient to probe quark distributions.

Considerable experimental difficulties had to be overcome in order to exploit the potential of this probe of nuclear structure. One needs to map out form factors to sufficiently high momentum transfers in order to bring out the details of nuclear charge and magnetization densities. Since form factors of heavy nuclei decrease rapidly as a function of momentum transfer, measurements of very small cross sections are imperative. Huge magnetic spectrometers of large solid angle, wide momentum acceptance, and intense electron beams are needed. The difficulty is increased by the need to isolate specific nuclear excitations. An energy resolution $\Delta E/E = 10^{-4}$ is barely sufficient for nuclear studies. Such experimental constraints led to the development (starting around 1970) of a completely new design for experimental facilities. The detection and data acquisition systems met these requirements in the mid-1970s, permitting for the past ten years the exploitation of the possibilities of electron scattering for nuclear studies.

## 2.2   The Nuclear Response

In electron scattering the incident electron transfers momentum $\mathbf{q}$ and energy $\omega$ to the nucleus, and is scattered to an angle $\theta$. In inclusive experiments, the final state of the nucleus is not known. In exclusive experiments, specific channels of the final state are selected and studied. When there are more than two particles in the final state, this necessitates the coincident measurement of some of the products of nuclear deexcitation.

Schematic views of the cross sections of electron-nucleus and electron-nucleon scattering as functions of the energy transfer $\omega$ for a fixed momentum transfer $\mathbf{q}$ are shown in Figure 1. The low energy, low momentum transfer region corresponds to conventional nuclear spectroscopy. The first peak is due to elastic scattering from the ground state, it is followed by the low-lying excited states and the giant multipole resonances. At larger energy transfers, the cross section increases and has a smooth $\omega$ dependence. This is the region of "continuum" excitation. The cross section reaches a maximum in the region $\omega \approx q^2/2M_{nucleon}$ corresponding to "quasielastic scattering" from individual nucleons. The nuclear quasielastic peak is followed at higher energy transfers by successive broad peaks due to the nucleon resonances. At the limit where both $q$ and $\omega$ become very large, one reaches the region of scattering by nucleon constituents. This is the region where evidence for pointlike particles, partons, has been found by scale invariance. In this region, a difference in the scaling behavior between iron and deuterium (4) was recently observed, the so-called EMC effect, named after the European Muon Collaboration (EMC) that first identified it.

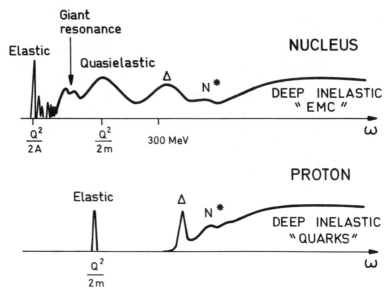

*Figure 1* Schematic shape of electron-nucleus and electron-nucleon cross sections as a function of the energy transfer $\omega$.

The comparison in Figure 1 between electron scattering from a nucleus and from the proton shows the striking difference arising from the influence of the nuclear medium.

## 2.3   *Formalism*

We present here only the basic definitions necessary for discussing electron scattering from nuclei in the one-photon-exchange approximation. More extensive treatments of the formalism can be found elsewhere (5–8).

For an unpolarized incident electron beam, the inclusive electron-nucleus cross section is the sum of two terms,

$$\frac{d^2\sigma}{dE'\,d\Omega} = \sigma_{\text{Mott}}\left[\left(\frac{q_\mu^4}{\mathbf{q}^4}\right)R_{\text{L}}(q,\omega)+\left(-\frac{1}{2}\frac{q_\mu^2}{\mathbf{q}^2}+\tan^2\frac{\theta}{2}\right)R_{\text{T}}(q,\omega)\right], \qquad 1.$$

where $\sigma_{\text{Mott}}$ is the point charge electron scattering cross section, given by

$$\sigma_{\text{Mott}} = \left[\frac{\alpha\cos\theta/2}{2E\sin^2\theta/2}\right]^2; \qquad 2.$$

$R_{\text{L}}(q,\omega)$ and $R_{\text{T}}(q,\omega)$ are the longitudinal and transverse response functions; and $q^\mu$ is the four-momentum of the exchanged virtual photon:

$$q^\mu = (\omega,\mathbf{q}) \qquad q = |\mathbf{q}|. \qquad 3.$$

Neglecting the electron mass, $q^\mu$ and $\omega$ are given by

$$Q^2 = -q_\mu^2 = -4EE' \sin^2 \frac{\theta}{2} \qquad \omega = E - E' \qquad\qquad 4.$$

$$q_\mu^2 = \omega^2 - q^2 \leq 0. \qquad\qquad 5.$$

$E$ and $E'$ are the energies of the incident and scattered electron; $\theta$ is the electron scattering angle (Figure 2), all quantities being measured in the laboratory frame.

As indicated in Equation 1, by selecting the kinematic conditions one can vary the polarization of the virtual photons and can separate the longitudinal and transverse response functions. One performs a combination of forward and backward measurements, keeping the value of the momentum transfer fixed by varying the incident electron energy. The response functions $R_L(q, \omega)$ and $R_T(q, \omega)$ contain the information on the structure of charge and current distributions in the nucleus.

## 3.  FEW-NUCLEON SYSTEMS

A basic goal of nuclear physics, the exact description of finite nuclei in terms of a realistic nucleon-nucleon interaction, is at present beyond the reach of theory. Drastic approximations have to be made to solve the nuclear many-body problem. It is only for few-nucleon systems, $A = 2$ and $A = 3$, that one has found exact solutions in a nonrelativistic framework. Thus, the experimental study of these systems is essential in order to explore the limits of the mesonic description of the nucleon-nucleon interaction. The electromagnetic properties of the two- and three-nucleon systems are of particular interest. Experimental progress has been considerable in the last five years. The form factors of the deuteron, of tritium, and of $^3$He have been accurately determined up to 1 $(GeV/c)^2$. These new

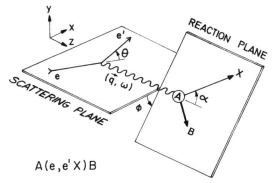

$A(e, e' X)B$

*Figure 2*   Kinematical definitions for coincident electron scattering in the one-photon-exchange approximation.

data offer a rich testing ground for theory and have thoroughly explored the character of meson exchange currents in few-nucleon systems.

## 3.1   Meson Exchange Currents

The importance of meson exchange currents in electromagnetic data was realized (9) shortly after the pion was discovered. Such currents are required by electromagnetic gauge invariance as well as other considerations based on current algebra and chiral symmetry. However, no conclusive evidence for their presence was available at that time. The large coupling constants associated with meson-nucleon vertices do not allow the derivation of a convergent diagrammatic expansion. The inclusion or omission of specific diagrams changes the interpretation of experimental data. Chemtob & Rho (10), by highlighting the role of chiral symmetry, have identified a clear hierarchy of dominant processes where the pion plays a central role. The description of the $\pi$ exchange current is constrained by model-independent theorems, valid for "soft pions" that have small momenta at the scale of the nucleon mass. The first investigations of the effects of mesons in nuclei have relied on these model-independent predictions. The classical example is the slight increase observed in the thermal neutron capture by the deuteron. Riska & Brown (11) showed that the 10% disagreement between experiment and theory was the signature of the $\pi$ exchange current. Various experimental results have now clearly demonstrated that low energy theorems are operative in nuclei (12).

It is crucial to investigate the validity of mesonic theory as a function of momentum transfer in order to find the momentum at which it begins to break down. This is precisely what has been accomplished by electron scattering experiments in the last decade.

## 3.2   Electrodisintegration of the Deuteron at Threshold

This M1 isovector transition is an admixture of two amplitudes, the $^3S_1$ and the $^3D_1$ components of the ground-state wave function of the deuteron coupled to the $^1S_0$ state of the n-p system. It is the inverse reaction of the neutron capture $n+p \rightarrow D+\gamma$. In thermal neutron capture, the pion has negligible momentum and the contribution of the $\pi$ exchange current is given in a model-independent way by low energy theorems. In the electrodisintegration of the deuteron at threshold, the spatial distribution of meson exchange currents can be explored with virtual photons of adjustable wavelength. The nucleonic and mesonic currents contributing to the cross section have strong destructive interferences that occur successively at different momentum transfers (13). Thus, measurements at specific momentum transfers isolate the contributions from different meson exchange processes (14, 15).

Cross sections for this reaction have been measured up to 28 fm$^{-2}$. Experimental data (13), averaged over the energy of the n-p system near threshold ($E_{np} \leq 3$ MeV), are shown in Figure 3, together with theoretical predictions using the Paris potential (15). The predictions take into account the effect of both nucleons and mesons. The purely nucleonic contribution has a deep minimum around $Q^2 = 12$ fm$^{-2}$, which results from a destructive interference between the $^3S_1$-$^1S_0$ and the $^3D_1$-$^1S_0$ amplitudes. Non-nucleonic degrees of freedom are essential for the interpretation of the data. In a large momentum transfer range, between 10 and 15 fm$^{-2}$, they account for nearly 100% of the experimental cross section. This process provides some of the most striking evidence of the presence of meson exchange currents in nuclei.

The experimental data are well described by theoretical predictions that allow for $\pi$, $\rho$, and $\Delta$ meson exchange currents. Up to $\sim 15$ fm$^{-2}$ there is almost no sensitivity to the choice of the nucleon-nucleon potential or the $\pi NN$ form factor. The theoretical predictions for higher momentum transfers strongly depend on the detailed structure of the currents and wave functions.

In order to achieve the best description of the data beyond 10 fm$^{-2}$, one must include the effects of the $\pi$, $\rho$, and the $\Delta$ meson exchange currents

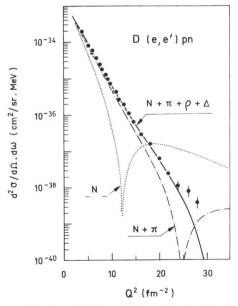

*Figure 3* Cross section for threshold electrodisintegration of deuterium at $\theta = 155°$ (13). The impulse approximation result (nucleon current only) (*dotted curve*) with the successive addition of pion exchange (*dot-dash*) and $\rho$ meson and isobar contributions (*solid curve*) are shown (15).

with hadronic form factors at the meson-nucleon vertex. The Dirac form factor $F_1$ must be used to fit the deuteron electrodisintegration data. The choice of the Sachs form factor $G_E$ would lead to a large discrepancy with experiment. The finite size of the meson-nucleon coupling plays an important role at the spatial scale probed by high momentum transfers. This size is accounted for by a hadronic form factor at the meson-nucleon vertex, with a cutoff parameter $\sim 1200$ MeV in the $\pi$NN form factor. This corresponds to a size of 0.5 fm for the hadronic interaction region. This value, which is smaller than the proton charge radius, appears to be consistent with the description of the nucleon in a two-phase model where charge and baryon number are fractioned between a quark core and a pion cloud (16). A radius of 0.54 fm is predicted for the core, while the radius of the cloud that includes the $\rho$ meson propagation is 0.91 fm. The difference between the electromagnetic and the hadronic form factors comes from the difference in the coupling of the photon and the pion to the nucleon. The photon, because of its large coupling to the $\rho$ meson, is sensitive to the cloud, while the pion is partially blind to the cloud and probes essentially the size of the nucleon core.

In order to avoid the use of phenomenological hadronic form factors, a different approach was recently proposed (17) in which meson exchange currents are derived from the nucleon-nucleon potential through the continuity equation. The advantage of this procedure is that hadronic form factors are automatically consistent with the nucleon-nucleon potential. This procedure yields results very similar to those of the perturbative approach. Therefore one is quite confident of the reliability of the mesonic description of the electrodisintegration of the deuteron at threshold.

The $\rho$ meson exchange and the $\Delta$ isobar current tend to cancel the effect of the $\pi$NN form factor. Short-range processes tend to cancel each other in this isovector process, and it seems that the only really significant contribution is due to the $\pi$ exchange current between two nucleons in a pointlike coupling (18). It is quite intriguing that meson exchange processes appear to be governed by chiral symmetry even at such large momentum transfer. This result suggests that chiral symmetry plays an important role for the description of meson exchange currents beyond the description of soft pions.

A few calculations have begun to investigate the role of quarks with phenomenological models. In the momentum transfer range measured at present, the role of mesons still predominates, quarks making essentially a negligible contribution (19, 20). Experimental data must now be extended to high momentum transfers to investigate such short-range processes.

The most surprising result of this experiment is that there is not yet any sign of breakdown of mesonic theory, even at 1 GeV/$c^2$, beyond its expected limit of validity.

## 3.3    *The Magnetic Form Factor of the Deuteron*

The magnetic form factor $B(Q^2)$ of the deuteron is of isoscalar nature. Therefore, the contribution of the pion exchange current derived from chiral symmetry, which plays a central role in isovector processes, vanishes. Thus, one has a unique opportunity to study the effects of other meson exchange currents. Recent experiments (21–23) have considerably extended the momentum transfer range of the measurements of $B(Q^2)$, providing a very discriminating test of theoretical calculations (Figure 4).

Gari & Hyuga (24) have shown that the $\rho\pi\gamma$ exchange current, which is the dominant isoscalar meson exchange current, increases the impulse approximation cross section by a factor of 3 at 30 fm$^{-2}$, in agreement with experiment. However, beyond 30 fm$^{-2}$, a smooth falloff is predicted, while the experimental data show a minimum around 50 fm$^{-2}$.

The effects of $\Delta\Delta$ and NN* (1440) components in the ground state of the deuteron have been investigated with a coupled-channel N-N interaction by Sitarski, Blunden & Lomon (25). They used N-N scattering data

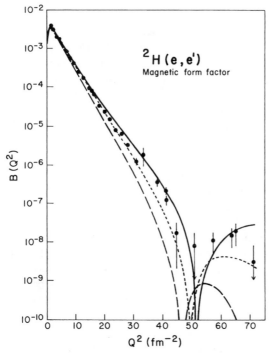

*Figure 4*    The magnetic form factor of the deuteron $B(Q^2)$. Result from an impulse approximation calculation (*dashed curve*) and the improvement resulting from the inclusion of meson exchange currents and isobar contributions (*dotted curve*) are also shown (25). The solid curve is the result of the Skyrme model (31).

up to 1 GeV, with the magnetic moment and the magnetic form factor of the deuteron as constraints. The form factors of Hoehler et al (26) have been used for the nucleon. The effects of meson exchange, as calculated by Gari & Hyuga (24), have also been included since they are assumed to be relatively independent of the details of the NN potential. This model predicts the existence of a diffraction minimum around 45 $fm^{-2}$. The best agreement with experiment up to 60 $fm^{-2}$ is obtained with NN(D state) = 5.45%, $\Delta\Delta \approx 1\%$ and NN* $\approx 0\%$.

Meson exchange currents are of relativistic order for an isoscalar process. The data have reached very high momentum transfers, and a nonrelativistic framework may no longer be reliable for such a light system as the deuteron. At present, it is not yet possible to draw a reliable conclusion on relativistic effects. Various relativistic approaches (27, 28) give completely different predictions. It is crucial to investigate the origin of their discrepancies and to find the appropriate relativistic framework for the description of the properties of the deuteron.

The parton model does not predict a diffraction minimum (29). The existence of a minimum around 50 $fm^{-2}$ shows that even at such high momentum transfers asymptotic behavior has not been reached. Hybrid models (19, 30) involving both nuclear and quark degrees of freedom are able to give a qualitative description of these data.

Nyman & Riska (31) have studied the isoscalar electromagnetic current operator in the Skyrme model. They have shown that this current is determined by the part of the effective QCD Lagrangian that accounts for chiral anomalies and therefore can be predicted at low energies in an essentially model-independent way. The isoscalar meson exchange currents are uniquely related to the baryonic current. It is then possible to derive the form factors of the deuteron from the isoscalar electric structure of the nucleons themselves. This approach gives a good description of the experimental data (see Figure 4).

## 3.4   The Trinucleon Systems

The simplicity of the deuteron makes it a natural starting point for the study of nucleon interactions in nuclei. The success of the impulse approximation, corrected for meson exchange currents, in describing its properties naturally leads one to ask whether this scheme can be as successful for more complex nuclei. In such systems short-range interactions play a different role; the deuteron is barely bound (1 MeV/nucleon instead of the 8 MeV/nucleon typical of heavy nuclei) and the two nucleons rarely find themselves at close proximity. Moreover, the influence of many-body forces (3-body forces in particular) or isospin-dependent effects cannot be examined (32). These aspects of the nuclear force can be studied in the trinucleon systems ($^3$He and $^3$H).

The calculation of properties of $A = 3$ nuclei with realistic forces is now reliable: Hamiltonians based on realistic two-body interactions yield almost identical results (33, 34) for the binding energies of $^3$He and $^3$H, results that are significantly lower than the experimental values (7.72 and 8.49 MeV). The difference between experiment and theory for the binding energy of the $A = 3$ system is now believed to arise largely from the three-nucleon force (33). Only the long-range (attractive) part of the three-body force, which is attributed to two-pion exchange, is well understood. The short-range part is believed to be repulsive and is adjusted phenomenologically. In a more ambitious approach (35) both the short- and long-range parts of the three-body force are derived microscopically by including meson exchange and isobar degrees of freedom. These calculations render support to the phenomenological approach by producing similar trends. The effect of the three-body force is to increase the binding energy by 1–3% of its potential energy. The resulting smaller root mean square (rms) radii are also in agreement with the experimental values. The detailed study of the charge and magnetic form factors of $^3$H and $^3$He can provide further insights on the structure and the momentum content of the ground state.

## 3.5  The Form Factors of $^3H$ and $^3He$

Elastic electron scattering cross section from $^3$H or $^3$He is a linear combination of the squares of the charge and magnetic form factors. Recent measurements (36, 42) separated both form factors up to $\sim 25$ fm$^{-2}$. Figure 5 provides a comparison of the experimental data for the charge form factors to the theoretical predictions of Hajduk et al (35). The impulse approximation, which accounts only for nucleon currents, does not describe the data correctly. The diffraction minimum is shifted by 3 fm$^{-2}$, while the amplitude of the calculated second diffraction maximum is too low by a factor of 2. Various combinations of two-body and three-body forces have been investigated (33, 37). None of them is able to account for the experimental charge form factors. The influence of the three-body force on their shape is very small. A larger negative contribution that vanishes at $q = 0$ and that increases up to the second diffraction maximum is needed. There is a term in the two-pion exchange three-nucleon interaction that gives such a contribution, but in order to explain the observed form factors its strength has to be increased beyond reasonable limits (37).

As in the case of the deuteron, the charge form factors of the three-nucleon system cannot be described without the explicit introduction of nonnucleonic degrees of freedom. A reasonable agreement is obtained for the charge form factors of $^3$He and $^3$H when the effects of $\Delta$ isobars and meson exchange currents are included (35, 38) (see Figure 5).

In a nonrelativistic framework, the relative importance of the meson

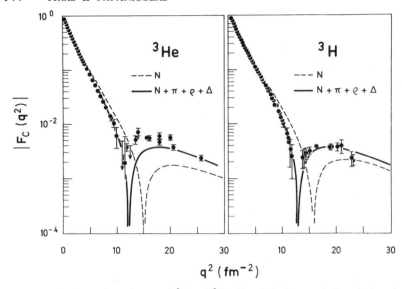

*Figure 5*   The charge form factors of ³H and ³He. The dashed curve depicts the impulse approximation results. The solid curve derives from a calculation in which meson exchange contributions are included (35).

exchange currents depends on the choice of the pion-nucleon coupling. In the pseudoscalar coupling the isoscalar and the isovector meson exchange contributions have approximately the same magnitude. In ³He the two contributions have the same sign, so the ³He form factor is not very sensitive to the choice of coupling. In ³H and for pseudoscalar coupling they contribute with opposite signs, which leads to an almost complete cancellation. Such a cancellation is not observed experimentally.

A description of the same data has also been attempted with quark models (39, 40). Results similar to those of traditional models employing pseudovector coupling are obtained. The photon is coupled to a quark-antiquark pair instead of the usual nucleon-antinucleon pair used for the pion exchange current. The ³He form factor has been also described with quark cluster models (40). These calculations are in reasonable agreement with experiment but have a large model dependence.

The major sources of uncertainty in the theoretical description of the trinucleon charge form factors are of relativistic origin (41). Meson exchange currents are relativistic corrections, while the three-body wave functions are calculated in a nonrelativistic framework. Unfortunately, at this point we lack the appropriate theoretical framework for incorporating in a consistent and complete fashion relativistic corrections.

The magnetic form factors of ³H and ³He (42) are shown in Figure 6. They correspond to an M1 isovector transition similar to the elec-

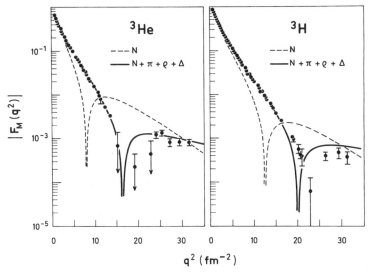

*Figure 6*    The magnetic form factors of ³He and ³He. The curves are defined as in Figure 5.

trodisintegration of the deuteron. The impulse approximation alone cannot explain the experimental data. In the region of $q^2 = 8$ fm$^{-2}$ for ³He and 12 fm$^{-2}$ for ³H, the cross section is entirely due to nonnucleonic processes because of destructive interference among the nucleonic amplitudes. The prediction of Strueve, Hajduk & Sauer (43), which takes into account the effect of both $\pi$ and $\rho$ meson exchange currents, agrees with the experimental data up to the diffraction minimum. In the region of the second diffraction maximum, there is a slight deviation from experiment. The calculation of Hadjimichael, Goulard & Bornais (38) is also in good agreement with the ³He and ³H magnetic form factors. Similar calculations have been performed by Riska (44) and by Maize & Kim (45). In the expression of the exchange current operators, one must use the Dirac form factor F1 to reproduce the experimental data, while there is strong discrepancy at large momentum transfers when the Sachs form factor $G_E$ is used. As in the case of deuteron electrodisintegration (18), the magnetic form factor of ³He can be well described by the impulse approximation corrected only by the contribution of soft pions. This result further highlights the significance of chiral symmetry.

Experiment and theory have made major advances in the three-nucleon problem. The two-body contribution is now theoretically well under control and thus enables us to establish the presence of nonnucleonic processes. Remaining uncertainties are at the level of relativistic effects and of the short-range part of the three-body force.

# 4. ELASTIC ELECTRON SCATTERING FROM HEAVY NUCLEI

The simplicity of few-nucleon systems is ideal for a detailed study of nucleon-nucleon interactions in a nucleus. However, these very light nuclei do not exhibit the special characteristic features of nuclear matter, such as saturation of the nuclear force and a constant binding energy per nucleon. In addition, collective behavior (an important degree of freedom in many-body systems) is totally absent. Such features can only be studied in heavy nuclei. There one is faced with the additional problem of isolating the properties of the bulk nuclear matter from those of the nuclear surface. Nuclear matter is the only form of matter that cannot be studied experimentally in bulk. Nuclear physicists are faced with the challenge of having to derive bulk properties from the behavior of droplets of nuclear matter, the atomic nuclei. The superb spatial resolution and penetrability of electron scattering facilitates the study of the interior of heavy nuclei, thus enabling the isolation of bulk properties from surface effects.

## 4.1 *Charge Distributions of Magic Nuclei*

The charge distribution of nuclei can be determined from the elastic electron scattering cross section. As late as 1975 the interior charge densities of medium and heavy nuclei had not been determined with sufficient accuracy to provide stringent tests for the emerging Hartree-Fock calculations (46). New electron scattering data, combined with the very precise measurements of muonic x-ray transitions, have since then determined the charge densities in the interior of many nuclei to an accuracy of 1%. This set of data, which is the culmination of an experimental effort spanning 35 years, offers a remarkably precise and well-understood nuclear structure observable. Typical is the case of $^{208}$Pb (see Figure 7), where the elastic cross section has been measured over 13 orders of magnitude (47). Presently no nuclear model can reproduce the cross section to the experimental accuracy. The observed discrepancies have provided a crucial impetus for the advancement of nuclear many-body theory.

The experimental charge densities for all doubly closed-shell nuclei (47–52; D. Goutte, private communication) are compared in Figure 8 with the prediction of a mean field (Hartree-Fock-Bogoliubov, HFB) calculation by Dechargé & Gogny (53) with a finite-range, density-dependent, effective force. The disagreement is most pronounced in $^{208}$Pb, which a priori seemed the most favorable case for a mean field description. In the mean field approach the interactions of a nucleon with all the other nucleons is approximated by an effective potential. This comparison shows that modern mean field calculations with effective interactions rather accu-

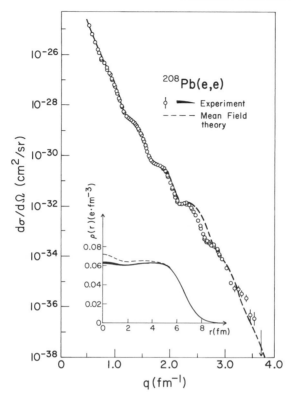

*Figure 7*   Elastic cross section from [208]Pb (47). The dynamic range of the measurements allows the reconstruction of the charge density distribution (*insert*). The thickness of the line in $\rho(r)$ depicts the experimental uncertainty. The mean field result in that of Dechargé & Gogny (53).

rately describe sizes and surface properties but systematically predict larger charge density fluctuations in the nuclear interior.

This discrepancy is today understood as arising largely from correlations whose effects cannot be incorporated into the mean field approach to the nuclear many-body problem. The effects of short-range correlations on the single-particle wave functions are partly accounted for by the introduction of density dependence in the effective nucleon-nucleon potential. The Pauli principle tends to diminish their significance since nucleons cannot scatter into occupied states, which explains why mean field theory is successful in dense systems such as nuclei. However, the effect of long-range correlations due to collective excitations such as the low-lying collective states and the giant resonances are not taken into account. Such correlations are most reliable when calculated consistently with the same force used to generate the Hartree-Fock (HF) ground state (54). They

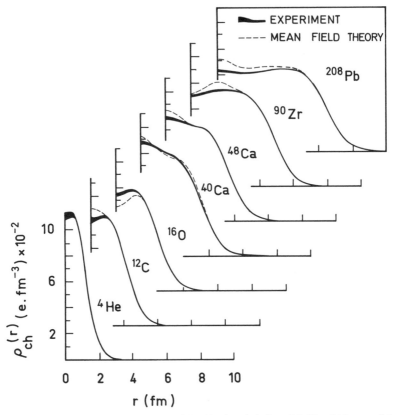

*Figure 8*    Charge density distributions of doubly closed-shell nuclei. The thickness of the
solid line depicts the experimental uncertainty. The mean field calculations are from (53).

have been estimated in the framework of the random phase approximation
(RPA) (55), which is equivalent to time-dependent Hartree-Fock in the
limit of small amplitudes. The large interior fluctuations predicted by
mean field theory are damped by the RPA correlations. However, such
corrections are not sufficient to reconcile theory and experiment.

It is then important to test whether the RPA provides a reasonable
description of the collective excitations of magic nuclei. The measured
transition charge densities (56, 58, 60; K. Seth, private communication; J.
Heisenberg, private communication) for the lowest octupole ($3^-$)
vibrations in magic nuclei are compared to RPA predictions (61) in Figure
9. These states are highly collective, exhausting a substantial fraction
of the energy-weighted sum rule (EWSR) (typically 30%). Most of the
transition charge is found concentrated in the surface. In light nuclei such

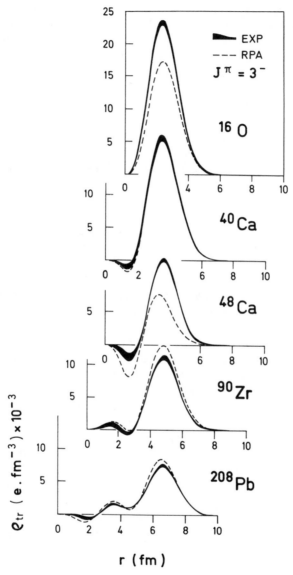

Figure 9  Transition charge densities for the first collective octupole vibrations of doubly closed-shell nuclei (56, 58, 60; K. Seth, private communication; J. Heisenberg, private communication). Same conventions as Figure 8. The theoretical predictions (53–55; Dechargé & Gogny, private communication) are obtained in a self-consistent RPA calculation.

as $^{16}$O, the absence of interior structure is striking. It is only in heavy nuclei, $^{90}$Zr and $^{208}$Pb, that fluctuations in the interior are observed. RPA indeed provides a reasonable overall description of collective excitations— at least at low excitation energies. In RPA, as in HF, the theoretical fluctuations are too large in the nuclear interior, which indicates a systematic problem of a more fundamental nature.

The inability of mean field theory to describe accurately the ground-state charge densities and the octupole transition charge densities of magic nuclei in the nuclear interior raises a fundamental question: "To what extent can a correlated wave function be approximated by an independent particle wave function?" A definitive answer has been obtained by isolating the charge distribution of a 3s proton and by studying single-particle distributions in the lead region.

## 4.2   Single-Particle Distributions

In the independent particle description, the charge distribution is simply the sum of the squares of the proton wave functions in the ground state. The narrow structure in the center of $^{208}$Pb is attributed to the two 3s protons occupying the valence orbit. The quenching of this structure in the measured charge density indicates that the 3s density distribution is significantly modified. Correlations could deform the radial structure of the 3s wave function or modify its occupation probability.

The detailed study of the interior of the charge distributions of $^{206}$Pb and $^{205}$Tl offers the possibility of learning about the shape and strength of the proton 3s orbital. Figure 10 shows the variation of the ratio of the cross sections of $^{205}$Tl to $^{206}$Pb (62, 63) as a function of the momentum transferred to the nucleus. The very special shape of the 3s orbit produces a narrow structure at high momentum transfer and thus allows the unambiguous identification of the contribution of the 3s proton. Mean field theory, regardless of the detailed assumptions of various models, correctly predicts a peak of large amplitude, totally different from the usual small fluctuations between neighboring nuclei. The phase and the shape of the calculated oscillations are in remarkable agreement with the experimental result, but an almost uniform reduction of 30 to 35% in their amplitude is observed. This peak, in the ratio of cross sections of $^{205}$Tl to $^{206}$Pb, is the signature of the shell model.

Figure 11 shows the charge difference between $^{206}$Pb and $^{205}$Tl (63, 64). The characteristic shape of the 3s orbit is observed. Experiment and theory also have the same remarkable similarity in configuration space as in momentum space. It is the first time that the spatial distribution of a particle in a quantum orbit has been isolated. The concept of an independent particle orbit has thus been demonstrated to be valid in the nuclear interior.

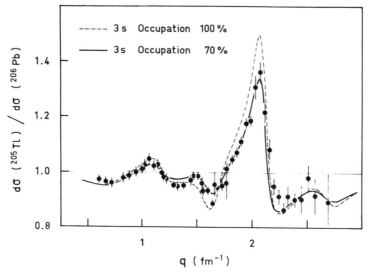

*Figure 10*   The ratio of elastic cross sections from $^{205}$Tl and $^{206}$Pb (62, 63). The peak at $q = 2$ fm$^{-1}$ is the signature of the 3s orbit. The curves are mean field predictions due to X. Campi.

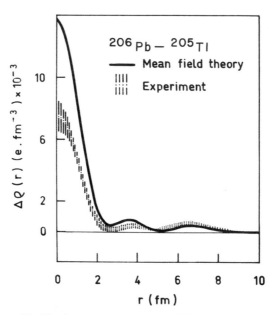

*Figure 11*   The charge density difference of $^{206}$Pb and $^{205}$Tl (63, 64).

One observes clearly a uniform reduction of the charge difference in the interior of the nucleus for $0 \le r \le 5$ fm. The fraction of charge that has left the center of the nucleus is redistributed at its surface.

Significant information on single-particle distributions and the limits of the mean field description is also provided by elastic magnetic scattering (1).

The measured form factor of $^{207}$Pb is shown in Figure 12. The orbital responsible for the magnetization density of $^{207}$Pb is the unpaired $3p_{1/2}$ neutron; its quantum number allows a single M1 multipole with high concentration of magnetization density at the center of the nucleus. The low frequency limit of this measurement is the magnetic moment of $^{207}$Pb, which agrees rather well with the Schmidt value. This agreement historically has been presented as one of the triumphs of the shell model. Electron scattering shows that, in fact, the modification of the $^{207}$Pb magnetization distribution due to correlations is substantial. A quenching of $0.7 \pm 0.1$ of the single-particle amplitude is observed at high momentum transfers.

The form factor of $^{205}$Tl arising from the unpaired $3s_{1/2}$ proton is quenched even further than that of $^{207}$Pb. This is due to the configuration mixing resulting from coupling to the low-lying $2^+$ states of $^{206}$Pb. When this configuration mixing is accounted for, a quenching identical to that of $^{207}$Pb is obtained at high momentum transfers ($q \approx 2.3$ fm$^{-1}$) (65). The same quenching factor ($0.7 \pm 0.1$) is also found for the magnetic form factor of $^{209}$Bi (66).

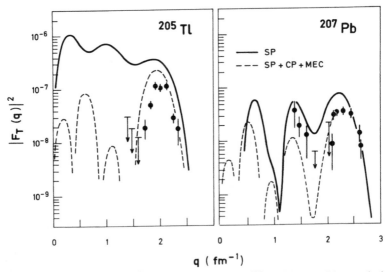

*Figure 12*    The elastic magnetic form factors of $^{207}$Pb and $^{205}$Tl (65). The solid curve depicts the single-particle predictions (53). The dashed curve includes in addition the effects of core polarization and meson exchange (67, 68).

The most complete available calculation in the mean field approach is that of Suzuki et al (67, 68), in which the effects of first-order core polarization, $\pi$ and $\rho$ exchange, and $\Delta$-h excitations have been explicitly accounted for. This calculation predicts the very strong modification of the form factor of $^{205}$Tl at intermediate $q$, which is attributed to first-order tensor core polarization (Figure 12).

The elastic magnetic scattering from $^{205}$Tl, $^{207}$Pb, and $^{209}$Bi, like all single-particle transitions studied by elastic or inelastic electron scattering in the Pb region, exhibits a systematic reduction from the mean field result regardless of the multipolarity or the character of the transition (3, 69). This reduction in the case of the lead region is of the order of 0.7 and was attributed to the effect of correlations that significantly modify the occupation probabilities. In this approach the quenching at high momentum transfer of the magnetization form factors of $^{205}$Tl, $^{207}$Pb, and $^{209}$Bi measures directly the discontinuity of occupation probabilities immediately below ($n_-$) and above ($n_+$) the Fermi sea. The deduced $Z$ factor ($Z = n_- - n_+$) for both neutron and proton shells is $Q = 0.7 \pm 0.07$. Mean field theory by definition produces $Z = 1$. An estimate (69) of the occupation numbers, shown in Figure 13, in $^{208}$Pb is obtained by adding to the microscopically calculated nuclear matter occupation numbers (70) the effects of RPA correlations (55). The resulting quenching factors are in agreement with the experimental results. Similar conclusions have been reached by Mahaux and coworkers (71) by examining the enhancement of the effective mass of nucleons at the Fermi energy.

Making (e, e'p) measurements on $^{206,208}$Pb and $^{205}$Tl provides comple-

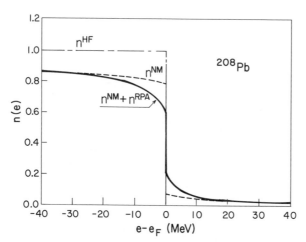

*Figure 13* An estimate of occupation numbers in $^{208}$Pb (*thick line*) derived from a nuclear matter calculation to which the effects of RPA correlations have been added (69).

mentary information on the momentum distribution of orbits as well as on their spectroscopic factors. The missing mass resolution of 100 keV achieved recently at NIKHEF-K (72) is a technological breakthrough that has made these experiments possible. An occupation probability of $0.8 \pm 0.1$ is derived for the $3s_{1/2}$ orbit of $^{208}$Pb, which confirms the interpretation of the elastic scattering measurements discussed earlier. Jaminon et al (73) have stressed that in a mean field approach it is not possible to reproduce simultaneously the charge and the momentum distribution. Additional information is then expected to emerge from the determination measurements of the momentum distribution of heavy nuclei through (e, e'p) measurements, especially at high momenta where the effect of correlations is predicted to be important.

## 4.3    Coupling of Single-Particle to Collective Modes

The interplay between single-particle and collective degrees of freedom not only is one of the most interesting consequences of collective behavior, it also allows the extraction of bulk properties through the study of single-particle properties. Recent precise measurements spanning a large dynamic range in momentum transfer have allowed us to test our theoretical understanding of such processes and to gain access to bulk properties such as the compressibility of nuclear matter.

The determination of the charge differences between $^{208}$Pb, $^{207}$Pb, $^{206}$Pb, and $^{204}$Pb from a combined analysis of muonic x-ray and electron scattering data measures the rearrangement of charge that results from the successive removal of neutrons from the doubly closed shell of $^{208}$Pb.

Mean field theory (53, 61) describes the isotopic density differences well. Excellent agreement is achieved despite the fact that the same theory does not successfully describe the charge distribution of each individual isotope. The isotopic density differences in the Pb region are mostly sensitive to three effects: the nuclear compressibility, the neutron pairing, and the electromagnetic spin-orbit contribution. In the case of the $^{207}$Pb-$^{208}$Pb difference, pairing plays no role at all; $^{207}$Pb is described by a neutron hole in $^{208}$Pb. Since the electromagnetic spin-orbit effect can be calculated reliably, quantitative information on the nuclear compressibility can be deduced from the $^{207}$Pb-$^{208}$Pb difference. Figure 14 shows a comparison between the experimentally derived density difference (74) and the results of two HFB calculations performed with two versions of Gogny's D1 force (53). These forces differ in their values of compressibility formula ($\kappa = 228$ MeV, $\kappa = 209$ MeV) but otherwise provide equally good fits to the set of observables used to adjust the force parameters. The data show a clear preference for $\kappa = 230$ MeV. This value of $\kappa$ is in agreement with the values derived from the analysis of breathing mode energies.

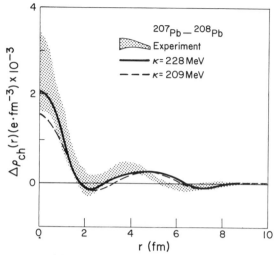

*Figure 14*   Experimental $^{207}$Pb-$^{208}$Pb charge density difference together with mean field predictions for two values of the nuclear compressibility (74).

# 5.   ELECTRON-NUCLEON SCATTERING IN THE NUCLEAR MEDIUM

Electron scattering from nuclei at energy transfers large compared to nucleon separation energies is dominated by interactions with quasifree nucleons. Recent inclusive measurements have studied quasielastic scattering and the excitation of the $\Delta$ isobar in a variety of nuclei.

## 5.1   *Quasielastic Scattering*

A prominent feature of the spectrum of inelastically scattered electrons by the nucleus is the result of quasielastic scattering (see Figure 1). It manifests itself as a broad peak at $\omega \approx (q^2/2m) + \bar{\varepsilon}$, where $\bar{\varepsilon}$ is the average binding energy per nucleon and $m$ its effective mass.

The simplest description of this process is given in terms of the non-interacting Fermi gas model, in which the nucleons are confined within the nuclear volume occupying levels up to Fermi momentum $k_F$. In this model, the experimentally observed quasielastic peak is viewed as providing direct information on $\bar{\varepsilon}$ and $k_F$. The position of the quasielastic peak determines $\bar{\varepsilon}$, while its width determines $k_F$. Earlier measurements (75) showed an excellent agreement with this simple picture. The shape of the quasielastic cross section was found by Moniz et al (76) to be very well described by a two-parameter fit. The values $k_F$ and $\bar{\varepsilon}$, derived for a variety of nuclei, were found in reasonable agreement with the widely accepted values.

More realistic shell-model calculations provided further credence to the noninteracting Fermi gas model by reproducing its results. In these calculations, the average binding energy $\bar{\varepsilon}$ is determined by a microscopic calculation, and final-state interactions are calculated by assuming that the ejected nucleon moves in the potential of the residual nucleus (77). The excellent agreement between experiment and theory suggested that there should not be any difficulty in describing separately the longitudinal and the transverse response functions. Furthermore these first measurements were interpreted as indicating that quasielastic electron scattering was well understood in the framework of the independent particle model.

## 5.2    Longitudinal and Transverse Response Functions

The agreement between the independent particle model and the inclusive quasielastic electron scattering cross section turned out to be completely fortuitous. The same model failed to account for the separated longitudinal and transverse response functions $R_L(q, \omega)$ and $R_T(q, \omega)$ for a variety of nuclei (78–83). Figure 15 shows the longitudinal and transverse response functions for $^{40}$Ca at a momentum transfer of 410 MeV/$c$. The parameters of the Fermi gas model $k_F = 250$ MeV/$c$ and $\bar{\varepsilon} = 36$ MeV were taken from a fit that successfully describes the unseparated cross section. The measured $R_L(q, \omega)$ shows significantly less strength at the peak as compared to the theoretical prediction.

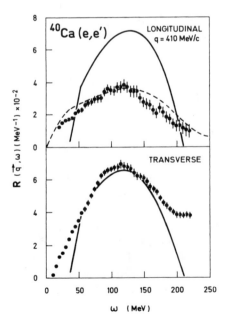

*Figure 15*    $^{40}$Ca longitudinal and transverse response functions at $q = 410$ MeV/$c$ (81). The solid curve is the Fermi gas prediction. The dashed curve is the result of a calculation for nuclear matter that treats correlations (97).

It was generally expected that the transverse response function would not be reproduced by calculations that neglect nonnucleonic degrees of freedom. Since meson exchange currents and delta isobar excitations are predominantly transverse, it turns out that its shape, up to $q = 400$ MeV/$c$, is not very sensitive to theoretical hypotheses and is reasonably well explained by nearly all standard calculations. Only at higher momentum transfers, $q = 550$ MeV/$c$, is a significant effect of nonnucleonic components found.

Similar results are observed for a variety of nuclei. In the momentum transfer range between 200 and 600 MeV/$c$, predictions based on the independent particle model are unable to account for the shape and the strength of the longitudinal response function. At present no theoretical model is able to describe simultaneously the longitudinal and the transverse response functions of nuclei heavier than $^{12}$C. The situation is different for $^3$He, for which both the longitudinal and the transverse response functions are well described by theory.

## 5.3    The Longitudinal Sum Rule

The longitudinal structure function $S_L(q)$ is defined in terms of the longitudinal response by the integral

$$S_L(q) = \frac{1}{Z} \int_{inel} \frac{R_L(q, \omega)\, d\omega}{|\tilde{G}_E(q_\mu^2)|^2}, \qquad\qquad 6.$$

where $Z$ is the charge of the target nucleus and $\tilde{G}_E$ is the appropriate free nucleon form factor (84). By integrating over all excitation energies of the nucleus, the following identity is obtained (84a, 85):

$$S_L(q) = 1 + \frac{1}{Z} \int_0^\infty e^{-iq(\mathbf{r}_1 - \mathbf{r}_2)} C_2(\mathbf{r}_1, \mathbf{r}_2)\, d\mathbf{r}_1\, d\mathbf{r}_2, \qquad\qquad 7.$$

where $C_2(\mathbf{r}_1, \mathbf{r}_2)$ is the two-body correlation function defined in terms of one- and two-body densities:

$$C_2(\mathbf{r}_1, \mathbf{r}_2) = \rho_2(\mathbf{r}_1, \mathbf{r}_2) - \rho_1(\mathbf{r}_1)\rho_1(\mathbf{r}_2). \qquad\qquad 8.$$

Ignoring relativistic corrections, conservation of electric charge requires that $S_L(q) \to 1$ for $q \to \infty$, where the contribution of correlation vanishes.

The asymptotic limit is generally assumed to be reached for $q > 2k_F$ when the effect of two-nucleon correlations is expected to be negligible. Figure 16 shows the behavior of $S_L(q)$ for $^3$He and $^{40}$Ca as a function of $q$, between 200 and 600 MeV/$c$. The experimental data are compared to the calculation of Schiavilla et al (85). While the asymptotic value is reached for $^3$He at $q = 500$ MeV/$c$, 40% of the sum rule is still missing for $^{40}$Ca and all heavier nuclei with the possible exception of $^{238}$U (83).

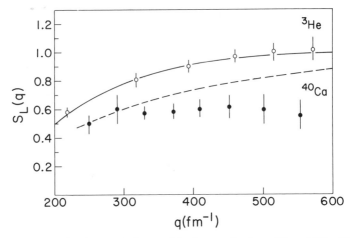

*Figure 16*    The Coulomb sum rule for quasielastic scattering from $^3$He and $^{40}$Ca. The solid curve results from an exact three-body calculation for $^3$He (85). The dashed curve derives from the same nuclear matter calculation (97) whose results are shown in Figure 15.

## 5.4    *The Missing Longitudinal Strength*

The discrepancy between theory and experiment has stimulated an intense theoretical activity. The prescription for the off-shell nucleon current operator is sufficiently well constrained by Lorentz and gauge invariance to be a negligible source of uncertainty (84). Different mechanisms have been invoked to explain the missing longitudinal strength; most notably relativistic effects, modification of the nucleon properties in the nuclear medium, and many-body correlations.

Relativistic effects have been investigated in some detail (86) in the mean field theoretical framework based on Dirac phenomenology. Calculations have been performed in the framework of the $\sigma$-$\omega$ model developed by Walecka and coworkers (87). The discrepancy between experiment and results from relativistic models is smaller than in their nonrelativistic counterparts. This is because the effective nucleon mass and the energy scale are modified by the presence of scalar $\sigma$ and vector $\omega$ fields, $m \to m^* = m + V_s$ and $E \to E^* = E - V_0$, with typical values $V_s = -400$ MeV and $V_0 = +300$ MeV for the scalar and vector potentials. The resulting effective mass $m^*$ enters differently in the charge and current operators. However, recent calculations (88) have shown that such relativistic effects may play a role, but they are not sufficient to explain quasielastic electron scattering data.

The hypothesis of a significant modification of the properties of the nucleon in the nuclear medium has received much attention. An increase of the nucleon radius (confinement volume) would quench both the longitudinal and transverse response functions. An increase of $\sim 20\%$ of the nucleon size reconciles theory and experiment for $^{56}$Fe (89). It also provides a possible explanation of the EMC effect. Celenza et al (90) have developed a model that predicts such an increase. This model is an attempt to make the synthesis of a soliton bag model for the nucleon and Dirac phenomenology. In addition to the standard quark fields, the $\sigma$, $\omega$, $\pi$, $\rho$ fields are used to build the soliton. The electromagnetic radii of the free nucleon are reasonably well described by this model. In the nuclear medium, the presence of the other nucleons modifies the $\sigma$ and $\omega$ fields. The size of the nucleon increases as a function of the nuclear density, becoming 20% larger for lead. This model reproduces the longitudinal response function measured experimentally, but its prediction for the transverse response function is not compatible with the experimental result (91).

The behavior of the nucleon form factor in the nuclear medium has been studied (A. Magnon, private communication) via the $^{40}$Ca(e, e'p) reaction for momentum transfers between 300 and 800 MeV/$c$. As in the inclusive process, the ratio of the longitudinal and transverse response functions is quenched by 35% with respect to theory. It has also been found that within experimental uncertainties the momentum transfer dependence of the electron-nucleon cross section in the nuclear medium is compatible to that of the free nucleon. This result disagrees with the predictions of Celenza et al (90). An increase of the nucleon radius larger than a few percent is excluded by this measurement. A similar conclusion was reached from a $y$-scaling analysis of $^{56}$Fe quasielastic data, which is discussed in the following section. A significant swelling of nucleons does not appear to be the explanation of the missing longitudinal strength.

Many-body correlations can be very significant in influencing the longitudinal response function but up until recently their effect at 400 MeV/$c$ was predicted to be small. Correlations due to collective excitations of the nucleus improve the agreement with theory at $q = 300$ MeV/$c$ (93–96), but their effect decreases at higher momentum transfer. At $q = 550$ MeV/$c$, these long-range correlations calculated in the framework of the random phase approximation become essentially negligible. Short-range and tensor correlations have been difficult to estimate in a reliable way (96, 97) because realistic calculations are possible only for few-nucleon systems and for nuclear matter. Calculations for finite nuclei are usually performed with effective interactions, which predict only a small effect for momentum transfers $q = 400$ MeV/$c$.

Fantoni, Pandharipande, and coworkers recently showed a large effect due to correlations in calculations performed with realistic nucleon-nucleon interactions in the few-nucleon systems (85) and in nuclear matter

(97). The effect of these correlations is to quench the amplitude of the quasielastic peak and to shift a significant fraction of its strength at higher excitation energies. The longitudinal response calculated for nuclear matter (97) is compared to experimental data for $^{40}$Ca at $q = 410$ MeV/$c$ in Figure 15. This theoretical prediction is in good agreement with experiment. The missing longitudinal strength is simply shifted beyond the maximum excitation energy measured experimentally. The longitudinal sum rule for nuclear matter is compared in Figure 16 to the experimental data for $^{40}$Ca (81, 82). The upper limit of integration of theoretical response was taken to be identical to the experimental one. Substantially better agreement with experiment is achieved by this calculation than by other theoretical approaches, but there is still some missing strength at high $q$. At 550 MeV/$c$, the $^{40}$Ca and $^{56}$Fe longitudinal response functions are overpredicted for excitation energies exceeding 200 MeV. This might be due to a number of approximations in the calculation. It is also quite possible that the excitation of the nucleon, which is not taken into account, is beginning to play a significant role.

The available data are still relatively limited in momentum transfer. It is important to extend these data to higher $q$ where the discrepancy is most serious. Further calculations with a realistic nucleon-nucleon interaction for finite nuclei are now crucial in order to reach a definitive conclusion on the capabilities and limitations of the traditional approach. The modification of the nucleon properties in the nuclear medium and relativistic effects appears to be less important than many-body correlations in explaining quasielastic electron scattering data.

## 5.5   Y-Scaling of the Quasielastic Response Function

The inclusive (e, e′) cross section in the quasielastic region has a shape and an amplitude that varies smoothly as a function of both the momentum transfer $q$ and the excitation energy $\omega$. In the impulse approximation, assuming a single-photon exchange, the cross section can be written as

$$\frac{d^2\sigma}{d\Omega\,d\omega} = \left( Z\frac{d\sigma_p}{d\Omega} + N\frac{d\sigma_n}{d\Omega} \right) R(q,\omega), \qquad\qquad 9.$$

where $d\sigma_p/d\Omega$ and $d\sigma_n/d\Omega$ are the proton and neutron cross sections.

West (98) has shown that if, in addition, one assumes a nonrelativistic Hamiltonian and neglects final-state interactions, then the response function $R(q, \omega)$ can be expressed in terms of a single variable $y$, for large energy and momentum transfer:

$$R(q,\omega) \to F(y) \qquad \text{for } q \to \infty, \qquad\qquad 10.$$

where

$$y \equiv \mathbf{k} \cdot \hat{q}; \qquad y \approx \frac{(\omega^2 + 2M_n\omega - q^2)}{2q}. \qquad\qquad 11.$$

After dividing by the electron-nucleon cross section, the experimental data measured for different values of $q$ and $\omega$ should lie on a universal curve (98). The scaling variable $y$ defined above is the component of the struck nucleon momentum $\mathbf{k}$ along the momentum transfer $\mathbf{q}$. The choice of $y$ depends on the theoretical assumptions used to describe final-state inter-actions, but Gurvitz, Tjon & Wallace (99) have shown that $y$-scaling is observed for a variety of possible $y$ variables.

The cross sections for $^3$He (see Figure 17) (100) and more recently those for $^4$He, $^{12}$C, $^{27}$Al, $^{56}$Fe, and $^{197}$Au (101) have been measured for a wide range of momentum transfers and excitation energies. The experimental data, when plotted as a function of $y$, lie on a universal curve in the quasielastic region (Figure 18$a$). The region of the quasielastic peak for which scaling behavior is observed is the region of small $y$, where $y < 0$. For $y > 0$, the response function does not scale. This can be understood easily, as $y > 0$ corresponds to an energy transfer sufficiently large to excite nucleon resonances that have form factors different from that of the free nucleon.

It is intriguing that at present no calculation based on a purely nucleonic

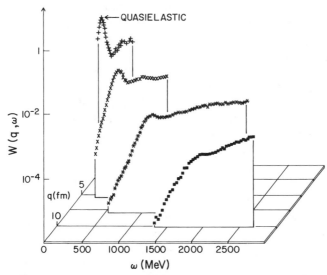

*Figure 17*    $^3$He(e, e$'$) response function in the quasielastic region for four different momentum transfers (101).

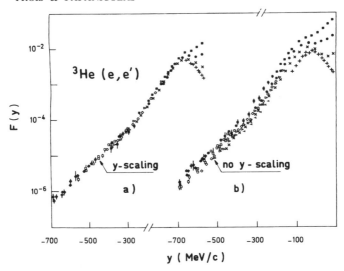

*Figure 18*   Scaling function $F(y)$ for $^3$He. The scaled data derive from the nuclear response functions shown in Figure 17 obtained by using (*a*) the free nucleon form factor and (*b*) a modified form factor corresponding to a size increase of the nucleon of 10% and simultaneously a mass decrease of 10% (104).

picture is able to reproduce the asymptotic scaling behavior of the experimental data. The theoretical spectral functions underpredict the data, which indicates a lack of high momentum components or nonasymptotic behavior where the effects of binding are still being felt (101a). Experimental information on the nucleon momentum distribution has been obtained at lower momentum transfers by measuring $^3$He(e, e'p) cross sections (102). These data are reasonably well reproduced by theoretical calculations (103). Therefore it is crucial to find if the problem lies in the breakdown of the theoretical description of the nucleon momentum distribution at very high momentum, or in the description of the reaction mechanism, as suggested by Ciofi degli Atti et al (99, 101a). The present $^3$He(e, e') data have a very small ($<15\%$) longitudinal component. We discussed in the previous section that it is the longitudinal response that provides the cleanest and most sensitive test to theoretical hypotheses. A precise separation of the longitudinal and transverse response functions at $q \geq 1$ GeV/$c$ for light and heavy nuclei may prove helpful in explaining the lack of strength observed at high nucleon momenta.

The $y$-scaling is very sensitive to the choice of the nuclear form factor, as its very definition clearly implies. If nucleons in the nuclear medium have properties different from the free nucleon, then the electron-nucleon cross section in the nucleus and the value of $y$ will be modified. The behavior of $y$-scaling has been investigated as a function of both the size and the mass of the nucleon (104). Figure 18$b$ shows that $y$-scaling in $^3$He is destroyed by a simultaneous 10% increase of the nucleon size and a 10% decrease of the nucleon mass. Sick (104) concludes for $^3$He that, if a variation of the radius is correlated with an inverse variation of the mass, the size of the nucleon increases by at most 3%, while if only the size of the nucleon is changed, the maximum increase possible is 6%.

A systematic study (101) of inclusive scattering on a variety of nuclei in tne quasielastic scattering region has been carried out recently over a large range of momentum transfers ($q = 2$–$12$ fm$^{-1}$). Day et al (101) showed that $y$-scaling is also observed for heavy nuclei. The maximum increase of the nucleon radius that is allowed in order to preserve $y$-scaling in $^{56}$Fe is $\sim 3\%$ (104a). This result further reinforces the view that an increase of the nucleon size in heavy nuclei, if present at all, is much smaller than the values that have been proposed ($\sim 15\%$) in order to explain the missing strength in $S_L(q, \omega)$ and the EMC effect.

## 5.6   Excitation of the $\Delta$ Resonance

Several inclusive experiments (79, 81, 82, 105) have mapped out the nuclear response for energy transfers $\omega$ up to 500 MeV. In these experiments the $\Delta$ resonance is as strongly excited as the quasielastic peak. This enables us to study the propagation of the $\Delta$ in the interior of nuclei. Since most of our knowledge of the behavior of the $\Delta$ resonance in the nucleus comes from pion-nucleus scattering, total photoabsorption and electron scattering provide an additional important test of the theoretical approaches developed for the $\Delta$-N interaction. While the pion-induced reactions are predominantly surface mechanisms, photons and electrons probe the entire nuclear volume. A further advantage of electrons is that $q$ and $\omega$ can be varied independently, providing a broader range of investigation of the $\Delta$-nucleus interaction.

O'Connell et al (105) recently studied the $\Delta$ resonance in kinematical conditions close to real photoabsorption, $q = \omega$ ($Q^2 = 0$). Data have been obtained on various light nuclei up to $^{16}$O. The cross section is found to vary linearly with $A$, which means that the cross section per nucleon is constant (Figure 19). This agrees with recent experimental results (106–108) obtained for real photoabsorption. The comparison with the electron-proton cross section in the $\Delta$ region shows the strong modification of the medium. This is due in particular to the Pauli blocking of the $\Delta$ decay, multiple scattering, and the damping of $\Delta$ propagation through the $\Delta$N $\rightarrow$ NN channel. Figure 19 gives the prediction of the $\Delta$-hole approach (109), which has been used successfully to describe photonuclear reactions (110). The pion absorption channel is described by a spreading potential taken from the analyses of pion-nucleus scattering. This calculation reproduces

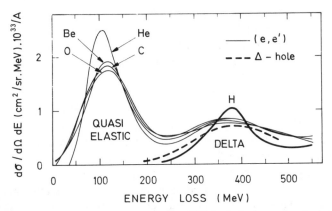

*Figure 19*  Inclusive cross-section measurements (105) in the isobar region for light nuclei. The dashed curve is the Δ-hole model prediction (109, 110). The thick curve corresponds to the electroexcitation of the free nucleon.

the medium effects reasonably well; however, about 15% of the strength observed experimentally is missing. The spreading potential fitted to pion scattering is not sufficient; more complicated corrections have to be included in the momentum transfer range probed by electron scattering. The failure of the model in the "dip" region between the quasielastic peak and the Δ resonance does not appear to be of fundamental nature. A complete theoretical description of this region is made difficult by the interference with 2p-2h and with higher order contributions from final-state interactions in the quasielastic process (93).

The momentum transfer dependence of the cross section in the Δ resonance region for $^3$He, $^{12}$C, $^{40}$Ca, $^{48}$Ca, and $^{56}$Fe has been studied by Laget (111, 113). He concludes that the resonance contribution seems to be reasonably well understood. His calculations are in better agreement for $^3$He; this indicates that the disagreement observed might be due to the approximations used for the many-body problem. Especially limiting are the uncertainties and approximations in treating theoretical two-body correlations. The basic mechanism of the excitation of the Δ in the interior of the nucleus appears well understood, but one needs now to perform coincidence experiments to separate the effects of correlations and final-state interactions.

## 6.  PERSPECTIVES FOR THE FUTURE

All the experiments discussed so far were performed at facilities designed in the late 1960s. They reflect both the technological advances and constraints of that era. Intense beams of high energy and tight emittance effectively employed by high resolution spectrometers made possible the measurements we discussed. On the other hand, the absence of certain classes of data, most notably coincident and polarization transfer data, is a reflection of the limitations of these facilities. Particularly limiting has been their low duty factor ($\sim 0.01$). A high duty factor is essential in coincidence experiments. The unavailability (until quite recently) of polarized electron beams and suitable polarized targets is responsible for the lack of exclusive data in the polarization degrees of freedom.

Both of these limitations were overcome in the early 1980s. Low energy electron scattering facilities with continuous beams (duty factor $\approx 1$) have been in operation for the past few years. Polarized beams of high intensity and high degree of polarization ($>0.3$) also became available at about the same time. Internal polarized targets have been demonstrated to be both feasible and capable of yielding reasonable luminosities when used together with the high intensity beams ($I > 100$ mA) of electron storage rings.

We briefly review in the following two sections some recent results that demonstrate the potential of the new experimental techniques. Recent coincident measurements have probed directly two-nucleon correlations. Experiments with low energy, continuous electron beams have explored phenomena ranging from subthreshold fission to giant resonances. Indicative of the significance attached to coincidence measurements is the fact that every major electron scattering facility under construction, regardless of its maximum energy, is of high duty factor ($>0.8$).

## 6.1  *Probing Two-Nucleon Correlations with Coincidence Experiments*

Investigation of two-nucleon correlations requires the coincident detection of two or three reaction products, as in the (e, e′N) or (e, e′2N) channels.

Potentially, two-nucleon knockout experiments (e, e′2N) yield the most direct information on the short-range behavior of the N-N interaction. Such experiments are today at the limits of operating facilities. However, recent results from the ongoing (e, e′p) experiments at low duty factor accelerators have provided evidence for the existence of such correlations.

Coupling of the virtual photon to correlated nucleon pairs was observed in the study of the three-body breakup in the $^3$He(e, e′p) reaction (112; C. Marchand, private communication). Figure 20 shows three missing energy spectra at recoil momentum values of 316, 401, and 458 MeV/$c$ respectively. The peak at 5.5 MeV corresponds to the one-body knockout reaction to a deuteron final state. A broad peak at higher missing energy values is values of missing energy in each spectrum, is due to the breakup of a p-n pair in $^3$He. It corresponds to the disintegration of a correlated p-n pair at rest. The curves shown in Figure 20 are the results of a calculation (113)

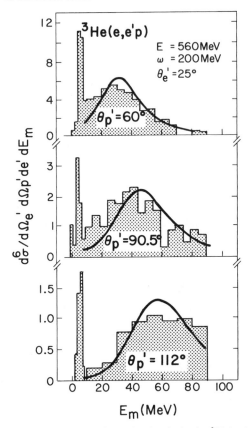

*Figure 20*  Evidence for the existence of correlated pairs in the $^3$He(e, e′p) spectra (112; C. Marchand, private communication). The secondary peak is the signature of virtual photon absorption on correlated N-N pairs; the solid curve is a theoretical prediction that includes final-state interactions and meson exchange currents (113).

for $^3$He based on Faddeev wave functions, a calculation that takes into account final-state interactions and meson exchange currents. Similar (e, e′p) measurements (114) in $^{12}$C also find a large increase in the high energy region of the missing mass spectrum.

The $^3$He(e, e′d)p experiment (115) has allowed the study of two-body mechanisms at low momentum transfers. Cross sections were measured in parallel kinematics with the momentum transfer fixed at $q = 380$ MeV/$c$ and the recoil momentum varying between 0 and 200 MeV/$c$. The experimental coincidence cross section is shown in Figure 21. The dotted curve is the expected cross section for the process in which the virtual photon couples to a single nucleon. The full curve is obtained if, in addition, the

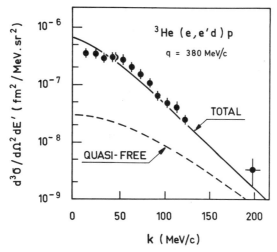

*Figure 21* ³He(e, e'd) cross section (115). The dashed curve results from a calculation of quasifree proton knockout. When photoabsorption on correlated n-p pairs, rescattering, and meson exchange effects are included (113), a good description of the data is achieved (*solid curve*).

coupling to a correlated p-n pair is allowed (in either the $T = 0$ or $T = 1$ state).

## 6.2 New Facilities: The First Results

Two major factors have so far limited the power of single-arm electron scattering: (*a*) Excitations in the continuum are broad and overlapping (e.g. giant multipole resonances), and their isolation from the elastic radiative tail, which often accounts for more than 90% of the observed cross section, is problematic. (*b*) The reconstruction of charge and current densities requires knowledge of individual multipole form factors; the single-arm cross section involves the incoherent sum of all multipole form factors allowed by the spin-parity selection rules. It is only in the case of scattering from spin-zero nuclei, and when one single multipole is allowed, that such reconstruction can be performed.

Both of these limitations are removed in coincident electron scattering. The coincidence requirement eliminates the contribution of the elastic radiative tail at the detection stage. The angular pattern of the decay products allows the model-independent determination of multipole strengths and therefore the reconstruction of transition currents and densities. Potentially more powerful is the flexibility that coincidence measurements offer for isolating and studying particular processes that are inaccessible in the inclusive channel.

Coincidence measurements will allow the study of modes of nuclear excitation whose investigation by electron scattering is still rather primitive. A good example is the case of giant resonances, the most important modes of nuclear collective motion. Bulk properties of nuclear and neutron matter, such as compression moduli, symmetry energies, and spin-isospin sound velocity, are derived from their study. Information on the energy dissipation of Fermi liquids can be obtained from the study of their damping (116, 117; J. Wambach, private communication).

The recent $^{208}$Pb(e, e′n) measurement (118) exhibits some of the advantages of coincidence electron scattering in the study of continuum excitations. The inclusive spectrum (see Figure 22) is dominated by the elastic radiative tail, but the coincidence spectrum is free of it. In this coincidence measurement, where no angular and energy information from the decay product is retained, the resulting spectra offer the same information as inclusive spectra after the subtraction of the radiative tail. Multipole strength functions were extracted by a multipole decomposition analysis of the combined (e, e′n) and (γ, n) data. The resulting monopole-quadrupole

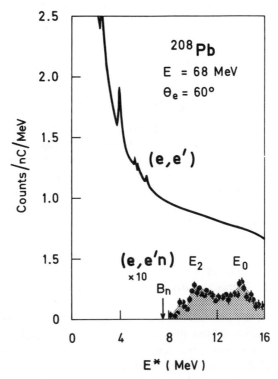

*Figure 22*  Single-arm and coincident (e, e′n) spectra from $^{208}$Pb (118). The coincidence condition removes the radiative tail, revealing the excitation of giant resonances.

strength function exhibits two resonant structures at 10.6 and 14.1 MeV, as evident in the coincidence spectrum (see Figure 22). The integrated strength between 9 and 12.5 MeV exhausts 65% of the energy-weighted quadrupole sum rule, the strength between 12.5 and 16 MeV will satisfy 130% of the isoscalar EWSR. These results are in good agreement with the HF-RPA predictions (117) and have definitively settled previous contradictory statements from different experiments. The ability of mean field theory to describe these very important collective modes of excitation of the nuclear many-body system is thus demonstrated to be as good as in the case of the nuclear charge densities and low frequency excitations discussed in Section 4.

As coincidence measurements become more exclusive, the power of the (e, e'x) probe to isolate particular channels and to yield microscopic information is greatly enhanced. In the study of the $^{28}$Si(e, e'p) reaction (119), the energy spectrum of the emitted particles allows the isolation of the decay to the ground state and various excited states of the daughter nucleus. The study of the angular distribution of the decay to any of these states then provides the most complete result of a two-arm coincidence measurement. The angular distribution of the $\alpha$ decay of $^{28}$Si to the ground state of $^{24}$Mg shown in Figure 23. It has allowed determination of multipole

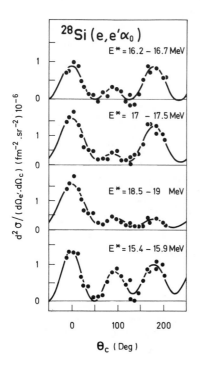

*Figure 23*   Angular distributions from the $^{28}$Si(e, e'$\alpha$)$^{24}$Mg coincidence measurements (119).

strengths. The derived E1 strength agrees with the corresponding photodisintegration experiment. The deduced E2/E0 strength amounts to 42%
of the E2 sum rule, twice the value derived from isoscalar probes, which
indicates the presence of strong isovector E2/E0 excitations.

In the (e, e′γ) reaction both the excitation and decay channels are
electromagnetic, and therefore free of final-state hadronic interactions. It
can be used in the study of both reaction (e.g. radiative corrections) and
structure questions. Its most obvious use is in the study of bound states
of non-zero-spin nuclei (120, 121). The angular pattern of the decay gamma
rays allows the determination of individual multipole form factors for
excited states of non-zero-spin nuclei, which in turn facilitates the reconstruction of the corresponding transition charge densities.

The first nuclear structure investigation utilizing (e, e′γ) was performed
on the 4.4-MeV state of $^{12}$C (120). This state had been previously studied
by inclusive inelastic scattering, and the two form factors involved ($F_T$,
transverse, and $F_L$, longitudinal) were isolated through the Rosenbluth
method. The (e, e′γ) reaction provides an alternative way of separating
them through the measurement of their interference and at the same time
reveals their relative sign. If $|F_T|^2$ were zero, a simple quadrupole pattern
relative to the $q$ axis would result. The effect of a nonvanishing $F_T$ is to
rotate the pattern with respect to the $q$ axis; the sense of rotation is
clockwise if $F_T$ and $F_L$ have the same sign, and counterclockwise if they
have opposite signs. A counterclockwise rotation was observed (see Figure
24), which means the relative sign was negative. This phase is in agreement

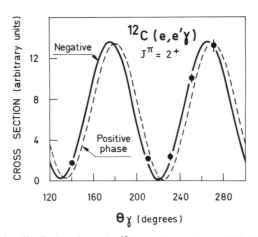

Figure 24   Angular distribution from the $^{12}$C(e, e′γ) experiment (120). It allows the separation of the two contributing multipoles (C2 and E2) and determination of their relative
phase.

with theory, which predicts that the transverse form factor is dominated by the convection current at low momentum transfers.

The increase in energy at electron accelerators with high duty factor will allow the study of these phenomena at a much wider dynamic range of momentum transfers. At higher energy, the same coincidence techniques, applied now to giant resonances, can be used for the investigation of nucleon resonances. The comparison of the transition currents of nucleon resonances studied in the proton and finite nuclei can provide precise and discriminating evidence of the influence of the nuclear medium on the intrinsic properties of the nucleon.

The drawbacks inherent to the standard setup of external beam–fixed targets are numerous: inability to study short-lived isotopes, degradation of the energy spectrum caused by straggling of particles emerging from the target, and difficulty in maintaining target polarization due to beam heating are among them. In a pioneering series of experiments at Novosibirsk, gas jet targets have been used in the VEPP-3 electron ring (122; S. C. Popov, private communication). The intense circulating electron beam ($I \approx 300$ mA) compensates for the very thin targets and results in reasonable luminosities.

Figure 25 shows the data obtained in the first internal target experiment reported, that of $^{16}O(e, e'x)$. Except for the internal target, the experimental arrangement is quite conventional: an electron spectrometer and decay particle detectors surround the interaction region. The quality of the data is impressive. Clearly, backgrounds are under control and worries regarding the feasibility of operating the ring and of obtaining high quality data in such an environment have been alleviated.

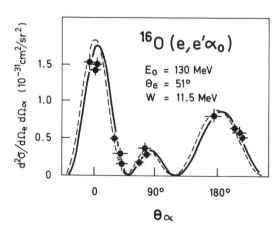

*Figure 25*   Coincidence electron scattering data from $^{16}O$ obtained in the depicted internal jet target setup of Novosibirsk (122).

Already the same group has reported electron scattering measurements from a polarized deuterium jet target (123). The possibilities offered by such measurements are numerous and exciting.

# 7. CONCLUSION

Recent electron scattering experiments have provided a wealth of new information on nuclear structure. The interpretation of the experimental results in the two- and three-body systems requires the explicit treatment of nonnucleonic degrees of freedom. Mesonic degrees of freedom are sufficient to account for the observed discrepancy between the nucleonic description and experiment up to $Q^2 = 25\,\mathrm{fm}^{-2}$. The role of the pion is well understood up to $\sim 15\,\mathrm{fm}^{-2}$. The modification of the intrinsic properties of the nucleon inside the nucleus appears to be less significant than originally suggested. The suppression of the longitudinal quasifree response appears to be induced mostly by many-body correlations. Discrepancies remain, which suggests that other effects are also contributing. Further experimental and theoretical investigations are needed.

The concept of nucleons moving in a mean field has been found to be valid throughout the nuclear volume by determining the charge distribution of the 3s orbit in lead. Yet the precise determination of ground and transition charge densities has shown that mean field theories using effective interactions cannot describe the interior structure of nuclei. Elastic charge and magnetic scattering and the study of a variety of other inelastic excitations in the lead region have shown that correlations reduce occupation probabilities by $\sim 30\%$ for valence orbitals.

In the last few years a noticeable transition to more exclusive measurements has been taking place. Coincident (e, e′p) studies in the quasifree region yield precise information on the momentum distributions of nucleons in the nucleus. These experiments recently provided for the first time direct evidence for two-body correlations. However, little is known about short-range processes.

The first results from continuous-beam electron accelerators have demonstrated, at low energies, the feasibility and the importance of coincidence measurements. Continuum excitations (especially giant multipole resonances) are now being studied with a precision previously restricted to bound states. The feasibility of isolating multipole form factors in mixed transitions through the (e, e′γ) reaction has been demonstrated. The study of nucleon resonances in nuclei, using these same coincidence techniques at higher energy, appears promising. The experimental method has been further widened by the use of internal polarized targets in coincidence measurements.

Quantum chromodynamics (QCD) has emerged as the fundamental theory of strong interactions, and in the asymptotic limit of short distances it successfully defines the desired behavior of any nuclear model, just as the traditional approach does in the limit of large distances. Yet, QCD remains intractable in the strong coupling regime, and the problem of confinement is still unsolved. A unified description of these two phases of hadronic matter within a general framework, based on fundamental theory, has emerged as one of the key tasks confronting nuclear physics (124).

Experimental data exploring the transition region between confinement and asymptotic behavior are needed to provide guidance to theory. The success of the previous generation of electron scattering facilities combined with the versatility and precision of the continuous-beam electron accelerators has led the nuclear community in the US to declare as its highest priority the construction of a multi-GeV electron accelerator suitable for coincidence and polarization studies. This goal is being realized by the construction of the Continuous-Beam Electron Facility (CEBAF) in Virginia (125). CEBAF will be augmented by lower energy, continuous-beam electron accelerators that will further explore the mesonic and collective degrees of freedom. Similar plans and construction projects are also being undertaken in Europe (125). The availability of these facilities in the early 1990s will bring the study of nuclei with electromagnetic probes to a new era.

### ACKNOWLEDGMENTS

We would like to express our deep appreciation for the many ideas, criticisms, and suggestions we have received from the electron scattering community and nuclear structure theorists. We are particularly indebted to our colleagues at Illinois and Saclay. An attempt to give even a partial list of people who have substantially contributed to this review would be prohibitively long. This work was supported in part by the National Science Foundation under grant NSF PHY 86-10493 and by a CNRS-NSF exchange program under grant NSF INT 83-12952.

*Literature Cited*

1. Donnelly, T. W., Sick, I. *Rev. Mod. Phys.* 56: 461 (1984)
2. Frullani, S., Mougey, J. *Adv. Nucl. Phys.* 14: 1 (1984)
3. Heisenberg, J., Blok, H. *Ann. Rev. Nucl. Part. Sci.* 33: 569 (1983)
4. Aubert, J. J., et al. *Phys. Lett.* 123B: 275 (1983)
5. Drechsel, D. *Prog. Part. Nucl. Phys.* 13: 64 (1984)
6. Ciofi degli Atti, C. *Prog. Part. Nucl. Phys.* 3: 163 (1980)

7. DeForest, T. Jr., Walecka, J. D. *Adv. Phys.* 15: 1 (1966)
8. Amaldi, E., Fubini, S., Furlan, G. *Pion Electro-production.* Berlin: Springer (1979)
9. Villars, F. *Helv. Phys. Acta* 20: 476 (147)
10. Chemtob, M., Rho, M. *Nucl. Phys.* A163: 1 (1971)
11. Riska, D. O., Brown, G. E. *Phys. Lett.* 38B: 193 (1972)
12. Rho, M. *Ann. Rev. Nucl. Part. Sci.* 34: 531 (1984)
13. Auffret, S., et al. *Phys. Rev. Lett.* 55: 1362 (1985); Simon, G., et al. *Nucl. Phys.* A324: 277 (1979) and references therein
14. Leidemann, W., Arenhövel, H. *Nucl. Phys.* A393: 385 (1983)
15. Mathiot, J. F. *Nucl. Phys.* A412: 201 (1984)
16. Brown, G., Rho, M., Weise, W. *Nucl. Phys.* A454: 669 (1986)
17. Riska, D. O. *Phys. Scripta* 31: 471 (1985); Buchmann, A., Leidemann, W., Arenhövel, H. *Nucl. Phys.* A443: 726 (1985)
18. Brown, G., Rho, M. *Comments Nucl. Part. Phys.* 10: 210 (1981)
19. Kisslinger, L. *Lect. Notes Phys.* 260: 432 (1986) and references therein
20. Yamauchi, Y., Yamamoto, R., Wakamatsu, M. *Phys. Lett.* 146B: 153 (1984)
21. Arnold, R. G., et al. *Phys. Rev. Lett.* 58: 1723 (1987)
22. Auffret, S., et al. *Phys. Rev. Lett.* 54: 649 (1985)
23. Cramer, R., et al. *Z. Phys.* C29: 513 (1985) and references therein
24. Gari, M., Hyuga, H. *Nucl. Phys.* A264: 409 (1976)
25. Sitarski, W. P., Blunden, P., Lomon, E. L. To be published
26. Hoehler, G., et al. *Nucl. Phys.* B114: 505 (1975)
27. Zuilhof, M. J., Tjon, J. A. *Phys. Rev.* C22: 2369 (1980)
28. Arnold, R. G., Carlson, C. E., Gross, F. *Phys. Rev.* C21. 1426 (1980)
29. Chemtob, M., Furui, S. *Nucl. Phys.* A454: 548 (1986)
30. Takeuchi, S., Yazaki, K. *Nucl. Phys.* A438: 605 (1985)
31. Nyman, E. N., Riska, D. O. *Phys. Rev. Lett.* 57: 3007 (1986)
32. Martino, J. *Lect. Notes Phys.* 260: 129 (1986)
33. Friar, J. L., Gibson, B. F., Payner, G. L. *Ann. Rev. Nucl. Part. Sci.* 34: 403 (1984) and references therein
34. Sasakawa, T., Ishikawa, S. *Few Body Systems* 1: 3 (1986)
35. Hajduk, C., Sauer, P., Strueve, W. *Nucl. Phys.* A405: 620 (1983)
36. Juster, F. P., et al. *Phys. Rev. Lett.* 55: 2261 (1985) and references therein; Beck, D. *PhD thesis*, MIT (1986)
37. Carlson, J., Pandharipande, V. R., Wiringa, R. B. *Nucl. Phys.* A401: 59 (1983)
38. Hadjimichael, E., Goulard, B., Bornais, R. *Phys. Rev.* C27: 831 (1983)
39. Beyer, M., Drechsel, D., Giannini, M. M. *Phys. Lett.* 122B: 1 (1983); Hoodboy, P., Kisslinger, L. *Phys. Lett.* 146B: 163 (1984)
40. Maize, M. A., Kim, Y. E. *Phys. Rev.* C35: 1060 (1987); Vary, J. P., Coon, S. A., Pirner, H. J. In *Few Body Problems in Physics*, Vol. II, ed. B. Zeitnitz. Amsterdam: North Holland (1986)
41. Tjon, J. A. *Nucl. Phys.* A446: 173c (1985)
42. Dunn, P., et al. *Phys. Rev.* C27: 71 (1983); Otterman, C., et al. *Nucl. Phys.* A436: 688 (1985); Cavedon, J. M., et al. *Phys. Rev. Lett.* 49: 986 (1982) and references therein
43. Strueve, W., Hajduk, C., Sauer, P. *Nucl, Phys.* A405: 620 (1983)
44. Riska, D. O. *Nucl. Phys.* A350: 227 (1980)
45. Maize, M. A., Kim, Y. E. *Nucl. Phys.* A420: 365 (1984)
46. Friar, J. L., Negele, J. W. *Adv. Nucl. Phys.* 8: 219 (1975)
47. Frois, B., et al. *Phys. Rev. Lett.* 38: 152 (1977)
48. McCarthy, J. S., Sick, I., Whitney, R. R. *Phys. Rev.* C15: 1396 (1977)
49. Emrich, J., et al. *Nucl. Phys.* A396: 401c (1983)
50. Sick, I., et al. *Phys. Lett.* 88B: 245 (1979)
51. Reuter, W., et al. *Phys. Rev.* C26: 806 (1982)
52. Cardman, L. S., et al. *Phys. Lett.* 91B: 203 (1980)
53. Decharge, J., Gogny, D. *Phys. Rev.* C21: 1568 (1968)
54. Gogny, D. *Lect. Notes Phys.* 108: 88 (1979)
55. Decharge, J., Sips, L. *Nucl. Phys.* A407: 1 (1983)
56. Buti, T. N., et al. *Phys. Rev.* C33: 755 (1986)
57. Deleted in proof
58. Wise, J. E., et al. *Phys. Rev.* C31: 1699 (1985)
59. Deleted in proof
60. Goutte, D., et al. *Phys. Rev. Lett.* 45: 1618 (1980)
61. Deleted in proof

62. Euteneuer, H., Friedrich, J., Voegler, N. *Nucl. Phys.* A298: 452 (1978)
63. Cavedon, J. M., et al. *Phys. Rev. Lett.* 49: 978 (1982)
64. Frois, B., et al. *Nucl. Phys.* A396. 409c (1983)
65. Papanicolas, C. N., et al. *Phys. Rev. Lett.* 58: 2296 (1987)
66. Platchkov, S., et al. *Phys. Rev.* 25. 2318 (1982)
67. Suzuki, T., et al. *Phys. Rev.* C26: 750 (1982)
68. Suzuki, T., Hyuga, H. *Nucl. Phys.* A204: 491 (1983)
69. Pandharipande, V. R., Papanicolas, C. N., Wambach, J. *Phys. Rev. Lett.* 53: 1133 (1984)
70. Fantoni, S., Pandharipande, V., R. *Nucl. Phys.* A427: 473 (1984)
71. Mahaux, C., et al. *Phys. Rep.* 120C: 1 (1985)
72. Quint, E., et al. *Phys. Rev. Lett.* 57: 186 (1986); 58: 1727 (1987)
73. Jaminon, M., Mahaux, C., Ngô, H. *Phys. Lett.* 158B: 103 (1985)
74. Cavedon, J., et al. *Phys. Rev. Lett.* 58: 195 (1987)
75. Whitney, R. R., et al. *Phys. Rev.* C9: 2330 (1974)
76. Moniz, E. J., et al. *Phys. Rev. Lett.* 26: 445 (1971)
77. Horikawa, Y., et al. *Phys. Rev.* C22: 1680 (1980)
78. Altemus, R., et al. *Phys. Rev. Lett.* 44: 965 (1980)
79. Barreau, P., et al. *Nucl. Phys.* A402: 515 (1983)
80. Deady, M., et al. *Phys. Rev.* C28: 631 (1983)
81. Meziani, Z. E., et al. *Phys. Rev. Lett.* 52: 2130 (1984); 54: 1233 (1985)
82. Marchand, C., et al. *Phys. Lett.* 153B: 29 (1985)
83. Blatchley, C. C., et al. *Phys. Rev.* C34: 1243 (1986)
84. DeForest, T. Jr. *Nucl. Phys.* A414: 347, 358 (1984)
84a. McVoy, K. W., Van Hove, L. *Phys. Rev.* 125: 1034 (1962)
85. Schiavilla, R., et al. *Preprint ILL-(NU)-86-60*, to be published
86. Do Dang, G., Van Giai, N. *Phys. Rev.* C30: 731 (1984)
87. Walecka, J. D. *Ann. Phys. NY* 83: 491 (1974); Serot, B. D., Walecka, J. D. *Adv. Nucl. Phys.* 16: 1 (1986)
88. Kurasawa, H., Suzuki, T. *Phys. Lett.* 154B: 6 (1985)
89. Noble, J. *Phys. Rev. Lett.* 46: 412 (1981)
90. Celenza, L. S., et al., *Phys. Rev.* C31: 232 (1985)
91. Traini, M. *Phys. Lett.* B176: 266 (1986)
92. Deleted in proof
93. Alberico, W., et al. *Ann. Phys.* 154: 356 (1984)
94. Cavinato, M., et al. *Nucl. Phys.* A423: 376 (1984)
95. Della Fiore, A., et al. *Phys. Rev.* C31: 1088 (1985)
96. Orlandini, G., Traini, M. *Phys. Rev.* C32: 320 (1985)
97. Fantoni, S., Pandharipande, V. *Preprint ILL-(NU)-86-52*, to be published
98. West, G. B. *Phys. Rep.* 18C: 264 (1975)
99. Gurvitz, S., Tjon, J., Wallace, S. *Phys. Rev.* C34: 648 (1986); Ciofi degli Atti, C. *Nucl. Phys.* A463: 127c (1987)
100. Sick, I., Day, D., McCarthy, J. S. *Phys. Rev. Lett.* 45: 871 (1980)
101. Day, D., et al. To be published
101a. Ciofi degli Atti, C., Pace, E., Salwe, G. *Phys. Lett.* 127B: 303 (1983)
102. Jans, E., et al. *Phys. Rev. Lett.* 49: 974 (1982)
103. Laget, J. M. *Phys. Lett.* 151B: 325 (1985)
104. Sick, I. *Phys. Lett.* 157B: 13 (1985)
104a. Sick, I. See Ref. 124
105. O'Connell, J. S., et al. *Phys. Rev. Lett.* 53: 1627 (1984)
106. Arends, J., et al. *Phys. Rev. Lett.* 98B: 423 (1981)
107. Ahrens, J., et al. *Phys. Lett.* 146B: 303 (1984)
108. Carlos, P., et al. *Nucl. Phys.* A431: 573 (1984)
109. Koch, J. H., Ohtsuka, N. *Nucl. Phys.* A435: 765 (1985)
110. Koch, J. H., Moniz, E. J., Ohtsuka, N. *Ann. Phys. NY* 154: 99 (1984)
111. Laget, J. M. *Phys. Rep.* 69: 1 (1981)
112. Marchand, C. *Phys. Lett.* 153B: 29 (1985)
113. Laget, J. M. In *New Vistas in Electronuclear Physics*, ed. E. L. Tomusiak, H. S. Caplan, E. T. Dressler, p. 361. New York: Plenum (1986)
114. Lourie, R. W., et al. *Phys. Rev. Lett.* 56: 2364 (1986)
115. Keizer, P. W. M., et al. *Phys. Lett.* 157B: 255 (1985)
116. Bertsch, G. F., Bortignon, P. F., Broglia, R. *Rev. Mod. Phys.* 55: 287 (1983)
117. Wambach, J. In *Proc. Int. Sch. Nucl. Struc.*, ed. V. G. Soloviev, P. Yu. Popov. Aluhsta, USSR (1986)
118. Bolme, G., et al. To be published; Cardman, L. S. In *Proc. Int. Sch. Intermediate Energy Nucl. Phys.*, ed. R. Bergere, S. Costa, C. Schaerf, p. 163. Singapore: World Sci. (1986)

119. Kihm, Th., et al. *Phys. Rev. Lett.* 56: 2789 (1986)
120. Papanicolas, C. N. *Phys. Rev. Lett.* 54: 26 (1985)
121. Ravenhall, D. G., et al. *Ann. Phys.* In press (1987)
122. Popov, S. C. *Proc. Workshop on the Use of Electron Rings for Nucl. Phys. Res., Univ. Lund,* 2: 150 (1982); Dmitriev, V. F., et al. *Nucl. Phys.* A464:

123. Dmitriev, V. F., et al. *Phys. Lett.* 157B: 143 (1983)
124. Walecka, J. D. In *Proc. Symp. Celebrating 35 Years of Electron Scattering,* ed. C. N. Papanicolas, L. S. Cardman, R. A. Eisenstein. *AIP Conf. Ser.* In press (1987)
125. Grunder, H. See Ref. 124

237 (1987)

*Ann. Rev. Nucl. Part. Sci. 1987. 37: 177–212*

# USE OF NEW COMPUTER TECHNOLOGIES IN ELEMENTARY PARTICLE PHYSICS

*Irwin Gaines and Thomas Nash*

Advanced Computer Program, Fermi National Accelerator Laboratory, Batavia, Illinois 60510

CONTENTS

## 1. INTRODUCTION

Elementary particle physics and computers have progressed together for as long as anyone can remember. The symbiosis is surprising considering the dissimilar objectives of these fields, but physics understanding cannot be had simply by detecting the passage of particles. It requires a selection of interesting events and their analysis in comparison with quantitative theoretical predictions. The extraordinary reach made by experimentalists into realms always further removed from everyday observation frequently encountered technology constraints. Pushing away such barriers has been

177

0163–8998/87/1201–0177$02.00

an essential activity of the physicist since long before Rossi developed the first practical electronic AND gates as coincidence circuits in 1930 (Figure 1) (1). This article describes the latest episode of this history, the development of new computer technologies to meet the various and increasing appetite for computing of experimental (and theoretical) high energy physics.

The experimenters' computing needs did not develop suddenly. The long-term growth in requirements is due to the increase in energy at which experiments are carried out. As the center-of-mass energy, $s^{1/2}$, increases, so do the complexity and, typically, the amount of data that must be analyzed. (Secondary particle multiplicities, total cross sections, and accelerator luminosities, albeit for different reasons, all tend to increase with energy.) As the haystack has grown, so has the needle shrunk. Point-like cross sections go as $1/m^2$. Furthermore, with the standard model becoming more established, other interesting physics to research tends to enter at lower cross sections.

The problem has deep causes and exposing them gives a hint of the future (see Section 4). Fundamentally, the task of selection and analysis remains the same whether using protractor and ruler for measurement in a cloud chamber or a $100 million detector at a colliding-beam interaction point. This observational, pattern-recognizing task was conceived by the human brain as an extension of the theoretical pattern recognition that is the creative core of science. Except for the quantity of data now involved, this is an activity better matched to the capability of a brain than that of

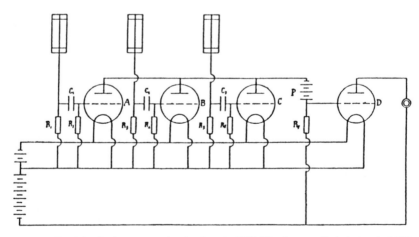

*Figure 1*   Particle physics computer technology in 1930. The first practical electronic AND gate, a Geiger Counter triple coincidence circuit. Pulses were "detected by a telephone" and scaled manually.

a computer. We have applied the brute force numerical abilities of computers to very nonnumerical problems. One acceptable definition of the fundamental goals of the research field known as Artificial Intelligence (AI) is the development of machines that will attack such nonnumerical problems in a more direct manner. Real success toward this key AI ambition has certainly been hard to come by, but its importance for high energy physics (and science in general) cannot be overestimated.

Experiment computing has traditionally been divided into the on-line tasks of monitoring and trigger selection (see Sections 1.1 and 2) and off-line reconstruction and analysis (see Sections 1.2 and 3). However, the distinction between on-line triggers and off-line reconstruction is becoming softer as parallel processors programmed in high level languages are applied more frequently in real time. Given adequate data rate capacity, a computer does not recognize whether it is being fed from a real-time buffer or a magnetic tape at a later time. The major concern on-line is in the validity of data and the criteria and programming that throw away data, forever. Intensive real-time high level filters are used today with reluctance and generally only in later runs of experiments. The need for such sophisticated on-line selection will become more and more compelling in the future. Artificial Intelligence concepts, such as program verification, again have fundamental relevance.

Formatting and calibrating data that is being recorded, and monitoring its validity, is a major on-line computing activity on which extensive personnel and computing resources are focussed. Many graduate students have cut their teeth on these tasks. The problems include checking that equipment is functioning properly, establishing and recording calibrations, and formatting and compacting data. Specialized hardware and software (including AI expert systems) are increasingly being applied. Covering this extensive subject would require much more space than we have available. Regrettably, we can only give it passing reference.

Theorists joined experimenters with large computing needs when lattice gauge theory was advanced by Wilson (2) as a means for numerically solving quantum chromodynamics (QCD). Widely accepted as the theory of the strong interactions, QCD cannot be solved perturbatively because of its large coupling constant. With existing algorithms and computers, precise lattice predictions of baryon masses, for example, seem tantalizingly out of reach; estimates like "70,000 Cray years" have been heard. Given their importance, the hope is that improved technology and algorithms will put these calculations on a sensible time scale. In Section 1.2 we introduce the lattice gauge problem from the computing perspective, and in Section 3.4 we describe processor efforts in this area.

Concluding the Introduction in Section 1.3, we discuss the place of elementary particle physics parallel computers in the modern world of computers. What seems like a surprising commonality in the approaches being taken to computing for experiment and theory is found to be no more than a manifestation of the underlying structural regularity of science.

## 1.1   *The Triggering Problem*

As early as 1933, some cosmic-ray experiments triggered cloud chambers on coincidences of pulses from Geiger tubes (3). Yet, for many years, triggering was not a common concern for experimenters. Some work was carried out by counting pulses from Geiger and scintillation counters; much of the rest depended on detectors with indiscriminate sensitivity. Emulsions recorded tracks from the moment they were poured until they were developed. There being no correlation of the sensitive time of these detectors with any specific interaction known to be occurring, the events they recorded were pot luck.

The first spark chambers revolutionized physics in 1959 (4). The chamber was fired and pictures taken only when an electronic signal indicated that an appropriate interaction was likely to have happened. The "trigger" signals were based on a simple coincidence and or-ing of the pulses coming from a number of photomultiplier tubes attached to scintillating material. The time available for these decisions was under 300 nsec so that the spark chamber pulse of $\sim 10,000$ volts could be formed in time. No longer was the leisurely tenth of a second time scale of cloud chamber triggers acceptable. The simple, but very high speed, electronics required to make these decisions was an outgrowth of the circuits used in a generation of "counter" experiments. By the time spark chambers took on a strong role, the standardization of "fast logic" electronics that could make a coincidence decision in 1–2 nsec was well underway.

Direct electronic spark chamber readout required the first use of on-line computers to write the data onto magnetic tape, still the primary data storage medium of particle physics. Spark chambers were typically fired less than ten times per second and the track multiplicities that the direct readout systems could support was limited. The data analysis from these experiments did not overwhelm computer centers already used to the load from bubble chambers. In the 1970s this situation changed as energies, multiplicities, and beam intensities increased and new detectors appeared that could support much higher rates. The irony was that, unlike spark chambers, these new detectors, multiwire proportional chambers (MWPCs), were untriggered devices. Yet triggering was necessary because the on-line computers could not keep up.

Triggers serve another essential purpose in experiments using drift chambers, a close relative of MWPCs in which the time for ions to drift toward a wire is measured to determine precisely the location of a passing particle. The time-to-digital converters (TDCs) that digitize this data require time reference signals. Similarly, analog-to-digital converters (ADCs) measure phototube pulse height, proportional to the shower energy in a calorimeter. The ADCs require a gate to define the time over which the phototube signal is to be integrated. The digitization process takes up valuable experiment live time (from a few to usually a few hundred $\mu$sec) during which the experiment cannot accept new events. Therefore, the signal to gate the digitizers must be made selectively, only when the probability of an interesting event being present is high.

Triggering electronics appear in large experiments in a variety of data rate environments and time domains as they carry out different functions. The triggers are intertwined with digitization, monitoring, and data acquisition electronics in an extraordinarily complicated system. The Collider Detector at Fermilab (CDF), as an example of this scale of effort, is spending over $1 million on triggers of all kinds, about $12 million on data acquisition electronics including digitizers, and over $2 million on large on-line data logging and monitoring computers and consoles (5).

Triggers are generally organized in a multilevel structure. Passing from each level is only as much data as following electronics, with tasks of increasing complexity, can handle. In this article, we classify triggers as "low level," "middle level," and "high level." In any specific experiment the distinction between these levels may be fuzzy. The levels may be compressed or expanded or bypassed, subject to the requirements and creativity of the experiment. "Low level triggers" determine when a basic interaction of interest has been detected at a rate that is appropriate to gate digitizers. An important limitation is that the selection can only be based on parallel information from individual subdetectors, never on global and tracking information, which is only available at later stages. A "middle level trigger" is, for our purposes, one that neither is used to gate digitizers nor is programmed in a high level language like FORTRAN. A trigger programmed in a high level language is here defined as a "high level trigger." The purpose of both middle and high level triggers is to reduce the amount of data ultimately stored for off-line analysis.

In sum, the broad reasons why modern experiments require triggers are as follows: to establish digitizer time references; to reduce the experiment dead time caused by on-line data recording; and to limit the amount of data that must be analyzed off line. Future experiments show no signs of becoming any less dependent on sophisticated real-time selection. In fact, for reasons like those affecting data analysis, the requirements for trigger reductions are becoming more severe with time. Data rates are increasing faster than the capacity of digitizing and data recording systems available at acceptable cost.

## 1.2 *The Off-Line Computing Problem*

An experiment's first off-line computing task is reconstructing raw detector signals into useful physics information about each interaction event. The raw data consists of digitized pulse sizes and times as well as arrays of bits indicating whether a wire or counter was hit. This must be transformed into the three-momentum, type, and originating (vertex) location of each particle and the error matrices for these quantities. The reconstructed data are the starting point for the real physics: analysis, in terms of theoretical or phenomenological models, of accumulated event distributions over relevant kinematical variables.

The two phases, reconstruction and analysis, pose very different problems. The first demands huge amounts of computation with human activity limited to monitoring of progress and quality. The second requires heavy, and one hopes efficient, human interaction and a far more moderate, though still significant, computing load. Development of new computers for experiments has been primarily directed at reconstruction. The analysis phase has been handled by commercial computers. Here, too, the situation is in sight of becoming intolerable because of inefficient use of the physicist's time, and attention is being directed at specialized solutions (Section 4).

The most common attribute of reconstruction software is that almost all such programs have been written specifically for one detector with its special geometry, physics, background, and rate considerations (and the whims of its physicists) in mind. Tracks are built up from position coordinates in wire chambers generally in a series of stages. Small curve segments, for example, may be projected into lists of other segments to find matches and produce larger segments. The process proceeds until all usable detector hits are accounted for. Energy in single calorimeter detector channels is identified with clusters of neighbors showing some energy and with nearby wire chambers. The tracks must be associated with Čerenkov counter and calorimetry information to establish particle type and account for as much interaction energy as possible. Least-squared fits throughout the program estimate the values of the required physical quantities.

This whole, extensive, handcrafted process is complicated by the need to consider a variety of subtle, detector-specific effects such as noise and inefficiency and ambiguities resulting from several particles hitting a small region. The programs, not surprisingly, are complex and long. They are generally unstructured, with frequent conditional branches and other features horrifying to computer scientists. Fixed-target experiments with reconstruction programs larger than 25,000 lines are the rule, and for big colliding detectors 100,000 lines are typical. Large memory banks are required just to manipulate lists of temporary data during matching and fitting operations. Calibration constants can consume several megabytes of memory. A 6-Mbyte memory requirement for a colliding-beam detector reconstruction program is no longer considered unreasonable by computer centers.

Reconstruction programs require extensive computation, primarily because their brute force pattern-recognizing approach requires large and deeply nested loops to test all reasonable combinations of detector hits for possible association into tracks or clusters for large numbers of events. Table 1 gives representative examples of the average computer time taken per event by a number of experiments, the number of events they have or anticipate having per calendar year of operation, and the total computer time required per calendar year. Throughout this article, we use the VAX 11/780 performance as a standard unit (VAXes). Relative performance of other machines are taken from Ref. (8). Clearly this amount of computing is not tenable with conventional computers within the limited budget of a basic science. This situation has prompted development of more cost-effective solutions by the high energy physics community.

In lattice gauge theory calculations the four space-time dimensions are mapped onto a grid of finite spacing (9). Monte Carlo methods are then used to evaluate expectation values of physically relevant quantities using Feynman's path integral formulation. For QCD, products of SU(3) matrices must be evaluated to determine how the action changes at each lattice step. This must be done for each SU(3) gauge field variable—corresponding to links in the lattice—to get a new configuration. With conventional algorithms, at least 10,000 such sweeps are required (and perhaps orders of magnitude more) to insure that the final configuration is stat-

**Table 1**  Representative computing-intensive experiments

| Experiment | Location | Events/CYr[c] | VAX-sec/Event | VAX-Yr[d]/CYr[c] |
|---|---|---|---|---|
| E691 | Fermilab fixed target | $1 \times 10^8$ | 7 | 30 |
| UA1 (6) | CERN $\bar{p}p$ | $5 \times 10^6$ | 60 | 20[b] |
| CDF[a] (6) | Fermilab $\bar{p}p$ | $10^7$ | 200 | 50–100 |
| L3[a] (7) | CERN LEP | $4 \times 10^6$ | >120 | 48–60 |
| Anticipated SSC experiment (7a) | | $10^7$ | 1000 | 1220 |

[a] Estimated.
[b] Constrained by limited computer resources.
[c] Calendar Year (CYr) includes typical beam on and off times.
[d] Includes analysis and simulations.

istically uncorrelated with the starting point. The total for a state-of-the-art $16^4$-site lattice is about 250 million floating-point operations per sweep or at least $2.5 \times 10^{12}$ per calculation. Depending on hardware utilization factors, this corresponds to 4–8 hours per calculation on a > \$10 million, > 400 Mflops (million floating-point operations per second), two-processor Cray XMP. Even this amount of computing provides crude calculations, accurate at best to 10%, and must be repeated frequently as different observables are studied.

The immediate, if limited, goal of this activity is a phenomenological understanding of the theory. Even for this, an order of magnitude more precision is required. In the long run, lattice gauge tantalizes with the opportunity to test with precision a fundamental theory of physics that cannot be tested otherwise. Much larger lattices will be required and quark loop effects will have to be included, both increasing computing requirements enormously. For example, a $1000^4$ lattice, which acceptably contains an entire proton, requires a factor of 16 million more time than current $16^4$ lattices—and significantly more yet when including quark effects. The truth is that the extrapolation of these requirements from present understanding is so large that accurate estimates are impossible. It is clear, however, that anticipated hardware improvements alone will not be enough. Much will have to come from better algorithms, and research into these is getting as much attention (and computer time) as phenomenological studies. Already new algorithms [such as the Langevin method (10)] are reducing the number of sweeps required to decorrelate. The large computing demand here has also led to the development of specialized processors, many of which have rather restrictive algorithm-specific architectures.

## 1.3    The Parallel Processors of Particle Physics

The very regular structures of the two critical high energy physics problems just described invite parallel computer solutions. Experimenters are using simple architectures with numbers of highly cost-effective, stripped down computers to process raw data. Single events are sent to individual processors, which pass the results to a host computer to record on tape. Until recently, the individual processors were all of a form called "emulators" because their hardware was arranged to "emulate" the instruction set of a large mainframe family of computers (generally IBM). This allowed the users of these machines to take advantage of the software tools developed for the commercial computers, in particular the compilers. A newer approach uses larger numbers of single-board computers based on commercial 32-bit microprocessors that are supported by their own FOR-TRAN compilers. Theorists have been building parallel computers, also often based on microprocessors in grid architectures, that naturally match their problem.

In order to understand where these particle physics computers fit in the modern world of computers, let's look at a computer taxonomy from a physicist's perspective. It is commonly implied that a computer must be either special or general purpose. A few, such as the low level triggers of high energy physics and military signal processors, are uniquely able to carry out one task. However, any programmable computer can execute different programs with different tasks. In this sense they are all general purpose. Even large commercial computers are designed to be efficient at specific tasks required by different portions of the marketplace, transactions or vector computations, for example. In this sense they are all special purpose. The physics machines, though driven by a special interest in a particular class of problems, are efficient at a large number of tasks. They fall somewhere in the middle of what is really a spectrum of generality.

Traditional computers are referred to as single instruction, single data stream (SISD) machines. Big vector computers like those made by Cray are SIMD machines since a single instruction operates on elements from multiple data streams. Truly parallel machines such as experimentalists' computers have multiple instruction streams operating independently on multiple data streams (MIMD). Some specialized lattice gauge processors, however, operate in lockstep with essentially a single instruction stream on all lattice points.

What probably distinguishes physicists' computers most is that they are explicitly parallel, with the individual processors primarily having their own local memory rather than sharing a global memory. The direction taken by most parallel computer research outside physics has been toward machines with many processors accessing a common, "global," memory through a complex switch that handles the necessary synchronization. Much computer science effort is directed at supporting implicit, automatic decomposition of algorithms onto parallel processors that share memory. The idea of identifying the structure of a problem and explicitly mapping it onto a parallel computer architecture has been much easier for physicists to accept than for computer scientists. In an extensive study (11), Fox demonstrated that most scientific problems can be explicitly mapped onto certain local memory gridlike architectures (hypercubes) so they make efficient (usually greater than 90%) utilization of the hardware. The interconnection of local memory processors is by nature simpler than global switches and, therefore, amenable to larger numbers of low-cost processors such as single-board computers. The willingness of physicists to accept explicit parallelism has been rewarded with access to what are the most cost-effective means of high level computing presently available, 32-bit microprocessor and floating-point arithmetic-integrated circuits. In a phrase, the computers of particle physics can be classified as primarily somewhat specialized, local memory, explicitly parallel computers.

## 2.    ON-LINE SPECIAL PROCESSORS

When designing triggers for most experiments, there is a conflict between two requirements, high speed and flexibility. This conflict is resolved with a hierarchy of increasingly complex triggers. Each successive level makes more detailed decisions requiring greater amounts of time on fewer events. Figure 2 shows the triggering hierarchy for the CERN UA1 experiment (12). Early levels of the triggers, while often blindingly fast and quite powerful, are truly specialized processors. They lack the programmability necessary to make complex physics decisions or to adjust to a variety of conditions. Often these low level triggers are only understood by a very small group of experts, which makes the event selection criteria inaccessible to most of the physicists in a collaboration. Nevertheless, the need for making increasingly complex decisions on more and more events has forced the use of such specialized and inflexible processors.

A major challenge will be to develop flexible systems that can be used at earlier trigger stages without sacrificing speed. In particular, middle level triggers function in a data rate environment too high for existing high level triggers. Many smaller experiments do not yet incorporate high level

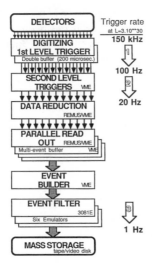

*Figure 2*  Overview of the UA1 multilevel trigger and data readout system. The numbers at right give the surviving event rate after each level of the trigger. Dead time is <10%.

triggers and stop at the middle (or even the low) level. However, the clear tendency is for more experiments to use high level triggers at the earliest possible place in the chain. The reason for this is not only the ease of preparing a trigger in FORTRAN. More importantly, it is the confidence in the validity and appropriateness of the algorithm that results from the far more extensive testing possible with high level systems. This allows more complex algorithms and deeper trigger reductions, which are requirements becoming more severe with time. Trigger reductions of $10^8$ to $10^9$ are anticipated for experiments on the proposed Superconducting Super Collider (SSC) (13). It is likely that such requirements virtually eliminate "middle level" triggers as we have defined them. Processors based on verifiable high level language code will have to take on these tasks.

Our coverage of lower level on-line processors is of necessity incomplete. Entire conferences are devoted to this topic, and we refer the reader to their proceedings for further information (14–28). We almost totally neglect the important use of processors for experimental monitoring, calibration, and data formatting and compaction. Such functions are an important part of all data acquisition systems usually performed by special processors, but space limitations require that we concentrate on the event-selecting processors.

## 2.1   Large Experiment Trigger Hierarchies

The trigger hierarchy is best understood by looking at two large colliding-beam experiments. ALEPH (29), a detector now under construction, will take data at the Large Electron-Positron Collider (LEP) at CERN beginning in 1989. The detector consists of an inner track chamber (ITC), a time projection chamber (TPC), electromagnetic and hadronic calorimeters, and muon chambers, with a total of about 500,000 digitizations per event. The beam-crossing rate at LEP is 40.5 KHz, allowing 25 $\mu$sec for trigger decisions before any dead time is incurred: The maximum event-recording rate is 2 Hz. The triggering system is designed to identify all $e^+e^-$ interactions while reducing background to an acceptable level, all with a minimum of dead time. Other trigger system constraints come from the detector. The TPC cannot be gated faster than 500 Hz, and the electromagnetic calorimeter requires 17 $\mu$sec to dump its charge and be ready for the next crossing.

A three-stage trigger scheme has been adopted. The low level trigger using ITC and calorimetry data will reduce the rate to below 500 Hz in about 3 $\mu$sec and will gate the TPC. The middle level trigger will use TPC trigger information to reduce the rate below 10 Hz in 50 $\mu$sec. The final high level trigger will be applied only after the entire event has been read out. It will use actual high level language reconstruction programs to bring the rate down below 2 Hz.

ALEPH also uses many on-line special processors besides the trigger devices. Read out controllers monitor, compact, calibrate, and format data from each subdetector. A trigger supervisor manages, for a variety of active triggers, the gate and dead-time signals of the subdetector systems. Finally, an event builder gathers data from the read-out controllers and assembles it into one contiguous bank for each event.

UA1 (12) is a large general-purpose detector that has been successfully taking data for several years at the CERN p̄p collider. The bunch spacing is 3.8 $\mu$sec, and the interaction rate is expected to reach 150 KHz. The trigger reduces the rate to 5 Hz recorded on tape. An important constraint is the massive amount of data from the central tracking chambers (CTC), which require 25 msec for data reduction and readout. Therefore, the first two levels of triggering cannot use CTC data and must reduce the rate to well below 40 Hz. Here again, a three-stage trigger is used (Figure 2).

The low level trigger uses data on hit muon chamber drift cells and analog sums of calorimeter channels. It reduces the rate to 100 Hz, a rate at which fully digitized events can be stored in a double buffer. The second-level trigger uses muon timing and digitized calorimeter data (with best available calibrations) to make more detailed physics selections. It requires 3 msec to reduce the rate to 20 Hz, at which the full CTC is read out. A special processor has been designed for the second-level calorimeter trigger. The data consist of 20,000 16-bit pulse height words and 20,000 addresses; the required processing includes pedestal subtraction, calibration in terms of transverse energy, comparison with a threshold, summation of transverse energy over regions of the calorimeter, and a weighted sum of energy over the full calorimeter. The task is split into two phases. Number-crunching hard-wired special-purpose devices do calculations at high speed in parallel. Their results are fed to pattern-recognition programs in standard microprocessors that identify detected electrons and jets and that trigger on missing and total transverse energy. Finally, the third-stage trigger uses six 3081E emulators programmed in FORTRAN to make the final physics selection.

## 2.2  Low and Midlevel Triggers

A good example of a modern low level trigger is the Level 1 trigger for the Collider Detector at Fermilab (CDF) (30), a large multipurpose p̄p detector due to take data in 1987. Typical of such triggers, it is required to make a decision in the less than 3.5 $\mu$sec between beam crossings to avoid dead time, and it must use data that is delivered separately from the normal data readout path. The fundamental component of this trigger comes from the electromagnetic and hadronic calorimeters. (Other parts are derived from muon chambers.) It exploits the projective geometry of the calorimeter by summing calorimeter towers into 15° azimuthal trigger sectors, 0.2 in rapidity wide. There are a total of 24 × 42 electromagnetic and hadronic trigger sectors. The trigger gets signals from dedicated outputs on the front-end sample and hold cards via dedicated cables. The signals first go to a "receive-and-weight" card that weights the energy by sin $\theta$, and then to "cluster sum" cards that calculate the number of towers with energies over several preset thresholds. The decision is then based on this number and the total transverse energy. Typical is the use of specialized dedicated hardware and the limited flexibility of this low level trigger. The calculations of calorimeter energy are refined in the CDF middle level trigger, which is more flexible and allows more sophisticated filtering.

Middle level triggers have more time available for complex trigger decisions, permitting systems with some programmability. These triggers still generally do not have the full event data available, since they tend to precede full event readout and process information from a separate path. Since events are usually not buffered while the middle level triggers are working, tight constraints on elapsed time can discourage use of the parallelism seen in high level triggers. They share some aspects of both low and high level processors and include hard-wired, microprogrammable, and data-driven systems. We describe examples of each type.

An elegant example of a hard-wired middle level trigger is the Mark II track finder used at SPEAR and PEP (31). This processor is designed to find curve tracks in a cylindrical geometry detector with an axial magnetic field. A low level pretrigger reduces the event rate to about 1 kHz. The middle level processor must then find tracks in about 30 μsec. Hits in the drift chamber feed a series of shift registers with variable delays, which allows a track with arbitrary curvature to be shifted into a straight line (Figure 3). Multiple curvature modules, each with different shift register delays, can look for tracks of different curvatures simultaneously. The shift register outputs then generate the address of a 2-bit RAM that determines whether or not that hit pattern corresponds to a good track. The data bits are artificially widened to generate a road. Required parameters, such as delays and widths, and RAMs are all programmable.

This device is an example of the commonly used technique of comparing the data from a particular event with a series of prestored patterns. The comparisons are made very rapidly since all calculations have been done in advance. In this case, the pattern of drift chamber hits is compared with prestored patterns for desirable tracks. Typical of middle level triggers, this one provides a limited degree of flexibility (what momentum tracks

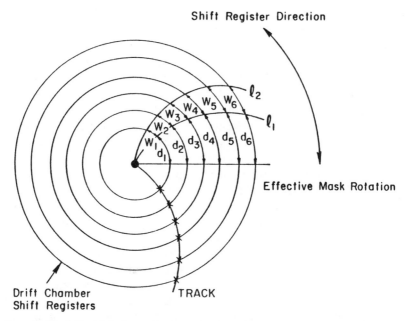

*Figure 3*   Mark II track finder mechanism. The effective mask for accepting a track is defined by the delays ($d_i$) necessary to shift the curved track into a straight line. Road widths are defined by the $w_i$.

are accepted, how many missing hits, how wide the roads, etc) without sacrificing speed.

Greater flexibility for midlevel triggers is possible with explicitly programmable devices. Special computers have been designed because conventional microprocessors are normally too slow. Bit slice technology, for instance, which requires many integrated circuits for a complete CPU, allows higher instruction execution speed. Bit slice processors are programmed in microcode, an extremely low level language that uses individual bits of instruction words to set the state of specific gates or multiplexors in the processor. Using microcode, a special instruction set may be prepared, tailored to the task at hand. To make the processor more efficient, designers may, for example restrict precision to 16-bit fixed point.

Examples of such devices are the CERN ESOP and XOP processors (32). The older ESOP processors have been used in several CERN experiments, including the European Hybrid Spectrometer, NA11, and R807, while the newer generation XOPs are used by UA2 and the LEP L3 experiment. The XOP processor is optimized for trigger computations of under 4000 instructions on 16-bit integers with execution times up to several msec.

XOP uses a very wide (160-bit) microinstruction word, concurrent execution of arithmetic operations, data address calculations, data accessing, condition checking, loop count checking, and next instruction evaluation. It supports such specialized instructions as bit search, population count, and loose compare. All this allows it to do trigger calculations 20 times faster than a 68000 microprocessor (or about two VAXes). Another microcoded processor is the M7, built at Fermilab (33).

All these processors share the disadvantage of lacking any high level language in which to program, thereby limiting the complexity of their algorithms. Similar microcoded special-instruction-set processors find extensive use in readout control and calibration applications. Here the lack of a high level language is less of a problem, since the calibration and readout programs are usually shorter and more stable than trigger programs. Examples are the MX used by CDF (34) and the BADC used at SLAC (35).

Data-driven processors represent an attempt to combine the speed of hard-wired devices with the programmability of special-instruction-set processor systems. Examples are the ECL-CAMAC system from Fermilab (36), CERN's MBNIM system (37), and the data-driven system developed at Columbia University (38). The latter ambitiously implements a full track reconstruction program with the processor.

These systems allow a trigger to be programmed, yet make maximum use of both parallelism and pipelining to insure that many processing steps take place at once. They also attempt to make each individual processing step as powerful as possible within a short time interval. They use a set of discrete modules that carry out processing steps simultaneously, each doing one operation every 25–50 nsec. Important are general-purpose table look-up modules that can be programmed (even in FORTRAN) with the tabulated result of an arbitrarily complicated calculation. A single 50-nsec step can in this way correspond to a huge, precomputed algorithm. Additional modules include stacks to store raw data and act as buffers between asynchronous loops, and even FORTRAN-like hardware DO loop indexers. These have been combined to implement powerful track-finding algorithms at speeds several orders of magnitude faster than on large main frame computers (39). The only drawback is the difficulty of "programming" such systems. Rather than working in a high level language, the user programs by recabling modules and loading precalculated results into the tables.

High level triggers typically use the full digitized event after it has been read out and buffered. These systems must make extremely large amounts of computing available so that many events can be examined in detail in a short time. The cheap and powerful high level language programmable processors described in the next section are natural solutions to this requirement, even being able to use the same programs as for off-line analysis. We describe in Section 3.3 how these are installed as high level triggers.

# 3.   HIGH LEVEL LANGUAGE EVENT PROCESSORS

By the mid 1970s, cost-effective microprogrammable devices were clearly useful for short trigger algorithms but far too difficult to program for any long trigger or reconstruction program. When it was recognized, probably first by Kunz (40), that the bit slice microprocessor technology used in such processors could be microprogrammed to execute the instructions of a commercial computer family, a new era of particle physics computing was started. Several such "emulating processors" were designed in this period (41), including Kunz's 168/E at SLAC, which was targeted for an IBM off-line environment; the NYU Courant Institute's PUMA, which emulated a Control Data 6600; and the CERN MICE processor intended for quick trigger programs and based on the PDP11 instruction set (42). Some were extremely popular with arrays of emulators computing a hundred times cheaper than big commercial machines. The 168/E, in particular, was followed by improved versions that are still being built today.

In the last year or so the emulators have felt strong competition in cost effectiveness and ease of use from systems with large numbers of single-board computers based on 32-bit microprocessors supported by FORTRAN 77 compilers. Improvements in integrated circuit technology—and the compilers—are likely to make the cost performance advantage so strong that such multimicroprocessors will take over from the traditionally popular emulators.

## 3.1   *Emulators*

Emulators have evolved steadily since their introduction. They were at first limited to important integer operations. The big-machine assembly language this permitted was a large jump from microprogramming. The emulation concept soon allowed another leap to FORTRAN using the excellent compilers prepared by the computer manufacturers. The 168/E supported most of the instructions required by IBM FORTRAN except for floating point. Even on-line programs for the MICE system were typically written half in FORTRAN. The popularity of MICE and the 168/E ultimately led to the development of floating-point processors for them so they would more completely emulate the instructions required by physics.

Both processors took advantage of the fact that neither trigger nor reconstruction programs require data input/output (I/O) during the execution of the kernels of their programs. It is sufficient to supply event information at the beginning of processing an event and retrieve results at the end. As a result, great simplifications can be had by not making direct, formatted FORTRAN I/O available directly from the processor. All I/O, including program and calibration constant downloading, is handled by a controlling computer. This "host" is connected by an appropriate interface to the processors.

The 168/E was originally designed to do off-line reconstruction of the huge data flow from the SLAC LASS spectrometer (40). The key components were eight of AMD's 4-bit microprocessor slices, the 2901. The 168/E's ultimate cycle time was 150 nsec and its performance was 50–75% of a 360/168 (or two to three VAXes). In 1980 the 168/E was reported to cost about $20,000 per processor fully loaded with 192 kilobytes of memory, including interfacing and assembly (but not testing) in a six-processor system (43).

While the 168/E was used for a variety of on-line triggers in addition to its original off-line applications, the MICE processor was always used only for on-line triggering. Its emulation of the PDP11 instruction set was accomplished with Motorola's 10800 series ECL bit slice at a micro cycle time of 105 nsec. It performed at speeds three times the most powerful computer in the PDP family, the PDP11/70 (or about one VAX), when programmed in FORTRAN or assembly language, and two to five times faster yet in microcode, which was often used in inner loops. A single processor installation including 4 kbytes of fast memory and a CERN Romulus interface cost $13,000 in 1980. The MICE processor was ultimately used in at least a half dozen CERN experiments, including the neutrino experiment WA1, the ISR experiment R704, and the Omega spectrometer.

The 168/E addressed the off-line problem in an ambitious and ultimately valuable way. This led to its becoming the most popular emulator, with about 50 processors built and installed at SLAC, CERN, DESY, Saclay, in Toronto and Tokyo and elsewhere. It has been used in "farms" of six or seven parallel machines. The new opportunity for complex high level triggers led to trigger applications at CERN and SLAC, and development of interfaces to CAMAC, Fastbus, and Unibus for activities as diverse as ISR experiments, the UA1 p̄p experiment, and the SLAC hybrid bubble chamber.

All emulators other than the 168/E and its successor the 3081/E have used fixed microcode for the execution of a defined, emulating, instruction set. They emulate at the machine instruction level. The 168/E was unique in employing software translation to microcode. The translator pass is transparent to users who see it only as an apparent delay in the IBM compiler-linker activity. The advantage of this approach is that it eliminates the need for complex hardware to decode the emulating instruction set into microinstructions. For the new 3081/E there is the added benefit that the translator can set up sophisticated, performance-enhancing, pipelines that would otherwise not be possible.

The 168/E's major weaknesses were identified by its designers and included the following: limited memory requiring extensive software and hardware to support overlays; lack of full 64-bit floating-point and byte addressing or manipulation instructions; and lack of adequate testing mechanisms (44). In addition, the 168/E had been designed as a one-of-a-kind installation for LASS. As a typical product of a high energy physics experiment, it used very tight timing and "clever," nonmodular, design techniques (45). These problems were recognized by Kunz and were professionally avoided in the beautifully modular 3081/E.

Also attempting to improve on the 168/E was the 370/E designed at the Weizmann Institute in Israel. This effort was led by Brafman, who had been involved in the 168/E design while visiting SLAC. Here the emphasis was on eliminating the software translation to microcode and emulating the IBM computers at the machine instruction level (46). The 370/E can run untranslated, unmodified IBM object code, including formated I/O, and the IBM operating system. This is felt to be a particularly desirable feature at DESY (47) and by others for whom the task of dividing a program into a host and processor part is perceived as too onerous. For highly parallel systems, where the I/O bandwidth can be saturated, the option of doing I/O directly from node processors gives naive users enough rope on which to hang themselves. Nevertheless, for low numbers of processors, as is the case with the 370/E, it is a decided convenience to be able to throw any old program directly into the emulator and have it run.

As may be seen in Figure 4 (*top*), the 370/E uses an architecture of multiple busses to which are connected several function units, integer, floating-point, multiplication, control, and interfacing. Separating functions in this manner increases the chip count somewhat but greatly reduces design, testing, and debugging time because the potentially complex control logic is much simplified. Although a bit slice 2901B is still used in the integer CPU, more functionality is now obtained with FAST and LS series TTL circuits than was the case with the earlier emulators. A multilevel pipeline prefetches instructions and allows microcoding machine language instructions preparatory to execution. The 150-nsec microcycle is narrowed to 100 nsec during multicycle shift, multiply, and divide operations. This emulator attains 60% of the speed of an IBM 370/168 (or 2.5 VAXes) and supports the full IBM address space. A system with two Mbytes of memory was reported as costing 45,000 DM (about $18,000) in 1985. A 20% faster prototype was completed at that time in Israel. At least 20 370/Es have been interfaced to a variety of DEC and IBM host computers at over ten locations in Israel, England, Germany, the US, and at CERN for the LEP experiments OPAL and DELPHI. Typical installations involve one or two units, but at Cornell there is a farm of six.

CERN people had seen the 168/E prototype running test programs at SLAC and started work at CERN in the Spring of 1978 on a copy. By the end of 1980 two systems were operational and being tested, off line by the European Muon Collaboration and as an on-line filter by the ISR Split Field Magnet group. In 1981 this careful CERN evaluation led to a collaboration with the original designers at SLAC to make a new version, subsequently named the 3081/E after the first of a new series of high

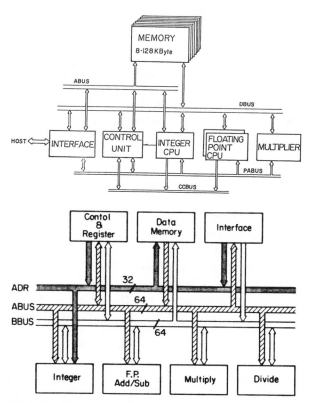

*Figure 4*    Block diagrams of two modern emulators: (*top*) 370/E; and (*bottom*) 3081/E.

performance IBM mainframes. The design work was divided equally between the two institutions.

Like the 370/E, this processor (45) is modular with separated functional units on several busses (Figure 4, *bottom*) and supports a more complete set of IBM instructions, including full double precision, and much larger memory, up to 14 Mbyte with 64K static RAMs, than its predecessor. Unlike their 370/E competition, the 3081/E designers retained software translation into microcode, which simplifies the design and can also automatically produce pipelined floating-point operations. The 3081/E runs at 1 to 1.5 times the speed of a 370/168 (or 4–6 VAXes), about twice that of the 168/E. Its cost with 4 Mbytes is now about $20,000, half attributable to the expensive fast static memory. The designers put a very strong emphasis on ease of building, debugging, and maintaining what was expected to be a frequently reproduced processor. Accordingly, the design was much simpler than the 168/E and conservative design rules were followed such as those requiring worst-case timing and multiple commercial sources for components.

A Common Interface board was designed to provide a means of communication with all the 3081/E busses and, for debugging purposes, control of its clock and state (47a). Execution may be halted by the interface on various condition traps. In various installations this interface has allowed connection to an impressive list of different busses, including IBM channels, CAMAC, VME, and FASTBUS, using NORD, IBM (mainframes and PCs), Apollo, Motorola 68000, and VAX computers as hosts. Major 3081/E facilities started production at CERN in the Fall of 1986 and a year earlier at SLAC. There are also 3081/Es at Saclay, Harvard, and in Italy.

## 3.2  Multimicroprocessors

The newly introduced 16-bit Motorola 68000 microprocessor was quickly adopted for particle physics applications in 1980. Though the speed of these devices continued to be a limiting factor, their extremely low cost encouraged their use in parallel systems. The Fast Amsterdam Multi Processor (FAMP) developed by Hertzberger and colleagues at NIKHEF was the most important early multimicroprocessor (48). It is still in use for high level triggering, having been a part of the triggers at the CERN ACCMOR spectrometer and UA1 experiments and the DESY JADE $e^+e^-$ spectrometer. Typically, three to seven processors in two levels, one processor acting as a supervisor, the rest as slaves, were operated in parallel on data from different detector regions of a single event. Each was a true single-board computer (SBC) with CPU and up to 16 kbytes of memory (with additional 128K extension boards available). Initially, programs were developed in assembly language, which is really only suitable for short algorithms. This was remedied with a UNIX$^{tm}$ operating system for FAMP under which high level languages like FORTRAN and C are available.

The big UA1 colliding-beam detector used 60 68000 SBCs for data readout in addition to the seven FAMP CPUs used for triggering. These were designed to be compatible with the 32-bit VME bus standard, which was new in 1982 and immediately recognized as appropriate for UA1 system by Cittolin and colleagues (49). Meanwhile at Bonn, a CAMAC Auxiliary Crate Controller based on the 68000 was developed to manage data acquisition (50). As part of this effort von der Schmitt wrote a FORTRAN compiler (RTF/68K) designed specifically for real-time applications.

For some time, such multimicroprocessors had only limited applications, mainly in middle level triggers and readout controllers, since the lack of power of the 16-bit CPUs and the incompleteness of the then available FORTRAN compilers limited the complexity of the algorithms that could be run. However, with the advent of the new generations 32-bit microprocessors and the development of full-fledged FORTRAN 77 compilers for these micros, the multimicroprocessors became competitive with emulators for use as off-line and on-line processor farms. In fact, the multiprocessors have already passed the emulators in cost effectiveness, and show potential for significant further improvements in cost effectiveness in the near future, as described below.

These developments became the basis for a major new multimicroprocessor effort aimed at off-line and highest level on-line computing by a new group at Fermilab. Named the Advanced Computer Program (ACP), it was formed in 1982 to confront the key particle physics computing problems, which by that time had been generally recognized as critical. The ACP's initial focus was primarily on event-oriented multiprocessing. Like most emulator activity, individual events were to be handled completely by separate CPUs (51).

There are many commercial producers of SBCs. Although competition has not yet driven these products to the status of a commodity, as has happened for memory chips, the ACP Multiprocessor was developed with an open, eclectic philosophy to take advantage of the strong competition in this area. The high speed, 32-bit ACP Branchbus connects up to 16 crates of SBCs per branch to each other and to the host computer. As the fundamental skeleton, the Branchbus and its optional $8 \times 8$ crossbar switch are the only features specified to remain the same in future variants of the ACP system. It is intended that, at any point in time, a system should use the most cost-effective node CPUs available from designers at high energy physics laboratories or, preferably, from commercial sources. Present ACP systems use VME standard crates interfaced to a Branchbus through a Branchbus-to-VME Interface (BVI), as shown in Figure 5. If motivated by SBC product availability in some other, "xBUS", standard, a "BxI" interface module can readily be designed. Because of this competitive, eclectic philosophy, the ACP system software is designed to be primarily resident in the host computer, with only a small node operating system that can be ported with relatively little effort to a new CPU.

The ACP has developed a new CPU based on the 68020, Motorola's 32-bit successor to the 68000, and another on AT&T's 32100, to demonstrate the technical and pricing requirements of nodes for such multiprocessors. Commercial SBC designs are still primarily aimed at controls applications, where they are used in small numbers, and are not yet acceptably cost effective. The ACP boards include two Mbytes of memory and

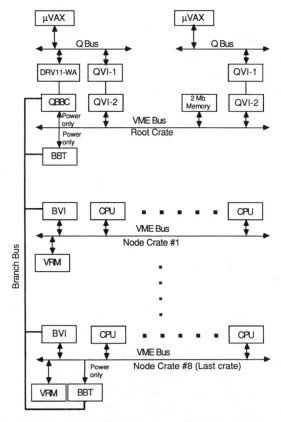

*Figure 5*    Block diagram of first ACP Multiprocessor in the Fermilab Computer Center.

the Motorola 68881 or AT&T 32106 floating-point coprocessor (FPU) appropriate to the CPU in use. Except for the CPU/FPU used, the boards are essentially identical. They are standard, double Eurocard, VME designs with all normally supported VME single-word transfer protocols. Unlike any other SBC available, they also support VME block transfers directly into memory. This permits extremely high rates of data transfer for on-line triggers. The system has been tested to transfer data error-free at 20 Mbytes/sec from a FASTBUS module through the Branchbus into CPU memory for over 48 hours.

These first ACP CPU modules currently cost under $1500 to produce and run FORTRAN reconstruction programs at about 0.7 VAXes. Including the low crate and interfacing overhead in large systems, their cost effectiveness is therefore now about $2000/VAX. Memory extensions up to six Mbytes (at under $200/Mbyte) can be located in slots in a single Eurocard crate immediately below the CPUs. The ACP plans at least one new generation of CPUs and is presently investigating at least eight microprocessor candidates, a much wider choice than was available when the first CPUs were designed. Several of these are reduced instruction set computers (RISC) and they will depend on good FORTRAN compilers for realization of their extraordinary performance potential. Design has been started on a new CPU based on the MIPs R2000 16.7-MHz microprocessor. Based on ACP reconstruction code benchmarks, this board is expected to run at 6 VAXes in FORTRAN. Another microprocessor, the Fairchild Clipper, approaches this performance and may also be supported. Since the cost of the new CPU will ultimately be similar to the present ones, the benchmarks demonstrate the possibility of attaining a cost effectiveness of better than $500/VAX in 1988.

Figure 5 shows in block form the first full-scale ACP Multiprocessor installed in the Fermilab Computer Center. It is a 140-node system, half based on 68020 and half on 32100 CPUs. Each VME crate contains up to 18 CPUs, a BVI, and a VME Resource Module (VRM) that handles arbitration. The BVIs act as master on VME and slaves on the Branchbus. The Branchbus Controller (BBC) is the Branchbus master and a slave on some other system, here the Qbus of a MicroVAX host computer. A VBBC is under design to allow direct VME control, and multiple mastership, of the Branchbus. The system may be hosted by one, two, or three Micro-VAXes that share a common memory on a special VME root crate to which they are interfaced through a Qbus-to-VME interface (QVI).

An application program destined to run on this system must, as for the 3081/E emulator, be divided into two parts. One, running in the host, handles all I/O. The other, running in each node, does the actual number crunching. The ACP provides system subroutines to communicate between the host and node programs. Routines exist to broadcast calibration constants at the beginning and to sum statistics at the end of a run. Individual events are sent, and results retrieved, asynchronously, by user-called send and get subroutines. Automatic floating-point and integer conversions, the latter in hardware, are available when required by different host and node CPU standards. The CERN ZEBRA data bank package (51a) is supported essentially transparently to the user. Converting programs to meet the multiprocessor requirements is relatively easy, aided by a full, multiprocess simulator that runs on a VAX. A visitor inexperienced with the system was able to bring up the Lund Monte Carlo program (51b) on real nodes in two days.

The initial system started running under Fermilab Computer Center operator control in July 1986, and ran for six months with no downtime on a huge backlog of data from the Tagged Photon Spectrometer experiment E691. With 100 processors (the remainder were assigned to other uses), the system performed at more than double the capacity of Fermilab's CDC 175 and 875 computers originally costing some $20 million. Omnibyte Corporation of West Chicago, Illinois, is selling all components of the ACP system at prices including initial testing and a two-year warranty. By March 1, 1987, this company had orders for over 240 processors from at least ten institutions (including SIN, Los Alamos, Brookhaven, Rutgers, Yale, and the Universities of Toronto and Montreal) and had shipped well over 100. This included an order for an 80 CPU second full system for the Fermilab Computer Center.

If there has been a weakness in the multimicroprocessor approach, it has traditionally been compilers, where there still remains an opportunity for a factor-of-two optimization. Effort in this direction is not encouraged by the commercial marketplace, which has been more interested in compatibility with PCs than performance. It is easier to produce an efficient compiler if the goals are limited as they have been in von der Schmitt's real-time compiler. However, for large programs a full FORTRAN 77 implementation, at minimum, is required. Some, such as the Absoft 68020 compiler, now are reasonably bug-free and have proven very satisfactory for writing new codes and for truly portable FORTRAN 77 programs. Converting from programs prepared under other compiler standards can be more time consuming. This is a problem not unique to microprocessors. It is commonly encountered in large experiment collaborations unless there is an agreement to outlaw exclusive VAX and IBM FORTRAN dialect features available on some home institution computers.

The ability to use full IBM FORTRAN is seen as a strongly desirable feature by many emulator users. Similarly many from the VAX environment have a strong preference for using MicroVAXes in a multimicroprocessor system even at a significant cost premium because it allows them to take advantage of VAX software. This was stated as a significant motivation by the D0 collaboration in their selection of a multi MicroVAX system developed by Cutts and Zeller at Brown University (52) for the highest level trigger on their Fermilab p̄p experiment. (After making their MicroVAX chipset available for prototyping, DEC decided not to release it for external designs, such as open architecture multiprocessors.) A 49-MicroVAX "farm" is planned with each node equipped with a special 256K dual port memory capable of absorbing data from an eight-data-cable parallel readout at the instantaneous 100-Mbytes/sec data rates required by the experiment. Another MicroVAX acts as the supervisor. Data is to be read out from the nodes, after event selection, via an Ethernet link to a host DEC 8600 at rates under a few hundred kbytes/sec. Higher rates will require using the dual port memory in write mode.

A 16-CPU farm has been used off line at Brown to generate GEANT Monte Carlo (52a) events badly needed by the experiment for final design decisions. The DEC ELN software toolkit is convenient for debugging and for handling data at the low rates required by a time-consuming large experiment Monte Carlo. Each MicroVAX, including 5 Mbytes of DEC memory and interfacing, costs about $16,000 (53). Thanks to the highly optimized VMS compiler, the MicroVAX chip, which is intrinsically not as fast as either the Motorola or AT&T circuits, runs slightly faster in FORTRAN, about 80–90% VAX. Therefore, the present cost effectiveness of this system is about $18,000/VAX. This is expected to improve by about 20% with the availability of new DEC board level products. Nevertheless, this is a considerable premium over other options, but advocates argue that the costs of "non-VAX operation" should not be forgotten (54). DEC has provided much support, including generous pricing, to this project, and one would expect it will be able to take advantage of successor MicroVAX chips rumored to be in production by 1988.

In another effort involving MicroVAXes, Siskind of NYCB Real Time Systems has used a Department of Energy Small Business Innovation Research grant to develop a FASTBUS MicroVAX that incorporates the DEC board level product as a piggybank in a multiboard, superfast (60 Mbyte/sec) FASTBUS interface (55). The SLAC SLD detector is planning on using 15 of these in a slower (12 Mbyte/sec) and cheaper TTL version for a high level trigger (56).

Each of the approaches to high level language processors has a strong advocacy. Emulators still provide higher performance individual processors, which can be essential in certain real-time applications. For those who have the luxury of living within the comfortable environment of one computer manufacturer, emulators, along with single-vendor multiprocessors, allow bit-for-bit comparison with popular main frames. On the other hand, open multimicroprocessor systems are more flexible, have a lower buy-in cost, and have hardware that is more easily made to work by nonexperts. They are more readily commercialized in an open competitive market. These systems already deliver at least as much performance per dollar as emulators, with much more cost-effective CPUs expected soon. The reader may sense in this summary the flavor of debate that typifies this field. What is important is that this intense interest and activity has led to devices that show promise of being able to handle the computing load of experiments for the foreseeable future.

## 3.3   On-Line Applications: The Interface Problem

Emulators and multiple microprocessor systems have both found application as high level triggers. Each individual processor can run large complex codes written in FORTRAN. The highly cost-effective arrays of such processors pioneered in off-line applications can be applied to obtain the massive amounts of processing power needed for on-line systems. Problems in applying them on line arise primarily from a lack of standardization by experiments.

CDF is using an array of ACP processors for their level-3 trigger (57). The system is required to accept events at 100 Hz and provide at least 50 VAX equivalents of processing power. The processor farm is fully integrated into the FASTBUS data acquisition system, through a FASTBUS-to-Branchbus Converter (FBBC). The FBBC will be used both for inputting events to the processors (from the FASTBUS Event Builders) and outputting from the ACP farm (by a VAX with a FASTBUS interface). The system is managed by the Buffer Manager, which controls the data acquisition system, via FASTBUS messages to the MicroVAX acting as host for the processor farm. On the other hand, the MEGA experiment at Los Alamos, also using an FBBC to attach ACP processors to a FASTBUS system, controls the flow of events to the processors by polling them directly from FASTBUS to determine which are ready to accept events (58; M. Oothoudt, private communication).

UA1 has used up to six 168Es (59) and recently 3081E emulators (60). The 168Es were originally interfaced to the CAMAC-based data acquisition system through the CAMFast interface, but to obtain higher performance a new PAX-Greyhound system was developed. The PAX module was a CAMAC sequencer that optimized CAMAC read out and interfaced to the Greyhound bus, a new bus used to interconnect the 168Es. A new memory board for the 168Es was also developed by the experiment. The 3081Es, which have recently replaced the 168Es, are interfaced to a new VME system. A VME event builder reads the events from dual port memories and broadcasts them (a nonstandard VME feature) to event task units, one of which is the farm of 3081Es. The Mark II at SLAC, on the other hand, attaches 3081Es directly to FASTBUS (61). As noted earlier, other experiments such as the D0 multiple MicroVAX system and the experiments using 370Es have also invented their own interfacing schemes.

These examples demonstrate the regrettable lack of standardization in the way these processors are interfaced. Experiments frequently find it necessary to develop new modules. It is hard to say whether this derives from the relative newness of these devices or from system designer's wishes to be different. One hopes that the future will bring greater uniformity in the way such on-line multiprocessors—and data acquisition systems in general—are implemented.

## 3.4 *Lattice Gauge Engines*

When confronted with the huge and important computing demands of lattice gauge calculations, theorists, encouraged by their experimentalist colleagues, began building their own parallel computers (62). All the main theory processors are programmed in high level languages, usually FORTRAN, and many are multimicroprocessors. Most supplement the FORTRAN engine with special devices that use high speed VLSI floating-point multiplier and adder chips to compute the intensive kernels of their programs.

Until recently, communication was thought to be required only between processing nodes working on adjacent parts of the lattice. This naturally led to gridlike architectural arrangements. At Cal Tech, Fox, a physicist, and Seitz, a computer scientist, joined forces to develop a hypercube of microprocessors (63). The hypercube is a good topology because it embeds the simple one- to four-dimensional grids typically used by scientific calculations and, in the worst case, communication path distances increase only as log $n$. The original system, called the Cosmic Cube, was based on 64 Intel 8086/87 16-bit processors and was completed in 1983. It did not emphasize performance but was used for a complete study of the applicability of local memory grid architectures to a wide variety of problems (11). This influential study alerted industry to the broad utility of such systems. It spawned commercial hypercube products from Intel, Ametek, N-Cube, and Floating Point Systems as well as a hypercube development effort at the University at Southampton using Inmos Transputer chips, which have excellent communication hooks for such purposes (64). An improved design using 68020s and, ultimately, daughter boards with Weitek floating-point chips is being developed by Fox in collaboration with the Jet Propulsion Laboratory. A 128-node system is planned, and was about half complete at the end of 1986 (65).

At the same time as the Cosmic Cube was developed, Christ and Terrano at Columbia constructed a 4 × 4 toroidal grid of nodes of Intel 80286/87 microprocessors coupled to a microprogrammed floating-point vector processor based on the then state-of-the-art TRW VLSI multiplier and adder chips (66; N. H. Christ, private communication). In this processor (Figure 6a) nodes operate in lockstep. After identical processing cycles, each transfers data synchronously to its nearest neighbor in a given direction. This is truly SIMD operation appropriate to the homogeneous problems then of interest to the designers. This processor was the first one using the maximally cost-effective floating-point chip that had just become available. It is able to reach 16 Mflops per node at a cost under $2500, still a very competitive figure, and has been used for important lattice calculations (67). A 64-node grid incorporating Weitek chips was completed in the

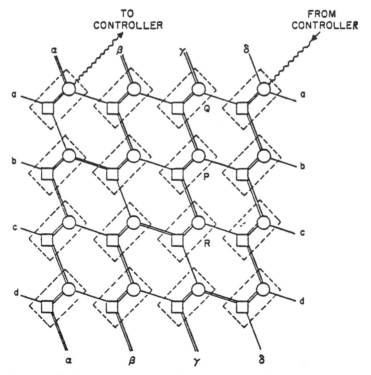

*Figure 6a*    Lattice gauge processor architecture: the Columbia 4 × 4 toroidal grid.

Spring of 1987, and a 256-node machine with 16 Gigaflop performance is planned.

An Italian group led by Cabibbo is developing another synchronous SIMD-like machine (called APE) of a very different architecture (Figure 6*b*) but also using Weitek chips to compute the lattice gauge kernels (68). Here a 3081/E is used to issue the simultaneous, SIMD, instructions to a bank of floating-point units (FPUs), each containing four multipliers and two adders so that the operation $y = a \times b + c$ on complex numbers is optimized. Each FPU has a maximum performance of 64 Mflops, so the planned 16-unit system could surpass a Gigaflop. Large memory is also important to big lattice calculations, and the design cleverly matches a Gigabyte of dynamic memory through a switch to the high speed FPUs. The switch can connect FPUs only to "neighboring" memory banks, but future design changes may relax this constraint. By the end of 1986 four FPUs were in place and the remainder were anticipated to be complete in 1987. [A similar project is the Space Time Array Computer (STAC) under development at Boston University (68a).]

At present the largest lattice gauge processor project, both in terms of Gigaflops and of dollars, is being undertaken at IBM Watson Labs by Weingarten et al (69). Targeted at eleven Gigaflops, the GF11 (Figure 6*c*) is also an SIMD machine with one high speed central controller issuing instructions, here to 576 floating-point processors again based on Weitek multiply and ALU chips. These 20-Mflop FPUs are flexibly interconnected through a three-stage Benes (shuffle) network of 24 × 24 crossbars. Used

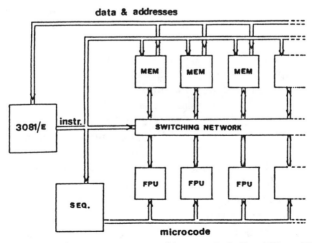

*Figure 6b*    Lattice gauge processor architecture: the Italian APE machine.

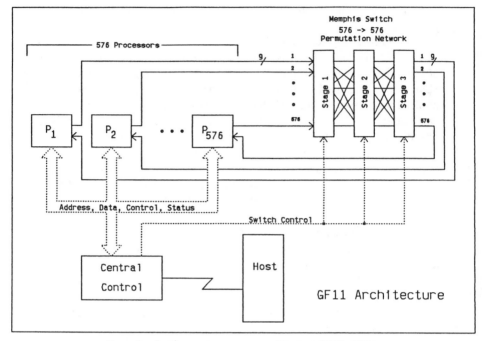

*Figure 6c*    Lattice gauge processor architecture: IBM's GF11.

typically as a 512-node grid for an $8^4$ lattice, the extra 64 processors are there to overcome the fundamental weakness of all grid processors. Unlike tree structures they are extremely fault intolerant, failing if a single node goes down—unless there are spares and a reconfigurable switch as in the GF11.

The ACP in a collaboration with Fermilab theorists is developing a lattice gauge processor with the architecture shown in Figure 6*d* based on ACP system modules. This structure is influenced by the more efficient nonlocal algorithms (the Langevin method) that have recently been proposed for lattice gauge calculations (10). It is just as easy for a processor to communicate to a far away node as to a nearest neighbor in this system. Truly MIMD, it also allows asynchronous algorithms that may someday be proposed. The nodes will be microprogrammed "array processor" boards giving the 5 Mflop per $1000 level of cost effectiveness typical of systems based on Weitek chips, but here in a less constraining, fault-tolerant architecture. The microcode will be prepared by experts and will look like subroutines to the theorist's FORTRAN program.

The lattice gauge processor story will not end here. We hope a com-

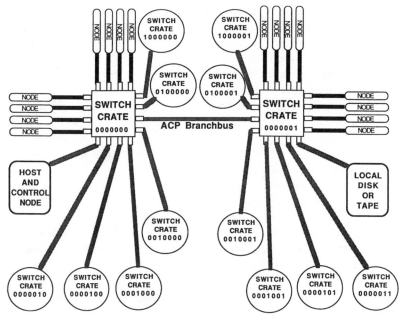

*Figure 6d*    Lattice gauge processor architecture: ACP Multiprocessor.

bination of dramatic new algorithm and hardware developments will make possible real QCD calculations before long. Given a realistic possibility of looking for fundamental QCD discrepancies or testing predictions of higher unified theories, funding for lattice gauge processors could reach a level that today would be surprising for a theorists' endeavor.

## 4.    UNSOLVED PROBLEMS FOR THE FUTURE

Technology projections are routinely overturned by unexpected inventions, but one must agree that at least the requirements of high energy physics appear certain and continuing. To date, new computer initiatives in physics have primarily applied industry-supplied components in suitably cost-effective ways. Hardware development will continue, and there is every expectation that it will meet the data rate and trigger reduction needs of experiments on 20-TeV colliding-beam machines. What is new is that demand for software innovation, not likely to be adequately met elsewhere, will put particle physics into the forefront of research in numerical algorithms and artificial intelligence.

At Fermilab the ACP is working on attaching devices based on video technology [such as the cheap, mass-produced Write Once Read Many times (WORM) optical disks or high-density video tape cartridges] directly to nodes within the ACP multiprocessor. This would allow systems originally conceived for reconstruction processing to reduce the analysis turnaround for a large experiment from weeks to half an hour. The impact this could have on particle physics is clearly enormous. However, such a dramatic hardware improvement will have to be accompanied by a similarly dramatic change in the way physicists interact with analysis programs. It would hardly be worth developing hardware that cuts the analysis iteration time to this level if, as now, it continues to take many days for an experimenter to prepare the next try.

Analysis programs go through much modification, frequently by many hands, as new ideas, techniques, and variables are tried. They are intrinsically big, messy, unstructured FORTRAN programs with scattered subroutine calls to histogram and plotting packages with long and obscure call lists. Making changes is, not surprisingly, a time-consuming mechanical endeavor. Tests, to make sure everything is right before committing to a full pass through the data, cause further delay. Something has to be done about this situation, and we think that the way is being shown by theorists and businessmen.

The obscurity of FORTRAN programs for doing science has been recognized in the theoretical context by Wilson et al (70). Their "Gibbs Project" at Cornell is attempting to address the problem by developing syntax for a language at a level higher than FORTRAN and more transparent to scientists. Programs will be organized like a text book, with chapters defining, in standard mathematical notation, the basic equations, boundary conditions, algorithms (e.g. Simpson's rule), starting values, etc of a problem. The project is presently using "human compilers" to establish a syntax sufficiently complete for an automatic compiler. High quality workstations for their research and, later, the science are considered a prerequisite by the Gibbs group.

Even before this effort by theorists, the business community had discovered the benefits of easily used software for everyday workplace tasks. Spreadsheets and efficient human interfaces, like that on the Apple Macintosh computers, are now a common part of every office environment. The human interface concepts based on "mouse" cursor control and desktop iconography were developed in AI research at the Xerox PARC Laboratory in Palo Alto in the early 1970s. We expect that successful personal computer software techniques, as well as Gibbs concepts, will be combined with a new generation of cheap, personal, science workstations to produce a spreadsheet friendly front end for experiment analysis. [Early work in this direction at SLAC and CERN was reported at the recent Asilomar

Conference on Computing in High Energy Physics (71, 72).] Software that allows an experimenter to request histograms with a few clicks of a mouse and to define the kinematic variables in standard mathematical notation will be an appropriate match to hardware engines that can process a large experiment data base in 30 minutes.

A subtle irony underlies the impressive worldwide work on experiment trigger hardware described in Sections 2 and 3.3. The high level language programmable processors are somehow less trusted, and less utilized, for large trigger reductions than are middle and low level devices, which are much harder to program. The issue has to do with trust in the correctness of the programs. Because of the difficulty in preparing them, programs for the lower level triggers are simpler and, thereby, more readily tested. High level systems are clearly capable of complex on-line programs and selection criteria as severe as off-line analysis that ultimately leads to publication of results. However, an event discarded on line is lost forever, and experiments are naturally very careful to test trigger programs thoroughly. Anyone who has wrestled with a large analysis or reconstruction program appreciates how long it takes before anything approaching certification of correctness can be made.

Interaction rates at 20-TeV hadron colliders are expected to be $10^8$ per second according to SSC design studies (13). Trigger reductions of $10^8$ will be required to achieve data logging rates of 1 Hz. Unbiased triggers that can accomplish this will be so complex that they will only be feasible in high level languages, even in environments we now see as middle and low level triggers. Developing high level processors that can handle the rates required will be accomplished by suitably parallel, probably tree structured architectures for data acquisition. In fact, there are high level processors existing today with the ability to handle data rates approaching 200 Mbytes/sec. The real problem will not be rates, but the trigger program confidence issue that is already making itself felt.

Modern software engineering techniques, such as "structured analysis, structured design" (SASD), are beginning to be applied in large experiments (72, 73). This work is a precursor to what will be essential: establishing techniques for developing and testing certified trigger code. Moreover, it will have to be possible to do this in almost real time, as the need of experiments for triggers that can be changed during a run is not likely to disappear. Rigorous certification of large and critical software projects is hardly a problem limited to high energy physics. In fact, this issue is at the heart of some important political debates about, for example, the Strategic Defense Initiative and nuclear reactor control systems. One can expect, therefore, that particle physics will not have to confront this problem alone. Nonetheless, the issue involves basic and unsolved difficulties of artificial intelligence research. For example, the desire will be to test not only whether the program is in fact carrying out the intentions of the designers, but whether those intentions are correct in the context of the detector parameters, known physics, and a miscellany of other variables.

It is a common trait of scientists to dismiss as "trivial" problems for which the solution path is understood. Development of new hardware for the computing-intensive problems of experimental and theoretical particle physics will follow paths well established today, though the effort can hardly be called trivial. We are not as sure about directions for attacking the lattice gauge algorithm, human interfacing, and software verification issues. These are also real and serious impediments to progress in particle physics. Finding approaches that hold promise of resolving them is therefore a pressing, and truly nontrivial, concern.

ACKNOWLEDGMENTS

We would like to thank our colleagues on the Advanced Computer Program for their stimulating support. M. Fischler and C. Quigg were helpful in areas with which we were not familiar. We thank the many individuals and groups working in this field for responding to our request for information. We apologize that space did not permit us to describe most of the work in any detail. Our choice of efforts to describe was based more on consideration of which would make the clearest examples than on any quality judgement. We thank L. Rauch for preparing the bibliography and manuscript. The authors are supported by funds from the US Department of Energy.

*Literature Cited*

1. Rossi, B. *Nature* 125: 636 (1930)
2. Wilson, K. G. *Phys. Rev.* D-3 10(8): 2445–59 (1974)
3. Blackett, P. M. S., Occhialini, G. P. S. *Proc. R. Soc.* A139: 699–726 (1933)
4. Fukui, S., Miyamoto, S. *Nuovo Cimento* 11(10): 113–15 (1959)
5. *Collider Detector at Fermilab (CDF) Activity Report.* Batavia: Fermilab, December 1986
6. Ballam, J., Chairman. *Report of Ad Hoc Committee on Future Computing Needs for Fermilab*, FERMILAB-TM-1230. 99 pp. December, 1983
7. L3 Collaboration. *Technical Proposal* May (1983). Submitted to CERN
7a. Newman, H. See Ref. 18

8. *Computing for Particle Physics, Report of the HEPAP Subpanel on Computer Needs for the Next Decade, August 1985*, DOE/ER-0234. p. 74. Washington: DOE (1985)
9. Hasenfratz, A., Hasenfratz, P. *Ann. Rev. Nucl. Part. Sci.* 35: 559–604 (1985)
10. Batrouni, G. G., Katz, G. R., Kronfeld, A. S., Lepage, G. P., Svetitsky, B., et al. *Phys. Rev.* D32(10): 2736–46 (1985)
11. Fox, G. C., Otto, S. W. *Phys. Today* 37(5): 50–59 (1984); Fox, G. The performance of the Caltech Hypercube in scientific calculations. In *Supercomputers—Algorithms, Architectures and Scientific Computation*, ed. F. A. Masten, T. Tajima, Univ. Texas (1985)

12. Dorenbosch, J. See Ref. 13, pp. 134–51 (1985)
13. Cox, B., Fenner, R., Hale, P., eds. *Proc. Workshop on Triggering, Data Acquisition and Off Line Computing for High Energy/High Luminosity Hadron-Hadron Coliders*. Batavia, Ill: Fermilab. 473 pp. (1985)
14. Michelini, A., Dobinson, R. W., Hoekemeijer, A., Innocenti, P. G., Jones, T., et al., eds. *Proc. Topical Conf. on the Appl. of Microprocessors to High Energy Phys. Exp.*, CERN 81-07. Geneva: CERN. 614 pp. (1981)
15. Istituto Nazionale di Fisica Nucleare Sezione di Padova, ed. *Proc. Three Day In-Depth Rev. on the Impact of Specialized Processors in Elementary Part. Phys., Padova, 1983*. Padova: INFN. 373 pp. (1983)
16. Donaldson, R., Kreisler, M. N., eds. *Proc. Symp. on Rec. Dev. in Computing, Processor and Software Res. for High-Energy Phys., Guanajuato, Mexico, 1984*. Batavia, Ill: Fermilab. 459 pp. (1984)
17. Hertzberger, L. O., ed. *Proc. Conf. on Computing in High Energy Phys.*, Amsterdam, 1985. Amsterdam: North-Holland. 431 pp. (1986)
18. Ash, W. W., ed. *Proc. Conf. on Computing in High Energy Phys.*, Asilomar State Beach, Calif. Amsterdam: North-Holland (1987). In press
19. IEEE Conf. on Real-Time Computer, Appl. in Nucl. Part. Phys., Santa Fe. *IEEE Trans. Nucl. Sci.* NS-26(4): 4365–4678 (1979)
20. IEEE Conf. on Real-Time Computer, Appl. in Nucl. Part. Phys., Oak Ridge. *IEEE Trans. Nucl. Sci.* NS-28(5): 3667–3928 (1981)
21. IEEE Conf. on Real-Time Computer, Appl. in Nucl. Part. Phys., Berkeley, 1983. *IEEE Trans. Nucl. Sci.* NS-30(5): 3721–4024 (1983)
22. IEEE Conf. on Real-Time Computer, Appl. in Nucl. Part. Phys., Chicago, 1985. *IEEE Trans. Nucl. Sci.* NS-32(4): 1260–1495 (1985)
23. Verkerk, C., ed. *Proc. on the 1984 CERN Sch. Computing, Aiguablava, Catalonia, Spain*, CERN 85-09. Geneva: CERN. 376 pp. (1985)
24. Verkerk, C., ed. *Proc. on the 1982 CERN Sch. Computing, Zinal, Valais, Switzerland*, CERN 83-03. Geneva: CERN. 350 pp. (1983)
25. Verkerk, C., ed. *Proc. on the 1980 CERN Sch. Computing, Vraona-Attiki, Greece*, CERN 81-03. Geneva: CERN. 408 pp. (1981)
26. Verkerk, C., ed. *Proc. on the 1978 CERN Sch. Computing, Jadwisin, Poland*, CERN 78-13. Geneva: CERN. 256 pp. (1978)
27. Macleod, G. R., Chairman. *Proc. on the 1976 CERN Sch. Computing, La Grande Motte, France*, CERN 76-24. Geneva: CERN. 283 pp. (1976)
28. Macleod, G. R., Chairman. *Proc. on the 1974 CERN Sch. Computing, Godoysund, Norway*, CERN 74-23. Geneva: CERN. 438 pp. (1974)
29. Videau, I. *IEEE Trans. Nucl. Sci.* 32(4): 1484–89 (1985)
30. Amidei, D., Campbell, M., Frisch, H., Grosso-Pilcher, C., Hauser, J., et al. *A Two Level Fastbus Based Trigger System for CDF*, CDF Note No. 510, Internal Document. Submitted to *Nucl. Instrum. Methods* (1987)
31. Brafmann, H., Breidenbach, M., Hettel, R., Himel, T., Horelick, D. *IEEE Trans. Nucl. Sci.* NS-25(1): 692–97 (1978)
32. Jacobs, D. A. *Computer Phys. Commun.* 26: 69–77 (1982); Bähler, P., Bosco, N., Lingjaerde, T., Ljuslin, C., van Praag, A., Werner, P. See Ref. 17, pp. 283–86
33. Droege, T., Gaines, I., Turner, K. J. *IEEE Trans. Nucl. Sci.* NS-25(1): 698–703 (1978)
34. Drake, G., Droege, T. F., Nelson, C. A. Jr., Turner, K. J., Ohska, T. K. *IEEE Trans. Nucl. Sci.* NS-33(1): 92–97 (1986)
35. Breidenbach, M., Frank, E., Hall, J., Nelson, D. *IEEE Trans. Nucl. Sci.* NS-25(1): 706–15 (1978)
36. Barsotti, E., Appel, J. A., Bracker, S., Haldeman, M., Hance, R., et al. *IEEE Trans. Nucl. Sci.* 26(1): 686–96 (1979)
37. Beer, A., Bourgeois, F., Corre, A., Critin, G., Huber, M. L., et al. *Nucl. Instrum. Methods* 160: 217–25 (1979)
38. Avilez, C., Borten, L., Christian, C., Church, M., Correa, W., et al. See Ref. 16, pp. 45–54
39. Martin, J., Bracker, S., Hartner, G., Appel, J., Nash, T. See Ref. 14, pp. 164–77
40. Kunz, P. F. *Nucl. Instrum. Methods* 135: 435–40 (1976); see also Hungerbühler, V., Mauron, B., Vittet, J. P. *Nucl. Instrum. Methods* 137: 189–92 (1976) for another project with these ideas at about the same time.
41. Verkerk, C. See Ref. 25, pp. 282–324
42. Halatsis, C., Joosten, J., Letheren, M. F., van Dam, A. *Proc. 7th Ann. Symp. on Computer Architecture, 1980, La Baule, France*. New York: IEEE Publ. 80CH1494-4C (1980)
43. Kunz, P. F., Fall, R. N., Gravina, M. F., Halperin, J. H., Levinson, L. J., et al.

*IEEE Trans. Nucl. Sci.* NS-27(1): 582–86 (1980)

44. Lord, E., Kunz, P., Botterill, D. R., Edwards, A., Fucci, A., et al. See Ref. 14, pp. 341–54

45. Kunz, P. F., Gravina, M., Oxoby, G., Trang, Q., Fucci, A., et al. See Ref. 15, pp. 83–100

46. Brafman, H., Fall, R., Gal, Y., Yaari, R. See Ref. 15, pp. 71–81

47. Notz, D. *A Data Processing System Based on the 370/E Emulator*, DESY 85-046. Hamburg: DESY. 17 pp. (1985)

47a. Paffrath, L., et al. *IEEE Trans. Nucl. Sci.* NS-33: 793–96 (1986)

48. Gosman, D., Hertzberger, L. O., Holthuizen, D. J., Por, G. J. A., Schoorel, M. See Ref. 14, pp. 70–82, 83–90

49. Cittolin, S., Demoulin, M., Haynes, W. J., Jank, W., Pietarinen, E., Rossi, P. See Ref. 16, pp. 413–27

50. Mertens, V., von der Schmitt, H. See Ref. 15, pp. 257–76

51. Bracker, S., Nash, T., Gaines, I. See Ref. 15, pp. 277–301; Gaines, I., Areti, H., Atac, R., Biel, J., Cook, A., et al. See Ref. 18, in press; Biel, J., Areti, H., Atac, R., Cook, A., Fischler, M., et al. See Ref. 18, in press. [A similar concept was conceived at UCLA and CERN where the VIRTUS processor was to use a CERN 68000 FASTBUS module, a combination which is not competitive in this application. See Ellet, J., Jackson, R., Ritter, R., Schlein, P., Yaeger, D., Zweizig, J. See Ref. 17, pp. 235–39; Müller, H. See Ref. 17, pp. 240–46]

51a. Brun, R., Goossens, M., Zoll, J. *ZEBRA User Guide*, CERN DD/EE/85-6. Geneva: CERN 141 pp. (1987)

51b. Sjöstruand, T. The Lund Monte Carlo for Jet Fragmentation and $e^+e^-$ Physics—JETSET version 6.2, LU TP 85-10. Lund Univ. 103 pp. (1985)

52. Cutts, D., Hoftun, J. S., Johnson, C. R., Zeller, R. T., Trojak, T., van Berg, R. See Ref. 17, pp. 287–91

52a. Brun, R., Bruyant, F., Maire, M., McPherson, A. C., Zanarini, P. *GEANT3 User's Guide*, CERN DD/EE/84-1. Geneva: CERN (1986)

53. D∅ *Management Plan and Cover Agreement Fermilab*, Internal Document. Batavia: Fermilab. 76 pp. (1985); see Ref. 54, p. 410

54. Johnson, T., Durham, T. *Parallel Processing: The Challenge of New Computer Architectures*, p. 410. London: Ovum (1986)

55. Siskind, E. J. See Ref. 16, pp. 281–84

56. Sherdan, D. J. *IEEE Trans. Nucl. Sci.* NS-32(4): 1479–83 (1985)

57. Beretvas, A., Carroll, J. T., Devlin, T., Flaugher, B., Joshi, U., et al. *Proc. 3rd Pisa Meet. on Advanced Detectors, Castiglione della Pescaia, Italy, 1986*. In press (1987)

58. Oothoudt, M., Naivar, F., Smith, W. *Use of the Fermilab Advanced Computer Project (ACP) for MEGA On Line High-level Triggering and Off Line Data Analysis*, MEGA Internal Document, Los Alamos, November 1, 1985

59. Carroll, J. T., Cittolin, S., Demoulin, M., Fucci, A., Martin, B., et al. See Ref. 15, pp. 47–70

60. Cittolin, S. See Ref. 17, p. 278

61. Rankin, P., Bricaud, B., Gravina, M., Kunz, P. F., Oxoby, G., et al. *IEEE Trans. Nucl. Sci.* NS-32(4): 1321–25 (1985); see also Ref. 47a

62. Pearson, R. B., Richardson, J. L., Toussaint, D. *Commun. ACM* 28: 385–89 (1985)

63. Seitz, C. L. *Commun. ACM* 28(1): 22–33 (1985); Brooks, E., Fox, G., Johnson, M., Otto, S., Stolotz, P., et al. *Phys. Rev. Lett.* 52(26): 2324–27 (1984); Otto, S. W., Stack, J. D. *Phys. Rev. Lett.* 52(26): 2328–31 (1984)

64. Hey, A. J. G., Jesshope, C. R., Nicole, D. A. See Ref. 17, pp. 363–69

65. Rogstad, D. H. *AMPlifier* (Jet Propulsion Lab., Pasadena, Calif.) 1(1): 5–6 (1986)

66. Terrano, A. See Ref. 15, pp. 135–53 (1983); Christ, N. H., Terrano, A. E. *IEEE Trans. Comp.* C-33(4): 344–50 (1984)

67. Christ, N. H., Terrano, A. E. *Phys. Rev. Lett.* 56: 111–14 (1986)

68. Bacilieri, P., Cabasino, S., Marzano, F., Paolucci, P., Petrarca, S., et al. See Ref. 17, pp. 330–37

68a. Brower, R. C., Giles, R. C., Maturana, G. See Ref. 17, pp. 339–44

69. Beetem, J., Denneau, M., Weingarten, D. *J. Stat. Phys.* 43: 1171–83 (1986)

70. The GIBBS Group (includes Bergmark, D., Demers, A., Gries, D., Lepage, P., Moitra, D., et al.) See Ref. 16, pp. 89–96

71. Burnett, T. See Ref. 18

72. Brun, R., Bock, R., Conet, O., Marin, J. C., et al. See Ref. 18

73. Kellner, G. See Ref. 18

74. Palazzi, P., Brazioli, R., Fisher, S. M., Zhao, W., et al. See Ref. 18

*Ann. Rev. Nucl. Part. Sci. 1987. 37: 213–41*

# NUCLEAR TECHNIQUES FOR SUBSURFACE GEOLOGY

*D. V. Ellis, J. S. Schweitzer, and J. J. Ullo*

Schlumberger-Doll Research, Ridgefield, Connecticut 06877-4108

CONTENTS

## INTRODUCTION

The technique of placing instruments into a well to measure continuously the properties of subsurface geological formations is known as logging and was invented in 1927 (1, 2). These first crude electrical devices localized possible hydrocarbon-bearing zones by measuring formation resistivity and spontaneously generated potentials in porous sedimentary formations. The main interest was the location of porous zones containing hydrocarbons and free of clay minerals, whose presence would inhibit the flow of fluids (and incidentally the generation of the spontaneous potential). Nuclear measurement techniques were first used in 1939 (3) and, since that time, have come to play a dominant role in the analysis of subsurface

213

0163–8998/87/1201–0213$02.00

geology. The developments in nuclear techniques have been motivated by a desire to generate a better understanding of the subsurface geology for the location and evaluation of oil, gas, and mineral resources (4–21), as well as for a better understanding of geological processes (22).

The earliest nuclear technique measured natural radioactivity, which is generally associated with clay minerals in sedimentary formations. It was immediately useful for the distinction of shaly from clean formations, much like the measurement of spontaneous potential. Subsequent developments, driven by needs of the petroleum industry, were related to the analysis of the fluids contained in sedimentary formations. The resistivity of a fluid-saturated rock depends on the nature of the saturating fluid (usually a conductive brine) and the volume fraction, or porosity, available to the fluid. Thus, a fundamental parameter for the interpretation of a resistivity measurement in terms of hydrocarbon/brine content is the porosity of the formation. Nuclear measurements provided the first quantitative measurements of in situ rock porosity.

Efforts were focused on measuring the bulk properties of rock using gamma rays and neutrons from radioactive chemical sources. Gamma rays, primarily through Compton scattering, are used to measure the bulk density of rocks. Neutron transport has become useful to determine the rock porosity since it is mainly influenced by hydrogen, which is usually associated with pore fluids. Following the development of pulsed accelerators (23, 24), other types of measurements became practical. The thermal neutron absorption cross section could be measured by analyzing the time dependence of neutron capture gamma rays following a pulse of high-energy neutrons. With the advent of improved gamma-ray detectors, it became practical to perform spectroscopic analyses of the gamma-ray flux reaching a detector, initially for natural radioactivity and then for gamma rays produced by neutron reactions with nuclei in the rock. The spectroscopic measurements have provided a basis not only for performing bulk geophysical measurements, but also for detailed geochemical analyses of the rocks and the fluids contained in them.

The difficulties in performing quantitative nuclear physics measurements in extended heterogeneous media required extensive calibrations under laboratory and field conditions. In recent years, the application of Monte Carlo and other analytical techniques has expanded the capabilities for performing quantitative analyses and has provided insight into the detailed physics involved in performing measurements in heterogeneous extended media. Before describing the details of particular techniques, we discuss the types of data that can be obtained, their relationship to the desired geophysical and geochemical parameters of the strata, and the advantages of and limitations imposed by making measurements within the well.

## Core and Logging Analysis

Petrophysical measurements routinely performed on core samples include the determination of porosity, permeability, rock type (lithology), mechanical properties, and the analysis of pore fluids. The analysis of core samples retrieved from a well might seem to eliminate the need for in situ logging analyses. However, the two types of analyses are complementary. In practice, coring, handling the cores, and laboratory analyses are expensive procedures and cannot be performed over an entire well. On the other hand, logging measurements can be made over the complete depth of the well. Furthermore, in many types of sedimentary rocks, only partial core recovery can be achieved (25). Thus, the core record is discontinuous and the location of retrieved samples may be highly uncertain. Often the extraction of the core sufficiently alters the sample so as to preclude meaningful measurements on it. Of a more fundamental nature is the heterogeneous nature of rocks (26, 27). A core sample is typically a few cm$^3$ in volume while in situ logging measurements average over volumes that are 2–3 orders of magnitude larger. Thus, logging measurements provide a more statistically valid sampling of the minimally altered geological strata.

Core analyses can be invaluable when used in conjunction with logging measurements. Cores can disclose fine details that are lost in the averaged logging measurements. Certain types of measurements, such as relative permeability, can only be performed on retrieved cores. In addition, core analyses can provide critical calibration data for certain types of logging measurements. The most complete analysis of geological strata can be obtained from a combination of logging measurements and core analyses.

## Nuclear Measurements and Geophysical Parameters

The use of nuclear techniques in subsurface geology is oriented toward the determination of geophysical and geochemical parameters. The actual quantities measured through nuclear techniques do not always correspond to the geological parameters desired. It is important to understand this difference. The geological parameters that are desired from nuclear measurements are generally as follows: the type and amount of fluids, permeability, lithology, clay identification, and, more recently, mineralogy (28). Information on these parameters can be obtained with nuclear techniques that measure gamma-ray transmission (which depends on Compton and photoelectric cross sections), neutron transmission (which is dominated by elastic scattering from hydrogen), the macroscopic thermal neutron absorption cross section, and elemental analysis through gamma-ray spectroscopy of natural or neutron-induced radioactivity.

It is obvious that the measurements do not provide the exact information desired on the geological parameters. For example, the neutron transmission measurement, while dominated by the amount of hydrogen present, also depends on the elastic, inelastic, and reaction cross sections for all other elements present. Thus, additional information is needed to relate the specific measurements obtained by nuclear techniques to the desired geological parameters.

NUCLEAR PARAMETER MIXING LAWS    The simplest way to understand how nuclear measurements and other information are combined to obtain geological parameters is to examine the relationship between microscopic cross sections for individual nuclei or atoms and their macroscopic equivalences, or mixing laws, which govern the actual measurements in extended heterogeneous media.

The measurement of density provides information on the amount of fluids present in the rock, since the bulk density can be written as

$$\rho_b = \phi\rho_f + (1-\phi)\rho_{ma}, \qquad\qquad 1.$$

which relates the bulk density, $\rho_b$, to the solid rock matrix of density $\rho_{ma}$ and the porosity or volume fraction, $\phi$, which contains a fluid of density $\rho_f$. From Equation 1, the porosity, $\phi$, can be determined from a measurement of bulk density, if the matrix and fluid densities are known. These are known with precision only if the fluid type and properties and the lithology are known. In practical terms, the density range of fluids is between 0.8 and 1.2 g/cm$^3$ although calcium chloride solutions may reach 1.4 g/cm$^3$) and most rock matrix densities of interest are between 2.60 and 2.96 g/cm$^3$.

In the region where the dominant mechanism for gamma-ray attenuation is Compton scattering, the transmission of gamma rays through the medium depends on the electron density, $n_e$ rather than $\rho_b$, the bulk density. The electron density is given by

$$n_e = \rho_b(Z/A)N_A, \qquad\qquad 2.$$

where $Z$ is the average atomic number, $A$ is the average atomic weight, and $N_A$ is Avogadro's number. For lower gamma-ray energies ($<100$ keV), the photoelectric cross section becomes significant. For geological material $Z$ is typically $<28$, so that gamma-ray energies are well above the K-absorption edge. In this energy region, the photoelectric absorption cross section per electron can be approximated (29) by

$$\tau = C\frac{Z^{3.6}}{E^{3.15}}, \qquad\qquad 3.$$

where $C$ is a proportionality constant. The average photoelectric absorption cross section per electron for a mixture of elements is given by

$$\langle \tau \rangle = \frac{\sum\limits_i Z_i m_i \tau_i / A_i}{\sum\limits_i Z_i m_i / A_i},$$    4.

where $m_i$ is the mass fraction of the $i$th element. Since many geological analyses are oriented toward volume fractions, it is useful to consider the quantity $U(\propto n_e \tau)$ whose scaling (30, 31) is analogous to Equation 1, i.e.

$$U = \phi U_f + (1 - \phi) U_{ma}.$$    5.

The macroscopic thermal neutron absorption cross section $\Sigma$ is the total capture cross section per cm$^3$. For a molecule $A_\alpha B_\beta$, $\sigma_{mol} = \alpha \sigma_A + \beta \sigma_B$, where $\sigma_{A(B)}$ is the microscopic thermal neutron absorption cross section for a nucleus A (B). Then for a collection of molecules, the macroscopic cross section is given by

$$\Sigma = \sum_i \frac{V_i \rho_i \sigma_{mol(i)} N_A}{M_i},$$    6.

where $M_i$ is the molecular weight of the $i$th type of molecule with volume fraction $V_i$. Frequently, $\Sigma$ is expressed in "capture units" (cu). The capture unit is 1000 times the macroscopic cross section (expressed in cm$^{-1}$). In these units water has a capture cross section of 22 cu and most rock material is in the range of 4–7 cu. The addition of NaCl to the formation water, or even impurities such as boron, leads to a rapid rise in the formation cross section, which in practical cases ranges from 10 to 80 capture units.

In gamma-ray spectroscopy measurements, there would seem to be no mixing laws, as such. However, when the yield of a particular gamma ray produced in a thermal-neutron-induced reaction is related to the elemental concentration producing it, it is necessary to know the macroscopic cross section (32). Unlike a thin-target experiment, in which the number of reactions is small compared with the number of beam particles and a reaction yield is directly proportional to the microscopic cross section, in a geological environment all of the thermal neutrons will be absorbed. Thus, in a thermal neutron capture reaction, the yield of gamma rays will not be proportional to just the concentration of the element times the cross section, but will be proportional to the ratio of the macroscopic cross section for the particular element to the total macroscopic cross section.

## Borehole Environment

The most fundamental limitation of the subsurface environment is the presence of matter everywhere in the region of the measurement. Thus nuclear physics techniques can use only neutral particles, i.e. neutrons and gamma rays.

Two additional constraints are placed on nuclear measurements by the borehole environment (33): geometry and contamination. The geometrical constraints can be visualized most easily. All equipment must be contained in a cylindrical housing to withstand the ambient pressures of 150 MPa (and temperatures to 150°C). The measurements are contaminated because wells are drilled with mud in the hole, which usually becomes part of the total measurement, and, since a pressure gradient is maintained between the hole and the surrounding rock, a mudcake is formed on the surface of the rock. In many circumstances, a steel casing is cemented into the well to ensure that the borehole does not collapse.

Since the source and detector must both be contained in the pressure housing, the detector must be shielded to ensure that it does not significantly respond to radiation transmitted through the tool. The inclusion of large amounts of shielding is not a simple solution, however, since detector counting rates decrease exponentially with source-detector spacing. Since measurements must typically be performed continuously at speeds of at least 5 cm/s, maximum counting rates are required to obtain data with good statistical precision. Furthermore, this geometry precludes the types of collimation that would be typically used in the laboratory to localize the region of a reaction or to minimize multiple scattering effects.

The presence of mud and mudcake introduces material into the measurement volume with significantly different properties than the rock. To obtain accurate measurements of rock properties, corrections must be made for the additional material. These corrections become so large in cased boreholes that some types of measurements are impractical. For tools with partial collimation, i.e. those that do not have uniform circumferential sensitivity, the distance between the detector and the borehole wall strongly affects the measured response. The following sections review both the physics of particular measurements and the relationship of measured quantities to geophysical and geochemical parameters.

# MEASUREMENT TECHNIQUES

## Application of Neutron Scattering

A measurement that was sensitive to neutron transport and moderation was introduced in the early 1940s to determine the hydrogen concentration of earth formations (4). The spatial distribution of epithermal or thermal neutrons resulting from the interaction of high-energy neutrons with a formation can be related to its hydrogen content. If the hydrogen (in water or hydrocarbons) is contained within the formation pore space, then the measurement yields porosity.

The conventional method of determining the neutron scattering properties of a formation employs a source of high-energy neutrons and monitors the resultant thermal or epithermal flux. The sources generally used in such a measurement are based on the $\alpha$-n reaction, such as Ra-Be or Am-Be. The neutron spectra from such sources are far from monoenergetic, but copious amounts ($\approx 10^8$ n/s) of neutrons ($\langle E \rangle \approx 4.5$ MeV) are produced. Although numerous neutron detectors are available, in the hostile environment of well logging simplicity and reliability are essential and are best provided by the use of gas counters such as those based on the $^3$He(n, p) reaction. To measure the epithermal neutron flux, a shield of thermal-neutron-absorbing material with a large cross section, such as cadmium, is used around the detector.

PARAMETERS FOR NEUTRON SCATTERING    At the neutron source energy, the primary interaction mechanism is elastic scattering. In a limestone formation the mean free path of source neutrons is about 8 cm, practically independent of the water content of the formation. However, the effect of hydrogen is noted at lower energies. For example, at 100 keV the mean free path ranges from 4 cm, in a pure limestone formation, to 2 cm in a 40% porous limestone containing water. Thus, the mean free path depends on porosity, and the logging measurement can be viewed as a determination of the moderating power of the formation. To describe this process some useful results from reactor physics are employed (34–36).

The transport of neutrons can be considered to depend on two processes: the loss of energy from the source energy to thermal energies, and the diffusion of neutrons at thermal energies until they are captured. Allen et al (37) have shown that the epithermal neutron flux from a point source can be characterized by an exponential dependence on the distance from the source with a length parameter given by the slowing-down length $L_s$. A simple model based on diffusion theory yields an epithermal neutron flux, $\Phi_{epi}$, at a distance $r$ from the source, of

$$\Phi_{epi} \propto \frac{e^{-r/L_s}}{D_{epi}r},$$ 7.

where $D_{epi}$ is an epithermal diffusion coefficient related to the transport mean free path. Kreft (38) calculated $L_s$ from a detailed knowledge of the cross sections for the elements in the volume of the measurement. The variation of the slowing-down length as a function of water-filled porosity in three types of sedimentary formations is shown in Figure 1.

For thermal neutrons, there is a comparable length parameter, the diffusion length $L_d$, such that the thermal neutron flux can be represented as

$$\Phi_{\text{th}} \propto \frac{L_d^2}{D(L_s^2 - L_d^2)} \frac{e^{-r/L_s} - e^{-r/L_d}}{r},$$    8.

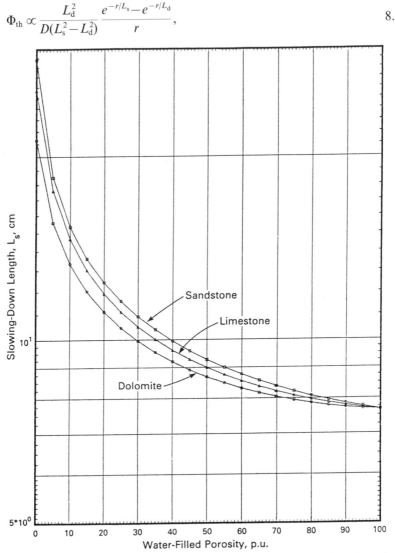

*Figure 1* The slowing-down length of three common rock types as a function of porosity. From Ellis (36).

where $D$ is the thermal diffusion coefficient. $D$ can also be calculated from the thermal neutron cross sections of the elements in the material and is related to $L_d$ by $D = L_d^2\Sigma$, where $\Sigma$ is the macroscopic thermal neutron absorption cross section for the material.

EPITHERMAL POROSITY MEASUREMENT   The preceding discussion indicates that a calculable parameter, $L_s$, can adequately represent the flux distribution in a formation, given the case of a point source in an infinite medium. However this is far from the reality of logging measurements, which must be made in a roughly cylindrical fluid-filled hole penetrating the formation in question. Some of the physical parameters of the formation and borehole that affect the response of a neutron porosity device include porosity, pore fluid type, salinity, rock type, borehole size, borehole fluid type, mudcake, and pressure. Of these, only one—porosity—is in fact the desired quantity. All the others can be related to a change in the local hydrogen density and thus can influence a measuring device that is sensitive to the concentration of hydrogen.

Figure 2 indicates the configuration of a typical neutron scattering device in the borehole. A first approximation to reducing the influence of the hydrogen-rich borehole on the measurement of the slowing-down properties of the formation is the use of two epithermal detectors. By comparing the counting rates of two detectors at different distances from the source, one obtains the shape of the epithermal flux distribution, which is more indicative of the slowing-down length of the formation.

Figure 3 shows the ratio of the counting rates of an experimental epithermal neutron scattering device as a function of the slowing-down length of some laboratory calibration formations. In this case all of the formations contain a simulated drill hole of 8″ diameter. There is some residual influence on the measured ratio for different borehole sizes, but this effect can be quantified by measurements in formations of different borehole sizes or by numerical simulation.

Curves such as those shown in Figure 3 allow the determination of the slowing-down length after correction for the borehole diameter. Normally the porosity is determined from the slowing-down length in a known rock type through the use of charts of the type shown in Figure 1. The variation of slowing-down length with porosity makes the measurement quite sensitive at low porosity, if the rock type is known. However, uncertainty in the rock type may introduce variations on the order of several porosity units. An additional complication is that important quantities of hydrogen may occur in formation rock, the most prominent example being the hydroxyls associated with clay minerals. Depending on the clay type, the apparent porosity of a formation can attain values of up to 60%, when in fact no fluid is present.

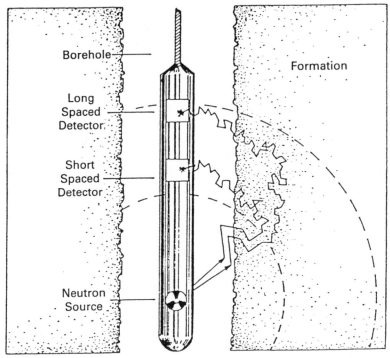

*Figure 2*    Schematic representation of a neutron porosity device in the borehole. It contains a chemical neutron source and two neutron detectors. The dashed lines are lines of constant radius from the source. From Ellis (35).

THERMAL NEUTRON DEVICES    As seen in the preceding section, the use of epithermal neutrons is a fairly uncomplicated way of determining the hydrogen content of a formation. Historically, thermal neutron detection has been used because of the much higher counting rates that can be obtained from a source of given activity.

Equation 8 suggests that the perturbation due to $\Sigma$ may be eliminated by placing the detector at a distance $r \gg L_d$, where the spatial thermal flux distribution is determined by $L_s$. In a practical neutron logging device this is generally not realizable because counting rates are insufficient at the distances required. As in the case of the epithermal device, a pair of thermal neutron detectors is used for determining the gradient of the thermal flux.

To characterize the laboratory data of the thermal neutron tool, which are at some variance with the predictions of Equation 8 because of the residual sensitivity to $\Sigma$, use is made of the migration length, $L_m$, $(L_m^2 = L_s^2 + L_d^2)$, which accounts explicitly for the additional perturbation

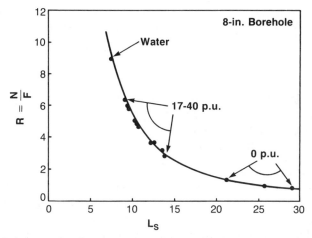

*Figure 3* Laboratory data from an epithermal neutron porosity device in rock formations over a wide range of porosity. The ratio of near to far detector counting rates is shown as a function of the appropriate slowing-down length of the test formations. From Ellis (35).

of neutron diffusion and absorption in the formation. To first order, representation of the tool response as a function of migration length allows the conversion of the measurements to porosity for given conditions of lithology and salinity (35). Empirical corrections (39) for formation and borehole water salinities are based on a series of laboratory formation measurements and numerical simulation.

## Applications of Gamma-Ray Scattering

Gamma-ray transmission through matter can be related to its bulk density (actually the number density of electrons) if the predominant interaction is Compton scattering. In the borehole environment, a classical transmission measurement is impossible. However, a gamma-ray transport measurement through a formation can be used to determine its density. The motivation for measuring formation bulk density comes from its direct relationship with formation porosity and from geophysical applications such as for determining the acoustic impedances of adjacent layers for seismic interpretation and for estimating overburden pressure.

BULK DENSITY AND GAMMA-RAY ATTENUATION    The interaction of gamma rays by Compton scattering depends only upon the number density of the scattering electrons. A scattering measurement can be strictly related to the bulk density, $\rho_b$, only if the ratio of $Z/A$ remains constant. For most elements $Z/A$ is about $\frac{1}{2}$ but there are several significant departures; hydrogen for example has a $Z/A$ ratio of nearly 1. For this reason it is convenient to define a new quantity, $\rho_e$, the electron density index, to be

$$\rho_e \equiv 2\frac{Z}{A}\rho_b. \qquad\qquad 9.$$

The bulk density and electron density index for the three major minerals (limestone, dolomite and quartz) are practically identical. However, for water there is an 11% discrepancy between the two (arising from the anomalous $Z/A$ value for H), which requires an additional correction.

DENSITY MEASUREMENT TECHNIQUE   A borehole device for measuring formation density consists of a source and a gamma-ray detector that is well shielded from the source. The gamma-ray source commonly used in density logging is $^{137}$Cs. Its 662-keV gamma rays are well below the threshold for pair production and insensitive to any significant photoelectric absorption. Although no unscattered gamma rays reach the detector, the intensity of multiply scattered gamma rays will still vary exponentially with the scattering material density. One explanation (31) for this behavior considers the multiply scattered gamma rays, detected far from the source, to have undergone most of their scattering in the formation close to the detector. These multiply scattered gamma rays are fed by a virtual source. It consists primarily of unscattered source-energy gamma rays that travel from the source, nearly parallel to the borehole wall, to reach the site of their last few collisions before detection. Their intensity depends on the probability of source gamma rays arriving at this site unscattered, and thus varies exponentially with formation density. Tittman (40, 41) has treated this as a diffusion problem and also concludes that the multiply scattered spectrum will vary exponentially with the formation density.

To compensate for the frequent occurrence of intervening mudcake, modern devices (42, 43) incorporate two detectors (generally both NaI) in a housing that shields them from direct source radiation. The device is forced up against the formation with a hydraulically operated arm. This arm also provides a measurement of the borehole diameter (along one axis).

The counting rates of the detectors vary exponentially with the electron density index of the formation. The formation density could be determined simply from an observed counting rate from either detector. However, when mudcake of unknown density and/or thickness is present on the borehole wall, each counting rate will be perturbed by an amount proportional to the product of the contrast in density, between the formation and the mudcake, and the mudcake thickness. This results in a discrepancy between the apparent formation density derived from the two detectors.

The discrepancy can be calibrated (43), using measurements with artificial mudcakes in laboratory test formations, into a correction for the apparent density measured by the farthest detector.

LITHOLOGY LOGGING    In modern density tools (43, 44), the shape of the low-energy portion of the scattered gamma-ray spectrum, which correlates with the formation photoelectric absorption parameters, is measured also. This permits one to differentiate the three major types of rock by their differing average atomic numbers. For this measurement it is convenient to define a new parameter, $P_e$, the photoelectric index, as

$$P_e \equiv \left(\frac{Z}{10}\right)^{3.6}, \qquad\qquad 10.$$

which is proportional to the photoelectric cross section per electron with the energy dependence suppressed. The attenuation of a flux $\Phi_0$ of gamma rays by photoelectric absorption alone can be written as

$$\Phi \propto \Phi_0 e^{-n_e P_e x}, \qquad\qquad 11.$$

where $n_e$ is the number density of electrons just as in the case of Compton scattering. The product $n_e P_e$, or $U$, is seen to have a volumetric mixing law, as noted earlier.

Qualitatively, the gamma-ray spectra observed with a logging device equipped with a window nearly transparent to low-energy gamma rays (such as Be) is shown in Figure 4. As the average atomic number (or $P_e$) of the formation increases, the lower-energy portion of the spectrum is progressively reduced. Thus, a measurement of this spectral shape at low gamma-ray energies can be calibrated as a function of $P_e$ (31). Once this calibration has been established, the value of $P_e$ can be determined continuously over the depth of the well to provide the gross formation lithology. For pure formations the values of $P_e$ are 1.83 for sandstone, 3.1 for dolomite, and 5.1 for limestone.

## Absorption Rate of Thermal Neutrons

Pulsed neutron sources are used in a device that responds to the macroscopic thermal neutron capture cross section, $\Sigma$. The value of $\Sigma$ depends on the chemical constituents of the rock matrix and the pore fluids, whose absorption cross section is usually dominated by chlorine. Thus, this measurement provides a means of identifying the formation fluid even in wells containing metallic casings where electrical measurements are useless.

MEASUREMENT TECHNIQUE    A logging tool to measure $\Sigma$ must include a pulsed neutron source and a gamma-ray detector. The measurement

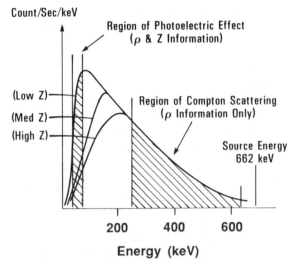

Count/Sec/keV

Region of Photoelectric Effect
($\rho$ & Z Information)

(Low Z)

(Med Z)

(High Z)

Region of Compton Scattering
($\rho$ Information Only)

Source Energy
662 keV

200        400        600

## Energy (keV)

*Figure 4*  Schematic illustration of the behavior of the multiply scattered gamma-ray spectrum detected by a density and lithology logging device. For illustration the formation density is held constant but the average atomic number of the rock formation is varied. Photoelectric absorption changes the shape of the lower energy portion of the spectrum. From Ellis et al (43).

sequence begins when a source of 14-MeV neutrons is pulsed for a brief period ($\approx 200$ $\mu$s). This forms a cloud of high-energy neutrons in the borehole and formation that becomes thermalized through repeated collisions. The neutrons are captured at a rate that depends upon the thermal absorption properties of the formation and borehole. The decay of the capture gamma-ray counting rate reflects the decay of the neutron population.

The reaction rate for thermal neutron absorption is given by the product of the macroscopic absorption cross section and the velocity of the neutron, $v$. The number of neutrons $N$ remaining at time $t$ is

$$N(t) = N(0)e^{-\Sigma vt}.\qquad\qquad 12.$$

The decay time constant, $1/v\Sigma$, is numerically equal to $4550/\Sigma$ [$\mu$s], when $\Sigma$ is in capture units.

The simple analysis of the capture gamma-ray time dependence misses an important aspect of actual measurements: the thermal neutron diffusion effect. At any observation point, the local thermal neutron density decreases because the neutrons are diffusing as well as being captured. To quantify the effect of the diffusion component on the local decay time constant it is necessary to use the time-dependent diffusion equation (45,

46). The result is that the apparent decay time of the local neutron population contains two components:

$$\frac{1}{\tau_a} = \frac{1}{\tau_{int}} + \frac{1}{\tau_{diff}},$$                    13.

where $\tau_{int}$ is the intrinsic decay time of the formation (i.e. that expected from global monitoring of absorption alone) and $\tau_{diff}$ is the diffusion time. The diffusion time depends on the distance from the source emission point and the thermal diffusion coefficient. The apparent $\Sigma$ of a formation will be greater than the intrinsic value because of the diffusion rate of the thermal neutron population in the vicinity of the detector.

As in most logging measurements, the response includes both borehole and formation signals. Since the borehole component is of little interest for the determination of the formation properties, techniques that decompose the capture gamma-ray time decay into formation and borehole components and correct for diffusion effects have been developed (47, 48).

Despite the complexity of the physics of the measurement and its engineering implementation, $\Sigma$ has a particularly simple mixing law. In the simplest case of a single mineral, the measured value, $\Sigma$, consists of two components, one from the matrix and the other from the formation fluid,

$$\Sigma = (1 - \phi)\Sigma_{ma} + \phi\Sigma_f.$$                    14.

To determine the water saturation, $S_w$ (i.e. the fraction of the pore volume $\phi$ containing water), the fluid component is broken further into water and hydrocarbon components:

$$\Sigma = (1 - \phi)\Sigma_{ma} + \phi S_w \Sigma_w + \phi(1 - S_w)\Sigma_h.$$                    15.

The $\Sigma_h$ of hydrocarbon is about 20 cu, nearly the same as fresh water. The ability of this measurement to distinguish hydrocarbon from water depends on the amount of chlorine in the water, which makes $\Sigma_w$ significantly larger than $\Sigma_h$. The presence of shale, which may contain thermal absorbers such as boron, seriously affects this simple interpretation scheme, but several methods for dealing with this problem have been developed (49).

## Natural Gamma-Ray Activity Measurements

Measurement of natural gamma-ray activity began as a supplement to the spontaneous potential for the identification of clay-free zones. The natural gamma-ray activity of sedimentary formations is dominated by the emissions of three isotopes with half-lives on the order of the age of the earth: $^{40}K$—$1.3 \times 10^9$ years, $^{232}Th$—$1.4 \times 10^{10}$ years, $^{238}U$—$4.4 \times 10^9$ years.

The decay of K produces a 1.46-MeV gamma ray. Thorium and uranium both decay through series of intermediate isotopes to stable isotopes of lead. Most gamma rays from the uranium series come from an isotope of bismuth while those of the thorium series are from thallium.

The largest source of formation radioactivity is potassium, which is fairly common in the earth's crust. By contrast, thorium- and uranium-bearing minerals are rare. In logging applications the uranium is frequently associated with the precipitation of uranium salts whose solubility accounts for its transport and frequent occurrence in organic shales. Thorium is frequently associated with heavy minerals such as monazite or zircon. Unlike potassium, which one expects to find at a mass concentration of a few percent, thorium and uranium may, at most, be expected to be tens of parts per million. Clay minerals, which are formed during the decomposition of igneous rocks, generally have a very high cation exchange capacity (50). This property enables clays to retain trace amounts of radioactive minerals that may have originally been components of the feldspars and micas that go into their production.

GAMMA-RAY DEVICES    The first gamma-ray devices measured only the total gamma-ray flux from the formation. These devices used Geiger counters or NaI detectors to measure the gamma rays above a practical lower limit (on the order of 100 keV). The total counting rate is a function of the distribution and quantity of radioactive material in the formation and is influenced by the detector size and efficiency. For this reason, calibration standards have been established at the University of Houston (51) by the American Petroleum Institute, and all total intensity gamma-ray logs are recorded in API units. The definition of the API unit of radioactivity comes from the calibration standards designed to simulate about twice the radioactivity of a typical shale. This formation, containing approximately 4% K, 24 ppm Th, and 12 ppm U, was defined to be 200 API units.

The response of a gamma-ray device, $GR_{API}$, is given by

$$GR_{API} = \alpha\,^{238}U_{ppm} + \beta\,^{232}Th_{ppm} + \gamma\,^{39}K_{\%}, \qquad\qquad 16.$$

where the subscripts refer to the mass concentration units of the isotope. The coefficients $\alpha$, $\beta$, and $\gamma$ depend on the actual detector used and the sonde design details.

Different types of shale have different total gamma-ray activity, depending on their respective Th, U, and K concentrations. With the development of improved spectroscopic-quality gamma-ray detectors, the gamma-ray tool was refined into a device for determining the actual concentrations of the three components. Calibration in standard formations, with known concentrations of K, U, and Th, permits determination of the indivdual mass concentrations in a formation as well as the total activity. To a good approximation, the gamma-ray intensity from a uniformly distributed source whose mass concentration is held constant is independent of the formation density even though the attenuation is a direct function of the formation density. Consequently, the $GR$ log responds directly to the mass concentration of radioactive elements (52).

Gamma-ray measurements also suffer to some degree from the borehole environment. Attenuation in the borehole mud is a function of the mud composition and the borehole size. Additional complications can arise from mud additives such as barite or KCl. Barium efficiently absorbs low-energy gamma rays from the formation. The potassium in KCl creates an unwanted source of radioactivity. Ellis (53) discusses a method for correcting spectral gamma-ray measurements for these effects.

## Neutron-Induced Reactions

Many geological properties can be determined from an analysis of the elemental concentrations (28, 50). However, measurements of individual gamma rays to determine a sufficient number of elemental concentrations require neutron sources to instigate reactions capable of producing characteristic gamma rays. Inelastic scattering, thermal neutron capture, (n, x) reactions, and delayed activity created by any of these reactions can be used. Detectors must be able to resolve the contribution of each element from the composite spectrum. In subsurface measurements it is sometimes straightforward to resolve the gamma-ray spectra but, because of competing reaction mechanisms or spectral interferences, it is sometimes difficult to relate gamma-ray intensity to the concentration of a particular element. If all of these problems can be overcome, the final difficulty remains to determine what fraction of the total intensity for a particular element is from the formation itself.

REACTIONS AND NEUTRON SOURCES    Neutron sources can be considered as two types: isotopic or accelerator sources. Because of the different properties of these sources, the types of reactions used to generate gamma-ray spectra are different. Isotopic sources, such as $^{252}$Cf or $^{241}$Am-Be, have been generally used for thermal neutron capture reactions (27, 54) or delayed activation produced by (n, $\gamma$) reactions (13, 21, 32, 55). The relatively low neutron energies of these sources (2.5–4.5 MeV) make it generally impractical to use these sources for (n, n') or (n, x) reactions. The development of compact pulsed accelerators (23, 24) that produced significant fluxes of 14-MeV neutrons from the $^2H(^3H, n)^4He$ reaction allowed sufficient yields to be obtained from (n, n') and (n, x) reactions. Since the 14-MeV neutrons are also moderated to thermal energies, the pulsed nature of the accelerators allows for multiple spectral measurements, gated at different times relative to the neutron burst, to heighten sensitivity to gamma rays produced by different types of reactions.

Inelastic scattering has been primarily used to determine the content of carbon and oxygen (12, 56–59). Most other elements are more efficiently detected through other types of reactions (60). Thermal neutron capture reactions have been used extensively to evaluate coal beds (54); to study mineral deposits, such as nickel (61) and mercury (62); and to determine most of the major constituents of sedimentary rocks (12, 63–65). The major use of delayed activity has been in mineral logging (13, 21, 27). Most of these applications have used isotopic chemical sources, as have the more general logging uses (32, 55). Future efforts for more complete elemental analysis will undoubtedly use delayed activity from both isotopic sources and 14-MeV accelerator-produced sources (21, 66, 67).

DETECTORS AND REACTION INTERFERENCES    Most spectrometers have been based on inorganic scintillators, primarily NaI(Tl) (12, 58, 59), though Ge detectors have also been used (13–19, 55), primarily for research purposes. Spectroscopic measurements based on the inorganic scintillators are generally limited to the detection of the most significant spectral contributions (68) because of their inherently poor energy resolution. Ge-detector-based spectrometers, with their dramatically improved energy resolution, have the capability of determining many more elements, including many that contribute only weakly to the total spectrum.

Regardless of the type of detector used in a spectrometer, two types of interferences are always present. These interferences are due to geological sediments containing all elements, though with dramatically varying concentrations. The first type is spectral interference, i.e. different elements produce gamma rays of similar energies, which cannot be resolved. The second type can have an isolated gamma ray, but the gamma ray can be produced by competing reactions, e.g. the 4.439-MeV gamma ray from $^{12}C(n, n')^{12}C^*$ can also be produced by the $^{16}O(n, n'\alpha)^{12}C^*$ reaction. These types of interferences are present in all analyses of geologic material (69), not just logging measurements. Means for dealing with the interferences can be optimized for any particular case; and they include improving system resolution (70), using time-dependent spectroscopy (71), and obtaining corrections from noninterfering reactions (72).

UNIQUE PROBLEMS IN SPECTROSCOPIC LOGGING    Two major problems exist in subsurface logging. The presence of the borehole introduces concentrations of at least some elements that are also present in the formation. Correcting the total measured concentration for the contribution from the borehole is a very difficult problem (73–76) whose solution is usually based on a model, which depends on the specific spectrometer, developed from extensive laboratory data. In addition, the neutron distribution changes substantially as a function of the elements present in the measurement volume. Again, the solution to this problem involves either a specific elemental normalization (77) or a model for the neutron flux that can be tied to measured parameters of the rock, fluid, and borehole (78).

Spectral and reaction interferences, as described above, further complicate these problems. It is important to remember that the spectrometer itself is unfortunately a portion of the measurement, and background contributions, especially in the case of trace elements, can be difficult to separate from the same elements present in the strata. This is apparent in the spectrum shown in Figure 5, where the gamma-ray peaks from neutron capture in Ni, Co, and Cr are from the cryostat containing the Ge detector and are shown together with peaks from Fe, Ti, and Si from the rock.

NUCLEAR DATA UNCERTAINTIES    The effect of nuclear data uncertainties, as well as the effects of interferences, on derived elemental concentrations can best be seen by considering the yield of a particular gamma ray, $Y_i$, that is observed when a neutron source is used to irradiate a geological formation. $Y_i$ is given by

$$Y_i = \sum_k \sum_j \int_{E_n} \int_t \int_V \Phi_n(\mathbf{r}, t, E_n) \sigma_{kj}(E_n) \eta_k(\mathbf{r}) b_{k,j,i} \Gamma(\mathbf{r}, E_{\gamma_i}) \, d^3\mathbf{r} \, dt \, dE_n, \qquad 17.$$

where $\Phi_n(\mathbf{r}, t, E_n)$ is the spatially varying, time- and energy-dependent neutron flux; $\sigma_{kj}$ is the reaction cross section for producing the nucleus $j$ through a neutron reaction on the nucleus $k$, whose concentration is $\eta_k$; $b_{k,j,i}$ is the probability of producing the gamma ray $i$ from the reaction $k(n, x)j$; and $\Gamma(\mathbf{r}, E_{\gamma_i})$ is the transmission and detection probability for the gamma ray $i$. Recent research in geochemical applications of nuclear data (28, 50) has emphasized the need for obtaining elemental concentrations from the spectroscopic measurements of $Y_i$. However, inverting Equation 17 to obtain the elemental concentrations $\eta_k$ is complicated by the interfering reactions $k, j$. The element $k$ can also be contained in the spectrometer or the borehole, which requires an independent correction. Finally, the neutron flux $\Phi_n$ depends on the cross sections for individual elements (79), and uncertainties in the cross sections themselves are also reflected in the accuracy with which $Y_i$ can be related to $\eta_k$ (66, 67). Naturally, for delayed activity the $Y_i$ also have a time dependence governed by the half-life of the parent nucleus.

Techniques for relating gamma-ray intensities to elemental concentrations have focused on approximate models (12, 73–76) that are calibrated with laboratory measurements, particular relationships between

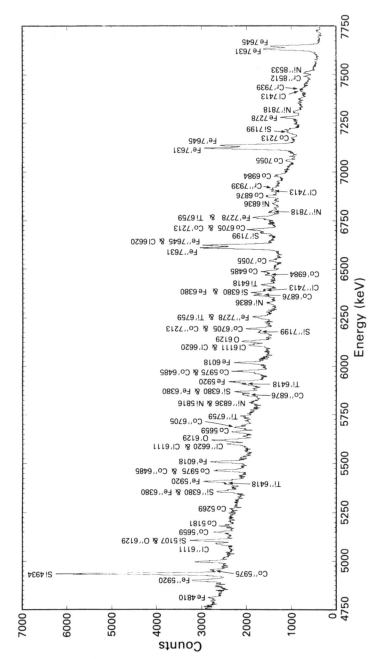

*Figure 5* Capture gamma-ray spectrum. The peaks from Cr, Co, and Ni are background from interactions in the germanium detector cryostat.

relative intensities of two gamma rays from two different elements (77), where one of the elements is well calibrated, or types of closure relationships (78) that allow an additional constraint to be applied to relative gamma-ray measurements. These techniques have been very useful for specific logging problems, as can be seen in Figure 6, which compares log-derived concentrations of Al, Si, and Ca with laboratory measurements performed on core samples from the well. But each new environmental situation requires a new approach to be developed from experimental and/or calculational studies.

Neutron-induced cross sections are generally not well enough determined to perform realistic, detailed calculations to relate gamma-ray yields to elemental concentrations (21, 66, 67). While thermal neutron cross sections are sufficiently well determined, high-energy neutron-induced reaction cross sections and gamma-ray production cross sections (60) are still inadequate. These data have a direct impact on the calculations that have become an increasingly important component of quantitative analysis of logging measurements.

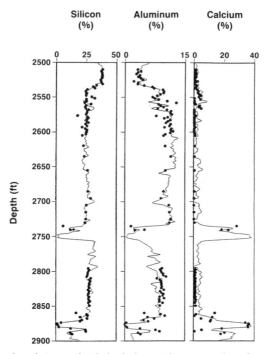

*Figure 6* Comparison between the derived elemental concentrations from spectroscopic logging measurements (*solid line*) and laboratory NAA results (*circles*) performed on core samples retrieved from the well.

## CALCULATIONAL TECHNIQUES

A very useful aid to furthering our understanding of nuclear logging measurements is the use of mathematical models to complement and extend experimental studies. Models are used to explore the feasibility of new techniques, support the design of new tools, and supply information on environmental effects that in many cases is otherwise unobtainable. With modern calculational methods and computers, this information can be provided with sufficient reliability and accuracy that the amount of experimental work can be greatly reduced and better focused.

The problem of calculating a nuclear tool response, in its simplest terms, is one of determining the flux of neutrons or photons reaching various detector locations. In general, the flux is a function of energy, spatial location, and direction of particle travel. The equation that governs the flux distribution at any point in space for a given source is the Boltzmann for radiation transport,

$$\Omega \cdot \nabla \Phi + \Sigma_t \Phi = \int dE' \int d\Omega' \, \Sigma_s(E' \to E, \Omega' \to \Omega)\Phi + S, \qquad 18.$$

where $\Omega \cdot \nabla \Phi$ is leakage (streaming) out of the phase space volume, $\Sigma_t \Phi$ describes the loss of particles due to interactions given by a cross section $\Sigma_t$, the double integral term is the scattering contribution to the given volume, and $S$ is the source term. This equation describes particle conservation at any typical point in phase space. Solutions to this equation can be obtained only for simple cases; ones in which the geometry is zero or one-dimensional and in which there is no angular or energy dependence to the cross sections, for example. In most logging problems the geometry is complicated (3-dimensional), has many interfaces, and does not usually include an axis of symmetry. The cross sections in general are complicated functions of energy and angle of scattering that do not lend themselves to a convenient mathematical description. Therefore, numerical methods have to be used to generate solutions to Equation 18 in logging problems. The two techniques most commonly used are Monte Carlo and deterministic numerical solutions based on discretization of the spatial, energy, and angular variables.

### Monte Carlo Methods

The Monte Carlo method is a statistical procedure whereby individual particle histories are simulated as a sequence of random events consisting of straight-line free flights each followed by a collision. These events are controlled by various probability distribution functions that depend on the details of the geometry and the interaction cross sections. This method is best suited for the analysis of complex geometries for which other numerical schemes are too approximate. However, because of its inherently statistical nature it is poorly suited for estimating fluxes over an extended range of spatial, energy, and angular variables. When 3-dimensional effects, time dependence, and/or cross-section energy details are important for accurate modeling of a logging tool, Monte Carlo techniques have to be used despite the computational costs.

Flexible, general purpose Monte Carlo codes have been available for some time, but their use in modeling nuclear logging tools is relatively recent (79–95). Although these codes were not developed with logging problems in mind, they possess capabilities that make them very useful for that purpose. In addition to the ease of describing complex 3-dimensional geometries, it is often the ability to do time-dependent calculations that has made Monte Carlo an exclusive choice for several logging applications.

The major drawback of Monte Carlo is the statistical uncertainty associated with each computed solution, an uncertainty that derives from the stochastic nature of the sampling techniques used. In an "analog" Monte Carlo simulation of a nuclear logging measurement, the detailed modeling of all the physical processes is not enough to produce practical solutions because of the excessive amount of computer time involved. As an example, a typical logging tool registers a response for one out of every 50 million source particles. This means that simulation of one second of logging time would require computer run times measured in CPU days, even on a modern supercomputer. This is true because most logging calculations are required to treat radiation transport over many mean free paths before reaching a given detector, which in turn makes adequate scoring statistics at particular detector locations difficult to obtain. The requirement of small statistical uncertainty is essential in calculations of nuclear logging responses because small changes in a detector response, sometimes no larger than a few percent, are important for the evaluation of formation properties. Hence, detector fluxes and, more importantly, differences in these fluxes produced by small environmental variations must be determined with a precision of 1% or lower in order to be useful.

Monte Carlo becomes practical for logging applications with the replacement of direct analog random walk models by nonanalog processes. The actual probability distribution functions are redefined or distorted so that events of interest occur far more frequently than in the analog case. This distortion is then compensated by altering the associated random variable estimator to remove any bias and ensure that a "fair game" is being played in the estimates of quantities of physical interest. These procedures, usually called importance sampling or variance reduction techniques, are essential to almost all Monte Carlo applications to logging.

The use of geometric splitting as a variance reduction technique was important in the development of a reasonably efficient Monte Carlo for gamma-ray scattering (86). Splitting involves defining spatial and energy boundaries with importance weights assigned to the volume and energy phase space elements. When a particle crosses a phase space boundary, it is split into multiple particles if the importance increases or is left unaltered or killed off if the importance decreases. Detailed importance maps must be defined for this purpose. Typical results obtained using this approach are illustrated in Figure 7, which shows a calculated pulse-height spectrum produced by the photon flux reaching a well-shielded long-spaced detector. Also shown in the same figure is an experimental photon pulse-height spectrum measured in a limestone block using the actual tool that was modeled. The spectra are normalized to the same total counting rate, and their close agreement shows the accuracy obtainable (and also required) with Monte Carlo. However, this was achieved only after considerable optimization of the importance sampling was done.

A variance reduction technique was also important for the calculation of time-dependent gamma-ray spectra observed following a burst of 14-MeV neutrons. Tsang & Evans (87) calculated the spectra from inelastic scattering events and thermal neutron capture at different detector locations and in different time windows. Calculated and measured spectra are compared in Figure 8. The chief difficulty with these calculations is that the spectra must be calculated from 1.5 to 7.5 MeV in 256 energy

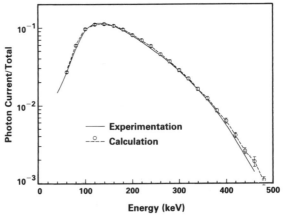

*Figure 7*  Comparison between a measured and calculated scattered gamma-ray spectrum from a $^{137}$Cs source in a limestone formation.

intervals. Such a discretization of the energy spectrum makes it very difficult to obtain adequate statistics at each energy. Variance reduction techniques in spatial, energy, angular, and time domains were used extensively to make the computer run times manageable for these calculations. These two examples (Figure 8) indicate that good spectral comparisons can be obtained and enhance our confidence in using these models to investigate the complicated responses in spectroscopy measurements.

A fair amount of success has been achieved in applying Monte Carlo to nuclear logging problems. The major difficulty remains the computational efficiency with which these calculations can estimate sometimes very small, but important, effects with acceptable and reliable statistical uncertainty. A substantial improvement in the efficiency of Monte Carlo calculations for logging applications will no doubt come from new computer architectures, with perhaps the greatest improvements coming from exploiting the inherent parallelism in Monte Carlo calculations by using multiple processors. Monte Carlo histories could be calculated independently and asynchronously on each processor, each accessing shared cross-section and geometrical data.

## Deterministic Methods

Unlike the Monte Carlo method, deterministic numerical solutions to Equation 18 do not suffer from statistical uncertainties. However, the treatment of the complex logging geometries is a greater difficulty for these methods. The description of the neutron or gamma-ray flux at detectors that may range from 10 to 20 mean free paths from the source makes it necessary to use accurate discretizations of the transport equation for numerical solution. The most widely applicable method, the discrete ordinates $S_N$ method (96), has until recently been limited to 2-dimensional geometries since the number of unknowns and computer storage requirements were prohibitively large in three dimensions. These limitations are being addressed by the availability of supercomputers in conjunction with the development of accurate, spatial coarse-mesh methods to solve the transport equation in three dimensions (97–99).

Deterministic transport methods have developed enough to complement the Monte Carlo method in many logging situations (100). However, some applications, such as gamma-ray spectroscopy, are probably beyond deterministic approaches simply because of the nuclear data representation that is required. For many environmental effects associated with formation composition (lithology, fluid type, temperature/pressure effects), deterministic models have yielded results that compare well with experimental data (80, 81, 83–85, 88, 91).

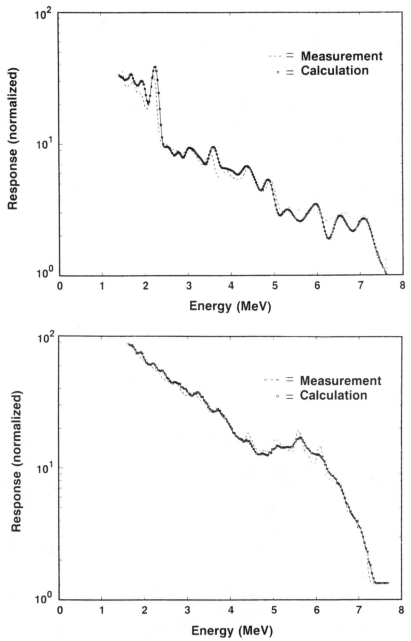

*Figure 8*   Comparison between measured and calculated gamma-ray spectra: (*top*) from thermal neutron capture, (*bottom*) from inelastic neutron scattering.

# CONCLUDING REMARKS

Nuclear physics techniques have been developed over the past forty years for in situ measurements of subsurface formations. The design of measurement devices, employing both neutrons and gamma rays, is dictated by the severe conditions of operating in a borehole environment. Their responses have been refined by extensive laboratory measurements in realistic borehole geometries and through mathematical simulation. Recent advances in measurement techniques have begun to provide bulk properties and elemental analyses of geological strata from which detailed geological information can be derived. Current and future efforts in both measurement and calculational techniques will be directed toward solving the fundamental problem of obtaining accurate geological properties from measurements performed in a complex extended medium. Improved cross-section data for neutron-induced reactions on geologically significant elements and improved computational techniques will lead to a better understanding of the physics of nuclear measurements in extended media and to overcoming the complex measurement calibration problem.

*Literature Cited*

1. Allaud, L., Martin, M. *Schlumberger, The History of a Technique.* New York: Wiley (1977)
2. Segesman, F. F. *Geophysics* 45: 1667 (1980)
3. Howell, L. G., Frosch, A. *Geophysics* 4: 106 (1939)
4. Pontecorvo, B. *Oil Gas J.* 40: 32 (1941)
5. Tittle, C. W., Faul, H., Goodman, C. *Geophysics* 16: 626 (1951)
6. Grosmangin, M., Walker, E. B. *J. Pet. Technol.* 9: 140 (1957)
7. Baker, P. E. *J. Pet. Technol.* 9: 289 (1957)
8. Voskoboinikov, G. M. *Bull. Acad. Sci. USSR, Geophys. Ser.* No. 3: 94 (1957)
9. Erozolimskii, B. G., Shkol'nikov, A. S. *Nuclear Geophysics.* Moscow: State Sci. Tech. Publ. House for Petroleum and Mining-Fuel Literature (1959)
10. Czubek, J. *Geophysics* 37: 1960 (1972)
11. Caldwell, R. L., Mills, W. R., Hickman, J. B. Jr. *Nucl. Sci. Eng.* 8: 173 (1960)
12. Hertzog, R. C,. *Soc. Pet. Eng. J.* 20: 327 (1980)
13. Senftle, F. E. In *Short Course in Neutron Activation Analysis in the Geosciences,* ed. G. K. Muecke. Toronto: Mineralogical Assoc. Canada (1980)
14. Clayton, C. G., Hassan, A. M., Wor-mald, M. R. *Int. J. Appl. Radiat. Isot.* 34: 83 (1983)
15. Lauber, A., Landstrom, O. *Geophys. Prospect.* 20: 800 (1972)
16. Tanner, A. B., Moxham, R. M., Senftle, F. E., Baicker, J. A. *Nucl. Instrum. Methods* 100: 1 (1972)
17. Senftle, F. E., Moxham, R. M., Tanner, A. B., Boynton, G. R., Philbin, P. W., Baicker, J. A. *Nucl. Instrum. Methods* 138: 371 (1976)
18. Nedostrup, G. A., Prokof'ev, F. N. *At. Energy* 35: 54 (1973)
19. Mellor, D. W., Underwood, M. C. *SPWLA 26th Ann. Logging Symp.* 2: Paper FFF. Houston, TX: Soc. Prof. Well Log Analysts (1985)
20. Goldman, L. H., Marr, H. E. *SPWLA 20th Ann. Logging Symp.* 2: Paper GG. Houston,· TX: Soc. Prof. Well Log Analysts (1979)
21. Michaelis, W. *NBS Special Publ.* 594: 615 (1980)
22. Anderson, R. N. *Nature* 316: 486 (1985)
23. Youmans, A. H., Hopkinson, E. C., Steward, R. M. *Tech. Symp. Soc. Petroleum Engineers, Dallas.* SPE1304G. Richardson, TX: Soc. Petrol. Eng. (1959)
24. Frentrop, A. H., Sherman, H. *Nucleonics* 18: 72 (1960)

25. Keelan, D. K. *Tech. Symp. Soc. Petroleum Engineers, Beijing.* SPE10011. Richardson, TX: Soc. Petrol. Eng. (1982)
26. Keelan, D. K. *J. Can. Pet. Tech.* 42: April-June (1972)
27. Nargolwalla, S. S., Seigel, H. *Can. Min. J.* 98: 75 (1977)
28. Herron, M. M. *Clay Clay Minerals* 34: 204 (1986); Herron, S. L. *SPWLA 27th Ann. Logging Symp.* Paper HH. Houston, TX: Soc. Prof. Well Log Analysts (1986)
29. Hubbell, J. H. *US Natl. Bur. Stand. Circ. 542* (1969)
30. Gardner, J. S., Dumanoir, J. L. *SPWLA 21st Ann. Logging Symp.* Paper N. Houston, TX: Soc. Prof. Well Log Analysts (1980)
31. Bertozzi, W., Ellis, D. V., Wahl, J. S. *Geophysics* 46: 1439 (1981)
32. Scott, H. D., Smith, M. P. *The Log Analyst* 14: 3 (1973)
33. Jordan, J. R., Campbell, F. In *Well Logging I.—Borehole Environment, Rock Properties, and Temperature Logging*, Dallas: SPE Monogr. Ser. (1984)
34. Duderstadt, J. J., Hamilton, L. J. *Nuclear Reactor Analysis.* New York: Wiley (1976); Henry, A. F. *Nuclear-Reactor Analysis.* Cambridge, Mass: MIT Press (1975); Glasstone, S., Sesonske, A. *Nuclear Reactor Engineering.* New York: Van Nostrand (1967)
35. Ellis, D. V. *First Break* 4(3): 11 (1986)
36. Ellis, D. V. *SPE Petroleum Production Handbook*, ed. H. Bradley, Ch. 50. Dallas: SPE. In press (1987)
37. Allen, L. S., Tittle, C. W., Mills, W. R., Caldwell, R. L. *Geophysics* 32: 60 (1967)
38. Kreft, A. *Nukleonika* 19: 145 (1974)
39. Gilchrist, W. A., Galford, J. E., Flaum, C., Soran, P. D., Gardner, J. S. *Tech. Symp. Soc. Petroleum Engineers, New Orleans.* SPE15540. Richardson, TX: Soc. Petrol. Eng. (1986)
40. Tittman, J. *Geophysical Well Logging.* Excerpted from *Methods in Experimental Physics, Vol. 24: Geophysics.* New York: Academic (1986)
41. Tittman, J., Wahl, J. S. *Geophysics* 30: 284 (1965)
42. Wahl, J. S., Tittman, J., Johnstone, C. W., Alger, R. P. *J. Pet. Technol.* 16: 1411 (1964)
43. Ellis, D., Flaum, C., Roulet, C., Marienbach, E., Seeman, B. *Tech. Symp. Soc. Petroleum Engineers, San Francisco.* SPE12048. Richardson, TX: Soc. Petrol. Eng. (1983)
44. Minette, D. C., Hubner, B. G., Koudelka, J. C., Schmidt, M. See Ref.

28, Paper DDD
45. Wahl, J. S., Nelligan, W. B., Frentrop, A. H., Johnstone, C. W., Schwartz, R. *J. Tech. Symp. Soc. Petroleum Engineers, Houston.* SPE2252. Richardson, TX: Soc. Petrol. Eng. (1968)
46. Ellis, D. V. *Well Logging for Earth Scientists*, Ch. 13. New York: Elsevier (1987)
47. Schultz, W. E., Smith, H. D. Jr., Verbout, J. L., Bridges, J. R., Garcia, G. H. *CWLS-SPLA Symp., Calgary.* Paper CC. Calgary: Can. Well Logging Soc. (1983)
48. Steinman, D. K., Adolph, R. A., Mahdavi, M., Marienbach, E., Preeg, W. E., Wraight, P. D. See Ref. 39, SPE15437
49. Hoyer, W. A., ed. *Pulsed Neutron Logging*, SPWLA Reprint Volume (1979)
50. Herron, M. M. *Int. J. Rad. Appl. Inst., Part E Nucl. Geophys.* 1: 197 (1987)
51. Belknap, W. B., Dewan, J. T., Kirkpatrick, C. V., Mott, W. E., Pearson, A. J., Rabson, W. R: *Drill. Prod. Prac.* Houston: API (1959)
52. Wahl, J. S. *Geophysics* 48: 1536 (1983)
53. Ellis, D. V. *SPWLA 23rd Ann. Logging Symp.* Paper O (1982)
54. Senftle, F. E., Tanner, A. B., Philbin, P. W., Boynton, G. R., Schram, C. W. *Mining Engineering (AIME)* 30: 666 (1978)
55. Everett, R. V., Herron, M., Pirie, G., Schweitzer, J., Edmundson, H. *Tech. Symp. Soc. Petroleum Engineers, Las Vegas.* SPE14176. Richardson, TX: Soc. Petrol. Eng. (1985)
56 Tittman, J., Nelligan, W. B. *J. Pet. Technol.* 12: 63 (1960)
57. Lock, G. A., Hoyer, W. A. *J. Pet. Technol.* 26: 1044 (1974)
58. Schultz, W. E., Smith, H. D. Jr. *J. Pet. Technol.* 26: 1103 (1974)
59. Chace, D. M., Schmidt, M. G., Ducheck, M. P. See Ref. 47, Paper X
60. Schweitzer, J. S., Hertzog, R. C., Soran, P. D. *Int. J. Rad. Appl. Inst., Part E Nucl. Geophys.* 1: 213 (1987)
61. Volkov, I. D., Ziv, L. A., Kostin, V. L., Postel'nikov, A. F. *Tr. Vses. Nauchno-Issled. Inst. Yad. Geofiz. Geokhim.* 13: 86 (1972)
62. Erkhov, V. A., Makarov, Y. I., Yaroslavtsev, V. F., Egorov, E. V., Sokolov, E. A., Ockhur, P. A. *Razved. Okhr. Nedr.* 2: 30 (1973)
63. Hertzog, R., Plasek, R. *IEEE Trans. Nucl. Sci.* NS-26: 1558 (1979)
64. Westaway, P., Hertzog, R., Plasek, R. E. See Ref. 23, SPE9461
65. Flaum, C., Pirie, G. *SPWLA 22nd Ann. Logging Symp.* 1: Paper H (1981)
66. Hertzog, R. C., Soran, P. D.,

Schweitzer, J. S. *Rad. Effects* 94: 49 (1986)

67. Hertzog, R. C., Soran, P. D., Schweitzer, J. S. *Int. J. Rad. Appl. Inst., Part E Nucl. Geophys.* 1: 243 (1987)
68. Grau, J. A., Schweitzer, J. S. *Int. J. Rad. Appl. Inst., Part E Nucl. Geophys.* 1: 157 (1987)
69. Pepelnik, R. *7th Int. Conf. Modern Trends in Activation Analysis, Copenhagen,* 1: 367. Roskilde, Denmark: Riso Natl. Lab. (1986)
70. Schweitzer, J. S., Manente, R. A. *AIP Conf. Proc.* 125: 824 (1985)
71. Mikesell, J. L., Senftle, F. E., Lloyd, T. A., Tanner, A. B., Merritt, C. T., Force, E. R: *Geophysics* 51: 2219 (1986)
72. Underwood, M. C., Dyos, C. J. *Appl. Radiat. Isot. Int. J. Appl. Instrum., Part A* 37: 475 (1986)
73. Schweitzer, J. S., Manente, R. A., Hertzog, R. C: *J. Pet. Technol.* 36: 1527 (1984)
74. Grau, J. A., Antkiw, S., Hertzog, R. C., Manente, R. A., Schweitzer, J. S. *AIP Conf. Proc.* 125: 799 (1985)
75. Roscoe, B. A., Grau, J. A. See Ref. 55, SPE14460
76. Grau, J. A., Roscoe, B. A., Tabanou, J. R. See Ref. 55, SPE14462
77. Senftle, F. E., Mikesell, J. L. *Int. J. Rad. Appl. Inst., Part E Nucl. Geophys.* 1: 227 (1987)
78. Fanger, U., Pepelnik, R. *Proc. Am. Nucl. Soc. Topical Meet., Austin.* Rep. CONF-720902: 245 (1972)
79. Case, C., Antkiw, S., Albats, P. *Rad. Effects* 94: 113 (1986)
80. Ellis, D., Ullo, J., Sherman, H. *Tech. Symp. Soc. Petroleum Engineers, San Antonio.* SPE10294. Richardson, TX: Soc. Petrol. Eng. (1981)
81. Ullo, J. See Ref. 80, SPE10295
82. Preeg, W., Scott, H. See Ref. 80, SPE10293
83. Case, C. *SPWLA 23rd Ann. Logging Symp.* 1: Paper L. Houston, TX: Soc. Prof. Well Log Analysts (1982)
84. Ellis, D., Case, C. *SPWLA 24th Ann. Logging Symp.* 1: Paper S. Houston, TX: Soc. Prof. Well Log Analysts (1983)
85. Sherman, H., Ullo, J., Robinson, J. See Ref. 84, 1: Paper R
86. Watson, C. See Ref. 43, SPE12051
87. Tsang, J. S. K., Evans, M. See Ref. 43, SPE12052
88. Ullo, J., Chiaramonte, J. See Ref. 43, SPE12137
89. Boyce, J., Carroll, J. *Tech. Symp. Soc. Petroleum Eng., Houston.* SPE13138. Richardson, TX: Soc. Petrol. Eng. (1984)
90. Smith, H., Shultz, W. *Tech. Symp. Soc. Petroleum Engineers, Houston.* SPE7432. Richardson, TX: Soc. Petrol. Eng. (1978)
91. McDaniel, P., Harris, J., Widman, D. H. See Ref. 84, 2: Paper LL (1983)
92. McDaniel, P., Harris, J. *Trans. Am. Nucl. Soc.* 45: 217 (1984)
93. Butler, J., Clayton, C. G. *SPWLA 25th Ann. Logging Symp.* Paper FFF. Houston, TX: Soc. Prof. Well Log Analysts (1984)
94. Sanders, L. G., Kemshell, P. B. See Ref. 93, Paper QQQ; Kemshell, P. B., Wright, W. V., Sanders, L. G. See Ref. 93, Paper PPP
95. Choi, H. K., Gardner, R. P., Verghese, K. *Trans. Am. Nucl. Soc.* 52: 354 (1986); Butler, J. *Trans. Am. Nucl. Soc.* 54: 148 (1987); Gartner, M., Dean, S. *Trans. Am. Nucl. Soc.* 54: 149 (1987); Case, C. R., Badruzzaman, A. *Trans. Am. Nucl. Soc.* 54: 150 (1987); Lehtihet, H. E., Altman, J. C., Quarles, C. A., Salaita, G. N. *Trans. Am. Nucl. Soc.* 54: 153 (1987); Petler, J. S. *Trans. Am. Nucl. Soc.* 54: 152 (1987); McDaniel, P. J., Lysne, P. S., Harris, J. M. *Trans. Am. Nucl. Soc.* 54: 154 (1987)
96. Carlson, B. G., Lathrop, K. D. In *Computing Methods in Reactor Physics,* ed. H. Greenspan, C. N. Kelber, D. Okrent. New York: Gordon & Breach (1968)
97. Badruzzaman, A., Xie, Z., Dorning, J., Ullo, J. *Proc. Topical Meet. Reactor Phys. Shielding, Chicago,* I: 170. La Grange Park, IL: Am. Nucl. Soc. (1984)
98. Badruzzaman, A. *Nucl. Sci. Eng.* 89: 281 (1985)
99. Badruzzaman, A. *Prog. Nucl. Energy* 18: 137 (1986)
100. Badruzzaman, A., Chiaramonte, J. *Proc. Int. Topical Meet. Adv. Reactor Phys. Math. Comput. Paris,* 3: 1333. La Grange Park, IL: Am. Nucl. Soc. (1987)

*Ann. Rev. Nucl. Part. Sci. 1987. 37: 243–66*

# ADVANCED ACCELERATOR CONCEPTS AND ELECTRON-POSITRON LINEAR COLLIDERS

## R. H. Siemann

Accelerator Division, Fermi National Accelerator Laboratory, Batavia, Illinois 60510; and Newman Laboratory of Nuclear Studies, Cornell University, Ithaca, New York 14853

CONTENTS

## INTRODUCTION

Many phenomena of high energy physics can be understood through the Standard Model of elementary particles and their interactions. Key experiments in the development of this model were made possible by advances in accelerator physics: stochastic cooling led to the discovery of the $W^\pm$ and $Z^0$ intermediate bosons, and electron-positron storage-ring development led to the discovery of the $\tau$ lepton and one of the two independent discoveries of the $J/\psi$. While the Standard Model has been extremely successful, phenomena including the quark and lepton mass spectra and the symmetry-breaking mechanism in the electroweak theory are not encompassed. Experiments at the TeV mass scale should give important information about these questions.

243

0163–8998/87/1201–0243$02.00

Either a linear or a storage-ring collider must be used to reach these high energies. There is substantial experience with storage rings, and they are feasible for multi-TeV proton beams. The Superconducting Super Collider design (1) has shown that the accelerator physics considerations for such a ring are not qualitatively different from those of previous facilities. For electron beams, storage rings are inconceivable because of large synchrotron radiation energy losses; as a result, linear colliders must be used. Although the linear collider concept (2, 3) was introduced in 1965, the Stanford Linear Collider (SLC) (4), which has a beam energy of 50 GeV and is expected to become operational in 1987, is the first such collider to be built. At present, linear colliders do not have the experience base of storage rings. In addition, as detailed in the next section, technological developments beyond the SLC are needed for a TeV-energy linear collider.

Given the sharp contrast between the need for these developments and the present feasibility of a proton storage ring, why pursue the novel acceleration techniques needed for linear colliders? There are two clear reasons. First, energy is not the only parameter of importance for experiments. Electron-positron colliders have proven superior for many types of experiments because of the well-determined quantum numbers of the initial state and low backgrounds. Therefore, a scientific case for an electron-positron collider covering the same mass range as a proton collider may develop. Second, the resultant technology may allow us to reach even higher energies than possible with proton storage rings. For these reasons there is substantial interest in novel acceleration techniques for electrons, and, just as in the past, advances in accelerator physics may be a key to our understanding of nature.

# SCALING LAWS FOR $e^+e^-$ LINEAR COLLIDERS

The requirements of high energy physics together with economic arguments lead to general relationships that are independent of specific acceleration mechanisms. Center-of-mass energy and luminosity are the primary high energy physics parameters. The cross section for the production of muon pairs sets the scale of interesting luminosities. At 2-TeV center-of-mass energy and a luminosity of $10^{33}$ cm$^{-2}$ s$^{-1}$ the event rate for $e^+e^- \to \mu^+\mu^-$ is 0.9 events per day. Luminosities greater than $10^{33}$ cm$^{-2}$ s$^{-1}$ are welcome!

Considerations of capital and operating costs are having substantial impact on the development of novel accelerator concepts. Estimates of capital costs are unreliable because most concepts are at a "proof-of-principle" stage, but a cost scaling analysis performed by Palmer (5) gives general guidance to cost-effective parameters. This analysis relates most closely to radiofrequency-driven accelerators, and a similar discussion is presented in the section on near-field accelerators. The beam power and the efficiency of converting "wall-plug" power to beam power determine the operating cost. While the conversion efficiency must be considered separately for different acceleration mechanisms, there is a direct trade-off between beam power and beam quality. This trade-off follows from the scaling laws developed below.

In terms of accelerator parameters the luminosity is given by (6; an earlier version, containing some errors, is 6a):

$$L = \frac{N^2 f H_D}{4\pi\sigma_x\sigma_y},$$    1.

where $N$ is the number of particles per beam bunch (assumed to be equal for the two beams), $f$ is collision frequency, $H_D$ is a luminosity enhancement factor caused by "disruption," and $\sigma_x$ and $\sigma_y$ are the rms beam sizes in the horizontal and vertical directions respectively. In addition to the hard collisions that produce elementary particles, the beams interact through their electromagnetic fields. This interaction is the dominant influence on beam behavior. It leads to beam focusing, which is called "disruption," and photon radiation, which is named "beamstrahlung," a name intended to evoke images of bremsstrahlung.

Disruption is characterized by a disruption parameter that measures the strength of the focusing provided by the oncoming beam. For small values of the disruption parameter there is an enhancement of luminosity, given by $H_D$, due to this focusing. At large values, particles undergo transverse plasma oscillations, a situation that is almost certainly unstable and to be avoided. Reasonable values for $H_D$ are thought to be between one and five (7, 8).

Beamstrahlung lowers the average center-of-mass energy and introduces a spread in this energy. The average energy loss has been calculated in two regimes: the "classical" regime (9), where the radiation spectrum is that of synchrotron radiation, with the critical energy much less than the beam energy; and the "quantum" regime (10)[1], where the critical energy is much greater than the beam energy and, as a result, the spectrum is cut off at the beam energy. These calculations give beamstrahlung parameters, $\delta_{cl}$ and $\delta_q$, that represent the average fractional energy losses and that have different dependences on bunch dimensions, number of particles, and beam energy (6). In addition, the distribution of center-of-mass energies is different in the two regimes.

---

[1] R. Blankenbeckler and S. Drell are studying the validity of the quantum beamstrahlung calculation at short bunch lengths; their results could affect Equation 3.

Expressions for the luminosity, beam power $P_b$, and the beamstrahlung parameters can be combined to give scaling relationships for linear colliders in the classical and quantum regimes (6):

$$\frac{L^2\gamma^3}{\delta_{cl}} = 7.2 \times 10^{-3} \frac{(\sigma_x+\sigma_y)^2\sigma_z H_D P_b^2}{r_e^3(m_ec^2)^2\sigma_x^2\sigma_y^2},\qquad\qquad 2.$$

and

$$\frac{L^2\gamma}{\delta_q^3} = 0.05 \frac{H_D(\sigma_x+\sigma_y)^2 P_b^2}{\alpha^4 r_e(m_ec^2)^2\sigma_x^2\sigma_y^2\sigma_z}.\qquad\qquad 3.$$

The terms $\alpha$, $r_e$, and $m_ec^2$ are the fine structure constant, the classical radius of the electron, and the rest energy of the electron; $\gamma$ is the beam energy in units of rest energy, and $\sigma_z$ is the rms bunch length. The parameters on the left-hand sides are determined by particle physics requirements, and once these are fixed the scaling laws relate beam power and beam dimensions.

Equations 2 and 3 are plotted in Figure 1, which shows several important features of these equations. First, since most novel acceleration techniques employ short bunches, it is probable that a very high energy collider would operate in the quantum regime. The transition from the classical to the quantum regime occurs at shorter bunch lengths as the energy decreases, and as a result, a lower energy collider using the same acceleration method may operate in a different beamstrahlung regime. Second, the beam radius

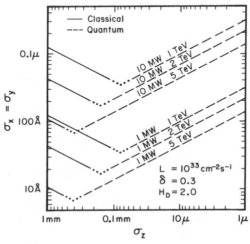

*Figure 1*   Linear collider scaling laws in the classical and quantum regimes assuming a round beam. Beamstrahlung is in the classical (quantum) regime at large (small) $\sigma_z$. Note that the $\sigma_z$ axis has an unconventional orientation.

is extremely small and directly proportional to the beam power and operating cost. Low beam power requires a small radius and, equivalently, good beam quality. Preservation of beam quality during acceleration is a critical issue discussed in more detail in the next section.

## BEAM QUALITY AND WAKEFIELDS

The acceleration and focusing of a particle can be described by a trajectory in six-dimensional phase space[2]. Liouville's theorem states that the phase-space density about any such trajectory is constant; consequently, the phase-space volume of the beam is constant although its shape may change. However, a distinction must be drawn between this statement and a more practical point of view. A beam may develop a complicated, filament-like structure enclosing empty regions of phase space. Theoretically, the phase-space filaments can be untangled, but this is impractical, and the average density of the beam has become lower. For a given intensity the useful measures of beam quality are the emittances, which are the projections of the occupied volume onto $(x,p_x)$, $(y,p_y)$, and $(z,p_z)$ planes. These emittances are constant at best.

In each of the transverse dimensions the spot size $(\sigma_x, \sigma_y)$ and the angular divergence $(\sigma_{x'}, \sigma_{y'})$ at the collision point are related to the emittance $(\varepsilon_x, \varepsilon_y)$ by

$$\sigma_i \sigma_{i'} = \frac{\varepsilon_i}{\gamma m_0 c^2}; \quad i = x, y. \qquad 4.$$

Focusing the beam to a small spot demands a strong final lens, and a common feature of ideas for making such a lens (11, 12; R. B. Palmer private communication) is a small beam size at the lens or equivalently a small angular divergence at the collision point. The combination of a small spot and a strong final lens call for small transverse emittances; for the spot sizes in Figure 1, these emittances need to be several orders of magnitude smaller than obtained to date.

The longitudinal emittance is the product of energy spread and bunch length. For a small collision spot, the chromatic and geometric aberrations of the interaction region optical system need to be corrected to high order. These corrections are difficult (13), and they place an upper limit on the energy spread. This limit, combined with the short bunch length required to operate in the quantum beamstrahlung regime, calls for a small longitudinal emittance.

[2] The phase-space coordinates are the three spatial coordinates and their conjugate momenta.

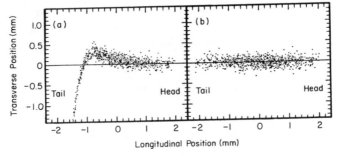

*Figure 2*   The results of a simulation of beam behavior in the SLC (14). In obtaining Figure 2*b*, Landau damping was used to prevent emittance blowup. (Reprinted with permission of K. L. F. Bane.)

Beam-generated electromagnetic fields called "wakefields" tend to increase the beam emittances. Figure 2*a*, the result of a computer simulation of beam behavior in the SLC (14), presents a graphic example. In this simulation the beam is injected with a 30-$\mu$m position error (with respect to the symmetry axis of the structure). As a result of this displacement, deflecting wakefields are generated. After acceleration, trailing particles have a large transverse displacement and the transverse emittance has increased. Transverse emittance growth can also be initiated by misalignment of accelerator sections or focusing elements (6). Ways to prevent this are as follows:

1. Remove misalignments and injection errors;
2. Use strong focusing along the accelerator;
3. Use Landau damping (14, 15). If the beam particles have a spread in focusing strength (with quadrupole magnets this is directly proportional to the beam energy spread), they do not move in phase, and the emittance blowup is strongly damped, as shown in Figure 2*b*.
4. Choose an accelerator with small wakefields.

The example in Figure 2 shows the effect of transverse wakefields. There are also longitudinal wakefields that have only a decelerating component[3]. These lead to an energy spread that is roughly inversely proportional to the bunch length. The longitudinal emittance, $\sigma_y \sigma_z$, is determined by the accelerator structure and can be reduced by the choice of accelerator (the fourth method above). The interplay between wakefields and accelerator properties is discussed in the next section.

---

[3] The transverse (longitudinal) wakefield is the transverse momentum (energy) change per unit length of accelerator.

# NEAR-FIELD ACCELERATORS

An accelerating electromagnetic wave must have a phase velocity equal to the beam velocity and a component of electric field in the direction of propagation. These waves can propagate in a variety of structures with the common feature of a periodic substructure with size and period comparable to the wavelength. The importance of nearby boundaries leads to the general classification of "near-field" accelerators. The most familiar example is the radiofrequency (rf) driven, disk-loaded, cylindrical waveguide.

For electron linear accelerators most experience is with 10-cm wavelength (S-band), room temperature structures. It would be natural to think of a very high energy collider based on this experience, but considerations of beam power and efficiency argue against it. Since these structures have a group velocity much less than the speed of light, a section of accelerator is "filled" with electromagnetic energy before the beam arrives. The beam extracts a fraction of this energy, and the remainder must be thrown away or saved for subsequent beam pulses.

The electromagnetic field energy per unit length is proportional to the square of a typical transverse dimension and the square of a typical field. The waveguide dimensions scale as the wavelength of the fundamental mode $\lambda$, and the accelerating gradient $G$ can be taken as a typical field. The energy extraction efficiency $\eta_b$ per bunch of $N$ particles is

$$\eta_b = A_b \frac{NG}{\lambda^2 G^2} = A_b \frac{N}{\lambda^2 G}.$$  5.

The proportionality constant, $A_b$, is $13.4 \times 10^{-8}$ V$-$m for the SLAC S-band linear accelerator (6).

The beam's fractional energy spread depends on $\eta_b$, $\sigma_z/\lambda$, and the phase of the bunch with respect to the accelerating wave. If the phase is chosen to minimize the energy spread, we have

$$\eta_b \propto \frac{\sigma_\gamma \sigma_z}{\gamma \lambda}.$$  6.

The proportionality constant depends on the structure (6). The choices for high efficiency are large longitudinal emittance, short wavelengths, or storage of rf energy for subsequent beam pulses.

Efficiency is one factor entering into a choice of wavelength; others are the dependence of limiting gradient on wavelength, transverse wakefields, and the availability of appropriate rf or laser power sources. The limitations on gradient from surface heating and electric field breakdown improve as the wavelength is reduced (6). However, with all other parameters fixed, the beam power is proportional to the gradient, and it may not be economical to operate near the limiting gradient.

Transverse wakefields favor long wavelengths. When $\lambda \gg \sigma_z$, the transverse wakefield within the bunch is linear with distance from the head. If the proportions of the accelerator are held fixed while the overall size is scaled to change the fundamental wavelength and the bunch length is kept a fixed fraction of the wavelength, then the transverse wakefield at the tail varies as $\lambda^{-3}$ (6). For weaker dependence on $\lambda$, $\sigma_z/\lambda$ must be reduced at the cost of increasing the beam power and decreasing $\eta_b$. As mentioned earlier, transverse wakefield effects can be damped with Landau damping, which comes from the energy spread of the beam. To the extent that Landau damping can be used, the longitudinal emittance and $\eta_b$ can be increased. While a detailed parametric study is needed for quantitative results, it is clear that short wavelengths are not favored by transverse wakefield considerations.

Any choice of wavelength is strongly influenced by the availability of an appropriate power source. The average power and peak power requirements depend on wavelength as $\lambda^2$ and $\lambda^{1/2}$ respectively. The former follows directly from the stored energy and the latter from the stored energy and the decrease in structure filling time at short wavelengths (6). These requirements are formidable, and development of power sources is needed at almost all wavelengths. Ultimately, considerations of power sources, economy, wakefields, fabrication tolerances, etc must be balanced in the design of an accelerator system. Since developments are still required before this can be done intelligently, a wide variety of approaches covering the wavelength range from $\lambda = 10$ cm to $\lambda = 10$ $\mu$m are under study. These are now described briefly.

Superconducting rf offers the advantage of being able to work at long wavelengths with a small number of continuous wave (cw) klystrons of proven design and a high efficiency for conversion of rf energy to beam energy. This high efficiency is obtained because the decay time of the rf energy is much longer than the time between beam pulses, and many beam bunches can be accelerated per rf pulse. This remains possible even if sufficient time for damping wakefields between bunches is required because external couplers can damp modes other than the fundamental without affecting the $Q$ of the fundamental mode (16).

To be practical for a high energy linear collider, the gradient and fundamental mode $Q$ must be increased from the presently obtained values of 5–10 MeV/m and 3–5 $\times$ 10$^9$. There is a fundamental limit to the gradient from the magnetic field at the surface of the superconductor. For Nb this limit is roughly 50 MeV/m, and for Nb$_3$Sn it is approximately 80 MeV/m

(17). In the past five years there has been substantial progress with the solution of multipacting and material defect problems (18), and at present the gradient is limited by field emission. This is under active study (17).

The fundamental mode $Q$ must be increased to make the cost of cooling the structure affordable. With the $Q$ raised to $5 \times 10^{10}$ and by operating with a modest duty cycle of 10–15%, the cryogenic power requirements can be reduced to the range of several hundred megawatts, which is beginning to become reasonable (19, 20).

In the design of a normal conducting linear accelerator there is a compromise between the peak power requirement and the structure efficiency defined as the inverse of the fraction of energy lost during the filling time. A factor-of-two increase above the minimum peak power increases the structure efficiency by about 2.5. Even with such an increase the average power at $\lambda = 10$ cm is prohibitive, and for $\lambda = 3$ cm it is acceptable only if multiple bunches can be accelerated with each rf pulse. It is unclear whether this is possible without interference between bunches due to wakefields. By reducing $\lambda$ to 1 cm it is possible to obtain reasonable efficiency with a single beam pulse per rf pulse (21). Small sections of a 1-cm wavelength disk-loaded waveguide have been constructed and tested (22); an average gradient of 180 MeV/m was produced. Future studies at this wavelength must include manufacturing tolerances, provision of water cooling, alignment, and potentially critical effects of wakefields.

For some representative parameters (21), the peak power requirement is 25–100 MW per meter of accelerator structure. The research and development work aimed at meeting this demand falls into one of two general categories: microwave power tubes and "two-beam" devices. Consider the rf power tubes first.

A peak power of over 150 MW has been obtained with good efficiency from a pulsed klystron at $\lambda = 10$ cm (23). However, this design cannot be used in the 1–3-cm wavelength range because the peak power capability of any given design falls with wavelength as $\lambda^2$. This is a consequence of space charge limits on current density combining with decreasing cathode area to reduce the beam current (21). For high power at short wavelengths the beam current must be increased by employing a ring or sheet beam that would interact with an appropriate cavity mode. For example, in the case of a hollow ring beam that mode is one with an electric field maximum at the beam radius. Preliminary work is beginning on the design of klystrons for these wavelengths employing ring and sheet current beams (24), but it is too early to have results.

The beam electrons in a gyroklystron undergo cyclotron motion and interact with a cavity TE mode. The cavities can be substantially larger than the wavelength, and as a result high beam currents can be used. A disadvantage of the gyroklystron is that the longitudinal component of the electron velocity must be kept large to counteract space charge forces, and the energy in this motion is not available for conversion into rf. As a result, the efficiency is limited to between 30 and 40%. Work is well advanced on the design and construction of a $\lambda = 3$ cm gyroklystron with a peak power output of 30 MW (25), and simulations are being used to study the scaling of this design up to 300 MW.

In the lasertron the cathode is a photoemitter. Striking this cathode with rf-modulated laser light makes the bunching cavities of the klystron unnecessary. Several experimental studies of the lasertron are underway (26, 27; J. LeDuff private communication). In the work being performed at SLAC (26), a lasertron for $\lambda = 10$ cm with a peak output power of 35 MW and an efficiency of 70% is under development. In the future it is expected that power levels of roughly 100 MW can be reached with approximately the same efficiency. The usefulness of this concept for very high energy linear colliders will depend on the results of studies of scaling to smaller wavelengths. As with the klystron, ring or sheet beam geometries could be used to obtain high power at short wavelengths.

Pulse compression schemes combine the power over a long pulse into a shorter, higher power pulse through the use of delay lines or energy storage cavities. These schemes can be important for increasing peak power or matching the pulse length of the tube to the accelerator fill time. An attractive idea for pulse compression, called binary pulse compression (28), uses delay lines, hybrid junctions, and phase reversals at the (low) drive power level. The delay lines are low loss lines running in a TE mode, and it is expected that losses will be sufficiently small to allow multiplication of peak powers by a factor of sixteen or more. Experimental tests are underway.

With any approach based on tubes, thousands of them will be needed. This unappealing prospect is avoided in two-beam accelerators in which rf energy is extracted from the "drive beam," a low energy, high current beam traveling parallel to the high energy accelerator (see Figure 3). The rf energy is replaced by reaccelerating the drive beam. The various two-beam accelerator concepts differ in the methods of energy extraction and reacceleration.

The original "Two-Beam Accelerator"[4] combined induction modules for acceleration with free electron lasers (FEL) for energy extraction (29). Induction modules offer efficient conversion of input power to beam power at a high repetition rate and moderate duty cycle (30). In the FEL section

---

[4] Andrew M. Sessler, the inventor of this approach, coined the name "Two-Beam Accelerator," which still refers to this specific two-beam concept.

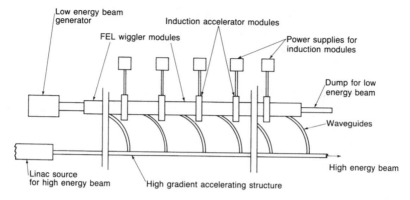

*Figure 3* Schematic of the "Two-Beam Accelerator" (29). This figure illustrates one of several possible two-beam concepts, which have the common features of a low energy drive beam running parallel to the high energy accelerator. (Reprinted with permission of A. Sessler.)

the beam oscillates transversely in the periodic magnetic field of a "wiggler" magnet. The ratio of wiggler wavelength to radiation wavelength is $\gamma_d^2$ ($\gamma_d$ is the drive beam energy) times a factor involving product of the magnetic field strength and wavelength of the wiggler. High efficiency is possible through the use of a tapered wiggler, a wiggler in which the wavelength or magnetic field changes to compensate for decreasing beam energy as rf is radiated (31).

Radiofrequency power generation with an FEL has been experimentally tested with impressive results (32): energy extraction efficiency was 34% and peak power was greater than 1.0 GW at $\lambda = 8.7$ mm. These results were obtained with a 3 m long tapered wiggler and a beam current and energy of 850 A and 3.5 MeV. Calculations and simulations studying other important aspects of this power source (including reacceleration, FEL sidebands and long-term beam stability, phase jitter, and rf extraction) are in progress with generally encouraging results (33). In addition, further experimental work, including development of a prototype Two-Beam Accelerator, is being planned.

A second two-beam approach combines superconducting rf for acceleration and bunched beam interaction with a resonant cavity, as in klystron, for energy extraction (34). The rf energy in the drive linear accelerator is supplied by large cw klystrons, and the drive beam pulses are short and spaced to match the repetition rate of the high energy accelerator. Superconducting rf allows efficient storage of energy between beam pulses; the required performance is that achieved routinely at present.

There are two other combinations of these acceleration and energy extraction methods: the "relativistic klystron" employing induction modules and klystron interaction (35), and a superconducting linear accelerator combined with an FEL (36). To date the optimal two-beam concept is not clear. Each concept has merits and drawbacks; for example, those with superconducting rf have the advantage of higher repetition rate and the disadvantage of generation and stability of intense beams in a multicavity linear accelerator. In all cases calculations and simulations are incomplete, and experimental work has been performed on only one of four ideas.

It was thought that a disk-loaded waveguide for $\lambda \ll 1$ cm would be impractical to fabricate, and other conductor configurations with accelerating fields were considered. Predominantly these were "open" configurations with one or more sides missing to allow for easier fabrication and coupling of energy from power source to accelerator. One of the first laser accelerator ideas was that of an optical grating as a near-field accelerator (37). The general properties of open structures have been studied (38), and a number of innovative ideas, including ink jets and etching, have been suggested for accelerator construction. Small sections of accelerator have been made by etching, and as this technology has developed, new ideas for structures have emerged (39). Among them is a row of connected holes resembling a disk-loaded waveguide.

Possible power sources for a short wavelength accelerator include FELs, $CO_2$ lasers, and the "microlasertron." FELs can operate at wavelengths as short as the ultraviolet, but there is a trade-off between extraction efficiency and gain that becomes more restrictive as the wavelength decreases (40). With the present rapid pace of FEL development, there should be substantially more experience with these power sources in the near future.

At $\lambda = 10$ $\mu$m it is natural to consider a $CO_2$ laser as the power source. This laser should have a high efficiency for conversion of wall-plug power to laser power, a high repetition rate equal to that of the accelerator, a pulse length in the picosecond range to match the accelerator filling time, and a favorable pulse extraction format. These have been achieved individually but not simultaneously (41). The results of a kinetics calculation for a high efficiency laser illustrate possible performance: a 3-ps output pulse length, and an efficiency of 20% obtained by extracting sixteen pulses spaced 50 ns apart (such multiple pulse extraction needs demonstration) (41). The efficiency of a single pulse is unacceptably low; either multiple beam bunches must be accelerated per laser pulse or output pulses must be combined with pulse compression. While the applicability to a full energy collider is unclear, a $CO_2$ laser will be the power source at an advanced accelerator test facility being planned by a collaboration between

Brookhaven National Laboratory and Los Alamos National Laboratory (42). Many ideas related to short wavelength accelerators will be tested at this facility.

The microlasertron is an adaptation of the lasertron for 1 to 10 mm wavelength (43). A large number of small cavities are constructed by etching, and a photocathode is incorporated into each of these. Modulated laser light strikes these photocathodes producing power. Though an individual cavity is not capable of high power, the large number compensates and, in fact, is an advantage because one cavity powers only a short section of accelerator. In principle, the efficiency can be close to 100%, but a number of detailed questions need study before we know the efficiency that can be realized in practice.

In addition to structures and power sources, other aspects of short wavelength accelerators have been considered in the past and need study in the future. The efficient coupling of energy from the power source to the accelerator is an example in which the solution is strongly dependent on the overall system. With the microlasertron, this coupling arises naturally, but with a laser, optical configurations leading to coupling must be considered in more detail. Overall, there has been progress in the development of short wavelength near-field accelerators with innovative solutions to the problems of small size.

# WAKEFIELD AND SWITCHED POWER ACCELERATORS

Wakefields, which usually limit performance, can be exploited for acceleration—a beam passing through a section of accelerator excites electromagnetic fields, and a second beam, injected with an appropriate time delay, is accelerated by these fields. In contrast with two-beam accelerators both beams travel in the same structure. For two point bunches traveling on collinear paths, energy conservation and linear superposition are sufficient to calculate both the stopping distance of the leading bunch and the acceleration gradient seen by a particle in the trailing bunch (44). While there is a trade-off between gradient and stopping distance, the transformer ratio $R$ is limited:

$$R = \frac{\text{gradient} \times \text{stopping distance}}{\gamma_1} = \frac{\Delta\gamma_2}{\gamma_1} \leq 2 - \frac{n_2}{n_1}, \qquad 7.$$

where $n_1$ and $\gamma_1$ ($n_2$ and $\gamma_2$) are the number of particles and energy of the leading (trailing) bunch. In this simple case the maximum energy increase is $2\gamma_1$. This surprising result also holds for any rigid, symmetric drive bunch (44). For $R > 2$ either the two beams must be noncollinear, or the drive beam must not be symmetric or rigid.

The DESY wakefield transformer (45, 46) shown in Figure 4 hopes to achieve a substantially better transformer ratio through the use of an annular driving beam coaxial to the trailing beam. For this configuration the transformer ratio is approximately $(r_b/r_a)^{1/2}$ where $r_a$ is the radius of the hole and $r_b$ the radius of the driving beam. There are potentially serious problems in the beam dynamics of the wakefield transformer. Azimuthal nonuniformity of the driving beam leads to transverse deflections of the accelerated beam, and the small hole, needed for a high transformer ratio, causes large wakefields in the accelerated beam itself. Simulations are being used to study these problems quantitatively (47).

An experimental test of the wakefield transformer is in progress (48). This experiment will serve as a proof of principle as well as identifying problems and directions for future research. The goals are a transformer

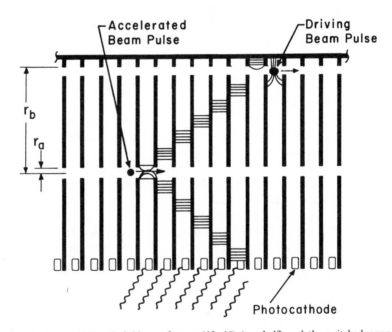

*Figure 4* The DESY wakefield transformer (45, 46) (top half) and the switched power accelerator (53), (bottom half) are a set of disks illustrated in section view. Fields generated at $r_b$ (by either an annular beam or light striking annular photocathodes) propagate toward the center and accelerate the high energy beam.

ratio of about ten and a gradient in excess of 100 MeV/m over a length of 0.42 m. The annular drive beam has been generated and accelerated up to 6 MeV. There have been some problems with the azimuthal uniformity and intensity of this beam, and these are under study (49). Construction of the apparatus is scheduled to be complete by the end of 1986, and results are anticipated shortly thereafter.

The limit $R \leq 2$ holds for a rigid driving beam, a beam in which the relative longitudinal positions of particles do not change. Wakefields within a beam that is not ultrarelativistic can cause mixing, a change of relative position. The "wakeatron" uses a proton drive beam with a mixing distance short compared to the stopping distance and thereby achieves a transformer ratio above the limit (50). [Early wakefield acceleration schemes were based on a proton beam (51), but the transformer ratio limit was not known at that time.] This increase in transformer ratio is at the expense of gradient unless a very small beam hole is used, and in that case beam stability is a major concern. A wakeatron with a transformer ratio of 10 and a gradient of 80 MeV/m has been discussed (52), and computer simulations are being used to study this concept in detail. Experimental work is being planned at the advanced accelerator test facility at Argonne National Laboratory.

The switched power accelerator (53) is closely related to the wakefield transformer, as illustrated in Figure 4. Instead of a drive beam, conductors are mounted between the disks of a transformer. These conductors are charged to a high voltage and then switched to ground, which creates an electrical pulse converging on the center. The transformer action of the radial transmission line increases the voltage substantially above the charging voltage. The switch is the key element, and a number of ideas have been discussed, including a laser/photocathode switch, a high pressure gas switch, and solid-state switches (54).

The laser/photocathode switches are being studied experimentally and with simulations. The goal of the experiments is a rugged cathode capable of high current density and with quantum efficiency of order 1%. Simulation results show that in a simple geometry about 10% of the stored energy before switching is converted to accelerating field energy (54). Azimuthal nonuniformity of the pulse caused by variation of the switch timing or efficiency would deflect the beam, and the sensitivity of the pulse to such variation is being measured with large-scale models. In addition, for a large transformer ratio the switched power accelerator has a small beam hole. As in the wakefield transformer this leads to strong wakefield effects. With all of these aspects of the switched power linear accelerator under active investigation, it should soon be possible to evaluate the promise of this idea.

# PLASMA ACCELERATORS

Plasma accelerators offer the prospect of accelerating gradients in excess of 1 GeV/m; typical parameters are given in Table 1. In these accelerators a plasma oscillation with a longitudinal, accelerating field is produced; particles from an external source are injected into the plasma and accelerated. Two widely discussed concepts, the plasma beatwave accelerator (55, 56) and the plasma wakefield accelerator (44, 57), differ in the mechanism producing the plasma oscillation. The wakefield accelerator is discussed first.

Maxwell's equations together with fluid equations describing the plasma are solved for a charged-particle beam traveling in a plasma. While the complete solution should be nonperturbative and include relativistic motion of the plasma electrons, the perturbation solution for a non-relativistic plasma (44) gives insight into the basic phenomena. Assuming a driving beam of $N$ particles in a disk with radial density distribution $\chi(r)$, i.e.

$$n_b = N\chi(r)\delta(z - v_b t),\qquad\qquad 8.$$

the plasma density behind the beam ($z < v_b t$) is

**Table 1**  Parameters for plasma accelerators from Ruth & Chen (58)[a]

| Parameter | Plasma beatwave accelerator ($CO_2$ laser) | | Plasma wakefield accelerator | |
|---|---|---|---|---|
| | 2 GeV/m | 0.94 GeV/m | 2 GeV/m | 0.94 GeV/m |
| $\omega$ (s$^{-1}$) | [$1.78 \times 10^{14}$] | [$1.78 \times 10^{14}$] | — | — |
| $N$ | — | — | [$5 \times 10^{10}$] | [$5 \times 10^{10}$] |
| $\gamma_i$ | — | — | [430] | [2040] |
| $\omega_p$ (s$^{-1}$) | $5.7 \times 10^{12}$ | $2.7 \times 10^{12}$ | $4.4 \times 10^{12}$ | $4.4 \times 10^{12}$ |
| $n_0$ (cm$^{-3}$) | $1 \times 10^{16}$ | $2.2 \times 10^{15}$ | $6.0 \times 10^{15}$ | $6.0 \times 10^{15}$ |
| $L$ (m) | [0.1] | [1.0] | [0.1] | [1.0] |
| $\alpha^b$ | [0.25] | [0.25] | 0.25 | 0.11 |
| $\delta$ | [$5\pi/16$] | [$5\pi/16$] | 0.042 | 0.018 |
| $\omega_p\tau$ | [1000] | [1000] | — | — |
| $\langle eE_z \rangle$ (GeV/m) | 2.0 | 0.94 | [2.0] | [0.94] |
| $R/(2\pi k_p)$ | 1.25 | 1.82 | [1.25] | [1.82] |
| $R$ (mm) | 0.41 | 1.3 | 0.54 | 0.78 |
| $W\tau$ (J) | 11 | 240 | — | — |
| $N\gamma_i m_e c^2$ (J) | — | — | 1.8 | 8.4 |

[a] Square brackets indicate chosen parameters; the other parameters are derived. The reference has a complete discussion.
[b] All symbols except $\alpha$ are defined in the text: $\alpha = n\ (r = 0)/n_0 - 1$, where $n(r)$ is the plasma density in Equation 9.

$$n(r) = n_0 + kN\chi(r)\sin(kz - \omega_\mathrm{p}t), \qquad\qquad 9.$$

where $n_0$ is the unperturbed density, $\omega_\mathrm{p}$ $[= (4\pi e^2 n_0/m_\mathrm{e})^{1/2}$ in cgs units] is the plasma frequency, and $k = \omega_\mathrm{p}/v_\mathrm{b}$.

The electrostatic potential produced by the perturbation is calculated using Poisson's equation. When $\chi(r)$ is a parabolic distribution of radius $R$ with $kR \gg 1$, the longitudinal and radial electric fields for $r \ll R$ are (44)

$$E_z = -\frac{8eN}{R^2}\left\{1 - \frac{r^2}{R^2}\right\}\cos(kz - \omega_\mathrm{p}t),$$

and                                                                                    10.

$$E_r = \frac{16eN}{R^2}\left\{\frac{r}{kR^2}\right\}\sin(kz - \omega_\mathrm{p}t).$$

Features of this solution are (58)

1. The density perturbation is a static ($v_\mathrm{group} = 0$) oscillation at the plasma frequency.
2. The number of particles, $N$, and the radius of the driving beam, $R$, determine the accelerating gradient.
3. For the chosen driving beam profile, $E_z$ depends on $r$; as a result the accelerated beam has an rms fractional energy spread of approximately $0.47(b/R)^2$, where $b$ is the accelerated beam radius.
4. For a range of one quarter of a plasma wavelength, the fields are accelerating and focusing.
5. The phase velocity is $v_\mathrm{b}$. Over a length $L$ where the driving beam energy changes from $\gamma_\mathrm{i}$ to $\gamma_\mathrm{f}$ there is a phase slippage (44) of

$$\delta = \frac{\omega_\mathrm{p}L}{2\gamma_\mathrm{i}\gamma_\mathrm{f}c} \qquad\qquad 11.$$

   between the beams. This limits the length of an acceleration stage.
6. The transformer ratio equals two (a consequence of assuming a disk in Equation 8); for a larger ratio the bunch must have an extended, nonsymmetrical shape.

The driven beam also produces a wakefield and, therefore, loses some energy to the plasma. The net energy change is the increase from acceleration minus this energy loss. The maximum efficiency for the transfer of energy from the plasma to the beam is approximately $(b/R)^2$, which is about twice the energy spread (58). The accelerated beam radius, $b$, is a function of the focusing and the beam emittance. The emittance is determined by the particle physics requirements as discussed in a previous section, and the focusing strength and gradient are related. Using these relationships the efficiency is found to be proportional to $\varepsilon_i(\omega_p/N\gamma)^{1/2}$. Is it possible to have simultaneously a high efficiency, a small energy spread, a small emittance, and a high gradient? The relationships among these quantities are dependent on $\chi(r)$, the driving beam radial profile (58). For $\chi(r)$ equal to a constant, the efficiency can be increased without increasing the energy spread, but the emittance cannot be reduced sufficiently without increasing the plasma frequency (58) and the accompanying emittance blowup due to multiple Coulomb scattering (59). While the importance of the radial profile of the driving beam is clear, to date there has not been a satisfactory positive answer to the question posed above.

The gradient, focusing strength, efficiency, and transformer ratio depend on the intensity and density distribution of the driving beam; this beam must be generated and propagated stably through the plasma. With the development of photoemitters combined with rapid acceleration, beams of the intensity needed may soon be feasible, but the problems of control and manipulation of the radial and longitudinal distributions have not been considered in detail (60). Beam stability is being studied with theory and simulations. The conclusions are (a) an ultrashort, relativistic beam can be longitudinally stable until it has lost more than 70% of its energy to the plasma (61), (b) transverse instabilities can be controlled with the transverse plasma temperature (61), and (c) self-focusing could make control of the radial profile difficult, particularly for the long driving bunches needed for a high transformer ratio (62). There has been progress in understanding, but more work is needed.

In the beatwave accelerator the driving beam is replaced by two laser beams differing in frequency by $\omega_p$ (55, 56). A single-frequency, plane wave cannot cause a net drift along the direction of propagation, but with two frequencies there is a modulation of the wave amplitude that results in a net force called the ponderamotive force. The plasma density perturbation grows linearly during the laser pulse, but can saturate as a result of secondary plasma modes (63) or relativistic effects (64). In the linear regime the density and the electric fields [for one particular $\chi(r)$] at the end of the pulse are given by Equations 9 and 10, with the replacement (58)

$$N \to \frac{W\omega_p^2\tau}{4\omega^2 m_e c^2}.$$

12.

The plasma frequency, the spot radius, and the laser pulse length ($\tau$), power ($W$), and average frequency ($\omega$) determine the gradient and focusing strength. With the laser pulse short enough to prevent exciting plasma modes other than the beatwave and the restriction that saturation be avoided, a rough rule of thumb is $\langle E_z(\text{V/cm})\rangle_{\text{max}} \approx [n_0(\text{cm}^{-3})]^{1/2}$.

The length of an acceleration section can be limited by physical optics, phase slippage, or pump depletion (loss of laser energy to the plasma). At low laser power the spot size is diffraction limited, and the section length should be $R^2\omega/c$, twice the Rayleigh length, for optimal use of the laser. At higher power a laser self-focuses in the plasma (65, 66), and in simulations a single-frequency, self-focused laser beam is stable when its radius is of order $c/\omega_p$ (61). Self-focusing in the beatwave system is being examined; these studies should include the feasibility of varying the radial distribution for good efficiency.

The beatwave propagates with a phase velocity equal to the velocity of an electron with $\gamma_p = \omega/\omega_p$ (55); the phase slippage is given by Equation 11 with $\gamma_i = \gamma_f = \gamma_p$. Increasing $\gamma_p$ increases the section length and the energy gain per section but decreases the gradient. Pump depletion is not a limit for the beatwave accelerator (67), but the surfatron, which avoids phase slippage (68), is limited by pump depletion to section lengths roughly equal to those of the beatwave accelerator.

The dominant factor in the overall efficiency of the beatwave accelerator is the laser efficiency. At 10-$\mu$m wavelength, a $CO_2$ laser can be used, and as discussed in conjunction with near-field accelerators, the short pulse efficiency of this laser is poor. Efficiencies of 7–10% are possible with KrF lasers, but these lasers are nonstorage lasers with a pump time of 100 to 1000 ns and an upper state lifetime of 5 ns (41). Therefore, energy must be extracted throughout the pump period, and the short pulse efficiency is not good. NdYag lasers are capable of short pulses, but these lasers run at a low repetition rate. At the present time a laser appropriate for a very high energy collider is not available.

The possible gradient of a plasma accelerator is unmatched by any other novel accelerator concept. Can this potential be realized? It is too early to answer definitively (69). Efficiency and control are factors outside the model above and to a significant degree outside the realm of simulations. In both accelerators efficiency is related to the driving beam profile, and crucial information about self-focusing and the stability of the profile is missing. The plasma and driving beam must be controlled to the degree required by the small focal spots; the solution of this problem depends on experimentation to give guiding tolerances and technological advances.

The first plasma wakefield experiments are to begin shortly at Argonne National Laboratory; these experiments are being performed in collaboration with the University of Wisconsin and the University of California, Los Angeles. The goals are the observation of plasma waves driven by an electron beam and acceleration of particles by those waves (70).

In the first successful observation of the beatwave, a $CO_2$ laser emitting light at 10.6 and 9.6 $\mu$m excited a 2 mm long hydrogen plasma of density

$n_0 \approx 10^{17}$ cm$^{-3}$ (71). The beatwave was diagnosed using Thomson scattering of ruby laser light. From the data it was concluded that the accelerating electric field was between 0.3 and 1 GeV/m and that the plasma density modulation was greater than 3%. In this experiment the beatwave saturated as a result of coupling to secondary electrostatic modes (63). These secondary modes are associated with a density ripple of the plasma ions and can be avoided with a laser pulse sufficiently short that the ions cannot move during the pulse (63).

The next goals of this group are the demonstration of acceleration of injected electrons and a study of interactions between these particles and the beatwave. To this end they are building a short pulse length laser to eliminate the saturation due to mode coupling, a small linear accelerator serving as the electron source, and a spectrometer for measuring energy changes. The experimental work with this system is beginning (62).

Beatwaves have also been inferred in a second experiment from the acceleration of injected particles (73). This experiment also used a $CO_2$ laser emitting at 9.6 and 10.6 $\mu$m, and both dry air and hydrogen gas were used. Typical plasma densities were $1 \times 10^{17}$ cm$^{-3}$. Electrons from a laser-illuminated Al target were injected into the plasma, and the preliminary results are that electrons injected at 0.6 MeV were accelerated to over 2 MeV in a 2 mm long plasma. This corresponds to a gradient of over 0.7 GeV/m.

A third experiment designed to observe beatwaves excited by 1-$\mu$m wavelength laser was unsuccessful because of a coincidental overlap with a rotational Raman spectrum line in $N_2$ (74). Time limitations have prevented repeating the experiment. An unanticipated outcome of the experiment was the observation of uniform plasmas produced by multiphoton ionization. To produce the beatwave, the laser frequency difference must match the plasma frequency, and the production of uniform plasmas has been a potential practical barrier. Multiphoton ionization is a promising direction. These experimental results are an encouraging beginning.

## COLLECTIVE IMPLOSION ACCELERATOR

The "collective implosion accelerator" concept (75), shown schematically in Figure 5, exploits the large electrostatic fields produced by an intense, relativistic beam traveling in a low density gas. The electron charging beam ionizes the gas, and the ionization electrons are ejected from the beam. Ion density builds up at a rate proportional to the beam and gas densities and can be controlled with the gas density. The beam has a sharp trailing end and leaves behind itself an ion column wake.

The energy loss of the charging beam goes into electrostatic field energy and kinetic energy flux of electrons hitting the wall. The ratio of electrostatic field energy to energy loss is $f_e/2$ where $f_e$ is the fractional neutralization, $f_e$ = (ion linear charge density at the tail of the beam)/(electron beam linear charge density). For a typical set of parameters (an electron beam with a current of 10 kA, a pulse length of 50 ns, and a gas density between $10^{13}$ and $10^{14}$ cm$^{-3}$) the deceleration rate is 20 keV/m, $f_e \approx 0.5$, and the electrostatic field energy is one quarter the energy loss.

A picosecond laser pulse and the beam being accelerated follow the charging pulse. Before the ion column can disperse, the laser photoionizes a region of gas of large radial extent. The photoelectrons flow toward the ion column, which produces an accelerating field with a gradient of several hundred MeV/m. The ion dispersal time is of the order of 1 ns. In addition to fixing the laser pulse delay, it specifies the sharpness of the electron beam tail; this, together with the rate at which the tail disperses, fixes the length of an acceleration section. Sections can be several hundred meters long before the tail needs to be resharpened.

Additional simplifying features include the following: the inrush of electrons is focusing as well as accelerating, the picosecond laser pulse energy requirements are modest, a preceding laser pulse can guide the charging beam (76), and the charging beam is self-focused (77). While all other aspects of the collective implosion concept have not been tested experimentally, the latter two features are used in routine operation of

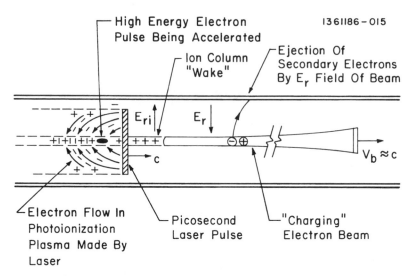

*Figure 5*   Schematic of the collective implosion accelerator concept (75). (Reprinted with permission of R. Briggs.)

high current electron accelerators. A disadvantage for linear colliders is that positron acceleration is not "natural," although a later phase of the plasma oscillation may be suitable. Overall, this accelerator concept combines a high gradient, a long section length, a high efficiency, and simplicity.

## SUMMARY AND CONCLUSION

Advanced accelerator concepts and their application to high energy physics have been reviewed. While the concepts selected do not make an all-inclusive list, they are those receiving substantial attention in the accelerator community. To be cost-effective for particle physics, small emittance beams and high acceleration efficiency are required. This is having a strong impact and is a stringent criterion for evaluation.

Experiments aimed at proving essential features are in progress or beginning soon. These experiments include the FEL power source, the wakefield transformer, the switched power accelerator, the beatwave accelerator, and the plasma wakefield accelerator. In addition, computer simulations are exploring issues related to using these concepts in complete accelerator systems.

Concepts that look promising in the proof-of-principle and simulation stages will soon be ready for detailed engineering. Prototype design to develop this engineering will play an increasingly important role in meeting the challenge of $e^+e^-$ collisions at very high energies.

ACKNOWLEDGMENTS

Papers by Perry Wilson (6), Wolfgang Schnell (34), and Ron Ruth and Pisin Chen (58) have been important to my understanding of advanced accelerator concepts. Chan Joshi reviewed the section on plasma accelerators. This paper was written at Fermilab, which is operated by Universities Research Association, Inc. under contract with the United States Department of Energy. Work at Cornell is supported in part by the National Science Foundation.

*Literature Cited*

1. Jackson, J., ed. *Superconducting Super Collider Conceptual Design*, SSC-SR-2020. Berkeley: SSC Central Design Group (1986)
2. Tigner, M. *Nuovo Cimento* 37: 1228–31 (1965)
3. Amaldi, U. *Phys. Lett.* B61: 313–15 (1976)
4. Erickson, R., ed. *SLC Design Handbook*. Stanford, Cal: Stanford Lin. Accel. Cntr. (1984)
5. Palmer, R. B. *SLAC-PUB-3678*. Stanford: SLAC (1985)
6. Wilson, P. B. *SLAC-PUB-3674*. Stanford: SLAC (1985)

6a. Wilson, P. B. *AIP Conf. Proc.* 130: 560–97 (1985)
7. Hollebeek, R. *Nucl. Instrum. Methods* 184: 333–47 (1981)
8. Fawley, W. M., Lee, E. P. *UCID-18584.* Livermore, Calif: Lawrence Livermore Lab. (1980)
9. Bassetti, M., Gygi-Hanney, M. *LEP Note 221.* Geneva: CERN (1980)
10. Himel, T., Siegrist, J. *AIP Conf. Proc.* 130: 602–8 (1985)
11. Palmer, R. B. *SLAC-PUB-3688.* Stanford: SLAC (1985)
12. Chen, P. *SLAC-PUB-3823.* Stanford: SLAC (1985)
13. Brown, K. L., Servranckx, R. V. *AIP Conf. Proc.* 127: 62–138 (1985)
14. Bane, K. L. F. *IEEE Trans. Nucl. Sci.* NS-32: 2389–91 (1985)
15. Balakin, V., et al. In *Proc. 12th Int. Conf. on High Energy Accel.*, ed. F. T. Cole, R. Donaldson, p. 119. Batavia, Ill: Fermi Natl. Accel. Lab. (1983)
16. Sundelin, R. M. *IEEE Trans. Nucl. Sci.* NS-32: 3570–73 (1985)
17. Kneisel, P. *AIP Conf. Proc.* 156: 145–60 (1987)
18. Peil, H. *IEEE Trans. Nucl. Sci.* NS-32: 3565–69 (1985)
19. Sundelin, R. M. *CLNS-85/709.* Cornell Univ., NY (1985)
20. Amaldi, U., et al. *CERN/EF 86-8.* Geneva: CERN (1986)
21. Schnell, W. *AIP Conf. Proc.* 156: 17–36 (1987)
22. Hopkins, D. B., Kuenning, R. W. *IEEE Trans. Nucl. Sci.* NS-32: 3476–80 (1985)
23. Lee, T. G., et al. *SLAC-PUB-3619.* Stanford: SLAC (1985)
24. Eppley, K. R., et al. *SLAC-PUB-4221.* Stanford: SLAC (1987)
25. Granatstein, V. L., et al. *IEEE Trans. Nucl. Sci.* NS-32: 2957–59 (1985)
26. Sinclair, C. *AIP Conf. Proc.* 156: 298–312 (1987)
27. Yoshioka, M. *AIP Conf. Proc.* 156: 313–21 (1987)
28. Farkas, Z. D. *SLAC-PUB-3694.* Stanford: SLAC (1985)
29. Sessler, A. M. *AIP Conf. Proc.* 91: 154–59 (1982)
30. Birx, D. L., et al. *IEEE Trans. Nucl. Sci.* NS-32: 2743–47 (1985)
31. Orzechowski, T. J., et al. *IEEE J. Quant. Electron.* QE-21: 831–44 (1985)
32. Orzechowski, T. J., et al. *Phys. Rev. Lett.* 57: 2172–75 (1986)
33. Wurtele, J., Sessler, A. M. *AIP Conf. Proc.* 156: 322–34 (1987)
34. Schnell, W. *CERN-LEP-RF/86-14.* Geneva: CERN (1986)
35. Marks, R. *LBL-20918/UC-34A.* Berkeley: Lawrence Berkeley Lab. (1985)
36. Amaldi, U., Pellegrini, C. *CERN CLIC Note 16.* Geneva: CERN (1986)
37. Palmer, R. B. *Part. Accel.* 11: 81–90 (1980)
38. Kroll, N. *AIP Conf. Proc.* 130: 253–70 (1985)
39. Palmer, R. B., et al. *AIP Conf. Proc.* 156: 234–52 (1987)
40. Slater, J. *AIP Conf. Proc.* 130: 505–17 (1985)
41. Lowenthal, D., Slater, J. *AIP Conf. Proc.* 130: 518–43 (1985)
42. Fernow, R. C. *AIP Conf. Proc.* 156: 37–45 (1987)
43. Palmer, R. B. *SLAC-PUB-3890-Rev.* Stanford: SLAC (1986)
44. Ruth, R. D., et al. *Part. Accel.* 17: 171–89 (1985)
45. Voss, G.-A., Weiland, T. *DESY M82-10.* Hamburg, DESY (1982)
46. Voss, G.-A., Weiland, T. *DESY 82-074.* Hamburg: DESY (1982)
47. Weiland, T., Willeke, F. See Ref. 15, p. 457
48. Weiland, T. *IEEE Trans. Nucl. Sci.* NS-32: 3471–75 (1985)
49. Bialowens, W., et al. *AIP Conf. Proc.* 156: 266–80 (1987)
50. Ruggiero, A. G. *AIP Conf. Proc.* 130: 458–74 (1985)
51. Perevedentsev, E. A., Skrinsky, A. N. In *Proc. 6th All-Union Conf. on Charged Particle Accel.,* p. 272 (1978)
52. Ruggiero, A. G., et al. *AIP Conf. Proc.* 156: 247–65 (1987)
53. Willis, W. *AIP Conf. Proc.* 130: 421–34 (1985)
54. Aronson, S. *AIP Conf. Proc.* 156: 283–97 (1987)
55. Tajima, T., Dawson, J. M. *Phys. Rev. Lett.* 43: 267–70 (1979)
56. Joshi, C., et al. *Nature* 311: 525–29 (1984)
57. Chen, P., et al. *Phys. Rev. Lett.* 54: 693–96 (1985)
58. Ruth, R. D., Chen, P. *SLAC-PUB-3906.* Stanford: SLAC (1986)
59. Montague, B. W., Schnell, W. *AIP Conf. Proc.* 130: 146–55 (1985)
60. Fraser, J. S., et al. *AIP Conf. Proc.* 130: 598–601 (1985)
61. Bingham, R. *AIP Conf. Proc.* 156: 389–94 (1987)
62. Joshi, C., et al. *AIP Conf. Proc.* 156: 71–104 (1987)
63. Darrow, C. et al. *Phys. Rev. Lett.* 56: 2629–32 (1986)
64. Rosenbluth, M. N., Liu, C. S. *Phys. Rev. Lett.* 29: 701–5 (1972)
65. Max, C. E., et al. *Phys. Rev. Lett.* 33: 209–12 (1974)
66. Forslund, D. W., et al. *Phys. Rev. Lett.* 54: 558–61 (1985)

67. Horton, W., Tajima, T. *AIP Conf. Proc.* 130: 179–84 (1985)
68. Katsouleas, T., Dawson, J. M. *Phys. Rev. Lett.* 51: 392–95 (1983)
69. Lawson, J. D. *AIP Conf. Proc.* 130: 120–29 (1985)
70. Rosenzweig, J., et al. *AIP Conf. Proc.* 156: 231–46 (1987)
71. Clayton, C. E., et al. *Phys. Rev. Lett.* 54: 2343–50 (1985)
72. Deleted in proof
73. Martin, F., et al. *AIP Conf. Proc.* 156: 121–33 (1987)
74. Dangor, A. E., et al. *AIP Conf. Proc.* 156: 112–20 (1987)
75. Briggs, R. J. *Phys. Rev. Lett.* 54: 2588–91 (1985)
76. Martin, W. E., et al. *Phys. Rev. Lett.* 54: 685–88 (1985)
77. Struve, K. W., et al. In *Proc. 5th Int. Conf. on High Power Part. Beams,* p. 408. Livermore, Cal: Livermore Natl. Lab. (1984)

*Ann. Rev. Nucl. Part. Sci. 1987. 37: 267–92*
*Copyright © 1987 by Annual Reviews Inc. All rights reserved*

# RELATIVISTIC EQUATION FOR NUCLEON-NUCLEUS SCATTERING

## Stephen J. Wallace

Department of Physics and Astronomy, University of Maryland, College Park, Maryland 20742

CONTENTS

## INTRODUCTION

Developments over the past 15 years have given credibility to a relativistic description of nucleon motion in the mean field provided by a nucleus.

267

0163–8998/87/1201–0267$02.00

The nucleon motion is described using the Dirac equation,

$$(E\gamma^0 - \gamma \cdot \mathbf{p} - M - \hat{U})\psi = 0, \qquad\qquad 1.$$

and the nucleon's interaction with the nucleus, $\hat{U} \approx S + \gamma^0 V$, is described mainly by a combination of an attractive scalar potential, $S$, and a repulsive vector potential, $V$. The use of the Dirac equation is motivated by the size of these potentials, each of which is about one third the nucleon mass. The traditional justification for the use of the Schroedinger equation is undermined when the interactions are so strong.

Meson-exchange models of the nucleon-nucleon (NN) interaction provide a qualitative explanation of the nuclear mean field. Exchanges of a scalar meson ($\sigma$) and a vector meson ($\omega$), both of isospin zero, yield scalar attraction and vector repulsion. However, this connection of the NN interaction to the nuclear mean field leaves much to be desired. The assumed 500-MeV scalar meson is not observed in nature as a particle, and meson-nucleon coupling constants enter essentially as phenomenological parameters.

Several years ago, the origin of the scalar attraction and vector repulsion was connected directly to the NN interaction by considering the optical potential for nucleon-nucleus scattering in a relativistic impulse approximation (1). The optical potential is the extension of the nuclear mean field to scattering energies, and the term relativistic impulse approximation refers to the prediction of the optical potential in the Dirac equation from the nuclear density and NN amplitudes for free-space scattering. Relativistic Fermi amplitudes constructed from phase-shift fits to NN experimental data were found to exhibit the characteristic scalar attraction and vector repulsion in isospin-zero states (2). Moreover, the optical potential calculated from the Fermi amplitudes and the nuclear density was found to provide an excellent description of spin observables in proton-nucleus elastic scattering (3, 4).

In this review, primary emphasis is placed on the relativistic approach to proton scattering based on NN amplitudes. Much of the current interest in relativistic nuclear physics stems from this approach and its successful predictions of spin observables in elastic proton scattering. Moreover, these aspects of the relativistic approach make contact with a large body of experimental work on nucleon-nucleus scattering and they can be tested in various ways.

There is a close connection between the relativistic optical potential and the mean field potential responsible for nuclear binding. To emphasize this connection, a unified discussion of relativistic approaches to the nuclear mean field and to the nucleon optical potential is presented. In both cases the potential used in the Dirac equation may be expressed as a product of

(a) a Feynman scattering operator, $\hat{M}$, for the NN interaction and (b) a nuclear density matrix $\hat{\rho}$, as follows,

$$\hat{U}(\mathbf{p}, \mathbf{q}) = -\frac{1}{4} \, \text{Tr}_2 \left[ (2\pi)^{-3} \int d^3k \, \hat{M}\left(\mathbf{p}, \mathbf{k} - \frac{1}{2}\mathbf{q} \rightarrow \mathbf{p} - \mathbf{q}, \mathbf{k} + \frac{1}{2}\mathbf{q}\right)\hat{\rho}(\mathbf{k}, \mathbf{q}) \right],$$

2.

where $\mathbf{p}$ is the nucleon momentum and $\mathbf{q}$ is the momentum transfer in the rest frame of the nucleus. The Feynman scattering operator $\hat{M}$ is expressed as an operator in the full Dirac space of two nucleons. Conventionally, particle 2 is designated the target nucleon and the trace is over Dirac indices of particle 2. The nuclear density matrix is an operator in the Dirac space of particle 2. It is determined by Dirac nuclear wave functions, $\psi_\alpha$, for the occupied single-particle states as follows,

$$\hat{\rho}(\mathbf{k}, \mathbf{q}) = \sum_\alpha \psi_\alpha\left(\mathbf{k} - \frac{1}{2}\mathbf{q}\right)\bar{\psi}_\alpha\left(\mathbf{k} + \frac{1}{2}\mathbf{q}\right).$$

3.

Only salient features of the relativistic mean field models are discussed because extensive reviews are available in two recent books. Serot & Walecka (5) provide a detailed account of the quantum hadrodynamics model initiated by Walecka (6). Celenza & Shakin (7) and Anastasio et al (8) review the relativistic Brueckner-Hartree-Fock approach initiated by Anastasio et al (9, 10). Recent conference proceedings provide further information on many of the developments in relativistic nuclear physics. Green (11) traces the history of the development of the Dirac-Hartree model, which was initiated by Miller (12) and Miller & Green (13). Clark et al (14) and Clark (15) review the Dirac phenomenology for proton scattering, which was initiated by Clark et al (16). Wallace (17, 18) reviews the development of the relativistic impulse approximation, IA1, which was initiated in the work of McNeil et al (1, 2) and the generalized impulse approximation, IA2, developed by Tjon & Wallace (19, 20) based on complete sets of Lorentz-invariant amplitudes (21, 22).

## RELATIVISTIC NUCLEAR STRUCTURE

Three basic motivations seem clear for pursuing relativistic models of nuclear structure. (a) Spin-orbit interactions are at the heart of the nuclear shell model and these may be understood rather simply in a relativistic Dirac-Hartree approach. (b) Nuclear saturation is a basic consequence of the nuclear force that causes all nuclei to have about the same density and binding energy per particle. A simple and intrinsically relativistic mechanism for saturation is found in a relativistic mean field approximation. Indeed, there is mounting evidence that the relativistic saturation mechanism may be predicted from meson-exchange models of the nucleon-nucleon (NN) interaction in nuclei. (c) Finally, there is a purely theoretical motivation. A relativistic description of nuclei is desirable, for example, to obtain wave functions suitable for the analysis of high momentum transfer reactions.

## Dirac-Hartree Shell Model

Dirac-Hartree calculations for the nuclear shell model were first performed by Miller (12) and Miller & Green (13). In this approach, single-particle wave functions are determined self-consistently by solving the Dirac equation with potentials generated by folding $\sigma$, $\omega$, and other meson-exchange interactions with the nuclear density. The $\pi$ meson plays no role because nucleon-exchange processes (Fock terms) are omitted. Miller & Green (13) and Miller (23) used generalized one-boson-exchange (OBE) interactions of the same form as those used earlier to fit NN scattering potentials. Phenomenological coupling constants were readjusted to obtain the desired nuclear shell model properties. Vertex form factors were included to regulate high momentum behavior. In subsequent work by Brockmann (24) and Horowitz & Serot (25), the analysis has been repeated with different assumptions regarding the meson-nucleon interactions and form factors.

Typically, these analyses yield a scalar potential $S \approx -400$ MeV and a vector potential $V \approx +300$ MeV inside the nucleus. Although scalar and vector potentials are both quite strong, they essentially cancel one another in the central part of the equivalent Schroedinger potential, $V_c \approx S + (E/M)V + (S^2 - V^2)/(2M)$. Therefore nuclear binding energies are comparable to those of the nonrelativistic shell model based on $V_c \approx -50$ MeV.

The most significant success of the Dirac-Hartree approach is the ability to fit single-particle energy levels, particularly the spin-orbit splittings that are essential to the nuclear shell model (23, 25). When opposing scalar and vector potentials are used in the Dirac equation, the equivalent Schroedinger-Pauli equation contains a spin-orbit interaction term, $V_{LS} \approx 2Mr^{-1}(S' - V')/(2M + S - V)$, where a prime denotes a radial derivative. This can be understood classically in terms of the Thomas spin-orbit interaction familiar from atomic physics. For atoms, the scalar potential, $S$, is zero and the vector potential, $V$, is attractive. The spin-orbit interaction raises the energy of electrons with angular momentum $j = \ell + \frac{1}{2}$. For nuclei, $S$ is attractive and $V$ is repulsive. The spin-orbit interaction lowers the energy of nucleons with angular momentum $j = \ell + \frac{1}{2}$.

Proton "vector" densities for closed-shell nuclei are in good agreement with electron scattering data (23, 25). Typically one has three distinct densities: scalar, vector, and tensor. Corresponding nuclear form factors are defined by Fourier transforms of these densities as follows,

$$\rho_S(q) = 4\pi \int_0^\infty dr\, r^2 j_0(qr) \left\{ \sum_\alpha \frac{2j+1}{4\pi r^2} [G_\alpha^2(r) - F_\alpha^2(r)] \right\}, \qquad 4.$$

$$\rho_V(q) = 4\pi \int_0^\infty dr\, r^2 j_0(qr) \left\{ \sum_\alpha \frac{2j+1}{4\pi r^2} [G_\alpha^2(r) + F_\alpha^2(r)] \right\}, \qquad 5.$$

$$\rho_T(q) = -4\pi m q^{-1} \int_0^\infty dr\, r^2 j_1(qr) \left\{ \sum_\alpha \frac{2j+1}{4\pi r^2} 4G_\alpha(r) F_\alpha(r) \right\}, \qquad 6.$$

where $G_\alpha(r)$ and $F_\alpha(r)$ are upper and lower components of Dirac radial wave functions, respectively, and $\alpha$ denotes quantum numbers of the occupied states. These densities are related to the density matrix of Equation 3 as follows,

$$\hat{\rho}(q) \equiv (2\pi)^{-3} \int d^3k\, \hat{\rho}(\mathbf{k}, \mathbf{q}) = \rho_S(q) + \gamma_2^0 \rho_V(q) - \frac{\boldsymbol{\alpha}_2 \cdot \mathbf{q}}{2m} \rho_T(q). \qquad 7.$$

In the Hartree model, direct meson-exchange Feynman amplitudes, $\hat{M}_H(q)$, are used and these depend only on the momentum transfer, $\mathbf{q}$. Therefore Equation 2 simplifies considerably because the NN amplitude factors out of the integral and one has

$$\hat{U}(q) = -\frac{1}{4} \operatorname{Tr}_2[\hat{M}_H(q) \hat{\rho}(q)]. \qquad 8.$$

Extension of the analysis by Miller (26) to include Fock terms encountered difficulties because of pion-exchange contributions. If one assumes a pseudoscalar $\pi$N coupling, then a very large correction results from the leading-order Fock term.

## Walecka Model

A mean field approximation for infinite nuclear matter was developed by Walecka (6) starting from a renormalizable quantum field theory of nucleons interacting with $\sigma$ and $\omega$ mesons. The physics is similar to that of the Dirac-Hartree model, but calculations are much simpler. Only plane wave functions need be considered and only forward scatterings ($\mathbf{q} = 0$) can occur. The scalar and vector potentials determined from Equation 8 are $S = -(g_\sigma^2/m_\sigma^2)\rho_S$ and $V = (g_\omega^2/m_\omega^2)\rho_V$, where $g_\sigma$ and $g_\omega$ are coupling constants and $m_\sigma$ and $m_\omega$ are meson masses. The vector density $\rho_V = 2k_F^3/(3\pi^2)$ is parameterized by the Fermi momentum $k_F$ since it is equal to the ground-state expectation value of the baryon density. Scalar density $\rho_S$ is determined self-consistently from

$$\rho_S = 2\pi^{-2} \int_0^{k_F} dk \ k^2 M^*/(M^{*2}+k^2)^{1/2} \qquad\qquad 9.$$

and $M^* = M-(g_\sigma^2/m_\sigma^2)\rho_S$. $M^*$ is the effective mass of a nucleon in the nuclear medium.

This model exhibits an intrinsically relativistic mechanism for nuclear saturation. At low density, the scalar attraction, $S$, dominates. As the density of the system increases, the attraction weakens relative to the repulsive vector potential, $V$. The origin of this effect is in Equation 9, from which one can show that $\rho_S/\rho_V$ and $M^*/M$ decrease toward zero as $k_F$ increases. At high density, the vector repulsion overcomes the scalar attraction. Consequently there is a saturation point at which the minimum energy is obtained as a function of density. Such a minimum is present in nonrelativistic models only because two-body correlations act to keep nucleons outside the range of the short-range repulsive force until the density becomes high. In the Walecka model, the minimum does not depend on correlations.

By making phenomenological choices for the two parameters of the Walecka model, $g_\sigma^2/m_\sigma^2$ and $g_\omega^2/m_\omega^2$, the nuclear medium saturates at 16-MeV/nucleon binding energy and density $\rho_{nm} \approx 0.16$ fm$^{-3}$ in agreement with inferences from the semiempirical mass formula. The effective nucleon mass turns out to be, $M^* \approx 0.6$ M, which signals the importance of relativistic effects in normal nuclear matter.

A large literature exists concerning development of the quantum hadrodynamics model, which stems from the work of Walecka and collaborators (5). In extending the mean field approximation to include Fock terms, Horowitz & Serot (27) encountered difficulties related to those found by Miller for finite nuclei based on pseudoscalar $\pi$N coupling. For pseudovector $\pi$N coupling, the Hartree-Fock analysis in nuclear matter produces results similar to those of the Hartree analysis (28).

## Relativistic Breuckner-Hartree-Fock Model

In 1980, Shakin and collaborators (9) published a relativistic Brueckner-Hartree-Fock model for infinite nuclear matter. The main ingredient of this approach is the relativistic NN amplitude, $\hat{M}_{BHF}$, in the nuclear medium where nucleons have an effective mass $M^*$. Two key points are (a) effects of nucleon correlations are included by solving an integral equation for matrix elements of $\hat{M}_{BHF}$ in $M^*$ spinor states (10), and (b) the one-boson-exchange potential is taken from a meson-exchange model that explains nucleon-nucleon scattering in the 0–300-MeV region (29). The effective mass $M^*$ is determined self-consistently. The principal result of this approach is that nuclear saturation is explained as a relativistic effect of the same character as is observed in the mean field approximation of Walecka. In this case all parameters are determined by the NN interaction.

Some technical approximations used in the analysis make the accuracy of the original calculation difficult to assess. For example, effects of the self-consistent mass $M^*$ were not included in the intermediate states of the NN integral equation. Also, couplings to negative-energy intermediate states were neglected. In principle, negative-energy couplings should have increased importance because $M^*$ is smaller than $M$.

The first of these approximations is eliminated in recent analyses by Brockmann & Machleidt (30) and Malfliet & ter Haar (31). Each of these analyses concludes that a simple explanation of nuclear saturation is attainable in lowest-order Brueckner-Hartree-Fock calculations. The compressibility is considerably improved over that of the Walecka model. Recently Horowitz & Serot (32) have pointed out that the explanation of nuclear saturation is likely to be fragile. The result can be changed significantly by including vacuum polarization corrections and it is sensitive to the meson-nucleon form factors assumed. Moreover, there is unknown sensitivity to the neglect of negative-energy intermediate states of the NN interaction.

While further work is warranted, recent progress establishes that the relativistic saturation mechanism is connected fairly directly to the meson-exchange description of the NN interaction.

# RELATIVISTIC NUCLEAR SCATTERING

Three motivations seem clear for relativistic models of nuclear scattering. (a) Spin observables are simply and accurately predicted in the relativistic approach. (b) Intermediate-energy scattering involves relativistic beams and it is very desirable to use a relativistic equation to incorporate kinematics correctly. (c) There is a theoretical motivation to unify the description of the optical potential with that of the nuclear mean field that is responsible for binding.

These motivations are not new. Until recently, theorists have opted to use the traditional Schroedinger-Pauli spin formalism for proton-nucleus scattering (33, 34). Relativistic approaches were held back by the lack of an evident starting approximation. A phenomenological approach based on the Dirac equation was introduced first (16). More recently, a relativistic impulse approximation has been found that does provide a useful starting point.

## Dirac Phenomenology

In parallel with the developments in relativistic nuclear structure, a Dirac phenomenology for proton-nucleus scattering was developed by Clark et al (16) and Arnold et al (35). The Dirac equation is solved at scattering energies using three-parameter Woods-Saxon potentials for real and imaginary parts of both the scalar and vector potentials, $S$ and $V$. Thus, four Woods-Saxon terms involving twelve fitting parameters are used to fit the proton-nucleus scattering data. In principle, there is an equivalence between the $S$ and $V$ phenomenology based on the Dirac equation and a phenomenology based on the Schroedinger equation with local central and spin-orbit potentials. Given the potentials of the Dirac equation, one may calculate the equivalent central and spin-orbit potentials to be used in the Schroedinger equation such that scattering amplitudes are identical (15).

Dirac $S$ and $V$ phenomenology determines that real parts of scalar and vector potentials are attractive and repulsive, respectively, in general accord with the Hartree model (14). For proton scattering near 200 MeV, Dirac phenomenology yields potentials that have a radial variation similar to that of the nuclear density, whereas the Schroedinger phenomenology requires complicated double Woods-Saxon shapes. Energy dependence of the Dirac potentials is simpler than that of the Schroedinger potentials.

The most significant result of Dirac phenomenology is in the description of spin observables. Scalar and vector potentials that fit the data on cross section $\sigma$ and analyzing power $A_Y$ for 500-MeV protons scattering elastically from $^{40}$Ca successfully predict the spin rotation function $Q$ (36). The $A_Y$ and $Q$ data at 500 MeV are poorly described by the Schroedinger equation using the impulse approximation of Kerman, McManus & Thaler (33) to predict the optical potentials theoretically (37). Qualitative analysis indicates that the nonrelativistic impulse approximation does not link $A_Y$ and $Q$ correctly at 500 MeV (38). Although a dozen parameters are varied to achieve the fit, Dirac $S$ and $V$ phenomenology succeeds in linking the spin observables $A_Y$ and $Q$ in just about the right way.

Subsequent analysis by Clark (15) and Kobos et al (39) showed that the Dirac phenomenology could equally well be based on scalar and tensor ($S$ and $T$) interactions in the Dirac equation, or on other possibilities. The $S$ and $T$ phenomenology is equivalent to the $S$ and $V$ one but it does not so successfully predict the $Q$ data when parameters are chosen to give a fit to just the cross section and analyzing power (39). Although one may infer that various equivalences make it difficult to arrive at definite conclusions from Dirac phenomenology, the $S$ and $V$ phenomenology seems to be most successful at giving a simple description of proton-nucleus scattering data (15). This conclusion is strengthened considerably by the relativistic impulse approximation.

## Relativistic Impulse Approximation: IA1

In 1983, a relativistic impulse approximation was developed by McNeil et al (1) using free NN scattering amplitudes determined by McNeil et al (2). In this approach, which is designated IA1, the Feynman scattering operator for NN scattering is expanded in terms of the Fermi covariants: $K_1 = 1$ (scalar), $K_2 = \gamma_1 \cdot \gamma_2$ (vector), $K_3 = \sigma_1^{\mu\nu}\sigma_{2\mu\nu}$ (tensor), $K_4 = \gamma_1^5\gamma_2^5$ (pseudoscalar), and $K_5 = \gamma_1^5\gamma_2^5\gamma_1 \cdot \gamma_2$ (axial vector), as follows,

$$\hat{M}(\mathbf{p}, \mathbf{q}) = -\kappa \sum_{i=1}^{5} F_i(s, t, u)K_i. \qquad 10.$$

Thus $F_1$ is a scalar amplitude, $F_2$ is a vector amplitude, and so on. Arguments $s$, $t$, and $u$ are the Mandelstam invariants and they may be expressed in terms of the proton momentum $\mathbf{p}$ and the momentum transfer $\mathbf{q}$ for on-mass-shell kinematics. Here $\kappa = -4\pi i p/M$ is a kinematic factor used to relate conventional scattering amplitudes to Lorentz-invariant Feynman amplitudes in a frame where the nucleon has momentum $\mathbf{p}$.

The use of Fermi covariants is natural for incorporating one-boson-exchange contributions to the amplitudes. Scalar amplitude $\kappa F_1$ may contain a scalar meson-exchange contribution, $g_\sigma^2/(q^2 - m_\sigma^2)$, and vector amplitude $\kappa F_2$ may contain a vector meson-exchange contribution, $-g_\omega^2/(q^2 - m_\omega^2)$. However, the Fermi amplitudes are calculated without reference to meson-exchange models. They are directly related to NN helicity amplitudes (20, 40), which are expressible in terms of empirical phase shifts (41). The helicity amplitudes, and therefore the Fermi amplitudes, are known for energies where sufficient NN data exist to constrain tightly the phase-shift analysis. At present, this means energies up to 800 MeV for pp scattering and up to 500 MeV for np scattering. Experimental programs at Los Alamos and Saclay are now extending the NN measurements to 1000 MeV and higher.

In the center-of-mass (cm) frame for NN scattering, conventional helicity amplitudes, $\phi(12 \to 1'2')$, are matrix elements of the Feynman operator $\hat{M}$, i.e.

$$-\kappa^{-1}\bar{u}_{1'}^+, \bar{u}_{2'}^+, \hat{M}(\mathbf{p}, \mathbf{q})u_1^+u_2^+ = \phi(12 \to 1'2'), \qquad 11.$$

where $u_1^+$ is a positive energy Dirac spinor. As illustrated by Equation 11, only the positive-energy matrix elements of the Feynman amplitude $\hat{M}$

are really controlled by the NN scattering data. Equation 10 provides a mechanism for extrapolating the NN data to the full Dirac space of two nucleons, and it makes possible the prediction of matrix elements of $\hat{M}$ in negative-energy states.

The optical potential in the relativistic impulse approximation is determined as in Equation 2 by using the free NN amplitude, $\hat{M}$, of Equation 10. One finds scalar, vector, and tensor terms in the potential,

$$\hat{U}(\mathbf{q}) = S(q) + \gamma^0 V(q) - \frac{\boldsymbol{\alpha} \cdot \mathbf{q}}{m} T(q), \qquad \qquad 12.$$

and each of these is determined in terms of the NN amplitudes and nuclear form factors, i.e. $S(q) = \kappa F_1(s, t, u)\rho_S(q)$, $V(q) = \kappa F_2(s, t, u)\rho_V(q)$, and $T(q) = \kappa F_3(s, t, u)\rho_T(q)$. Because of the factor $\mathbf{q}/m$, the tensor term is much less important than the scalar and vector terms.

The IA1 impulse approximation presumes that the five Fermi amplitudes, fixed in terms of NN scattering data, are sufficient to characterize the optical potential. In order to arrive at local potentials in coordinate space, one must localize the Fock contributions that are implicit in the antisymmetric Fermi amplitudes. Fock contributions involve momentum transfer comparable to the incident proton momentum and therefore they are suppressed by nuclear form factors at high energies (500–800 MeV). At lower proton energy, the Fock contributions are influential. A localization procedure (42a,b) commonly used for nonrelativistic optical potentials may be used to approximate Fock contributions to the optical potential.

The impulse approximation may be understood in terms of four assumptions or approximations to the definition of Equation 2:

1. factorization of the NN amplitude at $\mathbf{k} = 0$ from the nuclear density, as in Equation 8;
2. use of Dirac-Hartree wave functions for the nucleus;
3. use of the five Fermi-invariant NN amplitudes; and
4. localization of the Fock contributions.

One obtains the nucleon self-energy $\hat{U}$ from Equation 2 in all the relativistic approaches. In the Dirac-Hartree and Walecka models, an unsymmetrized one-boson-exchange amplitude, $\hat{M}_H$, is used. In the Hartree-Fock case, the antisymmetrized one-boson-exchange interaction, $\hat{M}_{OBE}$, is used. In the Brueckner-Hartree-Fock case, an antisymmetrized NN amplitude in the nuclear medium, $\hat{M}_{BHF}$, is used and in the relativistic impulse approximation the free NN amplitude, $\hat{M}$, which is also antisymmetrized, is used.

Figure 1 shows the predictions of the impulse approximation for real and imaginary parts of the scalar and vector strengths at nuclear matter

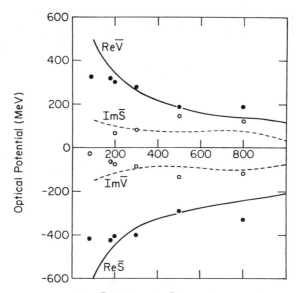

*Figure 1* Isospin average scalar ($\bar{S}$) and vector ($\bar{V}$) potentials at nuclear matter density as functions of the proton kinetic energy in MeV. Solid and dashed lines are the IA1 relativistic impulse approximation predictions (1). Filled and open circles are the phenomenological values for real and imaginary potentials, respectively, found by fitting proton scattering data using the Dirac equation (14).

density, $\rho_{nm} = 0.16$ fm$^{-3}$, as a function of proton energy (1). Assuming equal densities for protons and neutrons, the proton optical potentials $\bar{S}$ and $\bar{V}$ are based on the average of pp and pn amplitudes. The potential strengths of Dirac phenomenology are indicated by solid and filled points. One sees that the impulse approximation generally agrees with the strengths of $\bar{S}$ and $\bar{V}$ found by fitting the proton scattering data phenomenologically. Scalar attraction and vector repulsion emerge directly from the relativistic NN amplitudes, $F_1$ and $F_2$. Moreover, the imaginary potentials, which are absorptive for Im $\bar{V}$ and emissive for Im $\bar{S}$, agree with one of the key features of the $S$ and $V$ phenomenology.

The first applications of the impulse approximation to a finite nucleus were for 500-MeV protons scattering from $^{40}$Ca (3, 4). Figure 2 shows results of Clark et al (4) comparing the relativistic and nonrelativistic impulse approximations. The parameter-free Dirac impulse approximation provides a good description of the data, particularly the spin observables $P$ (polarization) and $Q$ (spin rotation). The parameter-free nonrelativistic impulse approximation fails to predict spin observables correctly (37). Even though the free NN interaction is used in both cases,

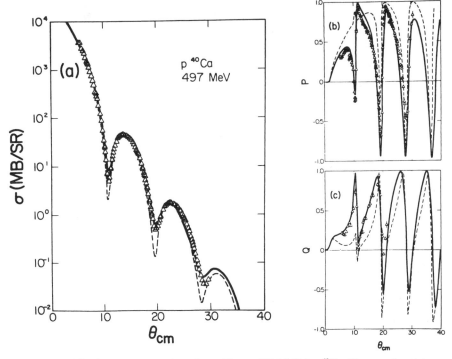

*Figure 2* Elastic proton scattering observables at 500 MeV for $^{40}$Ca. Cross section ($\sigma$), polarization ($P$), and spin rotation function ($Q$) data from (37) and (62) are compared with calculations from (4) of the nonrelativistic impulse approximation (*dashed lines*) and the IA1 relativistic impulse approximation (*solid lines*).

the potentials are not equivalent. By construction, the Dirac potentials are linear functions of density and consequently the equivalent Schroedinger potentials contain all powers of the density.

Subsequent calculations (43, 44) have indicated that the Dirac impulse approximation generally is superior for describing spin observables in the 300–800-MeV energy range, although the improvement over non-relativistic results is usually less obvious than in Figure 2.

## Virtual NN̄ Pair Contribution

One way to understand the basic new ingredient of the Dirac approach is to focus attention on the virtual NN̄ pair contribution that is implicit in solutions of the Dirac equation. Virtual pairs enter because the Dirac optical potential, $\hat{U}$, couples positive-energy and negative-energy amplitudes, $\psi^+$ and $\psi^-$, when the Dirac wave function is expanded as

$\psi = \psi^+ u^+(\mathbf{p}) + \psi^- u^-(\mathbf{p})$, where $u^\pm(\mathbf{p})$ are solutions of the free Dirac equation. Negative-energy amplitudes describe backward-in-time propagation of negative-energy solutions of the free Dirac equation. Equivalently, they describe forward-in-time propagation of positive-energy antinucleon states. When the scattering nucleon is far from the nucleus, negative-energy amplitudes vanish. They are present virtually in the region where the optical potential is strong, i.e. in the immediate vicinity of the nucleus. It is straightforward to eliminate negative-energy amplitudes to obtain an equation for the positive-energy amplitude, $\psi^+$, of the form

$$(E - E_p - U^{++} - U_{\text{pair}})\psi^+ = 0, \qquad\qquad 13.$$

where $U_{\text{pair}} = U^{+-}(E + E_p - U^{--})U^{-+}$ contains the effects of couplings to virtual-pair states. When $U_{\text{pair}} = 0$, this is essentially the Schroedinger-Pauli equation with relativistic kinematics.

Momentum-space calculations by Hynes et al (45, 46) have been performed to elucidate the role of pair effects. When couplings to $\psi^-$ are neglected, i.e. $U_{\text{pair}} = 0$, one recovers the Schroedinger result of Figure 2. When couplings of $\psi^+$ and $\psi^-$ are predicted from large scalar and vector potentials consistent with the impulse approximation, a systematic improvement in the prediction of spin observables $A_Y$ and $Q$ is found. The conclusion of Hynes et al (45) is that the key ingredient underlying the success of the Dirac approach is the implicit inclusion of virtual N̄N pair effects. A similar point was made regarding the success of Dirac phenomenology by Bawin & Jaminon (47).

A simple estimate of $U_{\text{pair}}$ can be made by neglecting $U^{--}$. One finds that $U_{\text{pair}}$ is proportional to the square of the difference between scalar and vector potentials, assuming $V \approx -S$,

$$U_{\text{pair}} \approx \frac{\boldsymbol{\sigma}\cdot\mathbf{p}}{2M} \frac{(S-V)^2}{2M} \frac{\boldsymbol{\sigma}\cdot\mathbf{p}}{2M}. \qquad\qquad 14.$$

Taking $S - V \approx -700$ MeV for nuclear matter, one finds $U_{\text{pair}} \approx 3(\rho/\rho_{\text{nm}})^{8/3}$ MeV, where $\rho_{\text{nm}}$ is nuclear matter density. This repulsive and strongly density-dependent effect accounts for the saturation mechanism that is present in relativistic models and is absent in nonrelativistic models (48). Indeed $U_{\text{pair}}$ is responsible for enhancement of the spin-orbit interaction in the Dirac approach.

# MESON THEORETICAL APPROACH

Scalar attraction and vector repulsion are key ingredients in the Dirac equation and lead to simple explanations of several aspects of nuclear physics, including spin-orbit splittings in the shell model, nuclear saturation, and spin-dependent proton-nucleus scattering. The scalar attraction and vector repulsion are seen to originate in the NN force. However the essential difference from nonrelativistic analyses is the virtual-pair effect. For proton scattering above about 300 MeV, this is predicted by an impulse approximation that uses only positive-energy NN data as input. Clearly such predictions cannot be regarded as being under control. In general, one needs more input to characterize fully the relativistic NN amplitude and the pair contributions to proton-nucleus scattering.

## Pseudovector versus Pseudoscalar $\pi N$ Coupling

At very low energy, overly strong scalar and vector potentials are predicted by the impulse approximation. This may be traced to the implicit use of pseudoscalar pion exchange in the representation of NN amplitudes (20). Isospin-averaged Fock contributions involving one-pion exchange are shown in Figure 3. Solid lines, showing $S$ and $V$ based on pseudoscalar $\pi N$ coupling, exhibit the same divergent behavior at low energy as occurs

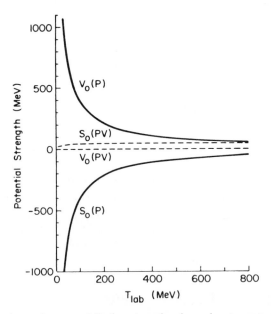

*Figure 3*  One-pion-exchange contributions to real scalar and vector potentials in nuclear matter (20). $P$ refers to pseudoscalar pion coupling and $PV$ refers to pseudovector pion coupling.

in the IA1 impulse approximation. Dashed lines show that much smaller contributions arise if pseudovector $\pi$N coupling is used.

The reason for this behavior is that the Fock contribution for pseudo-scalar pion exchange, $\hat{U}_{1\pi} \approx -\frac{3}{8}(1-\gamma^0)g_\pi^2/(2MT_{lab}+m_\pi^2)p_V$, has very large scalar and vector parts, of opposite sign, at low nucleon energy, $T_{lab}$. The structure of the phenomenological Fermi invariants causes the pion contribution to the optical potential, or the mean field potential, to enter in this fashion. Virtual-pair effects, estimated as in Equation 9, thus rule out the use of pseudoscalar $\pi$N coupling.

It is well known that pseudoscalar $\pi$N coupling causes intolerable pair contributions to NN scattering (49). The same thing occurs in the optical potential at low energy. A solution is to use pseudovector $\pi$N coupling to suppress $\pi$N$\bar{\text{N}}$ couplings. Since pseudovector coupling is not renor-malizable, one must also introduce phenomenological form factors to control the dynamics. A more fundamental solution within the context of renormalizable quantum field theories has long been sought based on imposing chiral invariance (5). However, candidate theories typically pre-dict strong many-body forces (50) and these have defied quantitative analysis at the level of the NN interaction. Therefore pseudovector $\pi$N coupling is usually assumed.

## Ambiguities in NN Amplitudes

Although Fermi covariants are natural from the point of view of meson-exchange models of the NN interaction, other choices for the repre-sentation of the Lorentz-invariant amplitudes are equally admissible (51). Generally one finds different virtual-pair couplings even when positive-energy matrix elements are held fixed. In principle, one needs a complete representation of the NN amplitude on the Dirac space of two nucleons in order to construct unambiguously the Dirac optical potential.

The choice of pseudovector vs pseudoscalar $\pi$N coupling exemplifies the ambiguity that can arise. For positive-energy matrix elements, such as in Equation 11, a pseudovector covariant, $P' = (4M^2)^{-1}\gamma_1^5\gamma_2^5 \cdot (p_1-p_1')\gamma_2 \cdot (p_2-p_2')$, and a pseudoscalar covariant, $P = \gamma_1^5\gamma_2^5$, are completely equiva-lent. However, $P$ and $P'$ have radically different matrix elements when negative-energy states are involved, as in pair contributions. This fact illustrates why it is ambiguous to characterize the NN amplitude from knowledge only of positive-energy matrix elements. In general, the em-pirical NN data are necessary but not sufficient input to characterize the Feynman amplitude, $\hat{M}$, needed in Equation 2.

Because of the obvious problem at low energy, a simple improvement over the IA1 assumption is to adopt a five-term representation of the NN amplitudes that forces the one-pion-exchange contribution to be pseudovector (20, 52). Essentially one replaces the pseudoscalar covariant with a pseudovector one. It is important to realize that this assumption does not remove the essential ambiguity of extending the NN amplitude defined on positive-energy states into an operator for the full Dirac space. In principle, the ambiguity one encounters can only be removed by use of a dynamical model for the relativistic NN amplitude.

## Relativistic NN Dynamics

At present, meson theory with appropriate cutoffs provides the only relativistic model for NN scattering dynamics that is (a) capable of describing the NN phase shifts over a broad energy range and (b) suitable for predicting the Feynman NN amplitude in the full Dirac space of two nucleons.

NN scattering dynamics within the meson exchange framework has been extensively analyzed in recent years (29, 53–58). Of particular relevance is the work Tjon and collaborators, in which the NN scattering is analyzed in a Bethe-Salpeter framework. Recently, the model has been extended to incorporate channel couplings between NN, NΔ, and ΔΔ states in order to allow for inelasticity in the NN channel; this has produced realistic results above pion production threshold (56–58). The model has also been extended to include negative-energy intermediate states in the NN channel (21, 22). Figure 4 illustrates the coupled integral equations of the van Faassen–Tjon model for NN scattering. Nucleons are indicated by horizontal lines and deltas are indicated by double lines. Various transition

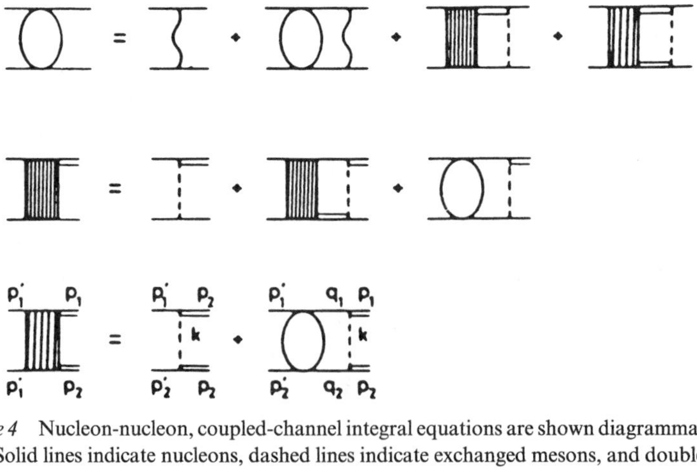

*Figure 4* Nucleon-nucleon, coupled-channel integral equations are shown diagrammatically (57). Solid lines indicate nucleons, dashed lines indicate exchanged mesons, and double solid lines indicate deltas.

amplitudes are indicated by ovals and rectangles. The interactions shown by wavy and dashed lines include $\pi$, $\varepsilon$, $\eta$, $\rho$, $\delta$, and $\omega$ meson exchange. Pseudovector $\pi$N coupling is used.

Calculations show that a reasonably accurate description of NN scattering data is achieved for the 0–1000-MeV range of laboratory energy. This is notable progress. One consistent set of meson-baryon couplings and cutoff parameters describes all the essential features of NN scattering over a broad energy range. Moreover, the dynamical equations used predict not only the positive-energy matrix elements but also a complete set of negative-energy matrix elements. Therefore, one has at hand a model capable of providing a dynamical basis for $\hat{M}$ and the implicit pair couplings of the Dirac optical potential. Currently, no other approach exists to fix $\hat{M}$ unambiguously or to include pseudovector $\pi$N coupling consistently.

The full Bethe-Salpeter analysis is approximated reasonably well by a three-dimensional quasi-potential equation provided some coupling constants are changed by 10–20% (56). Using this approximation, Tjon & Wallace (22) have calculated complete sets of NN helicity amplitudes in the cm frame of the NN system. Coupling constants used are consistent with those used by other workers (55) who generally solve equations involving couplings of only positive-energy NN states.

## Complete Sets of Lorentz-Invariant NN Amplitudes

For the optical potential, NN amplitudes are needed in the proton-nucleus cm frame. Since phase-shift analysis is practical only in the nucleon-nucleon cm frame, it is necessary to boost the amplitudes. This is conveniently done by introducing a Lorentz-invariant representation (21, 22, 59). The full Feynman amplitude is expanded in terms of Lorentz-invariant amplitudes times kinematical covariants formed out of the Dirac matrices and the four-momenta that characterize initial and final scattering states. In principle, many different representations of the Feynman amplitude are possible since the choice of covariants is not unique. All such representations are equivalent provided one expands in a complete and linearly independent set of covariants.

There are two parts to the problem of reconstructing $\hat{M}$ from solutions of the NN integral equations. First, one needs a complete set of kinematical covariants suited to expansion of the NN amplitude on the full Dirac space of two nucleons, with proper regard for parity invariance, time-reversal invariance, charge symmetry, and the generalized Pauli principle. The covariants selected must be linearly independent and the associated amplitudes must be free of kinematical singularities.

Second, one must determine the Lorentz-invariant amplitudes that multiply the kinematical covariants. In general the parity-invariant NN amplitude needed to construct the Dirac optical potential contains 128 terms for each isospin. However, not all 128 terms are independent. Charge symmetry and time-reversal invariance reduce the number of independent amplitudes to 56 for off-mass-shell kinematics and to 44 for on-mass-shell kinematics, again for each isospin (21, 22). The five invariants of Equation 10 are sufficient to describe all positive-energy matrix elements. Therefore these five invariants can be calculated directly from the phase-shift analysis of NN scattering. In the complete set of 44, there are 39 additional invariants that have vanishing matrix elements in positive-energy states. These additional amplitudes must be calculated from a meson-exchange model in order to have control over the Feynman scattering operator in the negative-energy sectors of the Dirac space of two particles.

In order to treat nucleon-exchange contributions to the optical potential in a reasonable fashion, it is necessary to separate the NN amplitudes into direct and exchange parts. Tjon & Wallace (22) provide a representation suited to this task based on using covariants that are even or odd with respect to particle exchange. Each of 44 independent amplitudes is fitted to a sum of Yukawa terms of the form $M(t) \pm M(u)$, which explicitly is symmetric or antisymmetric with respect to $t \to u$, where $t$ and $u$ are Lorentz-invariant Mandelstam variables. In the cm frame for NN scattering, $t = -2p^2[1 - \cos(\theta)]$ and $u = -2p^2[1 + \cos(\theta)]$, where $p$ is the cm momentum. Thus when $\theta \to \pi - \theta$, we find $t \to u$ and vice versa. Symmetry with respect to $t \to u$ is a Lorentz-invariant manifestation of the $\theta \to \pi - \theta$ symmetry that holds in the cm frame. The Yukawa fits force a separation of each amplitude into a direct part $M(t)$ and an exchange part $M(u)$.

Related work on the representation of the NN amplitudes is given in (21) and (59). In both these cases Lorentz-invariant amplitudes are not symmetric when $t \to u$ because the covariants employed do not have simple particle-exchange symmetry. A separation into direct and exchange parts is awkward, although in principle it is possible.

A straightforward method of expanding the NN amplitude (22) is to use covariant projection operators, $\Lambda^{\pm}$, to separate positive- and negative-energy sectors of the Dirac space as follows,

$$\hat{M}(p_1, p_2 \to p'_1, p'_2)$$

$$= \kappa \sum_{\rho'_1 \rho'_2 \rho_1 \rho_2} \Lambda_1^{\rho'_1}(\mathbf{p}'_1)\Lambda_2^{\rho'_2}(\mathbf{p}'_2)\hat{F}^{\rho'_1\rho'_2\rho_1\rho_2}(p_1, p_2 \to p'_1, p'_2)\Lambda_1^{\rho_1}(\mathbf{p}_1)\Lambda_2^{\rho_2}(\mathbf{p}_2). \quad 15.$$

Here $\rho_i = +$ or $-$, for $i = 1$ and 2, distinguishes positive- and negative-energy initial states and similarly, $\rho'_i = +$ or $-$ for final states. Thus Equation 15 has sixteen independent $\rho$-spin sectors corresponding to the values of $\rho'_1$, $\rho'_2$, $\rho_1$, and $\rho_2$. For each sector there are eight independent amplitudes in $\hat{F}^{\rho'_1\rho'_2\rho_1\rho_2}$. This accounts for 128 linearly independent terms and 128 invariant amplitudes. Because of symmetries, only 44 of the invariant amplitudes are actually independent on-mass-shell. All 128 amplitudes of the representation have been calculated in terms of the 44 independent ones (22).

## Generalized Impulse Approximation: IA2

It is straightforward to generalize the relativistic impulse approximation by using $\hat{M}$ of Equation 15 in Equation 2. This generalization is called IA2. One predicts the complete set of amplitudes needed to characterize the Feynman scattering operator by using the same meson couplings and integral equations that provide a good theoretical description of NN scattering data in the 0–1000-MeV energy range. Nuclear densities are fixed by the Dirac-Hartree model. These ingredients are combined to determine the Dirac optical potential with no free parameters being available to aid in fitting proton-nucleus scattering data. Thus a dynamical model is used in place of the assumption of five Fermi covariants. Virtual-pair couplings of the Dirac optical potential are given a foundation in the meson-exchange description of the nuclear force. This overcomes the basic ambiguity in the construction of the Dirac optical potential from NN amplitudes. The objective of incorporating pseudovector $\pi$N coupling in a consistent fashion is met by basing the IA2 analysis on the relativistic meson-exchange model of van Faassen & Tjon (56–58).

At intermediate energy, nucleon-nucleus elastic scattering is diffractive and forward peaked. In such circumstances, a local form of the optical potential proves to be reasonably accurate. Straightforward analysis leads to six nonvanishing terms in the IA2 optical potential. One of these is a space vector potential, $C(r)$, which may be absorbed into the remaining five to arrive at the reduced form,

$$\hat{U}(\mathbf{r}) = \tilde{S}(r) + \gamma^0 \tilde{V}(r) - i\boldsymbol{\alpha} \cdot \hat{\mathbf{r}}\tilde{T}(r) - [\tilde{S}_{LS}(r) + \gamma^0 \tilde{V}_{LS}(r)]\boldsymbol{\sigma} \cdot \mathbf{L}, \qquad 16.$$

in coordinate space (17, 19). Scalar spin-orbit and vector spin-orbit potentials, $\tilde{S}_{LS}$ and $\tilde{V}_{LS}$, did not appear in the IA1 analysis, nor was there a space vector potential, $C$.

Figure 5 compares the energy dependence of scalar and vector strengths that arise in the generalized impulse approximation potential (crosses) with strengths based on the original impulse approximation, IA1 (solid lines). The reduction of strengths relative to the original impulse approximation is consistent with expectations based on changing from pseudo-scalar to pseudovector $\pi$N coupling.

Figures 6 and 7 show results for cross section $\sigma$, analyzing power $A_Y$, and spin rotation $Q$ for elastic scattering of protons by $^{40}$Ca at 200 and

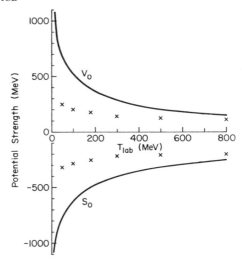

*Figure 5*   Scalar and vector strengths in nuclear matter are shown as functions of the proton kinetic energy in MeV (20). Solid lines show IA1 impulse approximation results and x's show IA2 impulse approximation results.

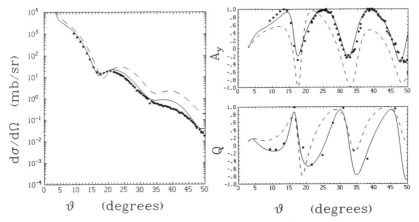

*Figure 6*   Elastic proton scattering observables at 200 MeV for $^{40}$Ca. Cross section ($d\sigma/d\Omega$), analyzing power ($A_Y$), and spin rotation function ($Q$) data from (60) are compared with the IA1 relativistic impulse approximation (*dashed lines*) and with the IA2 relativistic impulse approximation (*solid lines*).

500 MeV, respectively. In each figure a solid line shows the result of the generalized impulse approximation, IA2, and a dashed line shows the result based on the original impulse approximation, IA1. Circles show experimental data (37, 60, 62; P. Schwandt, private communication). A

very good description of 800-MeV data is obtained in both IA1 and IA2 approaches. The characteristic Dirac agreement with spin observables for proton-nucleus scattering is obtained with significantly smaller scalar and vector potentials at low energy in the IA2 approach.

## SUMMARY AND OUTLOOK

Relativistic approaches to nuclear binding potentials and nucleon optical potentials have progressed significantly in recent years. The attractive scalar potential and the repulsive vector potential in the Dirac equation are obtained directly from relativistic models of the NN interaction. The traditional method of constraining the many-body problem by organizing it around the basic NN interaction provides a successful starting point for the relativistic mean field and for the Dirac optical potential. Predictions of spin observables in intermediate-energy proton scattering are generally successful over a broad energy range based on an impulse approximation constructed from a complete set of Lorentz-invariant NN amplitudes. In the meson theoretical approach, virtual-pair effects included implicitly by use of the Dirac equation are given a foundation similar to that existing for pair currents in electromagnetic reactions.

In the phenomenological sense, relativistic optical potentials are quite successful. Although considerable progress has been made, the approach

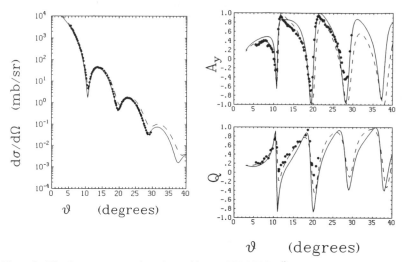

*Figure 7*  Elastic proton scattering observables at 500 MeV for $^{40}$Ca. Cross section ($d\sigma/d\Omega$), analyzing power ($A_Y$), and spin rotation function ($Q$) data from (37) and (62) are compared with the IA1 relativistic impulse approximation (*dashed lines*) and with the IA2 relativistic impulse approximation (*solid lines*).

based on meson theories that describe NN interactions is relatively new. Much more work is needed to fill out the theoretical picture. Moreover, several theoretical objections have been raised to the inclusion of virtual NN̄ pairs in nuclear physics.

## Is There a Unified Relativistic Dynamics?

One of the theoretical benefits to be gained from continuing development of the relativistic dynamics is the possible unification of descriptions of NN interactions, nucleon optical potentials, nuclear binding potentials, and meson-exchange currents. Relativistic meson-exchange models employing pseudovector $\pi$N coupling and meson-nucleon vertex form factors seem to provide a good starting description of each of these aspects of nuclear physics. With regard to improving the relativistic descriptions of elastic scattering of nucleons, three areas where progress is needed seem quite clear: (a) corrections to the first-order optical potential; (b) improvements of relativistic nuclear wave functions; (c) vacuum polarization corrections.

In multiple scattering theory, there are well-defined corrections to the first-order optical potential due to double scattering, triple scattering, and so on. A similar structure holds for multiple scattering of a Dirac wave by a collection of fixed nucleons. Thus approximate calculations of double scattering corrections are feasible. Corrections to the relativistic impulse approximation due to nuclear correlations first appear in the double scattering term of the optical potential, $U^{(2)}$. Lumpe & Ray (63) have reported calculations of $U^{(2)}$ based on retaining just the scalar and vector terms, $F_1$ and $F_2$, of Equation 10. The results for spin observables indicate very small changes at 500 MeV due to the double scattering corrections. This result provides the first demonstration that large scalar and vector interactions of opposite signs do not produce large double scattering corrections. Further work along these lines should be carried out.

Spin observables in elastic proton scattering are quite sensitive to the difference between the scalar and vector densities, $\delta\rho = \rho_V - \rho_S$. A notable feature of the meson theoretical optical potential IA2 is that smaller scalar and vector potentials arise than are commonly assumed in Dirac-Hartree analyses. This implies that nuclear wave functions based on the full NN interaction are likely to be significantly different from those based on one-meson-exchange interactions. For example $\delta\rho$ is likely to be smaller than the Hartree model predicts because there is less enhancement of lower component wave functions. A finite nucleus Brueckner-Hartree-Fock analysis is needed to develop wave functions consistent with the NN interaction.

Moreover, quantum field theories predict a significant alteration of $\delta\rho$

due to vacuum polarization effects. This effect has been estimated by Horowitz & Serot (64) and by Perry (65) but it has been omitted from most analyses of nucleon scattering and nuclear structure. Fully consistent calculations of relativistic effects in nuclei will require both better nuclear wave functions and some understanding of the vacuum polarization corrections.

## Are Virtual Pairs Avoidable?

Because the new ingredient of the relativistic approach is the virtual-pair contribution, one may question whether it is really needed (66). At energies considered in nuclear physics, the N$\bar{\text{N}}$ channel is closed and N$\bar{\text{N}}$ pairs are highly virtual. Consequently they are not subject to direct experimental tests. Logically one cannot exclude the possibility that effects in the Schroedinger dynamics may exist that can produce similar results. One possibility is modification of the NN interaction in nuclear matter. In the Schroedinger optical potential, effects of the nuclear medium significantly improve the description of spin observables in low-energy elastic scattering (67, 68).

It is important to stress that a proper relativistic framework fully includes the nonrelativistic dynamics. In addition, virtual-pair contributions are included. Therefore effects known to be important in nonrelativistic potentials must be included in relativistic ones in order to do a complete job.

Recent work by Horowitz & Murdock (69) finds that medium effects are important in the relativistic optical potential below 300–400 MeV. The analysis is based on an incomplete set of NN amplitudes and approximate estimates of medium modifications. In contrast, the IA2 impulse approximation does well at 200 MeV without medium effects when a complete set of NN amplitudes is used. Further work is needed to reconcile these results. Better ways of calculating medium effects are also needed in both relativistic and nonrelativistic formulations. Calculations appropriate to infinite nuclear matter are usually used to determine the NN interaction in the nuclear medium. The local density approximation used to apply the result to finite nuclei is suspect (70).

A second way of improving Schroedinger-Pauli spin calculations lies in the treatment of nonlocalities. Theis (71, 72) points out a possible nonlocality in the nonrelativistic optical potential, which, in the limit of zero-range NN interactions, produces the same results as the Dirac equation. A zero-range form of the nonrelativistic spin-orbit interaction, of the form $\sigma \cdot \mathbf{p} U_{\text{LS}} \sigma \cdot \mathbf{p}/(4M^2)$, produces nonlocal structure in the p-nucleus optical potential of the same form as is encountered in the $\pi$-nucleus optical potential. Lorentz-Lorenz corrections to the optical potential are needed to remove effects associated with short-range nucleon propagation between two nucleons. Since nuclear correlations prevent nucleons from coming too close together, the short-range propagation effects must be quenched. The upshot is that the Lorentz-Lorenz correction to the optical potential is equal to $U_{pair}$ of Equation 13! Thus in principle the same results for Dirac and Schroedinger-Pauli spin approaches are obtained in the limit of zero-range NN interactions.

Cooper & Jennings (73) point out that the Lorentz-Lorenz correction should be only about 20% of $U_{pair}$ for realistic ranges of the spin-orbit interaction. This is consistent with calculations of the correlation corrections by Lumpe & Ray (63). Moreover, Cooper & Jennings argue that the use of the Dirac equation removes sensitivity to short-range effects. Analysis shows that short-range structure is introduced in the nucleon propagator if one restricts propagation to positive-energy states of the free Dirac equation, as in "no-pair" theories (74). The corresponding Lorentz-Lorenz correction, for zero-range NN interactions, takes the same form as $U_{pair}$. For finite-range NN interactions, one has both a Lorentz-Lorenz correction and a correlation-quenched virtual-pair contribution. Remarkably, these add up to $U_{pair}$ as given by the Dirac approach. Thus the Dirac potential is insensitive to short-range effects, as suggested by Cooper & Jennings.

Such examples show that equivalent theoretical predictions can emerge from relativistic and nonrelativistic approaches. Comparisons with experimental data are very valuable, but one must keep in mind that they are less than conclusive for validating the inclusion of virtual-pair effects.

## What is the QCD Connection?

Whether or not virtual NN̄ pair effects are suppressed in ways not predicted by phenomenological meson-nucleon vertex functions is a difficult and important question in relativistic nuclear physics. Nonperturbative QCD models of the nucleon do not yet exist to provide a clear answer to this question. One needs a covariant model of a nucleon, a virtual antinucleon, and a meson in order to study the structure of the vertex function assumed in meson-exchange models.

Perturbative calculations in QCD predict a strong suppression of real NN̄ pairs at a time-like momentum transfer of 2 GeV or more, essentially similar to the suppression expected from the nucleon form factor. Such a suppression is observed in $e^+ + e^- \rightarrow p + \bar{p}$ experiments. Brodsky (75) argues that a similarly large suppression applies to virtual pairs in low-energy phenomena. The argument assumes that a time-like momentum transfer of $2M$ is involved whenever a virtual pair is produced, where $M$ is the nucleon mass. For some time orderings of a Feynman diagram, this is the correct momentum transfer flowing through the vertex of a Z-graph where a virtual pair is created or annihilated by a meson. However, the Z-graph diagrams included implicitly by using the Dirac equation are essentially different. They involve small, space-like momentum transfer and zero time-like momentum transfer. Such graphs are not suppressed significantly by the vertex form factors in a Lorentz-invariant field theory of mesons and nucleons. Only time will tell whether current approaches to relativistic nuclear physics are compatible with QCD.

ACKNOWLEDGMENT

This work was supported under US Department of Energy contract DE-ASO5-76-ERO5126.

*Literature Cited*

1. McNeil, J. A., Shepard, J. R., Wallace, S. J. *Phys. Rev. Lett.* 50: 1439–42 (1983)
2. McNeil, J. A., Ray, L., Wallace, S. J. *Phys. Rev.* C27: 2123–32 (1983)
3. Shepard, J. R., McNeil, J. A., Wallace, S. J. *Phys. Rev. Lett.* 50: 1443–45 (1983)
4. Clark, B. C., Hama, S., Mercer, R. L., Ray, L., Serot, B. D. *Phys. Rev. Lett.* 50: 1644–46 (1983)
5. Serot, B. D., Walecka, J. D. *Adv. Nucl. Phys.* Vol. 16, ed. J. Negele, E. Vogt. New York: Plenum. 327 pp. (1986)
6. Walecka, J. D. *Ann. Phys. (NY)* 83: 491–529 (1974)
7. Celenza, L. S., Shakin, C. M. *Relativistic Nuclear Physics.* Singapore: World Scientific. 238 pp. (1986)
8. Anastasio, M. R., Celenza, L. S., Pong, W. S., Shakin, C. M. *Phys. Rep.* 100: 327–92 (1983)
9. Anastasio, M. R., Celenza, L. S., Shakin, C. M. *Phys. Rev. Lett.* 45: 2096–98 (1980)
10. Anastasio, M. R., Celenza, L. S., Shakin, C. M. *Phys. Rev.* C23: 569–75 (1981)
11. Green, A. E. S. In *Antinucleon and Nucleon Interactions with Nuclei*, ed. G. E. Walker, C. D. Goodman, C. Olmer, pp. 143–58. New York: Plenum (1985)
12. Miller, L. D. *Phys. Rev. Lett.* 28: 1281–83 (1972)
13. Miller, L. D., Green, A. E. S. *Phys. Rev.* C5: 241–52 (1972)
14. Clark, B. C., Hama, S., Mercer, R. L. In *The Interaction Between Medium Energy Nucleons in Nuclei—1982*, ed. H. O. Meyer, AIP Conf. Proc. No. 97, pp. 260–87. New York: Am. Inst. Phys. (1983)
15. Clark, B. C. In *Relativistic Dynamics and Quark-Nuclear Physics*, ed. M. B. Johnson, A. Picklesimer, pp. 302–29. New York: Wiley (1986)
16. Clark, B. C., Mercer, R. L., Ravenhall, D. G., Saperstein, A. M. *Phys. Rev.* C7: 466–71 (1973)
17. Wallace, S. J. See Ref. 15, pp. 418–48
18. Wallace, S. J. *Prog. Theor. Phys. Jpn.* (*Suppl.*) 55: 99–112 (1985)
19. Tjon, J. A., Wallace, S. J. *Phys. Rev. Lett.* 54: 1357–59 (1985)
20. Tjon, J. A., Wallace, S. J. *Phys. Rev.* C32: 267–76 (1985)
21. Tjon, J. A., Wallace, S. J. *Phys. Rev.* C32: 1667–80 (1985)
22. Tjon, J. A., Wallace, S. J. *Phys. Rev.* C35: 280–97 (1987)
23. Miller, L. D. *Phys. Rev.* C14: 706–17 (1976)
24. Brockmann, R. *Phys. Rev.* C18: 1510–24 (1975)
25. Horowitz, C., Serot, B. D. *Nucl. Phys.* A368: 503–28 (1981)
26. Miller, L. D. *Phys. Rev.* C9: 537–54 (1974)
27. Horowitz, C., Serot, B. D. *Phys. Lett.* 109B: 341–45 (1982)
28. Horowitz, C., Serot, B. D. *Nucl. Phys.* A399: 529–62 (1983)
29. Holinde, K., Machleidt, R. *Nucl. Phys.* A256: 479–96 (1976)
30. Brockmann, R., Machleidt, R. *Phys. Lett.* 149B: 283–87 (1984)
31. Malfliet, R., ter Haar, B. *Phys. Rev. Lett.* 56: 1237–39 (1986)
32. Horowitz, C., Serot, B. D. *Nucl. Phys.* In press
33. Kerman, A. K., McManus, H., Thaler, R. M. *Ann. Phys. (NY)* 8: 551–635 (1959)

34. Glauber, R. J. In *Boulder Lectures in Theoretical Physics*, ed. L. G. Dunham, W. E. Brittin, pp. 315–414. New York: Interscience (1959)
35. Arnold, L. C., Clark, B. C., Mercer, R. L. *Phys. Rev.* C19: 917–22 (1979)
36. Clark, B. C., Mercer, R. L., Schwandt, P. *Phys. Lett.* 122B: 211–16 (1983)
37. Hoffmann, G. W., Ray, L., Barlett, M. L., Fergerson, R., McGill, J., et al. *Phys. Rev. Lett.* 47: 1436–39 (1981)
38. McNeil, J. A., Sparrow, D. A., Amado, R. D. *Phys. Rev.* C26: 1141–47 (1982)
39. Kobos, Á. M., Cooper, E. D., Johansson, J. I., Sherif, H. S. *Nucl. Phys.* A445: 605–24 (1985)
40. Goldberger, M. L., Grisaru, M. T., MacDowell, S. W., Wong, D. Y. *Phys. Rev.* 120: 2250–76 (1960)
41. Arndt, R. A., Roper, L. D., Bryan, R. A., Clark, R. B., VerWest, B. J., Signell, P. *Phys. Rev.* D28: 97–122 (1983)
42a. Brieva, F. A., Rook, J. R. *Nucl. Phys.* A291: 317–41 (1977)
42b. Brieva, F. A., Rook, J. R. *Nucl. Phys.* A297: 206–30 (1978)
43. Clark, B. C., Hama, S., Mercer, R. L., Ray, L., Hoffmann, G. W., Serot, B. D. *Phys. Rev.* C28: 1421–24 (1983)
44. Ray, L., Hoffmann, G. W. *Phys. Rev.* C31: 538–60 (1985)
45. Hynes, M. V., Picklesimer, A., Tandy, P. C., Thaler, R. M. *Phys. Rev. Lett.* 52: 978–81 (1984)
46. Hynes, M. V., Picklesimer, A., Tandy, P. C., Thaler, R. M. *Phys. Rev.* C31: 1435–63 (1985)
47. Bawin, M., Jaminon, M. *Nucl. Phys.* A407: 515–21 (1983)
48. Brown, G. E., Weise, W., Baym, G., Speth, J. *Comments Nucl. Part. Phys.* 17: 39 (1987)
49. Fleischer, J., Tjon, J. A. *Phys. Rev.* D21: 87–94 (1980)
50. Jackson, A. D. *Ann. Rev. Nucl. Part. Sci.* 33: 105–41 (1983)
51. Adams, D., Bleszynski, M. *Phys. Lett.* 136B: 10–14 (1984)
52. Horowitz, C. *Phys. Rev.* C31: 1340–48 (1985)
53. Holinde, K. *Phys. Rep.* C65: 121–88 (1981)
54. Lacombe, M., Loiseau, B., Richard, J. M., Vinh Mau, R., Cote, J., et al. *Phys. Rev.* C21: 861–73 (1980)
55. Machleidt, R. See Ref. 15, pp. 71–173
56. van Faassen, E. E., Tjon, J. A. *Phys. Rev.* C28: 2354–67 (1983)
57. van Faassen, E. E., Tjon, J. A. *Phys. Rev.* C30: 285–97 (1984)
58. van Faassen, E. E. *Relativistic NN Scattering With Isobar Degrees of Freedom*, PhD Thesis, Univ. Utrecht. 123 pp. (1984)
59. Picklesimer, A., Tandy, P. C. *Phys. Rev.* C34:1860–94 (1986)
60. Stephenson, E. J. *J. Phys. Soc. Jpn.* (*Suppl.*) 55: 316–21 (1985)
61. Deleted in proof
62. Rahbar, A., Aas, B., Bleszynski, E., Bleszynski, M., Haji-Saeid, M., et al. *Phys. Rev. Lett.* 47: 1811–14 (1981)
63. Lumpe, J. D., Ray, L. *Phys. Rev.* C35: 1040–59 (1987)
64. Horowitz, C. J., Serot, B. D. *Phys. Lett.* 140B: 181–86 (1984)
65. Perry, R. *Phys. Lett.* 182B: 269–73 (1986)
66. Negele, J. W. *Comments Nucl. Part. Phys.* 6: 303–19 (1985)
67. Rikus, L., von Geramb, H. V. *Nucl. Phys.* A246: 496–514 (1984)
68. Rikus, L., Nakamo, K., von Geramb, H. V. *Nucl. Phys.* A414: 413–55 (1984)
69. Horowitz, C. J., Murdock, D. *Phys. Lett.* 168B: 31–34 (1986)
70. Mahaux, C. See Ref. 14, pp. 20–41
71. Theis, M. *Phys. Lett.* 162B: 255–59 (1985)
72. Theis, M. *Phys. Lett.* 166B: 23–26 (1986)
73. Cooper, E. D., Jennings, B. *Nucl. Phys.* A458: 717–24 (1986)
74. Sucher, J. *Phys. Rev.* A22: 348–62 (1980)
75. Brodsky, S. J. *Comments Nucl. Part. Phys.* 12: 213–41 (1984)

*Ann. Rev. Nucl. Part. Sci. 1987. 37: 293–323*
*Copyright © 1987 by Annual Reviews Inc. All rights reserved*

# PROBES OF THE QUARK GLUON PLASMA IN HIGH ENERGY COLLISIONS

## K. Kajantie

Department of Theoretical Physics and Academy of Finland, Siltavuorenpenger 20 C, 00170 Helsinki, Finland

## Larry McLerran

Fermi National Accelerator Laboratory, P.O. Box 500, Batavia, Illinois 60510, USA

CONTENTS

0163–8998/87/1201–0293$02.00

# 1. INTRODUCTION

Quark gluon plasma (1) is the high-temperature high-density phase of matter described by the laws of quantum chromodynamics. At low temperatures and densities quarks, gluons, and color fields are confined to the interiors of strongly interacting particles, hadrons. At high temperatures and densities the hadrons overlap and lose their identity; quarks, gluons, and color fields are not confined within hadrons but can move over distances larger than the hadron size, 1 fm.

We expect that the early universe, when it was younger than about $10^{-5}$ s, was filled with quark gluon plasma (and, at least, photons and leptons). The possible observational consequences—relic cold strange quark matter (2, 3), energy density inhomogeneities (4, 5), black holes (6), gravitational radiation (2, 7), etc—are rather speculative and so far no observational evidence exists. Cold quark gluon plasma or quark matter could also exist in the present universe in the interiors of compact stellar systems. Here also no convincing observational evidence exists, although changes in cooling rates have been suggested as such (8). Information on possible quark gluon plasma can be gained, however, by studying ultrarelativistic nuclear collisions or very high multiplicity fluctuations in hadron-hadron collisions. What the signals could be is the subject of this review.

## 1.1 *General Properties of Ultrarelativistic Nucleus-Nucleus Collisions*

Even without the prospect of observing quark gluon plasma, ultrarelativistic nuclear collisions are very interesting and much theoretical work has been devoted to them (9–13). One has mainly tried to predict the rapidity and $p_T$ (or $E_T$) distributions of produced pions (possibly including neutrals) by extending models developed for hadron-nucleus collisions to nucleus-nucleus collisions: the multichain model (10), the additive quark model (11), the dual parton model (12), or the cascade model (13). Attempts have also been made to predict the nuclear stopping power, which describes how baryon number is distributed in the final state (14–16).

Although superficially these predictions have nothing to do with quark gluon plasmas, they are actually very important for the discussion of any plasma probes. Any thinkable plasma probes have a background arising from nonplasma mechanisms, and the above model calculations are a method of estimating this background. The model calculations based on particle physics concepts, for instance, are formulated in momentum space, while we expect quark gluon plasma to exist in space-time. The experiments thus must find effects that are not naturally described in momentum space, but follow naturally from a glob of matter flowing collectively in space-time and emitting various probes from its interior and surface.

In this context it is important to appreciate that quark gluon plasma is just the high-temperature high-density phase of QCD matter, and that it is also critical to observe the low-temperature, low-density phase (hadron gas) and the intermediate mixed phase. An additional complication arises because under present experimental conditions the mixed phase appears to dominate the important phenomena. The analysis would be simpler in the pure quark gluon plasma or the hadron gas phases.

There is now a major experimental effort under way at CERN and BNL to make and study ultrarelativistic nuclear collisions, as well as an effort at FNAL to study the extreme environment provided in high multiplicity fluctuations in $\bar{p}p$ collisions at the Tevatron. The first results from CERN came at the end of 1986 (17) and more will soon follow.

## 1.2  Review of Recent Lattice Gauge Theory Results

The only known way of performing first-principles nonperturbative computations on QCD matter is to use lattice Monte Carlo techniques (18–38). These, however, are only applicable to a rather limited set of static phenomena; there is no first-principles method to compute nonstatic phenomena. When discussing quark gluon plasma, it is usually assumed that the plasma behaves like a fluid, i.e. is in local thermal equilibrium (in contrast to a plasma described with the aid of kinetic theory). Thus it is essential to know the equation of state (EOS), which gives the pressure and energy, entropy, and net baryon-number densities in terms of temperature $T$ and chemical potential $\mu$. This is one important quantity that can be computed with the aid of lattice Monte Carlo, at least for baryon-number free systems, $\mu = 0$.

The fundamentals of finite-temperature lattice QCD are reviewed in (37) and more recent developments in (38). We discuss here only the question of scaling in the computations, which is a necessary condition for their validity and which was doubtful until recently, and the EOS, which is very important for practical purposes. Other notable recent developments are the determinations of static screening lengths for SU(3) (34) and SU(2) (35) and the first attempt to use lattice techniques to determine transport coefficients of QCD matter (36).

Finite-temperature QCD with only gluons, without quarks, has an order parameter (18–20) that identically vanishes in the confined hadron gas phase and is nonzero in the gluon plasma phase. On the lattice this is the expectation value $\langle L(x) \rangle$ of a trace of a product of SU($N_c$) matrices in the time-temperature direction at the spatial site $x$ (the Wilson-Polyakov loop); the trace arises because of thermal periodic boundary conditions.

A calculation performed on an $N_s^3 \cdot N_t$ lattice corresponds to a system with volume $(aN_s)^3$ and temperature $T = 1/aN_t$, where $a$ is the lattice spacing. Of course, the computer accepts only dimensionless numbers, and the distance $a$ is converted to dimensionless numbers by the equation

$$\frac{1}{a} = \Lambda_L (2.3\beta)^{0.42} e^{-1.2\beta} \equiv \Lambda_L f(\beta), \qquad \qquad 1.$$

where $\beta = 6/g^2$ is the dimensionless number fed into the computer, the expression for $f(\beta)$ follows from perturbation theory, and $\Lambda_L$ is a dimensionful quantity taken from experiment. One thus has

$$\frac{T}{\Lambda_L} = \frac{1}{N_t} f(\beta) \qquad \qquad 2.$$

and the determination of the critical temperature is made by finding at what value of $\beta$ the order parameter $\langle L(x) \rangle$ becomes nonvanishing. This is converted to a physical $T$ by Equation 2. However, the computation is only consistent if lattices of different $N_t$ give the same physical value of $T_c$. This is confirmed by the computations in (29, 30) and the result is shown in Figure 1. In this the solid curve corresponds to Equation 2 plotted for $T/\Lambda_L = 46.6$ and the points on the curve are from a computation (29) with $N_s^3 = 16^3$ and $N_t = 10, 12, 14$; the nonscaling points are from runs with smaller $N_s$ and $N_t = 2, 4, 6, 8$. From Figure 1 one can see how the common understanding of the scaling domain has changed; at first it was thought that even small lattices with $N_t = 2, 4$ would show scaling (slope parallel

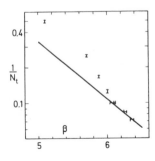

*Figure 1*  A test of whether or not lattice computations are sufficiently close to the continuum limit so as to be reliable for the computation of the deconfinement transition temperature. For lattice computations to be valid, the points have to lie on the solid line (see text).

to the solid line), but then larger lattices showed evidence for nonscaling. That the scaling sets in only for $N_t = 10$ is, of course, rather frustrating and implies that very large lattices will be required for detailed results on any phenomena.

Consider then the equation of state for $\mu = 0$, expressed as $\varepsilon = \varepsilon(T)$. At energy densities low compared to a scale of several hundreds of MeV/fm$^3$ to several GeV/fm$^3$, we presumably have a low-density gas of the ordinary constituents of hadronic matter, that is, mesons and nucleons. At densities very high compared to this scale, we expect an asymptotically free gas of quarks and gluons. At intermediate energy densities, we expect that the properties of matter will interpolate between these dramatically different phases of matter. There may or may not be true phase changes at these intermediate densities.

The result of a Monte Carlo simulation of the energy density is shown in Figure 2 (26). This is typical of the results of lattice Monte Carlo simulations. The precise values of the energy density are difficult to estimate as is the scale for the temperature. The figure does make clear the essential point, on which all Monte Carlo simulations agree, that the number of degrees of freedom of QCD matter changes by an order of magnitude in a narrowly defined range of temperature. There is apparently a first-order phase transition for SU(3) Yang-Mills theory in the absence of fermions, and a rapid transition that may or may not be a first-order transition for SU(3) Yang-Mills theory with two or three flavors of massless quarks.

The results are most accurate for pure gluon matter. Here the computations with lattices as large as $21^3 \cdot 14$ (31) reveal that the jump in energy density at the phase transition is $\Delta\varepsilon = 4.7T_c^4$, where $\Delta\varepsilon$ is in units of GeV/fm$^3$, with an error of about 40% in the numerical constant (about 10% error in $T_c/\Lambda_L$). To set the scale, note that for free gluon gas $\varepsilon = 5.3T^4$. The jump is thus large indeed.

The essential point, the large change in the number of degrees of freedom

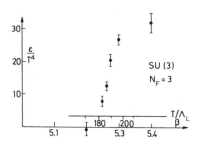

Figure 2  Energy density scaled by $T^4$ as a function of $T$. The data asymptote to a result not too far from the perturbative result. (Recall that an energy density of 1 GeV/fm$^3$ corresponds to an energy density of $\varepsilon/T^4$ of 5 at a temperature of 200 MeV.)

can be both physically understood and phenomenologically incorporated in terms of a simple bag equation of state. To formulate it, one simply gives the pressures in the quark gluon plasma phase and the hadron gas phase separately:

$$p_q(T) = g_q \frac{\pi^2}{90} T^4 - B, \qquad\qquad 3.$$

$$p_h(T) = g_h \frac{\pi^2}{90} T^4, \qquad\qquad 4.$$

where the effective numbers of degrees of freedom are estimated by assuming that quark gluon plasma consists of free quarks and gluons (one can also take $N_F = 2.5$ to simulate the effect of the strange quark mass) and hadron gas of free massless pions:

$$g_q = 2 \cdot 8 + 2.5 \cdot 2 \cdot 2 \cdot 3 \cdot \frac{7}{8} = 42.25, \qquad g_h = 3, \qquad\qquad 5.$$

and $B$ is a bag constant, vacuum energy density, which incorporates the effects of complicated QCD interactions. Including it makes the hadron phase the stable one at low $T$, i.e. $p_h > p_q$ for $T < T_c$, where the transition temperature $T_c$ is given by

$$B = (g_q - g_h) \frac{\pi^2}{90} T_c^4. \qquad\qquad 6.$$

Since it also holds that

$$\varepsilon(T) = Ts(T) - p(T) = Tp'(T) - p(T), \qquad\qquad 7.$$

it follows from Equations 3 and 4 that

$$\varepsilon_q(T_c)/\varepsilon_h(T_c) = 4g_q/g_h - 1/3 \qquad\qquad 8.$$

so that the large jump in the energy density is simply related to the large change in the number of degrees of freeedom. In theories with an arbitrary number of colors, $N_c$, the jump goes like $N_c^2$ (since color singlets appear below $T_c$ and gluons dominate above).

## 2.  SPACE-TIME PICTURES OF THE COLLISION

Collision processes in quantum theory are usually simplest to describe in momentum space. This is so since one must keep the momenta of the initial and final particles fixed. If one wants to study a glob of matter with a finite extent and finite lifetime, one must consider a space-time picture of the collision.

The space-time behavior of quantum electrodynamics was studied by Landau & Pomeranchuk (39) as early as 1953. The same was done for the parton picture of strong interactions by Gribov (40) and Bjorken (41), see also (42–46). The picture became particularly relevant when applied to the study of hadron-nucleus collisions (33, 34) since in these the nuclear radius provides one with a distance scale with which to measure space-time behavior. This has led to the inside-outside cascade model. In nucleus-nucleus collisions a further issue to discuss is thermalization.

## 2.1    Inside-Outside Cascade Picture

The physical foundation for the inside-outside cascade picture is very simple: one assumes that there is a hadronic time-scale $\tau_0$ defined so that partons created in a strong process can only reinteract when their proper time $\tau$ is larger than $\tau_0$. A natural QCD time unit is $\tau_0 \approx 1/\Lambda_{QCD} \approx 1$ fm/$c$; efforts have been made to improve this estimate by theoretical means (47–49). Experimentally, however, it has been impossible to distinguish between the inside-outside cascade model and various momentum-space models.

The physical consequences of this picture for hadron-nucleus collisions are also simple to see. When the hadron collides with the nucleus, it is immediately, within 1–2 fm from the surface, converted to a cloud of inactive partons. Only the partons with $v \cdot \gamma \tau_0 < 2R_A$, where $R_A$ = target radius, can reinteract within the nucleus and produce a cascade. In other words, partons with rapidity $y > \ln(4R_A/\tau_0) \equiv y_{cr}$ hadronize outside the nucleus. In this picture, the rapidity distribution $dN_\pi^{pA}/dy$ of produced pions thus contains an enhancement for $0 < y < y_{cr} \approx \ln 4A^{1/3}$ and behaves like $dN_\pi^{pp}/dy$ for $y > y_{cr}$.

Comparison with experimental data (48) confirms this highly simplified picture qualitatively but not quantitatively. There is a strong enhancement for small $y$ but also an enhancement relative to pp data at intermediate central region rapidities. This central region enhancement can be understood either in terms of reinteractions of the colliding proton with other target nuclei or in terms of color field effects (48). The highest available energy (200 GeV/$c$) corresponds to a total rapidity range of $0 < y < 6$ and with $y_{cr} \approx 3$ the separation of the different regions is not yet very clear.

At present the pA data are well understood with momentum-space models (multichain model, additive quark model, dual parton model, etc) and there is no need to introduce the space-time inside-outside cascade model. It is also not uniquely clear how the simplest version of this model should be refined. There is thus no way to fix the value of the essential parameter, $\tau_0$, of this model. A detailed analysis of these issues is carried out in (46). For the study of quark gluon plasma in AB collisions, the inside-outside cascade model nevertheless forms a very convenient starting point.

## 2.2  Thermalization

Thermalization is defined as the transition of the system to a state in which it consists entirely of bosons and fermions distributed according to the equilibrium distributions

$$n_B(x,p) = \frac{1}{e^{p \cdot u/T} - 1},$$    9.

$$n_F(x,p) = \frac{1}{e^{(p \cdot u - \mu)/T} + 1},$$    10.

where $T = T(x)$ is the temperature, $\mu = \mu(x)$ the chemical potential (anti-fermions have the sign of $\mu$ inverted) and $u^\mu = u^\mu(x)$ the four-velocity, each depending locally on $x^\mu = (t, x_T, z)$. If the system is locally thermalized, its behavior can be computed from the energy-momentum and baryon-number conservation equations $\partial_\mu T^{\mu\nu} = 0$ [equivalent to entropy conservation $\partial_\mu(su^\mu) = 0$] and $\partial_\mu(n_B u^\mu) = 0$ (see Section 3). This clearly is an enormous simplification. Note that the bosonic and fermionic degrees of freedom need not be those corresponding to gluons and quarks, although this is what one expects as a first assumption.

If the system does not thermalize, one may still discuss its behavior in terms of a single-particle distribution function $n(x,p)$ and its extensions to several particles, i.e. use kinetic theory. The question of thermalization and the effects of small deviations from it (dissipation, transport coefficients) have been discussed in (50–52) and kinetic theory in (53–56); for reviews, see (57, 58). In this context, kinetic theory is also related to the small-$x$ problem in QCD (59).

Consider now a central (zero-impact-parameter) ultrarelativistic collision of two large nuclei A and B. With thermalization, the sequence of events at zero transverse coordinate, as a function of time $t$ and longitudinal coordinate $z$, could be as shown in Figure 3. Before the collision, for $t < 0$, the partons in the system are distributed in the beam and target according to two (possibly smeared) delta functions: $n(z,p) \approx \delta(z \pm vt)$. After the collision there follows a preequilibrium phase during which individual parton-parton collisions start thermalizing the distributions. The entropy of the system increases from the initial value 0.

Figure 3 is drawn assuming that the system thermalizes in such a way

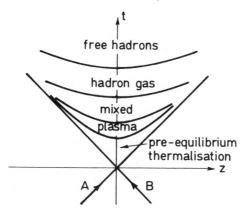

*Figure 3*  A collision of two large nuclei A + B in a space-time diagram. Here the two nuclei collide along the light cone trajectories marked A and B. Particles are produced in the region of the forward light cone. The various stages of the expanding matter at various times is marked on the figure.

that the initial temperature is large enough to place the system initially in the quark gluon plasma phase. In the inside-outside cascade picture, the proper time determines the reinteraction ability of the partons. Similarly the initial thermalization at $T_i$ takes place at a fixed proper time $\tau_i$ (60). The further evolution of the system is discussed in Section 3.

Thermalization here is discussed in an entirely phenomenological way; there exists no quantum theoretical treatment for it. At best one can present qualitative mean free path arguments to motivate it. For instance (47), the mean free path of a quark in QCD matter of energy density $\varepsilon$ would be

$$\lambda = (5, 0.5, 0.01)\,\text{fm for } \varepsilon = (0.15, 2, 200)\,\text{GeV/fm}^3. \qquad 11.$$

Another possibility is to use QCD finite-temperature perturbation theory to compute both gluon and quark mean free paths (50, 51), with the result

$$\lambda_g = 0.5 - 0.05\,\text{fm}, \quad \lambda_q = 2 - 0.2\,\text{fm for } \varepsilon = 1 - 1000\,\text{GeV/fm}^3. \qquad 12.$$

The uncertainties in these estimates are illustrated by the fact that different constant factors $T_0$ used in the logarithm in $\alpha_s \approx 1/\ln(T/T_0)$ in (50) and (51) gave rise to results differing by almost a factor two for the range of $T$ in question.

Estimates indicate that $\varepsilon$ of the order of 2 GeV/fm$^3$ should be attainable even in average collisions of two nuclei. Equations 11 and 12 indicate that for $A$ more than 100 the transverse size of the system ($\sim R_A$) is clearly larger than the mean free path and there is some motivation for thermalization. For oxygen ($A = 16$) the situation is very marginal and probably very atypical events with large $\varepsilon$ would be needed for thermalization.

One should emphasize that only qualitative arguments can be given for thermalization and it should basically be regarded as a practical first approximation to be tested experimentally.

## 2.3    *Cascade Simulation of Early Stages*

A numerical code simulating the thermalization process in an ultra-relativistic nucleus-nucleus collision has been developed by Boal (61). The initial momenta of the partons are taken from the standard structure functions of quarks and gluons, and in coordinate space they are randomly distributed within the nucleus so that the density of soft partons is lower than that of hard ones. Partons are then allowed to collide according to the most singular part of the $2 \to 2$ QCD cross sections. The code can be used, for instance, to follow how the energy density develops.

The limitations of codes of this type are seen from the fact that infrared singularities appear for small $x$ in the structure functions and in the forward direction for the scattering processes and have to be arbitrarily cut off. The physical reason for this is that partons are not the only degrees of freedom used to describe the process. Nonperturbative effects also have to be included. This can be done, for example, in the framework of color field models.

## 2.4    *Color Field Models*

Color field models (48, 62–67) assume that a color field $\mathscr{E}_1$ is formed between the receding nuclear disks after the collision. Note that the dimension of $\mathscr{E}$ is GeV$^2$ and that it is related to the total charge $Q$ by $Q = A_T \mathscr{E}_1$, where $A_T$ is the transverse area. Associated with $\mathscr{E}_1$ there is a time scale $\tau_0 = 1/\sqrt{\mathscr{E}_1}$. One thus has at one's disposal a new dimensionful parameter with which to model nonperturbatively the collision process.

In this model, there now are three questions to discuss: 1. magnitude of $\mathscr{E}_1$ and its process dependence, 2. conversion of the field to particles, 3. thermalization of the particles. Beyond that the process is described by fluid hydrodynamics. The color field models thus permit one to describe the entire sequence of events in the nuclear collision process, including the thermalization stage. There is some support for the model from hadron-nucleus collisions (48) but it has to be developed further and confronted with nuclear collision data to reveal its true potential.

# 3.  HYDRODYNAMICS IN ULTRARELATIVISTIC NUCLEAR COLLISIONS

With the possibly large energy densities achievable in ultrarelativistic nuclear collisions, and the large transverse extent of nuclei, it seems plausible that there is a viable hydrodynamic description of the collisions. Since the transverse extent of the system is large, we expect that, in the early phase, a good approximation is to treat the expansion as entirely longitudinal, that is along the beam axis. This expansion is particularly simple in the central region (60) and can be solved for analytically; it is the subject of Section 3.1. The situation in the fragmentation region is a bit more complicated (45, 68–71), since the space-time development of the matter is more complex and since the distribution of particles in rapidity is not uniform. The study of the fragmentation region is the subject of Section 3.2. At late times in the collision, the expansion becomes $3+1$ dimensional. To describe this expansion, $3+1$ dimensional hydrodynamic equations must be solved. This is a formidable task and has been accomplished so far only in the central region (47, 72–77). The general outline of the procedure for solving the $3+1$ dimensional hydrodynamic equations is the subject of Section 3.3, and the results of these computations are discussed in the sections on experimental probes. At late stages in the collision, the hadrons begin to decouple from one another and the hydrodynamic equations break down. It is at this time that the particle distributions become frozen into their final values. We discuss decoupling in Section 3.4.

We continue this introductory discussion with simple arguments that there should be a valid hydrodynamical treatment of the collision by first discussing time scales for expansion and comparing the expansion time, $\tau_E$, to the collision time $\tau_C$. If the system is to be well approximated as a perfect fluid that is adiabatically expanding, then $\tau_E \gg \tau_C$. In a perfect fluid expansion, the total entropy is conserved. To estimate the time of perfect fluid expansion we can thus use conservation of entropy. We take the entropy density to be proportional to $N_{\text{dof}}T^3$, where $N_{\text{dof}}$ are the number of particle degrees of freedom at the time of interest. The volume of the system is proportional to $V \approx t^d$ where $t$ is the time and $d$ is the dimensionality of the expansion. We therefore can relate the initial and final times to the entropy densities as

$$\left(\frac{t_f}{t_i}\right)^d = \frac{N_{\text{dof}}^i}{N_{\text{of}}^f}\frac{T_i^3}{T_f^3} \approx 10\text{--}10^4.$$ 

13.

At early time, the expansion is one dimensional and at later times becomes three dimensional. We estimate therefore that $t_f/t_i \approx 10\text{--}10^3$. Detailed $1+3$

dimensional hydrodynamic computations show that the final decoupling time is probably somewhere in the range of $t_f \approx 20\text{--}50 \text{ fm}/c$.

Large nuclei are clearly more favored systems for producing and studying a quark gluon plasma. This follows simply from the facts that the average energy density achieved is larger, and that the system is physically larger in transverse extent. As discussed in Section 2.1, collision lengths are then likely to be smaller than the size of the system. Experimental information on this has already been obtained at Bevalac energies (see Section 6).

## 3.1   *1 + 1 Dimensional Results for the Central Region*

The hydrodynamic equations in the central region are simplified by the observation that if the rapidity distribution is approximately flat, that is $y$ independent, then the description of this kinematic region should be approximately Lorentz invariant (60). We may introduce the space-time rapidity

$$\eta = \ln\left(\frac{t+z}{t-z}\right) \qquad\qquad 14.$$

and the proper time

$$\tau = \sqrt{t^2 - z^2}. \qquad\qquad 15.$$

The space-time rapidity equals the momentum-space rapidity for freely streaming particles that originated at $x = t = 0$. In the hydrodynamic model for the central region, it is also equal to the momentum-space rapidity of the fluid.

The Jacobian of the transformation above transforms $dt dz$ into $\tau d\tau d\eta$. Since the total entropy is conserved under perfect fluid expansion, we therefore find that the entropy density $\sigma$ is given by

$$\sigma(\tau) = \sigma(\tau_0)(\tau_0/\tau). \qquad\qquad 16.$$

To derive this result, we have used the fact that the entropy density is independent of rapidity. For an ideal gas, we therefore have

$$T = T_0(\tau_0/\tau)^{1/3}. \qquad\qquad 17.$$

The temperature falls very slowly with proper time, and more so at later times. Using a bag model equation of state, one finds that the time to go from an initial temperature of 300 MeV to a temperature of 150 MeV (in the pion phase) is measured in hundreds of femtometers. This time is so long that, for hydrodynamic simulations in $3 + 1$ dimensions of nuclear collisions, the effects of transverse expansion become important. This transverse expansion takes some time to set in because of the large but finite nuclear size, and has the effect of increasing the expansion rate.

We may now use this knowledge of the hydrodynamic equations to estimate the initially achieved energy density in terms of the final-state conditions. We first note that the hydrodynamic equations are entropy conserving and, to a good approximation, multiplicity conserving. The energy density in the initial configuration is given by the number of particles per unit volume times a typical energy per particle, which can be taken as the transverse mass at this time, $m_T^2 = m^2 + p_T^2$. The number of particles per unit volume is simply

$$\frac{N}{V} = \frac{1}{\pi R^2} \frac{dN}{dy} \frac{1}{\tau}.$$     18.

In the early stages of the collision when particles are forming, we expect that $m_T \approx 1/\tau$, a situation that is true in a variety of models of particle formation. The initial energy density is therefore

$$\varepsilon \approx \frac{1}{\pi R^2} \frac{dN}{dy} m_T^2.$$     19.

In this equation, $m_T^2$ is measured at the initial time. In general as the system expands, $m_T$ should monotonically decrease, and using experimental values in this equation should provide a lower bound on the energy density at formation.

## 3.2  1+1 Dimensional Results for the Fragmentation Region

Although the dynamics is more complicated in the fragmentation region, the methods described above may be generalized to include fragmentation (69–71). This may be done by providing sources for the hydrodynamic equations corresponding to the materialization of matter after the collision of the two nuclei. The source for the stress energy tensor may be related to the assumed initial distribution of particles. For the baryon-number currents, it is possible to treat the baryons as conserved through the entire scattering process (70). The source treatment for the stress energy tensor assumes that the sources correspond to the materialization of particles from pair production, and is consistent with the inside-outside cascade model.

There have been various numerical estimates of the energy and baryon-number density in the fragmentation region. Most recent treatments argue that the energy density in the fragmentation region probably approaches that in the central region. At asymptotically high energies, however, we expect that the high multiplicity in the central region will produce asymptotically higher energy densities.

The achieved baryon-number densities are quite controversial. The latest estimate shows that the baryon-number density may become up to ten times greater than that of nuclear matter (70), in sharp contrast with older results (69) where values only twice nuclear matter were found. The new result is in accord with previous order of magnitude estimates (45).

## 3.3    *1+3 Dimensional Results*

To handle the late stages of matter evolution, when the expansion becomes three dimensional, $1+3$ dimensional hydrodynamic equations must be solved. This is not too complicated in the central region, because there Lorentz invariance eliminates the rapidity as a variable. If we assume central collisions, there is azimuthal symmetry. The resulting hydrodynamic equation is effectively a $1+1$ dimensional problem.

The solution of the three-dimensional expansion of the system is important for computing transverse momentum and energy distributions. Such a computation can relate enhancements in these distributions to properties of the equation of state, which in fact determines the solution to the hydrodynamic equations. Such a solution also allows for detailed computations of the spectra of photons and dileptons, as well as strange particle production, as is discussed below.

For ideal gas equations of state, this problem is easily solved by the method of characteristics (72, 73). For a bag model equation of state, the method of characteristics is no longer applicable, since a shock front develops in the transverse rarefaction at the interface of the mixed phase with the hadron gas phase (74, 75). A variety of methods have been used to deal with this mathematical problem (74–76). The entire time evolution of the system from very high temperature to asymptotically low temperatures can now be solved in a minimum of computer time.

The treatment of the hadronization of the plasma via a mixed phase is subject to some criticism in these approaches. If there is truly a first-order transition between plasma and a hadron gas, such an approach may be invalidated by supercooling of the plasma, and various deflagration and detonation singularities may develop (78–81). If the transition is weak first order or second order, the treatment of the transition region as a mixed phase is probably quantitatively good.

There are a variety of uncertainties in these computations concerning the initial conditions. Only one unknown parameter, the multiplicity can be determined from experiment. The initial temperature affects the computation of dilepton distributions, and therefore the measurement of the dileptons may give a measure of this. Uncertainties in the shape of the initial matter distribution are not so important.

The treatment of decoupling follows the classical analysis of Cooper & Fry (82). It is assumed that at some fixed temperature, the hadrons decouple instantaneously from one another and become free-streaming noninteracting particles. Such an idealization leaves out much physics, such as entropy production at decoupling. If decoupling happens at an energy density much lower than that of the hadron gas at the phase transition temperature, then we expect that this sloppy treatment should provide a good approximation.

### 3.4   Decoupling and Pion Cascade in Late Stages

To handle the problem of decoupling properly and to test the assumptions underlying the hydrodynamic computations, a cascade simulation of the hadronic gas phase is useful. If the hadron gas is cool enough, then it is well approximated as a pion gas. The scattering of pions may be estimated from low-energy phase shift analysis.

Such a pion cascade is being developed by Bertsch et al (83). In addition to being better able to treat decoupling and to understand the quality of the hydrodynamic approximation, one might be able to compute the spectrum of low-energy photons and dileptons. In the low-energy region, these photons and dileptons are emitted by the bremsstrahlung process. Their distribution reflects the space-time history of the system at late times as the hadron gas freezes out.

## 4.   PROBES IN GENERAL

QCD matter formed in ultrarelativistic nucleus-nucleus collisions is at most a few tens of fm across and lives at most a few tens of $fm/c$. It is thus clear that no direct probes are feasible and that all diagnostics must be indirect and based on measuring the decay products of the matter. The general strategy of observing QCD matter and, in particular, its quark gluon plasma phase could thus be as follows:

1. Measure everything possible, e.g. differential distributions of all types of particles, and cross-correlate these with each other. Check if there is anything that cannot be understood in terms of the various models for particle production making no reference to space-time. Matter in local thermal equilibrium requires that there be some correlation that requires an explanation outside the context of simple particle production models.
2. Find evidence for collective flow irrespective of the phase the matter is in.

3. Find evidence for matter in the quark gluon plasma phase in particular.
4. If quark gluon plasma is found, diagnose its properties. How does it hadronize? What are the kinetic properties of the phase transition?

Potential experimental probes (with the physics they are sensitive to) are as follows:

1. Dileptons [quark gluon plasma, $T_i$ (for $1 < M < 3$ GeV), $T_{PT}$, collective flow (for $M \approx 1$ GeV), plasma expansion, impact parameter, resonance melting, low-mass pairs, and space-time evolution],
2. Photons (as dileptons but for $M$-dependent effects),
3. Jets (scattering cross section of quarks and gluons with plasma and hadron gas),
4. $\phi$ and $\psi$ production (quark gluon plasma),
5. Hadron $p_T$ distributions (equation of state, collective flow),
6. Strangeness (existence of hadron gas, dynamics of expansion),
7. Pion correlations (size and lifetime of the system).

All these probes should be measured together with the associated pion multiplicity $dN_\pi^{AB}/dy$ or the total transverse energy $E_T$ with the understanding that the larger these are, the more likely it is that the system will initially be in the quark gluon plasma phase. There should also be a separate trigger on small impact parameter collisions, like a cut in nuclear fragmentation or forward energy flow. The analysis of correlated variables will be complicated and one cannot argue that any of the probes will yield an unambiguous signal for the plasma. Using several different probes, it should, however, be possible to make a convincing case.

The size of the smaller of the colliding nuclei is also a very important parameter. The first experiments at ultrarelativistic energies will be done with rather small nuclei, such as oxygen, in the beam. It is quite likely that these will not yet reveal unambiguous matter effects. With the experience from Bevalac energies one might expect that $A \gtrsim 100$ will be needed before collective effects can be observed.

The various probes can also be characterized as volume or surface probes. With the former we have a probe (photon, dilepton, $\phi$, $\psi$, jet) formed in the inside of the plasma and leaking out of it to give information of the conditions inside. In the latter case the probe is either formed at the surface (hadron $p_T$, strange particles) or relates to the system as a whole (interference effects). Volume probes are clearly more direct.

## 5.    THE CORRELATION OF $p_T$ AND $dN/dy$

The correlation between $p_T$ and $dN/dy$ reflects properties of the equation of state of matter (84, 85). Measuring such a correlation is in principle straightforward.

## 5.1   *A Spherically Symmetric Example*

A correlation between $p_T$ and $dN/dy$ is easily seen from the example of a spherically expanding gas. We assume that at some initial time there is a spherically symmetric drop of hadronic matter of uniform density matter at rest. We then allow the system to expand hydrodynamically. We assume we know the volume of the initial system, $V_0$. We measure the total energy of all particles and the total multiplicity of particles in the final state. Since the system is slowly expanding at late times, the entropy of particles in the final state is known assuming the particles were produced thermally from a weakly interacting gas. Since energy and entropy are conserved in the expansion of a perfect fluid, the energy and entropy of the final state are those of the initial state. We can therefore experimentally measure the correlation between say $p_T$, which is proportional to $E/S$, and the energy density (86, 87). We can compare this to a theoretically predicted correlation determined by knowing the equation of state.

A plot of $E/S$ versus $\varepsilon$ is shown in Figure 4 for a bag model equation of state. The generic features of this curve are easy to understand. At low temperature, in the pion gas phase, and at high temperatures, in the plasma phase, we have $E/S \approx T$. The energy density in these two phases goes as $\varepsilon \approx N_{dof}T^4$. Since the number of degrees of freedom changes at the transition, there is a gap between these two limiting curves. The gap is filled by the region in which the plasma cools into a pion gas. This happens at a fixed $T$, and almost fixed $E/S$, for varying $\varepsilon$.

## 5.2   *Numerical Results for Head-On Collisions of Equal A Nuclei*

There are several problems when arguments like the above are applied to the more realistic expansion scenarios appropriate for central collisions of

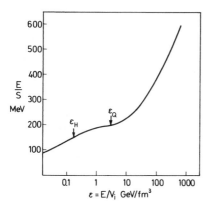

*Figure 4*   $E/S$, the energy per degree of freedom, vs $\varepsilon$, the energy density, in the MIT bag model.

heavy nuclei. First, $p_T$ is not conserved since longitudinal expansion causes the transverse momentum of individual particles to be converted into unobserved collective flow in the longitudinal direction. A correlation between $p_T$ and, say, multiplicity is therefore weaker than is the case for spherical expansion. It also depends more on the detailed numerical simulation of the hydrodynamic equations. Also, the initial conditions for the matter are not so well known. The final-state decoupling and perhaps a phase change may produce some entropy. Fortunately these problems do not appear to generate much dispersion in the numerical results for such a correlation (74). Finally, a severe limitation of present hydrodynamic simulations is that they are limited to the central region of collisions with zero impact parameter. The present computations may therefore only provide information on head-on collisions and their fluctuations. Since the number of particles is already large, the fractional fluctuations in the multiplicity for such head-on collisions is small.

There is also the potential problem of backgrounds from conventional processes such as minijets obscuring the $p_T$ enhancement from a quark gluon plasma (89). At energies typical of the SPS collider, production of minijets is presumably responsible for the high multiplicity events. In nuclear collisions at energies less than or equal to those proposed at RHIC, minijets are not expected to be a large background since the beam energy is low. Moreover, minijets should thermalize in the high multiplicity environment typical of central collisions of large nuclei, thus changing the

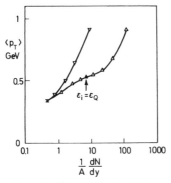

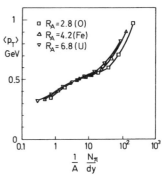

*Figure 5*  The results of a hydro-dynamic simulation for the correlation of $p_T$ vs multiplicity in head-on heavy-ion collisions for an ideal gas equation of state (*upper curve*) and a bag model (*lower curve*).

*Figure 6*  A scaling feature of the $p_T$ distributions as computed in hydrodynamic simulations. A graph of $p_T$ vs $dN/dy$ scaled by $1/A$ for a variety of $A$.

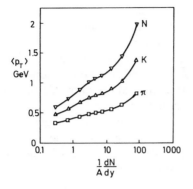

*Figure 7* Average $p_T$ vs $dN/dy$ for a variety of particles. Notice that the enhancement in $p_T$ for heavy particles is much greater than that for light. The pion enhancement is quite small.

initial conditions by making the matter initially a little hotter, but yielding a correlation between $p_T$ and $dN/dy$ that may be computed by hydrodynamics.

In Figure 5, the results of a hydrodynamic computation of $p_T$ vs $dN/dy$ are shown for an equation of state typical of the bag model and a pion gas equation of state. The large difference between these curves suggests that an experimental probe of this correlation can resolve various equations of state. A general feature is that the softer is the equation of state, the softer is the $p_T$. A quark gluon plasma produces lower $p_T$ particles at fixed multiplicity than does a pion gas.

In Figure 6, the same correlation is shown for head-on collisions of various nuclei. The curves scale approximately as a function of $1/A\ dN/dy$. A factor of $1/A^{2/3}\ dN/dy$ arises because the result must be proportional to

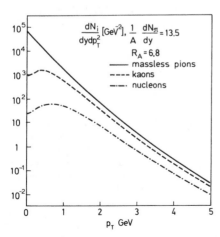

*Figure 8* The $p_T$ distributions for a variety of particles.

the multiplicity per unit area. An additional suppression by a factor of $A^{1/3}$ arises from the softening effects of longitudinal expansion.

As has been argued by Shuryak (84), heavy particles should show the effect of collective transverse expansion more strongly than light particles. This is shown in Figure 7, where $p_T$ is computed for pions, kaons, and nucleons as a function of multiplicity. The physical origin of this effect is that in fluid expansion, there is a collective fluid velocity $v$. Heavier particles have larger masses and therefore $p = mv\gamma$ is correspondingly larger.

In Figure 8, the $p_T$ distributions of pions, kaons, and nucleons are shown. The distribution of nucleons clearly shows the effects of collective flow, with the local maximum in $dN/dp_T^2$ at $p_T \approx 1$ GeV.

In Figure 9, an attempt is made to fit the experimentally observed correlation between $p_T$ and transverse energy per unit rapidity as seen in the JACEE collaboration (88). In the figure, $\varepsilon_i = 3/(4\pi)A^{-2/3} dE_T/dy$, where $A$ is the smaller $A$ of the nuclei involved in the collision. The JACEE data rise too rapidly to be explained by a quark gluon plasma. The data do seem to be fit by a pion gas model (dashed line), but the temperatures at which the system would be required to be in an ideal pion gas are quite large, and we consider this explanation unlikely. Either there is some nonthermal source of high $p_T$ particles in the JACEE data, something is wrong with the space-time picture of the collisions (86), or something is wrong with the data analysis.

# 6.  FLUID EFFECTS AND JETS

One of the most important phenomena to establish is that the matter produced in nuclear collisions behaves collectively as a fluid, and many of the probes discussed here ($p_T$ effects, dileptons) give information on this. This question has already been studied experimentally at Bevalac energies

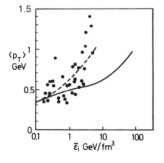

*Figure 9* An attempt to fit the JACEE cosmic-ray data with a bag model and ideal gas equation of state. The upper curve is the ideal gas, and the lower curve is the bag model. On the lower axis, energy density is defined to be $3/(4\pi) dE_T/dy$.

(90, 91). The data obtained also throw some light on the size of systems necessary for fluid dynamic effects to become important.

In collisions of nuclei of small impact parameter, single-particle collisions occur at large transverse momentum. The nuclei do not collectively flow in a given transverse direction unless there are subsequent rescatterings among the constituents of the nuclei. If these subsequent rescatterings do not occur, the transverse momentum of each particle is randomly oriented. To get collective flow, one needs rescattering, and this should be enhanced in collisions at small impact parameter, and collisions of large-$A$ nuclei.

In Figure 10, the flow angle is plotted for various measures of the impact parameter (large impact parameters at the top and small at the bottom of

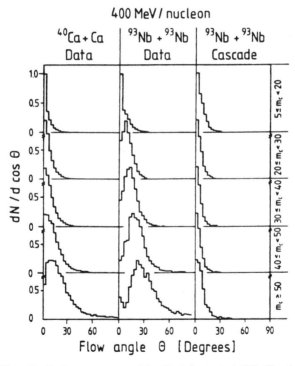

*Figure 10* Flow distributions as measured by Gustafsson et al (91). The three columns represent a light nucleus, Ca, a heavy nucleus, Nb, and a cascade simulation of Nb. Going from the top to the bottom corresponds to selecting decreasing impact parameter. Notice the nontrivial value of the typical flow angle in Nb collisions at small impact parameter, corresponding to a collective phenomenon (not seen in the simple cascade simulation presented here).

the figure) for two nuclei (small on the left and large on the right). The effect of flow is a correlation between the particle distributions and a nonzero average flow angle. The nuclei collectively flow through a nonzero collision angle. For a sum of uncorrelated single-particle collisions, the flow angle will be zero. The far right-hand column represents a simulation of Nb-Nb collisions showing that, within a simple cascade calculation, no flow angle arises. Little evidence of flow is shown for nuclei as large as calcium, and collective effects begin to become important for nuclei the size of niobium.

Another potential experimental probe of quark gluon plasma is the quenching of jets. The rescattering of jets after their production in a quark gluon plasma in principle provides a probe of the plasma and hadronic matter as the jet plows through the evolving system (92–94). The jets will scatter from the constituents of the plasma as well as from the constituents of hadronic matter that forms later. The degree of scattering is a measure of the quark-matter or gluon-matter cross section.

This scattering can dramatically change quantities such as the jet acoplanarity and can produce phenomena such as single jets. Theoretical predictions of jet acoplanarity for a variety of jet $p_T$ on a variety of nuclei have been performed (94). For nuclei with $A \approx 100$, and for jets of mass 10 GeV, the differences induced by the presence of a matter distribution are striking, and the rescattering removes the planar nature of the jets. Even at jet mass of 20 GeV, the difference is still significant, and the jets are remarkably acoplanar. In fact, at these masses the jets are probably largely extinguished.

The experimental measurement of this acoplanarity is very difficult. Particles with low rapidities along the jet axis, $y < 2$, must somehow be removed from the sample of particles contributing to the acoplanarity distribution. These low $p_T$ particles arise from conventional low $p_T$ processes and have little in common with the high $p_T$ particles associated with the jet.

## 7.   DILEPTONS

### 7.1   *General Properties; Correlation Between Mass and Time of Emission*

Dileptons from the plasma (95–106) are dominantly formed by the process $q + \bar{q} \rightarrow \gamma^* \rightarrow \ell^+ + \ell^-$ and from the hadron gas by $h + \bar{h} \rightarrow \ell^+ + \ell^-$; for small dilepton mass $M$, pairs emitted from soft virtual photons with a bremsstrahlung-type distribution are important. The rate per unit time and volume is easy to estimate and depends only on the temperature $T$ (and dilepton variables). In the hydrodynamical scenario (Section 3) the space-time history of the system is known and the predicted dilepton rate in nuclear collisions can be computed. Ideally one compares this prediction with the experimentally observed rate; agreement would verify the scenario and determine its parameters.

The general strategy of using dileptons as probes of quark gluon plasma is as follows. Measure the process $A + B \rightarrow \ell^+ + \ell^- + X$ as differentially as possible using as variables the mass $M$, rapidity $y$, $p_T$ of the dilepton, forward energy flow (as impact parameter trigger), associated hadron multiplicity $dN_h^{AB}/dy$, transverse energy $E_T$ of the event, etc. What is expected can be conveniently discussed in terms of $M$:

1. At large $M$ ($M > 3$ GeV, preferably $M > 10$ GeV) one observes the entirely nonthermal, single-collision Drell-Yan mechanism. The measured rate determines the structure functions of quarks in nuclei. There is no correlation between the pion and dilepton rates. The Drell-Yan pairs are emitted at times $t < 1/M < 0.1$ fm.
2. For $1 < M < 3$ GeV (the limits are very rough) and for events having large $dN_\pi/dy$, the dilepton rate is proportional to the square of $dN_\pi/dy$ and diagnoses the properties of quark gluon plasma. These pairs are also emitted very early in the course of the expansion (one has dominantly $M \approx 5T$ and $T_i$ is of the order of a few hundred MeV for times of the order of a few fm). Hardly any transverse flow has time to develop. The rapidity fluctuations of pions and dilepton are correlated.
3. For $M \approx 1$ GeV the pairs are emitted from the matter after it has cooled down to the transition temperature $T_{PT}$ or even below it to the hadron gas phase. The times involved may be of the order of tens of fm. Transverse flow will have time to develop.

The mass-time correlation is potentially a very useful property of the dilepton signal since it permits one to follow the time development of the system. In practice, it may be very difficult to unravel this signal in a clean form. In the interesting mass range between 1 and 3 GeV there are several backgrounds—bremsstrahlung, decays of charmed particles, preequilibrium emission, etc.

## 7.2   Use of Dileptons to Diagnose Quark Gluon Plasma

How dileptons of masses around 2 GeV diagnose properties of quark gluon plasma is illustrated by Figure 11. The figure is computed (100) for two different $1 + 3$ dimensional flows in central $U + U$ collisions having the same total entropy (same $dN_\pi/dy = 26 \cdot 238$) but different initial temperatures and initial times: $T_i = 350$ MeV, $\tau_i = 1.5$ fm, and $T_i = 500$ MeV, $\tau_i = 0.5$ fm; the transition temperature is assumed to be 200 MeV. The curve marked Mixed Phase gives the yield of dileptons from this fixed

$T = 200$ MeV; it is independent of $T_i$ and decreases quickly above $M = 1$ GeV. The quark phase contribution, on the other hand, depends on $T_i$ so that the larger $T_i$, the larger the rate for large masses. For this very large multiplicity, the Drell-Yan rate is clearly below the thermal rate.

Figure 11 also illustrates the complexity inherent in plasma diagnostics. When the data are plotted for various $dN_\pi/dy$, only the combination $T_i^3\tau_i$ is fixed. For each value of $dN_\pi/dy$ different values of $T_i$ are possible. It is thus not possible to determine the value of $T_i$ for each event separately, only the average value of $T_i$ for the events having a given value of $dN_\pi/dy$.

A further reservation is that preequilibrium emission is not yet included. It could affect the high-$M$ end of Figure 11.

## 7.3    Use of Dileptons to Diagnose Collective Flow

Above we discussed how the dependence of $\langle p_T \rangle$ on mass can be an indication of collective flow (which has to be transverse to be observable). A qualitatively similar mechanism operates for dileptons: if they are emitted from a matter flowing transversally, an increase in their transverse momentum is observed. The transverse flow has time to develop only if the system lives longer than $R_A/v_s \approx 10$ fm (as is confirmed by $1+3$ dimensional numerical computations). By then the system has cooled to temperatures $\sim 200$ MeV and pairs with $M \approx 1$ GeV are dominantly emitted. Transverse flow effects can thus be sought by looking at pairs with $M \approx m_\rho$.

A concrete example is shown in Figure 12. Here the dilepton rate for $M = 0.8$ GeV and $y \approx 0$ is shown as a function of the transverse mass $M_T = (p_T^2 + M^2)^{1/2}$ separately for the mixed phase ($T = 200$ MeV) and the quark gluon plasma phase with ($1+3$ dimensional hydro) and without

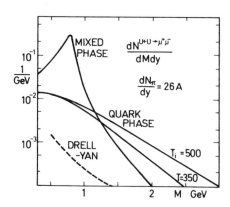

*Figure 11*    Mass distribution of dileptons at $y \approx 0$ from two flows with the same $dN_\pi/dy$ but different $T_i$.

(1+1 dimensional hydro) transverse flow. The flow parameters are as indicated in the figure. The quark gluon plasma phase has no time to develop transverse flow and is virtually unaffected by it. The mixed phase lives very long and ultimately develops a rapid transverse flow. The effect on the predicted rate is strong; the rate for a $p_T$ of a few GeV is changed by orders of magnitude.

The same effect can be seen in more detail by plotting the dilepton rate as a function of $M$ for $M_T = 1, 2, 3, 4$ GeV for example (105). If there is no transverse flow, the $\rho$ peak is clearly seen for small $M_T$, but at large $M_T$ the $M$-independent quark gluon plasma phase dominates and the $\rho$ peak disappears (dashed line in Figure 12). When transverse flow is included, the $\rho$ peak persists until large $M_T$. Therefore, this is the range to study to observe collective flow with dileptons.

## 7.4   Resonance Melting, $\psi$ Production

The vector meson resonances ($\rho$, $\omega$, $\phi$, $\psi$) are easy to observe in the dilepton spectrum. On the other hand, they clearly cannot exist in the quark gluon plasma phase, they melt away if the hadron gas is heated above the transition temperature (103). Naively one thus could expect their absence to be a signal for the plasma phase. However, they certainly are formed during the mixed and hadron phases (and couple to virtual photons that leak out of the system) and the problem becomes a quantitative one.

The situation with the $\rho$ and $\omega$ resonances was discussed above: they are abundantly formed during late stages of the process and the melting signal becomes a signal for collective flow. With the $\psi$ (and, to a lesser extent, the $\phi$) the situation may be different (106). The expected dominant production mechanism is two-gluon fusion $gg \rightarrow \bar{c}c \rightarrow \psi$. This would take

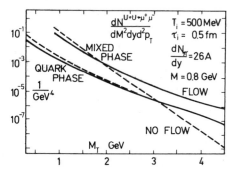

*Figure 12*   The transverse mass distribution of dileptons with $M = 0.8$ GeV $\approx m_\rho$ for a 1+3 dimensional flow with parameters as marked on the figure. The dashed line shows the result for a 1+1 dimensional flow with no transverse flow.

place very early, before or during the plasma stage but would then actually be hindered by the existence of the plasma. The magnitude of the $\psi$ peak above the dilepton background could thus be decreased if plasma existed. This is a very straightforward experimental quantity to study, but one must remember that the prediction of cross sections of processes involving charmed quarks has been notoriously difficult. In this case additional processes and surface effects could also give a sizable contribution.

## 8.  STRANGENESS

Strange particles seen in the detectors of a nuclear collision experiment are all formed at decoupling of the system—provided that an expanding system of matter is seen at all. They thus have no direct contact with a possible quark gluon plasma phase existing early during the history of the system and, if they exist at all, they are a very indirect probe of the plasma phase. On the other hand, if one could experimentally prove that the strange particles decouple from a hadron gas in local thermal equilibrium, they could serve as a signal for the existence of this system and as a probe of its properties. This in itself makes it worthwhile to study the production of strange particles carefully.

There has been much work done on the subject of strange-particle production in heavy-ion collisions (107–116). It was originally believed that in a quark gluon plasma in thermal equilibrium, there would be a larger strangeness abundance than in a corresponding hadron gas, and it was successfully argued that in a quark gluon plasma as produced in a heavy-ion collision, there would be time to produce an equilibrium abundance of strange quarks (107, 108). The first argument has been disputed (109–115), while the second argument has been supported by various computations.

While a hadron gas may be almost as rich in strange hadrons as is a quark gluon plasma, if the temperature in the plasma or hadron gas at which strangeness production decouples is sufficiently large, then there is quite likely to be a large strangeness abundance produced. We can see that this is the case by computing the strangeness abundance under the conservative assumption of a hadron gas. The result is quite sensitive to the assumed decoupling temperature for strangeness, but this is precisely why a measurement of this quantity is interesting.

Consider first strange mesons in the central region. In pp collisions the $K/\pi$ ratios are of the order of 0.1. For example, in 270 GeV + 270 GeV $p\bar{p}$ collisions the average number of $\pi^+ + \pi^-$ is 24 and that of $K^+ + K^-$ is 2.2. In thermal equilibrium the density of a boson with mass $m$ and degeneracy factor $g$ would be ($K_2$ is a Bessel function)

$$n = \frac{g}{2\pi^2} T m^2 \left[ K_2\left(\frac{m}{T}\right) + \frac{1}{2} K_2\left(\frac{2m}{T}\right) + \dots \right].$$    20.

Evaluating this at $T = 200, 160, 100$ MeV gives $(K^- + \bar{K}^0)/(\pi^+ + \pi^-) \approx 0.4$, 0.3, 0.1. Observing an enhanced $K/\pi$ ratio would thus be a signal for the kaons coming from hadron gas in thermal equilibrium and the magnitude of the ratio could be a probe of the decoupling temperature. Note that the mass dependence of $p_T$ discussed in Section 5.2 tests the same fact but also the collective flow of the system.

With the above numbers there is the risk that no firm conclusions can be drawn, even though the hadron gas system exists. If the decoupling temperature is close to 100 MeV, the $K/\pi$ ratio will be indistinguishable from its value in nonthermal pp collisions. This enhances the importance of performing a proper kinetic theory analysis of decoupling from the hadron gas.

One can also try to extend the above arguments to baryons in the central region. Again, at the p$\bar{\text{p}}$ collider the average numbers of $p + \bar{p}$, $n + \bar{n}$, $\Lambda + \Sigma^0 +$ antiparticles, $\Sigma^+ + \Sigma^- +$ antiparticles, and all $\Xi$'s are 1.45, 1.45, 0.53, 0.27, and 0.2, respectively. The leading baryon and antibaryon, one of each per event, were excluded here. In other words, for produced baryons, strange/nonstrange $= 1.0/2.9 \approx 0.3$. In thermal equilibrium a computation similar to the above gives strange/nonstrange $= 0.7, 0.5, 0.3$ at $T = 200, 150, 100$ MeV. Again the value 0.3 observed in nonthermal pp collisions is obtained at $T = 100$ MeV and higher values of the decoupling temperature may lead to enhanced strangeness. Because of the low baryon density, a proper kinetic theory computation would be even more important for baryons than for mesons. For baryons in the fragmentation regions the situation is still more complicated. This case is treated, for instance, in (116).

In summary, it is very hard to connect the strange particles observed with possible quark gluon plasma in a reliable way. On the other hand, it is very important to study their production experimentally, to see if anything unexpected takes place.

## 9.   HANBURY-BROWN-TWISS EFFECT

The Hanbury-Brown-Twiss effect arises from the interference of the matter waves of identical particles as they are measured in coincidence experiments. This effect arises since there are two possible paths of particles from emission to two coincidence detectors. If the amplitudes for this process are summed and squared, even for incoherent emission amplitudes, the result depends on the distance of separation of the emission regions. For relative particle momentum $k \leqslant R$, the detection probability is modified from its incoherent form.

The measurement of identical particles closely correlated in momentum therefore allows the possibility of measuring properties of the space-time evolution of matter produced in heavy-ion collisions (117–119). One can in principle measure the size and shape of the matter at the temperature when decoupling occurs, and perhaps verify the existence of an inside-outside cascade description.

The theoretical predictions of the Hanbury-Brown-Twiss correlation are complicated by a variety of factors. The interference may be obscured by final-state hadronic interactions, which are difficult to compute. The space-time profile of decoupling is not yet so well known and depends on details of the hydrodynamic simulations as well as the details of decoupling. Assuming that decoupling occurs at late times and large transverse sizes, $t, r_T \sim O(R)$, $R$ the correlation occurs only for very small relative momentum, and is very difficult to measure.

## 10. SUMMARY AND PROSPECTS

In this review we have attempted to present the prospects for making and detecting a quark gluon plasma in ultrarelativistic heavy-ion collisions. It appears that the prospect of making a plasma in collisions of very large nuclei at very high energy is quite good. The prospects for detecting and studying such a plasma are much less clear. Perhaps the most convincing evidence might come either from the measurement of collective effects in the matter expansion or from dilepton or photon signals. To get a reasonable dilepton yield, corresponding to a high initial temperature, probably requires very large beam energy. To observe collective effects probably requires very large $A$.

The current experiments at CERN and at BNL suffer from both small $A$ and fairly low energy—even at CERN the total rapidity range is 6 while one would hope for 10. Such experiments will give insight into both the conditions that prevail in high-energy heavy-ion collisions, and the experimental environment that will exist for such experiments. Still there is one characteristic of nuclear collisions these early experiments illustrate: the importance of geometric effects in minimum-bias events. In a majority of collisions only a part of the incident nucleus interacts, which is ineffective. For the observation of quark gluon plasma one has to trigger on those events in which the entire beam nucleus interacts. Possible triggers are an absence of beam fragments in the forward direction, a large $E_T$, or a large multiplicity.

*Literature Cited*

1. Collins, J. C., Perry, M. *Phys. Rev. Lett.* 34: 1353 (1975)
2. Witten, E. *Phys. Rev.* D30: 272 (1984)
3. Freedman, B. A., McLerran, L. *Phys. Rev.* D17: 1109 (1978)
4. Applegate, J. H., Hogan, C. J. *Phys. Rev.* D31: 3037 (1985)
5. Kajantie, K., Kurki-Suonio, H. *Phys. Rev.* D34: 1719 (1986)
6. Hogan, C. J. *Phys. Lett.* 113B: 172 (1983)
7. Hogan, C. J. *Mon. Not. R. Astron. Soc.* 218: 629 (1986)
8. Iwamoto, N. *Phys. Rev.* D28: 2353 (1983)
9. Bialas, A. In *Quark Matter Formation and Heavy Ion Collisions, Proc. Bielefeld Workshop, May 1982*, ed. M. Jacob, H. Satz. Singapore: World Scientific (1982)
10. Kinoshita, K., Minaka, A., Sumiyoshi, H. *Z. Phys.* C8: 205 (1981)
11. Bialas, A., Czyż, W., Lesniak, L. *Phys. Rev.* D25: 2328 (1982)
12. Capella, A., Pajares, C., Ramallo, A. V. *Nucl. Phys.* B241: 75 (1984)
13. Otterlund, I., Garpman, S., Lund, I., Stenlund, E. *Z. Phys.* C20: 281 (1983)
14. Busza, W., Goldhaber, A. S. *Phys. Lett.* 139B: 235 (1984)
15. Hüfner, J., Klar, A. *Phys. Lett.* 145B: 167 (1984)
16. Daté, S., Gyulassy, M., Sumiyoshi, H. *Phys. Rev.* D32: 619 (1985)
17. Bamberger, A., et al. *CERN preprint CERN/EP 86-194* (1986)
18. Polyakov, A. M. *Phys. Lett.* 72B: 427 (1978)
19. Susskind, L. *Phys. Rev.* D20: 2610 (1979)
20. McLerran, L., Svetitsky, B. *Phys. Lett.* 98B: 195 (1981), *Phys. Rev.* D24: 450 (1981)
21. Kuti, J., Polonyi, J., Szlachanyi, K. *Phys. Lett.* 98B: 199 (1981)
22. Kajantie, K., Montonen, C., Pietarinen, E. *Z. Phys.* C9: 253 (1981)
23. Engels, J., Karsch, F., Montvay, I., Satz, H. *Phys. Lett.* 101B: 89 (1981); *Nucl. Phys.* B205: 545 (1982)
24. Celik, T., Engels, J., Satz, H. *Phys. Lett.* 125B: 411 (1983); 129B: 323 (1983)
25. Celik, T., Engels, J., Satz, H. *Phys. Lett.* 133B: 427 (1983); *Nucl. Phys.* B256: 670 (1985)
26. Gavai, R., Karsch, F. *Nucl. Phys.* B261: 273 (1985)
27. Fukugita, M., Ukawa, A. *Phys. Rev. Lett.* 57: 503 (1986)
28. Kennedy, A., Pendleton, B. J., Kuti, J., Meyer, K. S. *Phys. Lett.* 155B: 414 (1985)
29. Gottlieb, S. A., et al. *Phys. Rev. Lett.* 55: 1958 (1985)
30. Christ, N., Terrano, A. *Phys. Rev. Lett.* 56: 111 (1986)
31. Gottlieb, S. A., Kennedy, A. D., Kuti, J., Toussaint, D., Meyer, S., et al. Univ. Calif. Preprint UCSD-IOPIO-271 (1987)
32. Redlich, K., Satz, H. *Bielefeld preprint BI-TP85/33* (1985)
33. DeGrand, T. A., DeTar, C. E. *Colorado preprint COLO-HEP-131* (1986)
34. DeGrand, T. A., DeTar, C. E. *Phys. Rev.* D34: 2469 (1986)
35. Kanaya, K., Satz, H. *Bielefeld preprint BI-TP86/16* (1986)
36. Karsch, F., Wyld, H. W. *Urbana preprint ILL-(TH)-86-51* (1986)
37. Satz, H. *Ann. Rev. Nucl. Part. Sci.* 35: 245 (1985)
38. Svetitsky, B. Proc. Quark Matter '86, *Nucl. Phys.* A461: 71c (1987)
39. Landau, L., Pomeranchuk, I. *Dokl. Akad. Nauk SSSR* 92: 535, 735 (1953); in *Collected Papers of Landau*, ed. D. ter Haar, papers 75, 76. New York: Gordon & Breach (1967)
40. Gribov, V. N. 1973. Space-time description of hadron interaction at high energies. VIII Leningrad Nucl. Phys. Winter Sch.
41. Bjorken, J. *Lectures at the DESY Summer Institute, 1975*, ed. J. G. Korner, G. Kramer, D. Schildnecht. Berlin: Springer (1976)
42. Casher, A., Kogut, J., Susskind, L. *Phys. Rev.* D10: 732 (1973)
43. Kancheli, O. V. *Pisma Zh. Eksp. Teor. Fiz.* 18: 469 (1973)
44. Levin, E. M., Ryskin, M. G. *Yad. Fiz.* 31: 429 (1980)
45. Anishetty, R., Koehler, P., McLerran, L. *Phys. Rev.* D22: 2793 (1980); McLerran, L. *Proc. 5th High Energy Heavy Ion Study*, LBL-12652. Berkeley, Calif: Lawrence Berkeley Lab. (1981)
46. Kisiliewska, D. K. *Acta Phys. Pol.* B15: 1111 (1984)
47. von Gersdorff, H., McLerran, L., Kataja, M., Ruuskanen, P. V. *Phys. Rev.* D34: 794 (1986)
48. Kerman, A., Matsui, T., Svetitsky, B. *Phys. Rev. Lett.* 56: 219 (1986)
49. Hwa, R., Kajantie, K. *Phys. Rev. Lett.* 56: 696 (1986)
50. Hosoya, A., Kajantie, K. *Nucl. Phys.* B250: 666 (1985)
51. Danielewicz, P., Gyulassy, M. *Phys. Rev.* D31: 53 (1985)

52. Gavin, S. *Nucl. Phys.* A435: 826 (1985)
53. Heinz, U. *Phys. Rev. Lett.* 51: 351 (1983); *Phys. Lett.* 144B: 228 (1984); *Ann. Phys.* 161: 48 (1985); *Phys. Rev. Lett.* 56: 93 (1986)
54. Elze, H.-Th., Gyulassy, M., Vasak, D. *Nucl. Phys.* B276: 706 (1986); *Phys. Lett.* 177B: 402 (1986)
55. Bialas, A., Czyż, W. *Phys. Rev.* D30: 2371 (1984); Z. *Phys.* C28: 255 (1985); *Acta Phys. Pol.* B17: 635 (1986)
56. Heinz, U. *Ann. Phys.* 168: 148 (1986)
57. Matsui, T. Proc. Quark Matter '86. *Nucl. Phys.* A461: 27c (1987)
58. Heinz, U. Proc. Quark Matter '86, *Nucl. Phys.* A461: 49c (1987)
59. Gribov, L. V., Levin, E. M., Ryskin, M. G. *Phys. Rep.* 100: 1 (1983)
60. Bjorken, J. *Phys. Rev.* D27: 140 (1983)
61. Boal, D. H. *Phys. Rev.* C33: 2206 (1986)
62. Ehtamo, H., Lindfors, J., McLerran, L. Z. *Phys.* C18: 341 (1983)
63. Biro, T. S., Nielsen, H. B., Knoll, J. *Nucl. Phys.* B245: 449 (1984)
64. Bialas, A., Czyż, W. *Nucl. Phys.* B267: 242 (1986)
65. Kajantie, K., Matsui, T. *Phys. Lett.* 164B: 373 (1985)
66. Gyulassy, M., Iwazaki, A. *Phys. Lett.* 165B: 157 (1985)
67. Gatoof, G., Kerman, A. K., Matsui, T. *MIT preprint CPT 1338* (1986)
68. Kajantie, K., McLerran, L. *Phys. Lett.* 119B: 203 (1982); *Nucl. Phys.* B214: 261 (1983)
69. Kajantie, K., Raitio, R., Ruuskanen, P. V. *Nucl. Phys.* B222: 152 (1983)
70. Gyulassy, M., Csernai, L. P. *Nucl. Phys.* A460: 723 (1986)
71. Chu, M.-C. *Phys. Rev.* D34: 2764 (1986)
72. Baym, G., Friman, B., Blaizot, J.-P., Soyeur, M., Czyż, W. *Nucl. Phys.* A407: 541 (1983)
73. Bialas, A., Czyż, W. *Acta. Phys. Pol.* B15: 229 (1984)
74. Kataja, M., Ruuskanen, P. V., McLerran, L. D., von Gersdorff, E. *Phys. Rev.* D34: 2755 (1986)
75. Friman, B., Kajantie, K., Ruuskanen, P. V. *Nucl. Phys.* B266: 468 (1986)
76. Blaizot, J. P., Ollitrault, J. Y. *Saclay preprint* (1986)
77. Chernavskaya, O. D., Chernavskaya, D. C. *Kiev preprint ITP-86-66* (1986)
78. Van Hove, l. Z. *Phys.* C21: 93 (1983); 27: 135 (1985)
79. Friman, B. L., Baym, G., Blaizot, J.-P. *Phys. Lett.* 132B: 291 (1983)
80. Gyulassy, M., Kajantie, K., Kurki-Suonio, H., McLerran, L. *Nucl. Phys.* B237: 477 (1984)
81. Danielewicz, P., Ruuskanen, P. V. *Phys. Rev.* D35: 344 (1987)
82. Cooper, F., Fry, G. *Phys. Rev.* D10: 186 (1974)
83. Bertsch, G., McLerran, L., Ruuskanen, P. V., Saarkinen, E. In preparation (1987)
84. Shuryak, E. V., Zhirov, O. *Phys. Lett.* 89B: 253 (1979); *Yad. Fiz.* 21: 861 (1975)
85. van Hove, L. *Phys. Lett.* 118B: 138 (1982)
86. von Gersdorff, H., Kapusta, J., McLerran, L., Pratt, S. *Phys. Lett.* 163B: 253 (1985)
87. Redlich, K., Satz, H., *Phys. Rev.* D33: 3747 (1986)
88. Burnett, T., et al. *Phys. Rev. Lett.* 50: 2062 (1983)
89. Halzen, F., Gaisser, T. K. *Phys. Rev. Lett.* 54: 54 (1985); Jacob, M., Landshoff, P. V. *CERN preprint CERN-TH.4562/86* (1986)
90. Buchwald, G., et al. *Phys. Rev. Lett.* 52: 1594 (1984); *Nucl. Phys.* A418: 625 (1984)
91. Gustafsson, H. A., et al. *Phys. Rev. Lett.* 52: 1590 (1984)
92. Bjorken, J. D. *FNAL preprint, Fermilab-Pub-82159-T* (1982)
93. Appel, D. *Phys. Rev.* D33: 717 (1986)
94. Blaizot, J. P., McLerran, L. *Phys. Rev.* D34: 2739 (1986)
95. Feinberg, E. L. *Nuovo. Cimento* 34A: 391 (1976)
96. Shuryak E. V., Zhirov, O. *Yad. Fiz.* 24: 195 (1976); Shuryak, E. V. *Phys. Lett.* 78B: 150 (1978); Shuryak, E. V. *Sov. J. Nucl. Phys.* 28: 408 (1978)
97. McLerran, L., Toimela, T. *Phys. Rev.* D31: 545 (1985)
98. Hwa, R., Kajantie, K. *Phys. Rev.* D32: 1109 (1985)
99. Kajantie, K., Kapusta, J., McLerran, L., Mekjian, A. *Phys. Rev.* D34: 2746 (1986)
100. Kajantie, K., Kataja, M., McLerran, L., Ruuskanen, P. V. *Phys. Rev.* D34: 811 (1986)
101. Pisut, J. *Acta Phys. Pol.* B17: 911 (1986); Csernai, V., Lichard, P., Pisut, J. Z. *Phys.* C31: 163 (1986)
102. Kajantie, K. Proc. Quark Matter '86, *Nucl. Phys. A.* To be published (1987)
103. Pisarski, R. *Phys. Lett.* 110B: 155 (1982)
104. Bochkarov, A. I., Shaposhnikov, M. E. *Nucl. Phys.* B268: 220 (1986)
105. Siemens, P., Chiu, S. A. *Phys. Rev. Lett.* 55: 1266 (1986)
106. Matsui, T., Satz, H. *BNL preprint BNL-38344* (1986)

107. Biro, T., Barz, H., Lukacs, B., Zimanyi, J. *Nucl. Phys.* A386: 617 (1982)
108. Müller, B., Rafelski, J. *Phys. Rev. Lett.* 48: 1066 (1982)
109. Redlich, K. Z. *Phys.* C27: 633 (1985)
110. Glendenning, N., Rafelski, J. *Phys. Rev.* C31: 823 (1985)
111. Kapusta, J., Mekjian, A. *Phys. Rev.* D33: 1304 (1986)
112. Matsui, T., McLerran, L., Svetitsky, B., *Phys. Rev.* D34: 783 (1986); *MIT preprint MIT-CTP-1344* (1986)
113. Kajantie, K., Kataja, M., Ruuskanen, P. V. *Phys. Lett.* 179B: 153 (1986)
114. Rafelski, J., Müller, B. *GSI preprint GSI-86-7* (1986)
115. McLerran, L. *Proc. Quark-Matter '86, Nucl. Phys.* A461: 245c (1987)
116. Koch, P., Müller, B., Rafelski, J. *Phys. Rep.* 142: 167 (1986)
117. Gyulassy, M., Kauffmann, S., Wilson, L. W. *Phys. Rev.* C20: 2267 (1979)
118. Pratt, S. *Phys. Rev. Lett.* 53: 1219 (1984); *Phys. Rev.* D33: 72, 1314 (1986)
119. Zajc, W. A. *Proc. RHIC Workshop, BNL-51921* (1985)

*Ann. Rev. Nucl. Part. Sci.* 1987. 37: 325–82

# HEAVY-QUARK SYSTEMS

## Waikwok Kwong and Jonathan L. Rosner

Enrico Fermi Institute and Department of Physics, University of Chicago, Chicago, Illinois 60637

## Chris Quigg

Fermi National Accelerator Laboratory, P.O. Box 500, Batavia, Illinois 60510

CONTENTS

325

0163–8998/87/1201–0325$02.00

# 1.  INTRODUCTION AND OUTLINE

## 1.1  *History*

1.1.1  ATOMS, NUCLEI, POSITRONIUM, AND OTHER PREHISTORY    Two-particle bound states have played a key role in teaching us about the fundamental interactions. The hydrogen atom was one of the fertile areas from which quantum mechanics sprang seventy years ago, and it continues to provide ever more precise tests of quantum electrodynamics. The deuteron was an essential laboratory for the development of the theory of nuclear forces. The electron-positron bound state, positronium, and corresponding bound states involving an electron and a muon are crucial in confirming our ideas about how processes may be described in quantum field theory. In much the same manner, bound states involving a heavy quark and antiquark have provided crucial confirmation of our present understanding of the strong interactions. These bound states and their theoretical implications form the subject of the present article. We also discuss, but to a lesser extent, bound states involving single heavy quarks.

1.1.2  LIGHT-QUARK SPECTROSCOPY    The strong interactions for many years resisted attempts at a quantitative treatment. A taxonomy of the hadrons, based on the quark model (1), developed in the mid-1960s. It succeeded in describing hundreds of strongly interacting particles in terms of three elementary spin-1/2 constituents: the quarks u, d, and s, for up, down, and strange. Masses, decay rates, and magnetic moments of the mesons and baryons were systematized (2–5), but the reasons for the many successes (and occasional failures) of the quark model were not satisfactorily understood.

1.1.3  COLOR AND ASYMPTOTIC FREEDOM  The requirement that quarks respect Fermi statistics led in the 1960s to the introduction of a new threefold degree of freedom for quarks. This property (5–7), which has now become known as *color*, was eventually recognized as a suitable strong-interaction charge on which to base a theory of interquark forces. The advent in the early 1970s of the color gauge theory (8) known as quantum chromodynamics, or QCD, brought the promise of a predictive theory of the strong interactions for the first time. Quarks were understood as interacting via field quanta called *gluons*, which interact with one another as well as with quarks. In electrodynamics, charge screening results in a growth of the effective charge at short distances. In contrast, in QCD the interactions among gluons (for which there is no analog in QED) lead to a diminution of the effective strong-interaction charge at short distances. This *asymptotic freedom* (9) of the strong interactions implies that perturbation theory becomes increasingly reliable at short distances, especially when these distances are small compared with the size (1 fm = $10^{-13}$ cm) of ordinary hadrons.

1.1.4  PSIONS AND CHARM  For the light (u, d, s) quarks involved in the hadron physics of the 1960s, QCD and asymptotic freedom have provided mainly qualitative insights. The complement to asymptotic freedom in QCD is that, at distances of 1 fm or more, the strong force becomes increasingly formidable, and quarks and gluons are *confined* (10). At such distances, confinement effects dominate the dynamics. Furthermore, quark pair production becomes important. Hadrons decay readily to other lighter hadrons and overlap and mix with one another, which frustrates precise spectroscopic descriptions.

In November of 1974, a remarkably narrow resonance (dubbed the J) was discovered (11) with a mass of 3.1 GeV/$c^2$, decaying to $e^+e^-$, in the reaction $p + Be \to e^+e^- + \dots$. Simultaneously, the resonance was discovered (12) in the direct channel in $e^+e^- \to$ hadrons (also $e^+e^-$, $\mu^+\mu^-$), and was named the $\psi$. The dual name J/$\psi$ has persisted. The initial evidence for it is shown in Figure 1. The cross section for its production in $e^+e^-$ annihilations, if integrated over center-of-mass energy $E$, yields (e.g. 13) its leptonic width $\Gamma_{J/\psi \to e^+e^-}$:

$$\int dE\, \sigma(e^+e^- \to J/\psi \to f; E) = \left(\frac{6\pi^2}{M_{J/\psi}^2}\right)\Gamma_{J/\psi \to e^+e^-}\Gamma_{J/\psi \to f}/\Gamma_{tot}, \qquad 1.$$

where the resonance may be observed in any final state f.

The J/$\psi$ and its excitations are now understood as the bound states of the charmed quark c and its antiquark (14–18). They are the first bound system of quarks to which QCD could be expected to apply, even approxi-

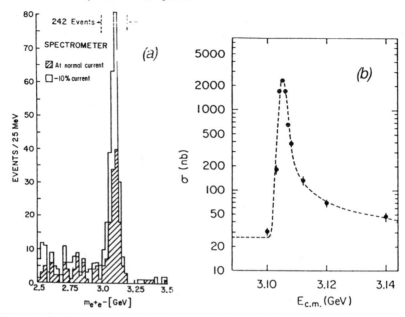

*Figure 1*  Initial evidence for the J/ψ. (*a*) Hadronic production in the reaction p + Be →
e⁺e⁻ + anything (11). (*b*) Hadronic formation cross section in electron-positron annihilations
(12).

mately, as a perturbative theory (16). The large mass of the c ($m_c \approx 1.5$
GeV/$c^2$) sets a mass scale high enough (and correspondingly implies a
bound-state size small enough) to approach the asymptotically free regime.
In analogy to positronium, the c$\bar{c}$ bound states were dubbed *charmonium*,
and heavy-quark–antiquark bound states have come to be known as *quarkonium*. Nonrelativistic potential models (18) successfully described and
predicted many properties of the new system.

Subsequently to the discovery of the J/ψ, hadrons containing a single
charmed quark were found (19). Their properties had been anticipated
theoretically to a large extent (14, 15, 18, 20, 21), and, in retrospect,
charmed hadrons had probably made their appearance several years earlier
in cosmic-ray interactions (22).

1.1.5  UPSILONS AND THEIR SIMILARITY TO PSIONS    The discovery in 1977
of the ϒ family of mesons was the first indication of the existence of a fifth
quark, the b (beauty or bottom), with mass $m_b \approx 5$ GeV/$c^2$ and charge
−1/3. The ϒ and two of its excitations were first observed in the reaction
p + (Cu, Pt) → μ⁺μ⁻ + ..., as shown in Figure 2 (23, 24). The ϒ family
was quickly identified as a set of b$\bar{b}$ levels. Comparison of b$\bar{b}$ and c$\bar{c}$ levels

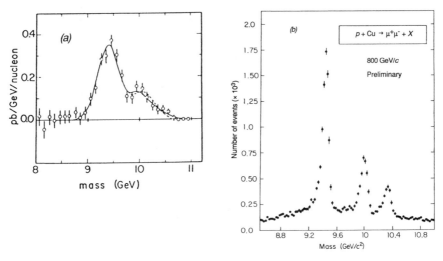

*Figure 2*    Evidence for the $\Upsilon$. (*a*) Initial muon-pair mass spectrum (23). (*b*) A recent spectrum in hadronic production (24).

showed that the interquark force was independent of the *flavor* of the quarks (25), as expected from QCD. (The flavor denotes the label u, d, s, c, b, ....) Hadrons containing a single b quark were identified in due course (26).

1.1.6    TOP    Our present understanding of the electroweak interactions implies that quarks must exist in pairs, differing in electric charge by one unit of the proton charge $|e|$. For many years, we knew of the pair u, d and of the unpaired s. The charmed quark was predicted as the partner of s to explain the absence of strangeness-changing, charge-preserving weak interactions (20), and was subsequently found. Within three years, however, the unpaired (and unpredicted) b appeared. Its hypothetical partner, the t quark (top or truth) has yet to make a definitive appearance, but it seems required to account for the absence of flavor-changing, charge-preserving weak decays of b (27), and we expect it will be found. The toponium ($t\bar{t}$) system should be a good new laboratory for precise hadron spectroscopy based on QCD.

## 1.2    Scope of This Article

We begin in Section 2 with a brief review of what is known experimentally about particles containing heavy quarks. We turn in Section 3 to the theoretical underpinnings of the spectroscopies of these particles. The J/$\psi$ family and other hadrons containing the charmed quark occupy Section

4. Section 5 is devoted to the spectroscopy of hadrons containing the b quark, and Section 6 to a comparison of the $\Upsilon$ and charmonium families. The still-to-be discovered top quark and the spectroscopy of its bound states are treated in Section 7. The spectroscopic methods that have been so successful for charmonium and the $\Upsilon$ family can be applied to the bound states of new strongly interacting constituents, such as quarks beyond the anticipated t, or spinless colored objects. Some methods for dealing with these exotic possibilities are discussed in Section 8. Conclusions are drawn in Section 9.

The length of the present article requires that we cover the topic selectively; some aspects are mentioned only briefly and others not at all. Some of these subjects are treated in greater depth in general reviews (2–5, 28–43). Earlier reviews in this series are collected in (44). The papers cited in (45–47) discuss specific quantum-mechanical techniques applicable to bound states of heavy quarks. An extensive compilation of information on the coupling strength in quantum chromodynamics has been made in (48). Space limitations prevent us from discussing the sum rule method in quantum chromodynamics (49–51). Weak decays of heavy quarks, mentioned briefly here, receive much more complete coverage in (52–55).

## 2.    REVIEW OF THE DATA

### 2.1    Preliminaries: Hadrons as Bound States of Quarks

2.1.1    MESONS    A quark-antiquark ($q\bar{q}$) meson is characterized by the total spin $S$ of the $q\bar{q}$ system ($S = 0$ or 1), the relative orbital angular momentum $L$, and the total angular momentum $\mathbf{J} = \mathbf{L} + \mathbf{S}$. The eigenvalues of $\mathbf{J}^2$ are $J(J+1)$, where $J$ can equal $L-1$, $L$, or $L+1$.

We use extensively the spectroscopic notation $n^{2S+1}L_J$, with $L = 0$ labeled by "S," $L = 1$ by "P," $L = 2$ by "D," and so on. The radial quantum number $n$ is equal to one plus the number of nodes of the radial wave function. Thus, the lowest group of S states are denoted 1S, the lowest P states 1P, the lowest D states 1D, etc.

The parity of a $q\bar{q}$ state is $P = (-1)^{L+1}$; the explicit factor of $-1$ arises from the opposite intrinsic parities of fermion and antifermion. A neutral $q\bar{q}$ meson is an eigenstate of the charge-conjugation operator, with eigenvalue $C = (-1)^{L+S}$. We often refer to a meson by the label $J^{PC}$.

2.1.2    BARYONS    In this article we are concerned only with the lowest-lying states of three quarks, with total orbital angular momentum $L = 0$. These baryons thus have positive parity, and spins $J = 1/2$ or $3/2$ equal to the total quark spin $S$.

## 2.2   Survey of States

We adopt the new nomenclature of the Particle Data Group (56) for hadrons involving heavy quarks, and indicate the traditional names in parentheses when needed.

We are concerned with mesons and baryons containing the quarks c, b, and t. Some examples of known particles containing c and b include the $c\bar{c}$ states (Table 1 and Figure 3), $b\bar{b}$ states (Table 2 and Figure 4), charmed hadrons (Table 3), and b-flavored hadrons (Table 4). This rich collection of states (56–64) was unknown until about a dozen years ago. In the rest

**Table 1**   The $c\bar{c}$ bound states

| $^1S_0$ $(0^{--})$ | Name | Mass (MeV/$c^2$)[a] | $\Gamma$ (MeV)[a] | |
|---|---|---|---|---|
| $1^1S_0$ | $\eta_c$ | 2981. ± 2 | 11±4 | |
| $2^1S_0$ | $\eta_c'$ | 3594. ± 5[b] | <8 (95% C.L.)[b] | |

| $^3S_1$ $(1^{--})$ | Name | Mass (MeV/$c^2$) | $\Gamma$ (MeV) | $\Gamma_{ee}$(keV) |
|---|---|---|---|---|
| $1^3S_1$ | $J/\psi$ | 3096.9 ± 0.1 | 0.063 ± 0.009 | 4.7 ± 0.3 |
| $2^3S_1$ | $\psi'$ | 3686.0 ± 0.1 | 0.215 ± 0.040 | 2.1 ± 0.2 |
| $3^3S_1$ | $\psi$ | 4030. ± 5 | 52. ± 10 | 0.75 ± 0.15 |
| $4?^3S_1$ [c,d] | $\psi$ | 4159. ± 20 | 78. ± 20 | 0.77 ± 0.23 |
| 4 or 5?$^3S_1$ [c] | $\psi$ | 4415. ± 6 | 43. ± 20 | 0.47 ± 0.10 |

| $1\ ^3P_J$ $(J^{++})$ | Name | Mass (MeV/$c^2$) | $\Gamma$ (MeV) |
|---|---|---|---|
| $^3P_0$ | $\chi_0$ | 3414.9 ± 1.1 | 13–21[e] |
| $^3P_1$ | $\chi_1$ | 3510.7 ± 0.5 | < 1.3 (90% C.L.)[f] |
| $^3P_2$ | $\chi_2$ | 3556.3 ± 0.4 | $2.6^{+1.4}_{-1.0}$[f] |

| $^3D_1$ $(1^{--})$ | Name | Mass (MeV/$c^2$) | $\Gamma$ (MeV) |
|---|---|---|---|
| $1^3D_1$ | $\psi$ | 3769.9 ± 2.4 | 25 ± 3 |

[a] All properties are from Ref. (56) unless otherwise indicated.
[b] Ref. (57).
[c] $^3S_1$ states above charm threshold may be substantially mixed with $^3D_1$ states.
[d] Possible $^3D_1$, if $\psi$(4415) is identified as $4^3S_1$.
[e] Crystal Ball (93).
[f] Ref. (76).

**Table 2**  The b$\bar{\text{b}}$ bound states

| $^3S_1(1^{--})$ | Name | Mass (MeV/$c^2$) | $\Gamma$(MeV)[a] | $\Gamma_{ee}$(keV)[a] |
|---|---|---|---|---|
| $1^3S_1$ | $\Upsilon$ | $9460.0 \pm 0.2$ | $0.043 \pm 0.003$ | $1.22 \pm 0.05$ |
| $2^3S_1$ | $\Upsilon'$ | $10023.4 \pm 0.3$ | $0.030 \pm 0.007$ | $0.54 \pm 0.03$ |
| $3^3S_1$ | $\Upsilon''$ | $10355.5 \pm 0.5$ | $0.0255 \pm 0.005$[b] | $0.40 \pm 0.03$ |
| $4^3S_1$ | $\Upsilon'''$ | $10577.5 \pm 4.1$ | $24. \pm 2$ | $0.24 \pm 0.05$ |
| $5^3S_1$ | $\Upsilon$ | $10864.8 \pm 7.9$ | $110. \pm 13$ | $0.31 \pm 0.07$ |
| $6^3S_1$ | $\Upsilon$ | $11019.1 \pm 8.6$ | $79. \pm 16$ | $0.13 \pm 0.03$ |

| $^3P_J(J^{++})$ | Name | Mass (MeV/$c^2$) |
|---|---|---|
| $1^3P_0$ | $\chi_{b0}$ | $9859.8 \pm 1.3$ |
| $1^3P_1$ | $\chi_{b1}$ | $9891.9 \pm 0.7$ |
| $1^3P_2$ | $\chi_{b2}$ | $9913.3 \pm 0.6$ |
| $2^3P_0$ | $\chi_{b0}'$ | $10230.5 \pm 2.3$[c] |
| $2^3P_1$ | $\chi_{b1}'$ | $10255.7 \pm 0.8$[c] |
| $2^3P_2$ | $\chi_{b2}'$ | $10268.6 \pm 0.7$[c] |

| $^1P_1(1^{+-})$ | Name | Mass (MeV/$c^2$) |
|---|---|---|
| $1^1P_1$ | $h_b$ | $9894.8 \pm 1.5$[d] |

[a] All properties are from Ref. (56) unless otherwise noted.
[b] Ref. (58).
[c] Based on inclusive photon spectra quoted in Ref. (58), and $\Upsilon''$(3S) mass quoted in Ref. (56).
[d] Ref. (59).

of this article we review what the study of these spectra has already taught us about fundamental physics, and what more we can hope to learn.

## 3.    THEORETICAL CONTEXT

### 3.1   *Quark Model Phenomenology for Hadron Masses and Widths*

Quarks as the constituents of hadrons were interpreted at first as convenient fictions, to be discarded once a more rigorous theory had been found, or as quasi-particles, to be used in phenomenological calculations

**Table 3**   Charmed hadrons

| $J^{PC}$ | Quark Content | Name | Mass[a] (MeV/$c^2$) | Lifetime ($10^{-13}$ sec) | Width[a] (MeV) |
|---|---|---|---|---|---|
| **Mesons** | | | | | |
| $0^{-+}$ | $c\bar{u}$ | $D^0$ | $1864.6\pm0.6$ | $4.3^{+0.2}_{-0.2}$[b] | |
| | $c\bar{d}$ | $D^+$ | $1869.3\pm0.6$ | $10.31^{+0.52}_{-0.44}$[b] | |
| | $c\bar{s}$ | $D_s$ (F) | $1970.5\pm2.5$ | $3.5^{+0.6}_{-0.5}$[b] | |
| $1^{--}$ | $c\bar{u}$ | $D^{*0}$ | $2007.2\pm2.1$ | | $<5$ |
| | $c\bar{d}$ | $D^{*+}$ | $2010.1\pm0.7$ | | $<2$ |
| | $c\bar{s}$ | $D^*_s$ (F*) | $2113.\pm8$ | | |
| $1^+/2^+?$ | $c\bar{d}$ | $D$ (2420) | $2426.\pm6$ | | $75.\pm20$[c] |
| **Baryons** | | | | | |
| $1/2^+$ | $cud$ | $\Lambda_c^+$ | $2281.2\pm3.0$ | $1.9^{+0.5}_{-0.3}$[b] | |
| | $cuu$ | $\Sigma_c^{++}$ | $M(\Lambda_c^+)+168.4\pm0.5$[d] | | |
| | $cdd$ | $\Sigma_c^0$ | $M(\Lambda_c^+)+165.8\pm0.7$[d] | | |
| | $cus$ | $\Xi_c^+$ (A+) | $2460.\pm25$ | $4.8^{+2.9}_{-1.8}$[e] | |
| | $css$ | $\Omega_c^0$ (T$^0$) | $2740.\pm20$ | | |

[a] Properties are from Ref. (56) unless otherwise indicated.
[b] Average values from Figure 8. The precision of these measurements is improving rapidly.
[c] Ref. (60).
[d] Ref. (61).
[e] Ref. (62).

**Table 4**   The b-flavored hadrons

| $J^{PC}$ | Quark Content | Name | Mass (MeV/$c^2$) | Lifetime[a] ($10^{-13}$ sec) |
|---|---|---|---|---|
| **Mesons** | | | | |
| $0^{-+}$ | $b\bar{u}$ | $B^-$ | $5277.9\pm1.1\pm3$[b] | $14.2\pm2.7$ |
| | $b\bar{d}$ | $B^0$ | $5281.0\pm0.9\pm3$[b] | |
| $1^{--}$ | $b\bar{u}$ | $B^{*-}$ | | $M(B)+52.0\pm4.5$[c] |
| | $b\bar{d}$ | $B^{*0}$ | | |

[a] Ref. (56). The precision of these measurements is improving rapidly.
[b] Ref. (63).
[c] Ref. (64).

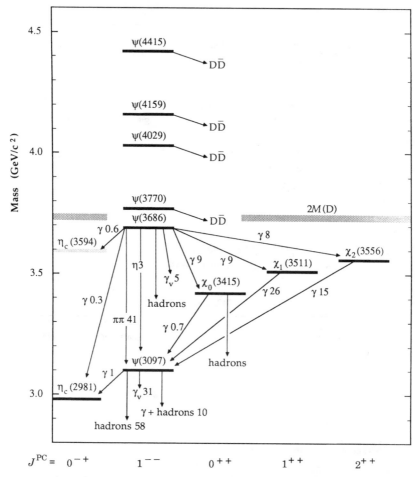

*Figure 3* Charmonium (c c̄) spectrum. Arrows denote decays and are labeled by branching ratios, in percent. The band at mass = 2M(D) denotes the flavor threshold, above which levels are broader than those below it.

of masses and widths based on effective Hamiltonians. We now regard them as fundamental entities, with interactions described by an underlying gauge theory. Before reviewing current understanding, however, it is helpful to recall some successes of early phenomenology with light quarks.

3.1.1 HADRON MASSES    The masses of the lowest s-wave mesons and baryons may be described to within 20 MeV by the following simple picture (3), motivated by an elementary treatment (21) based on QCD: (*a*)

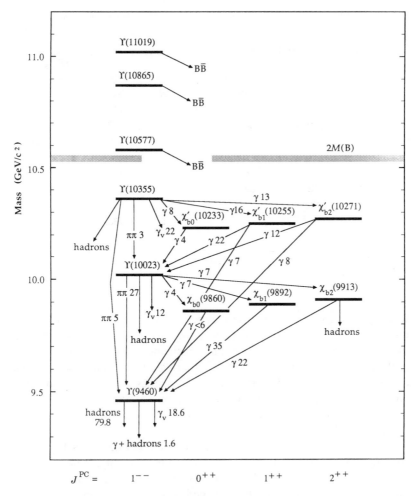

*Figure 4* Spectrum of the upsilon (bb̄) family. Arrows are labeled by branching ratios to specific channels, in percent. Levels above flavor threshold [band at mass = $2M(\text{B})$] are broader than levels below it.

Add all the quark masses $m_i$ in the hadron. (*b*) Add a term $\Delta M_{SS}$ for the spin-spin interaction of each quark-quark or quark-antiquark pair. This term is proportional to $\boldsymbol{\sigma}_i \cdot \boldsymbol{\sigma}_j/m_i m_j$, where $\boldsymbol{\sigma}_i$ is the Pauli spin operator for the *i*th quark. This picture totally neglects the kinetic energies or differences in binding energies of the quarks. It leads to the mass predictions shown in Table 5. Why does it work so well?

3.1.2  BARYON MAGNETIC MOMENTS  By coupling the spins of quarks

**Table 5**   Naive model of light-quark meson and baryon masses described in text

| $J^P$ | Meson | Mass (MeV/$c^2$) | | $J^P$ | Baryon | Mass (MeV/$c^2$) | |
|---|---|---|---|---|---|---|---|
| | | Calculated[a] | Measured | | | Calculated[b] | Measured |
| 0⁻ | $\pi$ | 140 | 138 | 1/2⁺ | $N$ | 939 | 938 |
| | $K$ | 485 | 496 | | $\Lambda$ | 1114 | 1116 |
| | $\eta$ | 559 | 549 | | $\Sigma$ | 1179 | 1195 |
| | $\rho$ | 780 | 776 | | $\Xi$ | 1327 | 1318 |
| 1⁻ | $\omega$ | 780 | 783 | 3/2⁺ | $\Delta$ | 1239 | 1232 |
| | $K^*$ | 896 | 892 | | $\Sigma^*$ | 1381 | 1385 |
| | $\phi$ | 1032 | 1020 | | $\Xi^*$ | 1529 | 1533 |
| | | | | | $\Omega$ | 1682 | 1672 |

[a] $m_u = m_d = 310$ MeV/$c^2$; $m_s = 483$ MeV/$c^2$; $\Delta M_{SS} = (160$ MeV/$c^2) \times \sigma_i \cdot \sigma_j (m_u^2/m_i m_j)$.
[b] $m_u = m_d = 363$ MeV/$c^2$; $m_s = 538$ MeV/$c^2$; $\Delta M_{SS} = (100$ MeV/$c^2) \times \sigma_i \cdot \sigma_j (m_u^2/m_i m_j)$.

suitably, one can determine the magnetic moments of the hadrons containing them. For example, the magnetic moment of the proton (uud) turns out to be $\mu_p = (4\mu_u - \mu_d)/3$. That of the neutron is $\mu_n = (4\mu_d - \mu_u)/3$, so if $\mu_d = -(1/2)\mu_u$, as implied by the quark charges, then $\mu_n = -(2/3)\mu_p$, in good agreement with experiment. The values of the u and d quark moments $\mu_u$ and $\mu_d$ extracted from $\mu_p$ and $\mu_n$ are remarkably close to those of Dirac particles with the same masses as found in the hadron mass calculations. Why should this be so?

The above examples are just two of the results of light-quark phenomenology that we still hope to see established more firmly. When we apply similar ideas to heavy-quark physics, we begin to understand why they might hold. Our insight has been provided to a large extent by quantum chromodynamics, which we now discuss briefly.

## 3.2   Quantum Chromodynamics

3.2.1   COLOR   Just 300 MeV/$c^2$ above the neutron and proton lie three-quark states totally symmetric in spin and flavor (e.g. the $\Delta^{++}$ resonance, composed of three u quarks coupled to $J = 3/2$) that have a spatially symmetric (ground-state) wave function. If the wave functions of the $\Delta$'s are to obey Fermi statistics, they must be antisymmetric in something else, called *color*. Baryons are composed of quarks of three different colors, and baryon wave functions are antisymmetrized in color. Color is a type of charge, coupled to a field (the *gluon* field) just as electromagnetic charge is coupled to the photon. The gauge theory of the interactions of gluons and colored quarks is known as quantum chromodynamics, or QCD.

The gluons couple both to each other and to quarks. The quark-gluon coupling contributes a term

$$L_{\text{int (quark-gluon)}} = -g_s \bar{\psi} \gamma_\mu \sum_a T^a A^{\mu a} \psi \qquad \qquad 2.$$

to the interaction Lagrangian, where $g_s$ is the (strong) coupling constant, $T^a (a = 1, \ldots, 8)$ are $3 \times 3$ matrices in color space, and $A^{\mu a}$ are eight gauge fields of colored gluons. The $T^a$ may be expressed in terms of the familiar Gell-Mann matrices $\lambda^a$ of SU(3) as $T^a = \lambda^a/2$.

The quark-gluon interaction bears strong parallels to the electron-photon interaction in quantum electrodynamics. The gluon is massless, as is the photon. The Born term for the quark-quark or quark-antiquark interaction is thus of the familiar Coulomb $(1/r)$ form, at least at short distances.

The gluon self-coupling results in a slow decrease of the effective coupling strength with decreasing distance. The distance scale is conveniently expressed in terms of its Fourier conjugate variable $Q$, a characteristic momentum. By calculating the first quantum corrections to the color Coulomb potential, we find that the strong-interaction analog of the fine-structure constant $\alpha_s \equiv g_s^2/4\pi$, can be parametrized as

$$\alpha_s(Q^2) = \frac{12\pi}{(33 - 2n_f) \ln [Q^2/\Lambda^2]}. \qquad \qquad 3.$$

Here $n_f$ is the number of fermion flavors with mass below $Q$, and $\Lambda$ is a characteristic scale, measured in various processes (48) to be of order 200 MeV. The decrease of $\alpha_s$ with increasing $Q^2$ may be contrasted with the growth of the electromagnetic coupling at short distances, which occurs when a test charge penetrates the vacuum polarization cloud that screens a charge at large $r$. Because of confinement, there is no meaningful Thomson (long-distance) limit for QCD, so there is no "natural" scale on which to define $\alpha_s$. For electromagnetism, on the other hand, it is conventional to define the charge in terms of its long-distance behavior.

3.2.2 SHORT DISTANCES    Single gluon exchange at short distances leads to a Coulomb-like interaction

$$V(r) = -\frac{4}{3} \frac{\alpha_s(r)}{r}, \qquad \qquad 4.$$

for a quark-antiquark pair bound in a color singlet. [The factor of 4/3 comes from the group theory of SU(3).] We illustrate the interaction

(Equation 4) symbolically in Figure 5*a*. As a result of Equation 3, the coupling $\alpha_s(r)$ varies logarithmically with $r$, so that at very short distances, gluon exchange becomes weaker. This property, known as *asymptotic freedom*, is responsible for the quasi-free behavior exhibited by quarks in hadrons probed at very short distances by deeply inelastic scattering.

3.2.3  LONG DISTANCES   Equation 4 is dramatically modified in QCD at momentum scales smaller than $\Lambda \approx 200$ MeV, i.e. at distances of about 1 fm or more. Chromoelectric lines of force bunch together into a tube of approximately constant cross-sectional area, as illustrated in Figure 5*b*. By Gauss' Law, this leads to a constant, distance-independent force, or a potential

$$V(r) = kr, \qquad\qquad\qquad 5.$$

where the force constant $k$ is about 0.14–0.18 GeV$^2$, or 0.7–0.9 GeV/fm. [The value of $k$ may be deduced from the spectrum of light-quark mesons and baryons with high orbital excitations (65).] An indefinitely rising

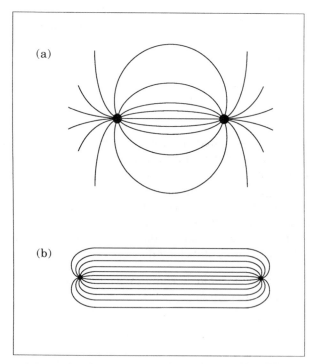

*Figure 5*   Chromoelectric lines of force for the interquark interaction at (*a*) short and (*b*) long distances.

potential such as Equation 5 permanently confines quarks so that they cannot be produced as separate entities. Up to now, no isolated quark has been observed. The form in Equation 5 receives theoretical support from calculations based on a space-time lattice approach to QCD (65a).

## 3.3 Interpolating Potentials

3.3.1 POWER LAWS  Systems with radii larger than 1 fm, such as high orbital excitations of the light-quark hadrons, represent one extreme limit in QCD, that of long distances. In the complementary limit of short distances, a Coulomb interaction is operative. The systems of heavy quarks (c and b) that we discuss here are more compact than 1 fm, but not so small as to be described by a Coulombic interaction alone. An effective potential

$$V(r) = Ar^{v} \qquad (-1 < v < 1) \qquad\qquad 6.$$

thus can interpolate between the short-distance and long-distance behavior of QCD. A comparison of $c\bar{c}$ and $b\bar{b}$ states (see Section 6) yields an effective power $v$ close to zero (45, 66).

3.3.2 QCD-MOTIVATED INTERPOLATIONS  Potentials motivated by perturbative QCD but incorporating the expected linear behavior at large separations have been proposed. The Fourier transform of a $1/r$ potential at small $r$ behaves as $1/Q^2$ for large momentum transfer $Q$, while that of a linear potential for large $r$ behaves as $1/Q^4$ for small $Q$. An expression embodying both limits that reproduces the expected logarithmic variation of the strong coupling constant for large $Q^2$ is (67, 68)

$$V(Q^2) = \frac{16\pi}{(33 - 2n_{\mathrm{f}})Q^2 \ln[1 + Q^2/\Lambda^2]}. \qquad\qquad 7.$$

## 3.4 Spin-Dependent Effects

The spin-independent features of quarkonium spectroscopy are well described by the potentials just noted, or variations on them. However, various phenomena in quarkonium systems are sensitive to spin dependences in the interquark interaction. For example, they depend on whether this interaction is of the form $V_{\mathrm{V}}(r)$ that would arise from the exchange of a vector particle (a single gluon) or of the form $V_{\mathrm{S}}(r)$ that would arise from an effective scalar exchange (in QCD, a collective phenomenon involving many gluons, such as a rotating flux tube). The vector interaction comes from the Fourier transform of a transition matrix element

$$M_{fi}^V = [\bar{u}(p_f')\gamma_\mu u(p_i')] V_V(Q^2) [\bar{u}(p_f)\gamma^\mu u(p_i)], \qquad\qquad 8.$$

while the scalar interaction comes from

$$M_{fi}^S = [\bar{u}(p_f')u(p_i')] V_S(Q^2) [\bar{u}(p_f)u(p_i)]. \qquad\qquad 9.$$

In principle other effective interactions are also possible. Each gives rise to characteristic spin-spin, spin-orbit, and tensor forces. These are most easily found by expanding Equations 8 and 9 to order $\beta^2$, where $\beta$ is a characteristic quark velocity in units of $c$. In practice this amounts to expansion in inverse powers of quark masses. A detailed review of this method is contained in (28).

3.4.1 SPIN-SPIN INTERACTIONS    The hyperfine electromagnetic inter-action between a proton and an electron leads to a 1420-MHz level splitting between singlet and triplet states of atomic hydrogen. In light-quark systems, a similar spin-spin force due to single gluon exchange between quarks generates the splittings between the masses of the pion and the $\rho$ resonance, the nucleon and the $\Delta$ resonance, the $\Sigma(1193)$ and the $\Sigma(1385)$ hyperons, and so on. The spin-spin interaction is of the form

$$V_{SS}(\mathbf{r}) = \frac{\boldsymbol{\sigma}_1 \cdot \boldsymbol{\sigma}_2}{6m_1 m_2} \nabla^2 V_V(r) \qquad\qquad 10.$$

for a quark of mass $m_1$ and an antiquark of mass $m_2$ with spins described by the Pauli matrices $\boldsymbol{\sigma}_1$ and $\boldsymbol{\sigma}_2$, respectively. The expectation value of $\boldsymbol{\sigma}_1 \cdot \boldsymbol{\sigma}_2$ is $+1$ for a state with total quark spin $S = 1$ (triplet), and $-3$ for $S = 0$ (singlet). Only $V_V(r)$ contributes to the spin-spin interaction. If we take Equation 4 for the vector interaction, and neglect the effect of $\nabla^2$ on the slow variation of $\alpha_s(r)$ with $r$, we obtain a spin-spin interaction

$$V_{SS}(\mathbf{r}) = \frac{8\pi\alpha_s \boldsymbol{\sigma}_1 \cdot \boldsymbol{\sigma}_2}{9m_1 m_2} \delta^3(\mathbf{r}). \qquad\qquad 11.$$

Because of the $\delta$ function, this expression has nonzero matrix elements only between S states. The absence of appreciable spin-spin splitting in states with $L > 0$ thus is a crucial test of the short-range Coulomb-like nature of the force between quarks. Such tests are just now becoming feasible for $c\bar{c}$ and $b\bar{b}$ states. As a result the value of $\alpha_s$ in Equation 11 is an effective short-distance value, $\alpha_s \approx \alpha_s(Q^2 = 4m_1 m_2)$. There are also small corrections to the $\delta^3(r)$ form. In practice these lead to mass shifts of no more than a few MeV for p-wave $c\bar{c}$ and $b\bar{b}$ levels.

3.4.2 SPIN-ORBIT INTERACTIONS    Spin-orbit forces between quarks are present for both vector and scalar interactions, but in different form.

Denoting the relative orbital angular momentum of a $q\bar{q}$ pair by $\mathbf{L}$, and its total spin by $\mathbf{S}$, we find for quarks of equal mass $m$:

$$V_{LS}(r) = (\mathbf{L} \cdot \mathbf{S})(3 d V_V/dr - d V_S/dr)/(2m^2 r). \qquad 12.$$

For $({}^3P_2, {}^3P_1, {}^3P_0)$ states, $\langle \mathbf{L} \cdot \mathbf{S} \rangle = (1, -1, -2)$. The vector contribution contains effects of both explicit spin-orbit interactions and Thomas precession, while only the Thomas precession is present for the scalar interaction.

3.4.3 THE TENSOR FORCE    A vector interaction leads to a tensor force of the form

$$V_{\text{tensor}} = \frac{S_{12}}{12 m_1 m_2}\left(\frac{1}{r}\frac{d V_V}{dr} - \frac{d^2 V_V}{dr^2}\right), \qquad 13.$$

where $S_{12} \equiv 2[3(\mathbf{S} \cdot \hat{\mathbf{r}})(\mathbf{S} \cdot \hat{\mathbf{r}}) - \mathbf{S}^2]$ has nonzero matrix elements only for $L \neq 0$.[1] Its expectation values in $({}^3P_2, {}^3P_1, {}^3P_0)$ states are $(-2/5, 2, -4)$.

3.4.4 MODEL-INDEPENDENT DISCUSSIONS    Parametrizations of spin-dependent effects in quarkonium have been given that are more general than those discussed here. We refer the reader to the literature for details of analytic (70) and lattice-QCD (65a) treatments.

## 3.5   Spin-Independent Relativistic Corrections

An expansion of Equations 9 and 10 in inverse quark masses also yields spin-independent relativistic corrections. These introduce a flavor (or quark-mass) dependence into the effective interaction, even if none was present before. A typical such correction comes from expanding the total energy $(\mathbf{p}^2 + m^2)^{1/2}$ to higher order in $\mathbf{p}^2$, yielding the term $-(\mathbf{p}^2)^2/8m^3$, where $m$ is the quark mass. As we see in Section 6, a typical expectation value of the kinetic energy $\mathbf{p}^2/2m$ is several hundred MeV for a heavy-quark–heavy-antiquark bound state. A typical relativistic correction to quarkonium energy levels is then (several hundred MeV)$^2/m$, or about a hundred MeV for $c\bar{c}$ and several tens of MeV for $b\bar{b}$.

## 3.6   Coupled-Channel Effects

As the mass of quarkonium state approaches the threshold for decay to pairs of flavored mesons, important corrections to the form of the inter-quark potential may arise from communication with open-flavor channels (18). These coupled-channel effects may lead to irregularities in an other-

---

[1] A simple method for evaluating $S_{12}$ may be found in (69).

wise orderly progression of masses, leptonic widths, and other properties of the quarkonium levels given by the one-channel potential model.

## 3.7   Dipole Transition Rates

Just as in atomic physics, one can calculate electromagnetic transition rates using simple quantum mechanics. In the dipole approximation, we have

$$\text{or} \quad \frac{\Gamma(n^3S_1 \to n'^3P_J + \gamma)}{\Gamma(n^3P_J \to n'^3S_1 + \gamma)} = \frac{4\alpha e_Q^2 E_\gamma^3}{27}(2J_f + 1)|\langle f|r|i\rangle|^2 \qquad 14.$$

for the simplest electric dipole transitions, where $J_f$ is the spin of the final state, and the matrix element involves normalized radial wave functions. In all of these expressions $e_Q$ denotes the quark change in units of the proton charge $|e|$. The simplest magnetic dipole transitions occur between $^3S_1$ and $^1S_0$ states, and are described by the rate expressions (18)

$$\text{or} \quad \frac{\Gamma(^3S_1 \to {}^1S_0 + \gamma)}{\Gamma(^1S_0 \to {}^3S_1 + \gamma)} = 4\alpha e_Q^2 E_\gamma^3(2J_f + 1)|\langle f|j_0(E_\gamma r/2)|i\rangle|^2/3m_Q^2, \qquad 15.$$

where the matrix element $\langle f|j_0(E_\gamma r/2)|i\rangle$ of a spherical Bessel function between radial wave functions reduces to one or zero in the long-wavelength limit, depending on the principal quantum numbers of the initial and final states. Finite-size corrections to Equation 14 also involve matrix elements of the appropriate spherical Bessel functions. Here the three-dimensional wave function $\Psi(r)$ is expressed in terms of spherical harmonics and (normalized) radial wave functions by

$$\Psi(r) = Y_{LM}(\theta, \phi)R_{nL}(r). \qquad 16.$$

We normalize this wave function in such a way that the integral of its square over all 3-space is 1, so that

$$\int_0^\infty r^2 \, dr [R_{nL}(r)]^2 = 1. \qquad 17.$$

## 3.8   Annihilation Decays of Quarkonium States

Many quarkonium decays proceed via the annihilation of the heavy quark and antiquark, into photons and/or gluons. For s-waves, the probability of this annihilation is proportional to the square $|\Psi(0)|^2$ of the wave function at the origin. For higher partial waves $L \neq 0$, the $L$th spatial derivative of the radial wave function $R_{nL}(r)$ at $r = 0$ governs the annihilation. Perturbative QCD expressions for annihilation decay rates in quarkonium are shown in Table 6. The decaying quarkonium state is taken to be a color singlet. [We always suppose SU(3) to be the color gauge group.]

In deriving perturbative results involving gluon emission, it is assumed that the concept of an on-mass-shell gluon makes sense. In fact, the gluon must "dress" itself before emerging as hadrons. Gluons materialize as distinct jets of hadrons only when their energies exceed a few GeV, and it is for such gluons that the perturbative results probably start to be reliable.

**Table 6**  Lowest-order expressions and first-order QCD corrections for decay processes of $c\bar{c}$ and $b\bar{b}$.

| Process | Rate[a] | Correction factor | Equation number |
|---|---|---|---|
| $n\,{}^3S_1 \to e^+e^-$ | $\dfrac{16\pi}{3} N_c\,\alpha^2\,e_Q^2\,\lvert\Psi(0)\rvert^2\big/M_n^2$ | $1 - 16\,\alpha_s/3\pi$ | 18 |
| $\to \gamma\gamma\gamma$ | $\dfrac{16\,(\pi^2-9)}{9} N_c\,\alpha^3\,e_Q^6\,\lvert\Psi(0)\rvert^2\big/m_Q^2$ | $1 - 12.6\,\alpha_s/\pi$ | 19 |
| $\to ggg$ | $\dfrac{40\,(\pi^2-9)}{81}\,\alpha_s^3\,\lvert\Psi(0)\rvert^2\big/m_Q^2$ | $1 + 4.9\,\alpha_s/\pi$ for $J/\psi$ <br> $1 + 3.8\,\alpha_s/\pi$ for $\Upsilon$ | 20[a] |
| $\to gg\gamma$ | $\dfrac{32\,(\pi^2-9)}{9}\,e_Q^2\,\alpha\,\alpha_s^2\,\lvert\Psi(0)\rvert^2\big/m_Q^2$ | $1 - 0.9\,\alpha_s/\pi$ for $J/\psi$ <br> $1 - 1.7\,\alpha_s/\pi$ for $\Upsilon$ | 21[a] |
| $n\,{}^1S_0 \to \gamma\gamma$ | $4\pi N_c\,e_Q^4\,\alpha^2\,\lvert\Psi(0)\rvert^2\big/m_Q^2$ | $1 - 3.4\,\alpha_s/\pi$ | 22 |
| $\to gg$ | $\dfrac{8\pi}{3}\,\alpha_s^2\,\lvert\Psi(0)\rvert^2\big/m_Q^2$ | $1 + 10.6\,\alpha_s/\pi$ for $\eta_c$ <br> $1 + 10.2\,\alpha_s/\pi$ for $\eta_b$ | 23[a] |
| $n\,{}^3P_2 \to \gamma\gamma$ | $\dfrac{12}{5} N_c\,e_Q^4\,\alpha^2\,\lvert R'_{nP}(0)\rvert^2\big/m_Q^4$ | $a$ | 24[b] |
| $\to gg$ | $\dfrac{8}{5}\,\alpha_s^2\,\lvert R'_{nP}(0)\rvert^2\big/m_Q^4$ | $(1 + 8.4\,\alpha_s/\pi)\,a$ for $\chi$ <br> $(1 + 11.7\,\alpha_s/\pi)\,a$ for $\chi_b$ | 25[a] |
| $n\,{}^3P_1 \to q\bar{q}g$ | $\dfrac{8}{9\pi}\,n_f\,\alpha_s^3\,\ln(2m_Q\langle r\rangle)\,\lvert R'_{nP}(0)\rvert^2\big/m_Q^4$ | not known | 26[c] |
| $n\,{}^3P_0 \to \gamma\gamma$ | $9N_c\,e_Q^4\,\alpha^2\,\lvert R'_{nP}(0)\rvert^2\big/m_Q^4$ | $(1 + 5.5\,\alpha_s/\pi)\,a$ | 27 |
| $\to gg$ | $6\,\alpha_s^2\,\lvert R'_{nP}(0)\rvert^2\big/m_Q^4$ | $(1 + 20.4\,\alpha_s/\pi)\,a$ for $\chi$ <br> $(1 + 21.2\,\alpha_s/\pi)\,a$ for $\chi_b$ | 28[a] |

[a] $N_c$ is the number of quark colors. Rate expressions that do not contain a factor of $N_c$ are for $N_c = 3$. For $N_c = (6, 8)$, the rate expressions in Equations 21, 23, 25, and 28 should be multiplied by a factor of $(25/2, 27/2)$, and Equation 20 by $(49/2, 0)$.

[b] Naive absorption of a $1/v$ term into $\lvert R'_{nP}(0)\rvert^2$ gives (73) $a = 1 - 16\alpha_s/3\pi$.

[c] $n_f$ is the number of light-quark flavors.

As mentioned at the end of Section 3.2.1, there is no "natural" scale on which to define $\alpha_s$, so the form of the perturbative expansions in this parameter depends on the mass scale chosen for $\alpha_s$. The first-order corrections in Table 6 are all based on evaluating $\alpha_s$ at the mass of the decaying state, in the $\overline{\text{MS}}$ (modified minimal subtraction) renormalization scheme (48, 71). A widely employed alternative prescription (72) for evaluating $\alpha_s$ at a more physically motivated (and generally smaller) scale leads in most cases to smaller first-order corrections. Predictions for ratios of rates (73) are expected to be more reliable than individual predictions involving $|\Psi(0)|^2$ or $|R'_{nP}(0)|^2$, which are subject to uncertainties (74) in the definition of the nonrelativistic wave function. We now comment about individual processes.

3.8.1    ONE VIRTUAL PHOTON    The decay of a $^3S_1$ quarkonium state into a lepton pair proceeds via a single virtual photon, as long as the initial mass $M_n$ is sufficiently small that the contribution of a virtual Z can be ignored. The $Z^0$ contribution is taken into account in the discussion of $t\bar{t}$ bound states (Section 7). Equation 18 in Table 6[2] holds for electromagnetic decay into any final fermion-antifermion pair $f\bar{f}$ if multiplied by $e_f^2$, by the number of colors of the fermion $f$ (three for quarks), and by a kinematic correction $(1+2m_f^2/M_n^2)(1-4m_f^2/M_n^2)^{1/2}$ for $m_f \neq 0$.

3.8.2    TWO PHOTONS
*S states*    Because of charge-conjugation invariance, an s-wave quarkonium state can annihilate into two photons only from the spin-singlet state. The rate is given by Equation 22. An analogous expression (with a different higher-order correction) holds for para-positronium.

*P states*    The first derivative of the p-wave radial wave function at $r = 0$ governs the annihilation amplitude. The states that can decay to two real photons are $^3P_0$ and $^3P_2$. The rates are given by Equations 24 and 27.

3.8.3    THREE PHOTONS    As for positronium, the decay of the spin-triplet (ortho-) state leads to (at least) three photons. The rate for quarkonium is given by Equation 19.

3.8.4    TWO GLUONS
*S states*    Charge-conjugation invariance prevents the spin-triplet s-wave quarkonium state from decaying to two gluons. The rate for the spin-singlet state is given by Equation 23. Note the large $O(\alpha_s)$ corrections.

*P states*    The rates for two-gluon decay of $^3P_2$ and $^3P_0$ states are given in Equations 25 and 28. In lowest order, a common group-theoretic factor

---

[2] Equations 18–28 are in Table 6.

governs the ratio of two-photon to two-gluon widths for the $^3P_2$, $^3P_0$, and $^1S_0$ states: $\Gamma(\gamma\gamma)/\Gamma(gg) = 9e_Q^4\alpha^2/2\alpha_s^2$.

3.8.5 THREE GLUONS The hadronic decays of a color-singlet $^3S_1$ quarkonium state must proceed via at least three gluons. A single (virtual) gluon is forbidden by color symmetry, and two gluons are forbidden by charge-conjugation invariance. The three-gluon rate is given by Equation 20.

Equation 20 for the triplet state has an extra factor of $\alpha_s$ and a greatly reduced coefficient (because of the three-gluon decay) relative to Equation 23 for the width of the singlet state, in agreement with the narrowness of the $J/\psi$ ($\Gamma_{tot} = 63 \pm 9$ keV) in comparison with the $\eta_c$ ($\Gamma_{tot} = 11 \pm 4$ MeV). The corresponding ratio of $3\gamma$ to $2\gamma$ decay rates for the $^3S_1$ and $^1S_0$ states of positronium is $4(\pi^2-9)\,\alpha/9\pi \approx 1/1115$.

3.8.6 TWO GLUONS + PHOTON The ratio $\alpha/\alpha_s$ may be measured by comparing the rates for quarkonium annihilation into $\gamma gg$ and $ggg$. The result for the $\gamma gg$ decay is given by Equation 21. Corrections to the shape of the photon spectrum have also been published (75). The mass spectrum of the two-gluon final state can be affected not only by details of how gluons turn into hadrons, but also by final-state interactions between the gluons. Consequently, the process is popular in searching for quarkless hadrons ("glueballs").

3.8.7 ONE REAL AND ONE VIRTUAL GLUON A $J = 1$ particle cannot decay to two transversely polarized identical spin-1 particles. As a corollary of this result, the $^3P_1$ state of quarkonium cannot annihilate into two real gluons. However, the process can occur if one of the gluons is virtual and materializes into a quark-antiquark pair. The resulting rate is given by Equation 26.

## 3.9 Hadronic Production Mechanisms

The processes involved in hadronic production of quarkonium states still are only partly understood. We illustrate some possible mechanisms in Figure 6.

3.9.1 TWO-GLUON FUSION AND SUBSEQUENT ELECTROMAGNETIC DECAY When two hadrons collide, a gluon from one hadron can combine with a gluon from the other to form a C-even quarkonium $\chi(^3P_J)$ state. This state can then decay via photon emission to a $^3S_1$ state (Figure 6a).

3.9.2 QUARK-ANTIQUARK ANNIHILATION A light quark q and antiquark $\bar{q}$ can combine into a virtual gluon that decays into a heavy $Q\bar{Q}$ state. The

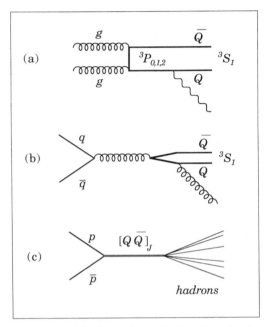

*Figure 6* Possible mechanisms for hadronic production of quarkonium states: (*a*) two-gluon fusion and subsequent electromagnetic decay; (*b*) quark-antiquark annihilation; (*c*) direct-channel production of a quarkonium state in proton-antiproton annihilations.

$Q\bar{Q}$ pair can then radiate one or more soft gluons to reach a color-singlet quarkonium state (Figure 6*b*).

3.9.3 PROTON-ANTIPROTON ANNIHILATIONS   When protons and anti-protons collide with a center-of-mass energy equal to the mass of a quarkonium state of total spin $J$, as shown in Figure 6*c*, that state can be produced with peak cross section

$$\sigma(\bar{p}p \to [Q\bar{Q}]_J) = \pi(2J+1)B([Q\bar{Q}]_J \to p\bar{p})/k^{*2}, \qquad 29.$$

where $k^*$ is the magnitude of the center-of-mass 3-momentum, and $B([Q\bar{Q}]_J \to p\bar{p})$ is the branching ratio for a quarkonium state of spin $J$ to decay to $p\bar{p}$: more generally, $B(X \to Y) \equiv \Gamma(X \to Y)/\Gamma(X)$. This method has already been used in Experiment R-704 at the CERN ISR (76) to produce the $\chi(^3P)$ states and to uncover candidates for the charmonium $1^1P_1$ and $1^1S_0$ states in collisions of stored antiprotons with a hydrogen gas jet. A follow-on experiment (E-760) at the Fermilab antiproton accumulator ring (77) should achieve very high accuracy for charmonium masses and yield total widths with a precision of $\pm 300$ keV for the $\chi$ states.

## 3.10    Inverse Scattering

The interquark potential may be estimated without appeal to theoretical biases about short-distance and long-distance behavior using the inverse-scattering formalism. This procedure permits the construction of potentials with any desired spectrum of s-wave levels using only the level positions and the squares of the corresponding wave functions at the origin, $|\Psi(0)|^2$, as obtained using Equation 18 from leptonic widths (25, 78). These potentials provide excellent approximations to radial quarkonium potentials in the range of distances actually probed by the known levels. The method has recently been simplified by an appeal to supersymmetric quantum mechanics (79).

# 4.    THE J/$\psi$ FAMILY AND CHARM

## 4.1    Hadronic and Radiative Decays of J/$\psi$

Decays of the J/$\psi$ provide rich information on hadrons containing light quarks. The initial state has well-defined mass, spin, parity, isospin, SU(3), and charge conjugation. It is copiously produced; nearly fifteen million J/$\psi$'s have been accumulated in experiments at SPEAR (SLAC) (80) and DCI (Orsay) (81). We give a sample of this information, referring to the literature for details.

4.1.1    HADRONIC DECAYS OF J/$\psi$    The hadronic decays of the J/$\psi$ are expected to proceed mainly via three-gluon emission (Equation 20). The ratio of rates for three-gluon and lepton pair emission (Equation 18) is

$$\frac{\Gamma(J/\psi \rightarrow ggg)}{\Gamma(J/\psi \rightarrow \ell^+\ell^-)} = \frac{10(\pi^2-9)}{81\pi e_c^2}\frac{\alpha_s^3(M^2)}{\alpha^2}[1+10.3\alpha_s(M^2)/\pi], \qquad 30.$$

if we set the J/$\psi$ mass $M$ equal to $2m_c$. The large $O(\alpha_s)$ correction (71) corresponds to a substantial fraction of the total rate. The gluons are not nearly energetic enough to appear as distinct jets. As a result, perturbative QCD is only a qualitative guide to the decay rate of J/$\psi$ into hadrons. If we were to ignore the QCD correction in Equation 30, we would infer $\alpha_s(M^2) \approx 0.19$ from the observed hadronic width.

When the gluons in J/$\psi$ decay materialize into a small number of hadrons, one may hope to learn about the properties of those hadrons. The Mark III collaboration has analyzed processes of the type

$$J/\psi \rightarrow (0^- \text{ meson}) + (1^- \text{ meson}), \qquad 31.$$

which proceed via the graphs of Figure 7. The $1^-$ mesons include $\rho$ and $\omega$ (composed of $[u\bar{u} \pm d\bar{d}]/\sqrt{2}$) and $\phi$ (composed of $s\bar{s}$). The $0^-$ mesons

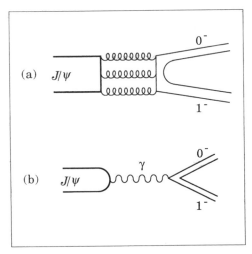

*Figure 7* Graphs describing production of $0^- 1^-$ pairs in $J/\psi$ decays: (*a*) three-gluon intermediate state; (*b*) one-photon intermediate state.

include $\pi$, $\eta$, $\eta'$, and $\eta(1440)$ ($\iota$). To the extent that Figure 7*a* dominates the decay, production of a $0^-$ meson opposite an $\omega$ tells the nonstrange quark content of that meson, while production opposite a $\phi$ tells the strange quark content. The rate may be compared with that for $J/\psi \rightarrow \rho\pi$, in which the quark content of both mesons is assumed known. The result is that while the $\eta$ is almost exclusively a quark-antiquark state, with roughly half strange and half nonstrange quarks, the $\eta'$ has some room (about 1/3 in probability) for a gluonic admixture (40, 80).

4.1.2  RADIATIVE DECAYS  Perturbative QCD, in the form of Equations 20 and 21, predicts that (71)

$$\frac{\Gamma(J/\psi \rightarrow \gamma gg)}{\Gamma(J/\psi \rightarrow ggg)} = \frac{16}{5} \frac{\alpha}{\alpha_s(M^2)} [1 - 5.8\alpha_s(M^2)/\pi].$$    32.

The observed ratio of $10 \pm 4\%$ for charmonium (Ref. 82, as quoted in last of Refs. 71) corresponds to a value of $\alpha_s(M^2) = 0.29 \pm 0.08$ if we ignore the radiative corrections (which, however, are substantial).

Many interesting final states may be reached in $J/\psi$ radiative decays (43). These are expected mainly to be those states either composed of, or with substantial couplings to, a pair of gluons. We summarize these states in Table 7. The radiative decay to a neutral pion may probe its small admixture of zero isospin. The decays to $\eta$ and $\eta'$ probably are sensitive to the relative gluonic admixtures in these two particles. The $f_2(1716)$ ($\theta$) and $X(2232)$ ($\xi$) were first seen in radiative $J/\psi$ decays, and the $\eta(1440)$ ($\iota$)

**Table 7**    Discrete states reached in radiative decays of $J/\psi$

| State[a] | Mass (MeV/$c^2$) | $J^P$ | Final state(s) | $J/\psi$ branching ratio (%) | Remarks |
|---|---|---|---|---|---|
| $\pi^0$ | 135 | $0^-$ | All | 0.004±0.001 | $^1S_0$ $q\bar{q}$ state |
| $\eta$ | 549 | $0^-$ | All | 0.086±0.008 | " |
| $\eta'$ | 958 | $0^-$ | All | 0.42 ± 0.05 | " |
| $f_2$ ($f$) | 1274 | $2^+$ | All | 0.16 ± 0.02 | $^3P_2$ $q\bar{q}$ state |
| $\eta$ (ι) | 1440 | $0^-$ | $K\bar{K}\pi$ <br> All | 0.46 ± 0.07 <br> >0.69±0.11 | Glueball candidate |
| $f_2'$ ($f'$) | 1525 | $2^+$ | All | 0.016±0.005 | $^3P_2$ $q\bar{q}$ state |
| $f_2$ ($\theta$) | 1716 | $2^+$ | $\eta\eta$ <br> $K\bar{K}$ <br> $\pi\pi$ | 0.026±0.011[b] <br> 0.096±0.014[b] <br> 0.020±0.004[b] | Glueball candidate |
| $X$ ($\xi$) | 2232 | $2^+$? | $K^+K^-$ <br> $K_sK_s$ | 0.0084±0.0032 <br> 0.013±0.007 | Interpretation unknown |
| $\eta_c$ | 2981 | $0^-$ | All | 1.27 ± 0.36 | $^1S_0$ $c\bar{c}$ state |

[a] Where different from the Particle Data Group nomenclature, the common name of the state is shown in parentheses. Branching ratios are from (56) unless otherwise noted.
[b] Averages quoted in (42).

(seen earlier in $p\bar{p}$ annihilations) also is quite prominent. The $\eta(1440)$ and $f_2(1716)$ are candidates for gluonic bound states. The nature of the X(2232) is uncertain, but it appears to have $J \geq 2$ (83). Other final states produced opposite a photon in $J/\psi$ radiative decays (80, 81) include $K\bar{K}\pi$, $\eta\pi\pi$, $\rho\rho$, $\omega\omega$, $\phi\phi$, and $\gamma\rho$. All of them but the last couple to two gluons.

## 4.2    Hadronic Decays of $\psi'$

4.2.1    THE DECAYS $\psi' \to \psi\pi\pi$ AND $\psi' \to \psi\eta$    Over half of the decays of $\psi'$ consist of hadronic transitions to the $\psi$, accompanied by the low-mass isoscalar systems $\pi\pi$ and $\eta$. Semiquantitative treatments of these processes in terms of two-gluon emission by the charmed quarks exist (84).

4.2.2    DECAYS OF $\psi'$ TO HADRONS CONTAINING LIGHT QUARKS    An intriguing puzzle (85) is the relative suppression of the rates for certain hadronic $\psi'$ decays: $B(\psi' \to \rho\pi)/B(\psi \to \rho\pi) \leq 0.6\%$ (90% confidence level); $B(\psi' \to K^*\bar{K})/B(\psi \to K^*\bar{K}) \leq 2.07\%$ (90% C.L.), while the corresponding ratios for $2\pi^+2\pi^-\pi^0$ and $3\pi^+3\pi^-\pi^0$ are $9.5 \pm 2.7\%$ and $13 \pm 7\%$, respectively. Nodes in the $\psi'$ radial wave function could be responsible for this peculiar behavior (86).

## 4.3  $\chi$ States

4.3.1  ELECTRIC DIPOLE TRANSITIONS    The only known $^3P_J$ states of charmonium are the $\chi_0$, $\chi_1$, and $\chi_2$ shown in Figure 3. They are presumably the lowest $^3P$ states, and the only narrow ones. They were first discovered in radiative decays from the $2^3S_1$ level, the $\psi(3686)$. The $\chi$ states themselves decay in part by radiative transitions to the $J/\psi$, with branching ratios indicated in Figure 3. Measured transition rates are compared with theoretical predictions (56, 87–89) in Table 8.

Purely nonrelativistic estimates of the rate for $\psi(3686) \to \gamma\chi$ are high by a factor of two to three. The exact position of the node in the $2^3S_1$ wave function strongly affects the dipole matrix element $\langle 1P|r|2S \rangle$. Relativistic distortions of the 2S and 1P wave functions substantially reduce this matrix element (88, 90). Differences still remain among various relativistic

**Table 8**    E1 transitions in $c\bar{c}$ systems

| Transition | Photon energy (MeV) | Experimental branching ratio (%)[a] | Partial decay rate (keV) | | | |
|---|---|---|---|---|---|---|
| | | | Experiment[b] | Theory[c] | | |
| | | | | MR(87) | MB(88) | GRR(89)[d] |
| $2S \to 1^3P_2$ | 128 | $7.8 \pm 0.8$ | $17 \pm 4$ | 41 (39) | 27 (27) | 24 (33) |
| $1^3P_1$ | 172 | $8.7 \pm 0.8$ | $19 \pm 4$ | 48 (51) | 31 (40) | 35 (49) |
| $1^3P_0$ | 261 | $9.4 \pm 0.8$ | $20 \pm 4$ | 37 (54) | 19 (45) | 44 (64) |
| $1^3P_2 \to J/\psi$ | 429 | $14.8 \pm 1.7$ | $385^{+212}_{-155}$ | 609 (495) | 347 (362) | 502 (753) |
| $1^3P_1 \to J/\psi$ | 389 | $25.8 \pm 2.5$ | $< 355$ | 460 (368) | 270 (250) | 369 (562) |
| $1^3P_0 \to J/\psi$ | 303 | $0.7 \pm 0.2$ | $147 \pm 38$ | 226 (174) | 128 (121) | 171 (253) |

[a] From compilation of (56).
[b] Based on total widths (76, 93) of $2.6^{+1.4}_{-1.0} < 1.3$, and 13–21 MeV for $^3P_{2,1,0}$ states, respectively.
[c] Nonrelativistic predictions are shown in parentheses below the relativistically corrected values.
[d] 1985 values. Substantial changes occur in 1986 work.

treatments, but the overall agreement is satisfactory. Relativistic effects are much less important for electric dipole transitions in the $b\bar{b}$ system.

The angular distributions in the radiative 2S → 1P and 1P → 1S charmonium decays are in accord with expectations that the lowest multipole (electric dipole, or E1) dominates the transitions. Studies at a slightly higher level of sensitivity than those already performed (91–93) may uncover small magnetic quadrupole (M2) contributions. The magnetic moment of the charmed quark as measured in these transitions should agree with that found in magnetic dipole, or M1, transitions.

4.3.2 TWO-GLUON AND TWO-PHOTON DECAYS   Recent measurements in $p\bar{p}$ annihilations have provided the results (76) $\Gamma(\chi_2 \to \gamma\gamma) = 2.9^{+1.3}_{-1.0} \pm 1.7$ keV, $\Gamma_{tot}(\chi_2) = 2.6^{+1.4}_{-1.0}$ MeV, and $\Gamma_{tot}(\chi_1) < 1.3$ MeV. The predicted ratio (73)

$$\Gamma(\chi_2 \to \gamma\gamma)/\Gamma(\chi_2 \to gg) = (8\alpha^2/9\alpha_s^2)(1 - 8.4\alpha_s/\pi),\qquad\qquad 33.$$

when combined with these measurements, would imply $\alpha_s = 0.21 \pm 0.09$ if the QCD correction were ignored. Since this correction is more than 50%, however, the convergence of the perturbation expansion cannot be guaranteed. The observed two-gluon width of the $\chi_2$, when compared with predictions of Equation 25 and potential models (94) for the wave function, entails $\alpha_s > 0.3$. The ratio $\Gamma(\chi_1 \to gq\bar{q})/\Gamma(\chi_2 \to gg) \approx 1.7\alpha_s$ implied in Table 6 [using $\langle r \rangle = 3.2$ GeV$^{-1}$ obtained in a typical potential model (87)] only constrains $\alpha_s < 0.6$ on the basis of the data quoted above. Substantial improvements in the above data are expected in forthcoming experiments (77).

## 4.4   Spin-Singlet S and P States

Spin-triplet charmonium states are produced copiously in $e^+e^-$ annihilations by formation of $^3S_1$ levels and subsequent E1 radiative transitions to $^3P_J$ levels, but the spin-singlet states have been more elusive. The first candidate for a $^1S_0$ $c\bar{c}$ level (95) appeared nearly five years after the discovery of the J/ψ, while only hints (76) of a $^1P_1$ state exist.

4.4.1 THE $^1S_0$ STATES   The $\eta_c$, mentioned in Table 1 and shown in Figure 3 as the lowest $J^{PC} = 0^{-+}$ level, is seen in magnetic dipole (M1) transitions from both the J/ψ and the ψ′. In Table 9 we summarize the observed partial widths for these transitions and the matrix elements $\langle f | j_0(E_\gamma r/2) | i \rangle$ implied by these widths through Equation 15. Also shown are matrix elements calculated on the basis of nonrelativistic wave functions. The discrepancy between the two values for J/ψ decay indicates that mixing effects and relativistic corrections appear to be important in these tran-

**Table 9**   M1 transitions in charmonium

| Decay | Partial width (keV) | Photon energy (MeV) | Overlap $\langle f \mid j_0(E_\gamma r/2) \mid i \rangle$ Experiment[a] | Theory |
|---|---|---|---|---|
| $J/\psi \rightarrow \eta_c \gamma$ | $0.80 \pm 0.25$ | 114 | $0.64 \pm 0.10$ | 0.9975 |
| $\psi' \rightarrow \eta_c' \gamma$ | 0.4 - 2.8 | $91 \pm 5$ | 0.64 - 1.67 | 0.9925 |
| $\psi' \rightarrow \eta_c \gamma$ | $0.60 \pm 0.17$ | 638 | $0.042 \pm 0.006$ | 0.066 |
| $\eta_c' \rightarrow J/\psi \gamma$ | — | 463 | — | 0.037 |

[a] Extracted from Equation 15 with $m_c = 1.8$ GeV/$c^2$.

sitions. However, there does not yet appear to be unanimity regarding the magnitude of these effects (18, 96–98).

The two-photon width of the $\eta_c$ has recently been measured in $\gamma\gamma$ collisions (via $e^+e^- \rightarrow e^+e^-\eta_c$), yielding $\Gamma(\eta_c \rightarrow \gamma\gamma) = 15.0 \pm 6.3$ keV (42), and in $\bar{p}p \rightarrow \eta_c \rightarrow \gamma\gamma$, yielding $5.7 \pm 2.6 \pm 3.7$ keV (76). The average (42) is $\Gamma(\eta_c \rightarrow \gamma\gamma) = 9 \pm 4$ keV. The partial width predicted from the ratio of Equations 22 and 18, in the approximation $M(^3S_1) = 2m_Q$, is (73)

$$\Gamma(\eta_c \rightarrow \gamma\gamma) = (4/3)\Gamma(J/\psi \rightarrow e^+e^-)(1 + 1.96\alpha_s/\pi), \qquad 34.$$

or about 7 keV for $\alpha_s = 0.22$ (42). The predicted ratio

$$\Gamma(\eta_c \rightarrow \gamma\gamma)/\Gamma(\eta_c \rightarrow gg) = (8\alpha^2/9\alpha_s^2)(1 - 14\alpha_s/\pi), \qquad 35.$$

when combined with the new $\gamma\gamma$ width and the total width of the $\eta_c$ (93) of $11.5 \pm 4.5$ MeV, would lead to a value of $\alpha_s = 0.26 \pm 0.08$ in the absence of the QCD correction term. With the renormalization scale set at the mass of the $\eta_c$, the QCD radiative corrections are too large to permit a reliable perturbative estimate of the $\eta_c$ width.

If we scale results from $\eta_c$ to $\eta_c'$ using the ratios of $|\Psi(0)|^2$ obtained from the corresponding $^3S_1$ leptonic widths, we would expect the total width of $\eta_c'$ to be a few MeV. [The process $\eta_c' \rightarrow \eta_c\pi\pi$ is not expected to be a major contributor to the $\eta_c'$ total width; the partial width is estimated (84) to be $\approx 0.1$ MeV.]

Further $p\bar{p}$ experiments (77) may be able to produce the $\eta_c'$ and detect it via $\eta_c' \rightarrow \gamma J/\psi$. This is a "hindered" $2S \rightarrow 1S$ transition, as is the observed decay $\psi' \rightarrow \gamma\eta_c$. For $m_c = 1.8$ GeV/$c^2$, using the matrix element in Table 9, we estimate $\Gamma(\eta_c' \rightarrow \gamma J/\psi) \approx 0.5$ meV, or a branching ratio $B(\eta_c' \rightarrow$

$\gamma J/\psi) \approx 10^{-4}$. A more optimistic estimate of about $10^{-3}$ was obtained in the last of Refs. (18).

4.4.2 THE $^1P_1$ STATE    The same $p\bar{p}$ experiment that observed the $\eta_c$ also has candidate events (76) for the $^1P_1$ charmonium state near the mass $M(^1P_1) = [5M(^3P_2)+3M(^3P_1)+M(^3P_0)]/9 \approx 3522$ MeV/$c^2$ expected if the spin-spin force (Equation 10) were negligible for P states. As mentioned in Section 3, this would confirm our notion that the spin-spin force is indeed due to a Coulomb-like interaction. Fermilab experiment E-760 (77) will be able to search more conclusively for this state.

## 4.5    D-Wave States and Prospects for Further Observations

The $\psi(3770)$, which lies about 40 MeV above $D\bar{D}$ threshold and decays mainly to $D\bar{D}$, is a candidate for the lowest $^3D_1$ level of charmonium. It has a larger leptonic width than one would expect for a pure D state, for which $\Gamma \propto |R''(0)|^2$. This is probably the result of an admixture of $2^3S_1$ induced by the spin-dependent tensor force and by coupling to decay channels (18). The total width of $\psi(3770)$, $25 \pm 3$ MeV, is such that any electromagnetic or hadronic transitions to lower $c\bar{c}$ states are expected to occur with branching ratios of no more than a few percent (18).

One also expects $^3D_2$, $^3D_3$, and $^1D_2$ $c\bar{c}$ levels not far from the $\psi(3770)$. In one model (87) (which ignores coupled-channel effects, however) the predicted masses of these states are 3.81, 3.84, and 3.82 GeV/$c^2$, respectively, for a $^3D_1$ level at 3.77 GeV/$c^2$. The parities of all these levels are negative. Now, a $J^P = 2^-$ particle cannot decay to two $0^-$ ones, so the $^3D_2$ and $^1D_2$ $c\bar{c}$ states expected at 3.81 and 3.82 GeV/$c^2$ cannot decay to $D\bar{D}$. Moreover, they are predicted to lie 50 or 60 MeV/$c^2$ *below* the $D^*\bar{D}$ threshold of 3.87 GeV/$c^2$. In the absence of a kinematically allowed strong decay into a pair of flavored mesons [such as makes the $\psi(3770)$ so broad], their transitions to lower $c\bar{c}$ states may be observable. In particular, the $^3D_2$ state may decay to $\gamma + {}^3P_{2,1}$ or to $\pi\pi + J/\psi$, while the $^1D_2$ state may decay to $\gamma + {}^1P_1$. Both states can be formed, in principle, in the direct channel of $p\bar{p}$ reactions.

## 4.6    States Above Threshold

Above flavor threshold, $c\bar{c}$ states are no longer exceptionally narrow, as shown in Table 1. The levels at 4.03 and 4.415 GeV/$c^2$ are good candidates for $^3S_1$ states; a level at 4.16 GeV/$c^2$ may be either a $^3S_1$ or $^3D_1$ state, or a mixture. The irregularities in the leptonic widths of the states above the charm threshold probably reflect the effects of many hadronic channels to which such states can couple, about which little is known. It would be of particular interest to find levels that couple appreciably to $D_s\bar{D}_s$ (formerly known as $F\bar{F}$).

## 4.7    Charmed Mesons and Baryons

We summarized some information about charmed mesons and baryons in Table 3. Here we add a few details.

4.7.1    LIFETIMES    Comparison of weak leptonic and semileptonic processes has led to the idea of Cabibbo universality, according to which, in modern language, the intrinsic strengths of the charged-current couplings to quarks and leptons are identical. This idea is embodied in the $SU(2)_L \times U(1)_Y$ electroweak theory, in which the charged-current couplings are determined by weak-isospin quantum numbers. Mixing among quark generations then gives couplings expressed in terms of the Cabibbo angle as $\cos\theta_C$ and $\sin\theta_C$ for the $u \leftrightarrow d$ and $u \leftrightarrow s$ transitions, in units of the electron $\leftrightarrow$ neutrino coupling. It has been known for many years from the study of kaon and hyperon decays that in nonleptonic processes strangeness-changing ($\Delta S = 1$) interactions occur with effective strength 1, not $\sin\theta_C$. The amplification by a factor of 20 is only partially understood (99) as the work of the strong interactions. Since the discovery of the $J/\psi$, the hope has persisted that the study of charmed-particle decays would lead to new insights into the nature of this nonleptonic enhancement. Let us examine some of the issues.

In Figure 8 we compare lifetimes for charmed hadrons measured in a number of experiments (55) over the past few years. Several conclusions may be drawn from these values.

1. The decay rates of hadrons containing charmed quarks are in crude accord with a free-quark picture based on the decays

$$c(1.6\,\text{GeV}/c^2) \rightarrow s(0.5\,\text{GeV}/c^2) + \begin{cases} e^+\nu \\ \mu^+\nu \\ u(0.3\,\text{GeV}/c^2)\,\bar{d}(0.3\,\text{GeV}/c^2) \end{cases}, \qquad 36.$$

where we assume (20) that the weak $c \rightarrow s$ transition proceeds with the same strength as $u \rightarrow d$. The charmed-quark lifetime would be about $10^{-12}$ s in this model. Its branching ratios to $e^+\nu$ or to $\mu^+\nu$ would each be about 20%, with nonleptonic decays making up the other 60%. In fact (56), $B(D^+ \rightarrow e^+ + \text{anything}) = 18.2 \pm 1.7\%$, close to the naive expectation, and the $D^+$ lifetime is just about $10^{-12}$ s.

2. Charmed quarks appear to decay more rapidly in the $D^0$, $D_s(F^+)$, and $\Lambda_c^+$ than in $D^+$. In every case, the semileptonic decay rate $\Gamma_{SL} \equiv \Gamma(c \rightarrow s\ell\nu)$ is in accord with the free-quark model (Equation 36), as required by

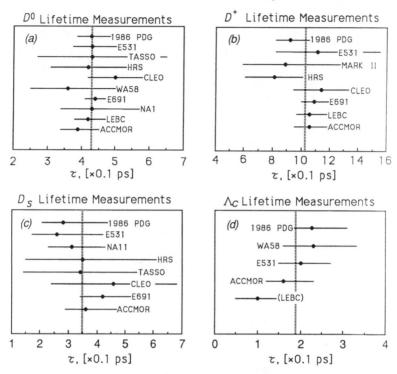

*Figure 8* Compilation of charmed-particle lifetime measurements. (*a*) D⁰; (*b*) D⁺; (*c*) Dₛ; (*d*) Λc.

Cabibbo universality; it is the nonleptonic decay rates that are enhanced by about a factor of $2\frac{1}{2}$ to 4. Using the branching ratios quoted in (56) and lifetimes from (55), we find $\Gamma_{SL}(D^+, D^0, \Lambda_c^+) = (1.98 \pm 0.34, 1.63 \pm 0.32, 2.37 \pm 1.03) \times 10^{11}\ s^{-1}$, compatible with the semileptonic rate of $1.8 \times 10^{11}\ s^{-1}$ one would expect for a charmed quark of $1.6\ GeV/c^2$ in the decay process (Equation 36).

3. The differences in nonleptonic decay rates have been ascribed to several possible mechanisms, reviewed in (52–55). The D⁺ semileptonic branching ratio is close to the free-quark prediction. Thus, there is probably neither a net enhancement nor suppression (100) of D⁺ nonleptonic decays. It appears that the environment of the charmed quark (100–102) can play a significant role in its nonleptonic decay rate. The study of final states in Dₛ and in charmed-strange baryon decays can be expected to shed further light on the mechanism of nonleptonic enhancement.

4.7.2 MASSES The simple model of hadron masses that was so successful for light-quark systems (Section 3.1.1) leads to the predictions shown in Table 10 for the masses of hadrons containing one charmed quark. The model is satisfactory for ordering the levels qualitatively. From the quantitative deviations, we begin to see dynamical effects. Thus, the predicted spin-averaged mass of the $D_s$ and $D_s^*$ is too high. The $c\bar{s}$ system has a greater reduced mass than the $c\bar{u}$ or $c\bar{d}$ system, which leads to increased binding effects that are ignored in our simple model. The predicted $c\bar{s}$ hyperfine splitting is too small in comparison with that of D and D*; we have ignored the increase of $|\Psi(0)|^2$ with increasing reduced mass. On the other hand, the model successfully predicts $M(D^*) - M(D) = (m_s/m_c) [M(K^*) - M(K)]$.

**Table 10**   Charmed hadron masses in a model with additive and hyperfine contributions

| $J^P$ | Meson | Mass (MeV/$c^2$) | |
|---|---|---|---|
| | | Predicted[a] | Observed[b] |
| $0^-$ | $D = c\bar{u},\ c\bar{d}$ | 1882 | 1867 |
| | $D_s = c\bar{s}$ | 2088 | 1971 |
| $1^-$ | $D^* = c\bar{u},\ c\bar{d}$ | 2002 | 2009 |
| | $D_s^* = c\bar{s}$ | 2164 | 2113 |

| $J^P$ | Baryon | Mass (MeV/$c^2$) | |
|---|---|---|---|
| | | Predicted[c] | Observed[b] |
| $1/2^+$ | $\Lambda_c^+ = udc$ | 2281 (input) | $2281 \pm 3$ |
| | $\Sigma_c = uuc,\ udc,\ ddc$ | 2438 | $2448 \pm 4$[d] |
| | $\Xi_c = usc,\ dsc$ | 2505[e] | $2460 \pm 25$ |
| | | 2604[f] | |
| | $\Omega_c = ssc$ | 2775 | $2740 \pm 20$ |
| $3/2^+$ | $\Sigma_c^* = uuc,\ udc,\ ddc$ | 2502 | |
| | $\Xi_c^* = usc,\ dsc$ | 2658 | |
| | $\Omega_c^* = ssc$ | 2818 | |

[a] $m_c = 1662$ MeV/$c^2$; other parameters as in Table 5.
[b] Masses are from (56) unless noted otherwise.
[c] $m_c = 1705$ MeV/$c^2$; other parameters as in Table 5.
[d] Ref. (60).
[e] State antisymmetric under exchange of s and u.
[f] State symmetric under exchange of s and u.

The $\Lambda_c$ was probably the first charmed particle observed (103) after the discovery of the $J/\psi$. The $\Sigma_c$, first seen some time ago (56, 103), has recently been confirmed in $e^+e^-$ annihilations (60). Candidate events for the charmed-strange baryons have been seen rather recently in experiments with good capability for short-track detection (62). The two $\Xi_c$ states listed in Table 10 mix with one another slightly as a result of spin-spin interactions. Since masses in a two-state mixing problem repel one another, the lower state is expected to be somewhat lower than shown in the table. The agreement for all known charmed baryons is certainly adequate for such a crude model.

Simple quark model considerations based on summing u-d mass differences, Coulomb self-energies, and hyperfine interactions (3, 4, 104) lead to the following electromagnetic mass differences, where experimental values (56, 61) are shown in brackets:

$$M(D^+) - M(D^0) = 4.3[4.7 \pm 0.3]\,\text{MeV}/c^2; \tag{37a.}$$

$$M(D^{*+}) - M(D^{*0}) = 4.2[2.9 \pm 1.3]\,\text{MeV}/c^2; \tag{37b.}$$

$$M(\Sigma_c^{++}) - M(\Sigma_c^0) = 2.6[2.5 \pm 1.0]\,\text{MeV}/c^2; \tag{37c.}$$

$$M(\Sigma_c^{++}) - M(\Sigma_c^+) = 2.2\,\text{MeV}/c^2. \tag{37d.}$$

Ignoring the hyperfine electromagnetic interaction, the authors of (104a) find $M(D^+) - M(D^0) = 6.7$ MeV/$c^2 \approx M(D^{*+}) - M(D^{*0})$; $M(\Sigma_c^{++}) - M(\Sigma_c^0) = -6$ MeV/$c^2$; and $M(\Sigma_c^{++}) - M(\Sigma_c^+) = -2$ MeV/$c^2$.

4.7.3 HADRONIC AND ELECTROMAGNETIC DECAYS

$^3S_1$ *states*    The D* states can decay both to $\pi$D and to $\gamma$D. The relative rates into these channels appear comparable, though one process is strong and the other electromagnetic. Quark models (18, 105) predict the rates for these processes in terms of others, such as hadronic and electromagnetic decays of the K*(892), with reasonable success. [One should note the prediction $B(D^{*+} \to \gamma D^+) = 1$–4%, to be compared with the value in (56) of $17 \pm 11\%$.]

*The* $\Sigma_c$    Quark model techniques similar to those just mentioned lead us to expect $\Gamma(\Sigma_c \to \Lambda_c \pi) \approx 2$ MeV, $\Gamma(\Sigma_c^+ \to \Lambda_c \gamma) \approx 90$ keV. We use a kinematic factor $p_{\text{cms}}^3 E(\Lambda_c) E_\pi / M(\Sigma_c)$ suggested in (18) to relate the first process to $\Sigma(1385) \to \Lambda \pi$, via a symmetry prediction for the transition matrix elements given in (106).

4.7.4 P-WAVE HADRONS CONTAINING CHARMED QUARKS    The p-wave excitations $^3P_2$, $^3P_1$, $^1P_1$, and $^3P_0$ of the c$\bar{\text{u}}$ and c$\bar{\text{d}}$ systems (we call them D** here) are expected to lie somewhere between 2.2 and 2.5 GeV/$c^2$. The corresponding c$\bar{\text{s}}$ excitations should be about 100–150 MeV/$c^2$ heavier.

Each D** has decay modes characteristic of its spin and parity: $^3P_2 \rightarrow$ ($\pi D^*$ or $\pi D$); ($^3P_1$ or $^1P_1$) $\rightarrow \pi D^*$; $^3P_0 \rightarrow \pi D$. The $^3P_1$ and $^1P_1$ states are expected to mix with one another, in the manner of the two axial-vector kaons (56) $K_1(1280)$ and $K_1(1400)$. In the limit in which the charmed quark is much heavier than the u and d quarks, the mass eigenstates correspond to definite values (1/2 or 3/2) of the light quark's total (orbital + spin) angular momentum (105).

A candidate for a p-wave charmed meson, the D(2420), decaying to $\pi^- D^{*+}$, has been reported in one experiment (61, 107). It lies $416 \pm 6$ MeV/$c^2$ above the $D^{*+}$, and has a width of $\Gamma[D(2420)] = 75 \pm 20$ MeV. If the particle is real, it could be either a $1^+$ or a $2^+$ state. If $J^P = 1^+$, the $\pi D$ decay is forbidden, whereas if $J^P = 2^+$, the $\pi D$ decay rate should be about 1.5 times the $\pi D^*$ rate.

The study of $\pi D$ resonances could be a very fruitful source of information on the light-quark–heavy-quark bound state (108). In the limit that one quark is very heavy, this system becomes the relativistic one-body problem of QCD.

# 5.   THE UPSILON FAMILY AND
b-FLAVORED STATES

The family of $b\bar{b}$ bound states is our best example so far of a "hadronic hydrogen atom." Many s- and p-wave levels have been discovered, more are likely to be seen, and there is even the prospect of finding some of the predicted d-wave states. We mention in this section what insights various levels of the upsilon family have provided.

## 5.1   Hadronic and Radiative Decays of $\Upsilon$

5.1.1   PERTURBATIVE QCD PREDICTIONS   All of the $\Upsilon$ decays observed so far proceed via $b\bar{b}$ annihilation. Virtual photons lead to lepton and quark pairs, while decays that can be interpreted as proceeding via $\gamma gg$ and $ggg$ also have been observed. The partial widths of the $\Upsilon$ are summarized in Table 11.

The predicted ratio of strong and leptonic decay rates (with $M_\Upsilon$ taken as $2m_b$),

$$\Gamma(\Upsilon \rightarrow ggg)/\Gamma(\Upsilon \rightarrow \mu^+\mu^-) = 10(\pi^2 - 9)[\alpha_s(M_\Upsilon^2)]^3/9\pi\alpha^2, \qquad 38.$$

is subject to a large perturbative correction (71) of the form $[1 + 9.1\alpha_s(M_\Upsilon^2)/\pi]$. If we ignore this correction and use the experimental value $\Gamma(\Upsilon \rightarrow ggg)/\Gamma(\Upsilon \rightarrow \mu^+\mu^-) = 28.7 \pm 2.3$ implied by Table 11, we obtain $\alpha_s(M_\Upsilon^2) \approx 0.17$. The experimental error is insignificant compared

**Table 11** Partial decay modes of the $\Upsilon$

| Decay mode | Branching ratio (%) | Partial width (keV) |
|---|---|---|
| $e^+ e^-$ | $2.82 \pm 0.31$ | $1.22 \pm 0.05$ |
| $\mu^+ \mu^-$ | $2.78 \pm 0.22$ | $1.20 \pm 0.05$ |
| $\tau^+ \tau^-$ | $2.76 \pm 0.2^a$ | $1.19 \pm 0.05^a$ |
| $\gamma^* \to q\bar{q}$ | $10.3 \pm 1.0^a$ | $4.42 \pm 0.55^a$ |
| $\gamma g g$ | $1.62 \pm 0.43^b$ | $0.70 \pm 0.19$ |
| $g g g$ | $79.8 \pm 1.4$ | $34.4 \pm 2.5$ |
| | $100$ | $43.1 \pm 3.1$ |

[a] Calculated value.
[b] From ratio $\Gamma(\gamma gg)/\Gamma(ggg) = 2.03 \pm 0.53\%$ (109).

with the theoretical one. If the energy scale $Q^* = 0.157 M_\Upsilon$ for evaluating $\alpha_s$ is adopted (72), the radiative correction $[1 - 14\alpha_s(Q^{*2})/\pi]$ is even larger.

The process $\Upsilon \to \gamma g g$ can be separated from background for photons of greater than about half their maximum energy (109, 110). The observed rate (see Table 11), when combined with the predicted ratio

$$\Gamma(\Upsilon \to \gamma gg)/\Gamma(\Upsilon \to ggg) = (4/5)[\alpha/\alpha_s(M_\Upsilon^2)][1 - 5.5\alpha_s(M_\Upsilon^2)/\pi] \qquad 39.$$

then implies $\alpha_s(M_\Upsilon^2) = 0.29 \pm 0.08$ if we ignore the substantial radiative correction. Taking account of this correction, Csorna et al (109) quote $\alpha_s([0.157 M_\Upsilon]^2) = 0.23$–$0.5$ depending on the assumed photon spectrum.

5.1.2 NEW NARROW BOSONS  In principle the radiative decays of the $\Upsilon$ can give rise to new narrow bosons, such as Higgs bosons (111), bound states of scalar colored objects (112), or axions (113). Searches for all of these objects in $\Upsilon$ decays have proved negative so far (58, 114, 115). The level of sensitivity of Higgs boson searches is just now approaching the expected electroweak limit, corresponding to a branching ratio in the decay $\Upsilon \to \gamma + H$ of a few parts in $10^4$.

5.1.3    THE $\eta_b(^1S_0)$ STATES    The $^3S_1$ b$\overline{b}$ state (the $\Upsilon$) is expected to decay to a lower-lying $^1S_0$ b$\overline{b}$ state (the $\eta_b$) via a magnetic dipole transition at a rate given by Equation 15: for $m_b = 4.9$ GeV/$c^2$,

$$\Gamma(\Upsilon \to \gamma\eta_b) = 45\,\text{eV} \times (E_\gamma/100\,\text{MeV})^3. \qquad 40.$$

Estimates (40, 87–89, 116) of the $\Upsilon$-$\eta_b$ mass difference range between 30 and 100 MeV/$c^2$. Present limits (58) imply only $M(\Upsilon) - M(\eta_b) < 168$ MeV/$c^2$.

We estimate using Equation 15 and nonrelativistic wave functions (see also 117) the hindered transition rate $\Gamma(\Upsilon' \to \gamma\eta_b) = 3$–6 eV, corresponding to a branching ratio of about $10^{-4}$. However, corrections (96–98) to Equation 15 appear to be appreciable for this transition.

## 5.2    Hadronic Decays of Higher $\Upsilon$ Levels

5.2.1    THREE-GLUON DECAYS OF $\Upsilon(2S)$ AND $\Upsilon(3S)$    The total widths of $\Upsilon(2S)$ and $\Upsilon(3S)$ may be estimated from the measured leptonic widths (see Section 1.1.4) and leptonic branching ratios $B_{\ell\ell}$, of which $B_{\mu\mu}$ is the most precisely measured. For the 2S and 3S states, a theoretical estimate of $B_{\mu\mu}$ may usefully be obtained as follows. All decays of $\Upsilon(2S)$ and $\Upsilon(3S)$ except for radiative (E1) transitions and $\pi\pi$ emission to lower $\Upsilon$ levels are expected to proceed via b$\overline{b}$ annihilation. The rates for these annihilation processes are all proportional to $|\Psi_{nS}(0)|^2$. Accordingly, we may write (58)

$$B_{\mu\mu}[\Upsilon(nS)] = B_{\mu\mu}[\Upsilon(1S)]\{1 - B_{\pi\pi}[\Upsilon(nS)] - B_{E1}[\Upsilon(nS)]\}. \qquad 41.$$

In this manner we obtain values of $1.52 \pm 0.13\%$ and $1.72 \pm 0.20\%$ for the 2S and 3S levels, to be compared with measured values (56, 58) of $1.80 \pm 0.44\%$ (2S) and $1.53 \pm 0.36\%$ (3S), respectively. Using the more precise results based on Equation 41, we estimate the total and partial widths and branching ratios for the 2S and 3S levels quoted in Tables 12 and 13. The branching ratios to $\gamma$gg and ggg have been obtained by subtraction, and we have assumed the same $\gamma$gg/ggg ratio as for the 1S level (Table 11). We then find $\Gamma(\text{ggg})/\Gamma(e^+e^-) = 28.5 \pm 2.8$ (2S); $28.4 \pm 4.4$ (3S), in close agreement with the corresponding ratio of $28.7 \pm 2.3$ for the 1S level. The implied values of $\alpha_s$ thus are all about 0.17 in the absence of QCD corrections.

5.2.2    DECAYS $nS \to mS + \pi\pi$    The decays of excited $\Upsilon$ levels to lower ones can proceed via the emission of two gluons, which then materalize into a $\pi\pi$ system. This mechanism also appears to be responsible for the process $\psi' \to J/\psi + \pi\pi$, which accounts for over half the $\psi'$ decay rate. Rates and spectra for these processes have been estimated in (84).

**Table 12**  Partial decay modes of the $\Upsilon(2S)$ level

| Decay mode | Branching ratio (%) | Partial width (keV) |
|---|---|---|
| $e^+ e^-$ | $1.52 \pm 0.13$ | $0.537 \pm 0.033$[a] |
| $\mu^+ \mu^-$ | $1.52 \pm 0.13$ | $0.537 \pm 0.033$ |
| $\tau^+ \tau^-$ | $1.51 \pm 0.13$ | $0.534 \pm 0.033$ |
| $\gamma* \to q\bar{q}$ | $5.6 \pm 0.5$ | $1.97 \pm 0.12$ |
| $\gamma \chi_{b0}$ | $4.31 \pm 0.96$[a] | $1.52 \pm 0.38$ |
| $\gamma \chi_{b1}$ | $6.73 \pm 0.86$[a] | $2.38 \pm 0.39$ |
| $\gamma \chi_{b2}$ | $6.57 \pm 0.87$[a] | $2.32 \pm 0.39$ |
| $\Upsilon \pi^+\pi^-$ | $18.75 \pm 0.99$[a] | $6.62 \pm 0.78$ |
| $\Upsilon \pi^0 \pi^0$ | $9.37 \pm 0.50$ | $3.31 \pm 0.39$ |
| $\gamma g g$ | $0.88 \pm 0.23$ | $0.31 \pm 0.09$ |
| $g g g$ | $43.28 \pm 1.94$ | $15.28 \pm 1.74$ |
| | $100$ | $35.3 \pm 3.7$ |

[a] Based on compilation of (56). All other values are calculated on the basis of Equation 41 as explained in the text.

The $\pi\pi$ mass spectra are peaked toward the high end in $\psi' \to J/\psi + \pi\pi$, $\Upsilon(2S) \to \Upsilon(1S) + \pi\pi$, and $\Upsilon(3S) \to \Upsilon(2S) + \pi\pi$, but near its center in $\Upsilon(3S) \to \Upsilon(1S) + \pi\pi$ (59). A likely explanation of these differences in spectra would be a broad s-wave dipion resonance coupled to a pair of gluons (118).

5.2.3  THE DECAY $\Upsilon(3S) \to {}^1P_1 + \pi\pi$  One expects a ${}^1P_1$ $b\bar{b}$ state ("$h_b$") at the center of gravity of each group of ${}^3P_J$ levels, as in Section 4.4.2. For the lowest $b\bar{b}$ states, this corresponds to $M({}^1P_1) = [5M(\chi_{b2}) + 3M(\chi_{b1})$

**Table 13**   Partial decay modes of the $\Upsilon(3S)$ level[d]

| Decay mode | Branching ratio (%) | Partial width (keV) |
|---|---|---|
| $e^+ e^-$ | $1.72 \pm 0.20$ | $0.402 \pm 0.031$[a] |
| $\mu^+ \mu^-$ | $1.72 \pm 0.20$ | $0.402 \pm 0.031$ |
| $\tau^+ \tau^-$ | $1.71 \pm 0.20$ | $0.400 \pm 0.031$ |
| $\gamma^* \to q\bar{q}$ | $6.3 \pm 0.5$ | $1.47 \pm 0.11$ |
| $\gamma \chi_{b0}(2P)$ | $5.3 \pm 2.3$[b] | $1.24 \pm 0.54$ |
| $\gamma \chi_{b1}(2P)$ | $11.7 \pm 3.0$[b] | $2.73 \pm 0.80$ |
| $\gamma \chi_{b2}(2P)$ | $12.8 \pm 3.3$[b] | $2.99 \pm 0.88$ |
| $\Upsilon(1S)\,\pi^+\pi^-$ | $3.47 \pm 0.34$[c] | $0.81 \pm 0.14$ |
| $\Upsilon(1S)\,\pi^0\pi^0$ | $1.74 \pm 0.17$ | $0.41 \pm 0.07$ |
| $\Upsilon(2S)\,\pi^+\pi^-$ | $2.1 \pm 0.5$[c] | $0.49 \pm 0.14$ |
| $\Upsilon(2S)\,\pi^0\pi^0$ | $1.05 \pm 0.25$ | $0.25 \pm 0.07$ |
| $h_b(1^1P_1)\,\pi^+\pi^-$ | $0.37 \pm 0.15$[c] | $0.09 \pm 0.04$ |
| $h_b(1^1P_1)\,\pi^0\pi^0$ | $0.19 \pm 0.08$ | $0.04 \pm 0.02$ |
| $\gamma gg$ | $0.88 \pm 0.23$ | $0.21 \pm 0.06$ |
| $ggg$ | $48.86 \pm 5.1$ | $11.42 \pm 1.99$ |
| | 100 | $23.37 \pm 3.26$ |

[a] Based on compilation of (56).
[b] Ref. (58).
[c] Ref. (59).
[d] Except as indicated in footnotes $a$–$c$, all values are calculated on the basis of Equation 41, as explained in the text.

$+ M(\chi_{b0})]/9 = 9.9002 \pm 0.0007$ GeV/$c^2$. In decays of $\Upsilon(3S)$ to $\pi\pi + ($anything$)$, a weak $(2.5\sigma)$ signal has been seen (59) at $M(^1P_1) = 9.8948 \pm 0.0015$ GeV/$c^2$ at a branching ratio $B[\Upsilon(3S) \to \pi\pi h_b(1^1P_1)] = 0.37 \pm 0.15\%$. As for charmonium, the close agreement between the predicted and observed masses would indicate that the spin-spin splitting is very small in the lowest p-wave $b\bar{b}$ levels. This, in turn, would be further evidence for the Coulomb-like nature of the force leading to the spin-spin interaction between quarks.

The branching ratio for $h_b(1^1P_1) \to \gamma + \eta_b$ is expected to be around 50%, and the photon energy is expected to be around $500 \pm 30$ MeV. The $1^1P_1$ level thus may be a useful source of the lowest $b\bar{b}$ state.

## 5.3  $\chi_b$ States and Electric Dipole Transitions

Two sets of p-wave $b\bar{b}$ levels have been discovered: a triplet around 9.9 GeV/$c^2$ (which we call $\chi_b$, or 1P) and a triplet ($\chi_b'$, or 2P) around 10.25 GeV/$c^2$. The properties of these levels are noted in more detail in Table 2. The spin-parity assignments of some of these levels are assumed from theory, but all tests performed so far (119) on the $\chi_b$ levels are compatible with the values of $J^P$ shown and with the dominance of electric dipole transitions.

5.3.1  FINE-STRUCTURE SPLITTING    The masses of the $\chi_b$ levels provide valuable information on the spin dependence of the interquark force. In the presence of vector and scalar interactions, the masses of $^3P_J$ levels $\chi_J$ are determined by the spin-dependent interactions in Sections 3.4.2 and 3.4.3. The spin-orbit and tensor terms probe different combinations of vector and scalar potential.

A simple model (88–90, 97, 120, 121) based on a short-range vector interaction $V_V(r) = -(4/3)\alpha_s/r$ and a long-range scalar interaction $V_S(r) = kr$ is compatible with present data. [A more extensive discussion is given in (122).] The vector interaction is expected on the basis of single-gluon exchange, while an effective scalar interaction can arise from a rotating flux tube (70, 121), as mentioned in Section 3.4. If we define the parameter (123)

$$R \equiv [M(\chi_2) - M(\chi_1)]/[M(\chi_1) - M(\chi_0)], \qquad\qquad 42.$$

we expect in such a model $R = 0.8\ (1 - 5\lambda/16)/(1 - \lambda/8)$, where $\lambda \equiv k\langle 1/r\rangle/(\alpha_s\langle 1/r^3\rangle)$. For a purely Coulombic force ($\lambda = 0$) one would have $R = 0.8$. If $M(\chi_2) > M(\chi_1) > M(\chi_0)$ (so the scalar interaction is not dominant), the model predicts $R < 0.8$.

We show in Table 14 the predictions of this model for $\alpha_s = 0.374$, $k = 0.18$ GeV$^2$, and $m_b = 4.9$ GeV/$c^2$. From both sets of p-wave levels, there is evidence for a non-Coulombic part of the interaction, and the deviations from $R = 0.8$ are as expected for a scalar long-range interaction. The question of whether there is a small long-range effective vector part of the interaction (120, 122) is still unresolved (40). Arguments (87, 123–126) in favor of a dominantly vector-like confining part of the potential must be evaluated in light of the successes of the scalar confinement model.

Predictions for $b\bar{b}$ fine-structure parameters (87–89, 120) are discussed in (40). The quality of new experimental data will permit distinctions among these models.

5.3.2  ELECTRIC DIPOLE TRANSITION RATES    If the quarkonium wave functions are known, one can calculate rates for electric dipole transitions

**Table 14**   Predictions for $\Delta M \equiv M(^3P_J) - \langle M(^3P_J) \rangle$ of a scalar confinement model

| State | $\Delta M(\chi_b[1P])$, MeV/$c^2$ | | $\Delta M(\chi_b'[2P])$, MeV/$c^2$ | |
|---|---|---|---|---|
| | Predicted[a] | Observed[b] | Predicted[a] | Observed[c] |
| $^3P_0$ | −39.3 | −40.4 ± 1.4 | −29.4 | −29.6 ± 2.4 |
| $^3P_1$ | −8.6 | −8.3 ± 0.8 | −6.5 | −4.4 ± 1.0 |
| $^3P_2$ | 13.0 | 13.1 ± 0.7 | 9.8 | 8.5 ± 0.9 |
| $\dfrac{M(\chi_2) - M(\chi_1)}{M(\chi_1) - M(\chi_0)}$ | 0.71 | 0.67 ± 0.06 | 0.71 | 0.57 ± 0.06[d] |

[a] Ref. (127).
[b] Ref. (56).
[c] Ref. (58).
[d] Based on combination of inclusive and exclusive photon spectra in (58).

among levels. These may be compared with measured transition rates provided the total widths of the decaying states are known, since it is their branching ratios that are measured directly.

One way to learn the wave functions of $b\bar{b}$ states is to construct the interquark potential directly from bound-state data (25, 78, 79, 127). This permits the evaluation of the necessary dipole matrix elements in Equations 14; all approaches (18, 87–89) agree on their values to 10%. The total widths of p-wave levels are estimated theoretically as the sum of electric dipole and hadronic partial decay rates; the latter are based on Equations 22, 26, and 23 for the $^3P_{2,1,0}$ levels, respectively.

We show the results of a comparison of predicted (127) and observed E1 transition rates in Table 15. For just one process out of 14 is the partial decay rate more than two standard deviations away from its predicted value: the transition $2^3P_1 \rightarrow \gamma + \Upsilon$. Relativistic $[O(E_\gamma/m_b)]$ deviations from the simple dipole formula (Equations 14) for the hard photon ($E_\gamma = 763$ MeV) in this transition are estimated (87–89) to be at the 20–30% level. Notice the small predicted 3S → 1P rates; the above approaches do not agree on the exact value of the matrix element, but nearly all agree it is very small.

## 5.4   D-Wave Levels

A key test of the very existence of a potential for quarkonium systems [which could be questioned (128) if gluonic degrees of freedom are impor-

**Table 15**  E1 transitions in b$\bar{b}$ systems

| Transition | | Photon energy (MeV) | Experimental branching ratio (%) | Partial decay width (keV) | |
|---|---|---|---|---|---|
| | | | | Experiment | Theory[a] |
| 2S | $\rightarrow$ $1^3P_2$ | 109.5±0.6 | 6.57 ± 0.87 | 1.97 ± 0.55[b] | 2.1 |
| | $1^3P_1$ | 130.7±0.7 | 6.73 ± 0.86 | 2.02 ± 0.55[b] | 2.2 |
| | $1^3P_0$ | 162.3±1.3 | 4.31 ± 0.96 | 1.29 ± 0.43[b] | 1.4 |
| $1^3P_2$ | $\rightarrow$ $\Upsilon$ | 442.9±0.6 | 22.0 ± 4.2 | 29.2 ± 5.6[c] | 38 |
| $1^3P_1$ | $\rightarrow$ $\Upsilon$ | 422.5±0.7 | 35.0 ± 8.0 | 22.4 ± 5.1[c] | 33 |
| $1^3P_0$ | $\rightarrow$ $\Upsilon$ | 391.7±1.3 | < 6 | < 22.9[c] | 26 |
| 3S | $\rightarrow$ $2^3P_2$ | 86.5±0.7 | 12.8 ± 3.3 | 3.3 ± 1.1[d] | 2.8 |
| | $2^3P_1$ | 99.3±0.8 | 11.7 ± 3.0 | 3.0 ± 1.0[d] | 2.5 |
| | $2^3P_0$ | 124.2±2.3 | 5.3 ± 2.3 | 1.4 ± 0.7[d] | 1.6 |
| | $1^3P_2$ | 432.8±0.8 | - | - | 0.025 |
| | $1^3P_1$ | 453.2±0.9 | - | - | 0.017 |
| | $1^3P_0$ | 483.8±1.4 | - | - | 0.006 |
| $2^3P_2$ | $\rightarrow$ $\Upsilon'$ | 242.3±0.8 | 15 ± 8 | 17 ± 9[e] | 18.7 |
| | $\Upsilon$ | 776.8±0.7 | 16 ± 6 | 19 ± 7[e] | 9.8 |
| $2^3P_1$ | $\rightarrow$ $\Upsilon'$ | 229.7±0.9 | 24 ± 10 | 14 ± 6[e] | 15.9 |
| | $\Upsilon$ | 764.8±0.8 | 7 ± 3 | 4.2 ± 1.8[e] | 9.3 |
| $2^3P_0$ | $\rightarrow$ $\Upsilon'$ | 205.0±2.3 | 4 ± 3 | 13 ± 10[e] | 11.3 |
| | $\Upsilon$ | 741.5±2.3 | <3 (90% c.l.) | <10[e] | 8.5 |

[a] See (127).
[b] Based on total width of 2S level quoted in (56).
[c] Based on calculated total widths of (132, 64, 382) keV for $J = 2, 1, 0$.
[d] Based on total width of 3S level quoted in (58).
[e] Based on calculated total widths of (116, 60, 336) keV for $J = 2, 1, 0$.

tant] is its ability to predict as yet unseen aspects of the spectrum correctly. The d-wave b$\bar{b}$ states provide such a test.

5.4.1  EXPECTED MASSES  Nearly all potential models agree that the lowest d-wave b$\bar{b}$ levels have centers of gravity around 10.16 (1D) and 10.44 (2D) GeV/$c^2$. Predictions of the fine-structure splitting differ somewhat, but most authors agree that it should be smaller than in the P states. For example, in (127) the masses (10149, 10156, 10161) and (10434, 10440, 10444) $MeV/c^2$ are obtained for the $1^3D_{1,2,3}$ and $2^3D_{1,2,3}$ levels in the simple model described above. Other models yield results differing by only a few MeV (18, 87, 89).

5.4.2 SIGNATURES IN $e^+e^-$ ANNIHILATIONS  The most promising way to observe the d-wave levels appears to be to study electromagnetic transitions of the $\Upsilon(3S)$ level, as produced in $e^+e^-$ annihilations. The expected rate of production of d-wave levels may be gauged in terms of the transitions $\Upsilon(3S) \to \gamma + \chi'_b \to \gamma + \gamma + \Upsilon$ (1S and/or 2S), which have already been seen. It is likely that in the present sample of more than $10^5$ $\Upsilon(3S)$ decays, at least $10^3$ 1D states have been produced. The challenge is to separate them from background.

The photons expected in the transitions $\Upsilon(3S) \to \gamma_1 + 2^3P \to \gamma_1 + \gamma_2 + 1^3D$ are expected to have a total energy lying in a narrow band: $E_{\gamma_1} + E_{\gamma_2} = 244 \pm 10$ MeV, as a result of the small fine-structure splitting in the D states. It is probably necessary to observe some signature of D-state decay: either an additional photon $\gamma_3$ with energy around 250 MeV resulting from a transition to a $1^3P$ state, or a pion pair+(missing mass = $M_\Upsilon$). These signatures are shown in Figure 9a.

An energy scan in $e^+e^-$ annihilations with extremely high sensitivity and resolution could in principle excite the $^3D_1$ states. Their leptonic widths are expected to be very small (87): 1.5 eV for the $1^3D_1$ (about $10^{-3}$ of the $1^3S_1$ leptonic width, corresponding to an expected branching ratio of $3 \times 10^{-5}$), and 2.7 eV for the $2^3D_1$. Mixing with S states, which probably affects the observed $1^3D_1$ state in charmonium [the $\psi(3770)$], is expected to be much less significant for the $1^3D_1$ and $2^3D_1$ $b\bar{b}$ levels, since (in contrast to the $c\bar{c}$ state) they lie well below flavor threshold. Thus, coupled-channel effects are not expected to be nearly as important. (Tensor force mixing of $^3S_1$ and $^3D_1$ $b\bar{b}$ levels was taken into account in the above estimates.)

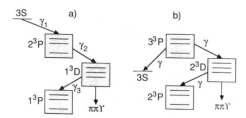

*Figure 9*  Electromagnetic transitions giving rise to $1^3D_{1,2,3}$ states (*a*) from the $3^3S_1$ $b\bar{b}$ level; (*b*) from the $3^3P$ levels, capable of giving rise both to the $3^3S_1$ state [the $\Upsilon(3S)$] and to $2^3D$ levels.

5.4.3 HADRONIC PRODUCTION    It is likely that one important mechanism for $^3S_1$ $\Upsilon$ production in hadronic experiments, such as those leading to the spectrum in Figure 2b, is the hadronic production of p-wave states $\chi_{0,2}$ (since these couple to two gluons), followed by E1 decays. (The $\chi_1$ may also be produced in this manner if one of the gluons is virtual.) The presence of the $\Upsilon(3S)$ in the effective mass distribution of Figure 2b then may signal the production of 3P b$\bar{\text{b}}$ states in hadronic experiments. These states can decay not only to $\Upsilon(3S)$, but also to d-wave (particularly $2^3D$) levels. The expected transitions are shown in Figure 9b. It may be possible to pick up the $2^3D$ states through their $\pi\pi\Upsilon$ decays. The $1^3D$ levels shown in Figure 9a could arise similarly from hadronically produced $2^3P$ states. High-resolution studies of soft photons in hadronic experiments could shed light on decay schemes such as those shown in Figure 9b.

## 5.5    States Above Threshold

Just as for the charmonium levels, b$\bar{\text{b}}$ states are highly stable only if strong decays into pairs of flavored mesons are kinematically forbidden. Thus, the levels above the shaded line labeled by $2M(B)$ in Figure 4 are broader than those below it, as one may see from Table 2. In light of the measured masses of B and B* states, the $\Upsilon(4S)$ should decay only to pairs of $0^-$ mesons (B) each containing a b quark or antiquark and a nonstrange antiquark or quark. The decays of the higher levels (129) may be especially good sources of other b-flavored hadrons (130), such as vector mesons, baryons, and particles containing strange quarks. We turn to a brief description of such states.

## 5.6    Flavored Mesons and Baryons

5.6.1    THE B(0⁻) STATES    Mesons containing a single b quark (Table 4) have been reconstructed from their decays into charmed mesons (26, 131). The simple model (3, 4, 104) described earlier for charmed-meson electromagnetic mass differences predicts $M(B^0) - M(B^+) = 5.7$ MeV/$c^2$, to be compared with the measured values (131) of $2.0 \pm 1.1 \pm 1.0$ MeV/$c^2$ (CLEO) and $2.4 \pm 1.6 \pm 1.0$ MeV/$c^2$ (ARGUS). A prediction of 4.4 MeV/$c^2$ is obtained in the last of Refs. (18).

5.6.2    VECTOR MESONS (B*)    The B*-B hyperfine splitting is expected to be proportional to $|\Psi(0)|^2/m_u m_b$, where $|\Psi(0)|^2$ is the square of the wave function at the origin. This wave function is expected to depend primarily on properties of the light quark, and should not be much affected if we replace the heavy b quark by another (say, c). In this manner we are led to expect that

$$M(B^*) - M(B) = (m_c/m_b)\,[M(D^*) - M(D)]$$
$$\approx (1/3)\,[M(D^*) - M(D)]$$
$$\approx 50 \text{ MeV}/c^2,  \hspace{4em} 43.$$

which is well verified. We predict $M(B^{*0}) - M(B^{*-}) = 5.8$ MeV/$c^2$, whereas the last of Refs. (18) gives 4.4 MeV/$c^2$.

5.6.3  PREDICTED STRANGE AND BARYONIC STATES   Additional states containing a b quark may be predicted with the help of the naive quark model described in Section 3. We estimate $m_b - m_c$ by comparing the spin-averaged masses of the B's and D's:

$$m_b - m_c = \frac{3M(B^*) + M(B)}{4} - \frac{3M(D^*) + M(D)}{4} = 3.34 \text{ GeV}/c^2. \hspace{2em} 44.$$

The resulting predictions for masses of the lowest b-flavored states (cf 3; see also 132) are shown in Figure 10. The $B_s$ meson is expected to be very useful in studies of mixing of neutral mesons (as in the K-$\bar{K}$ system) and $CP$ violation (130). Indeed, recent experiments (133, 133a) suggest mixing for both $B_s$-$\bar{B}_s$ and $B^0$-$\bar{B}^0$ systems.

# 6.  COMPARISONS OF QUARKONIUM SPECTRA

The spectra of bound states of different flavors of heavy quarks appear to be described by the same potential. In this section we recapitulate some

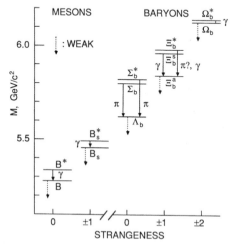

*Figure 10*   $L = 0$ mesons and baryons containing a single b-flavored quark (adapted from 3).

arguments in favor of this assertion, describe some elementary power-law properties of the potential, show how heavy quarks can provide new information at short distances, and mention expectations at larger distances (near flavor threshold).

## 6.1    Flavor-Independence of Potential

When both $c\bar{c}$ and $b\bar{b}$ bound-state data became available, it was possible to construct potentials directly from these two sets of data using the inverse scattering method (25). Two such potentials based on the most recent set of masses and leptonic widths are compared in Figure 11. They agree extremely well in the range of distances where data exist on both sets of bound states. Improved leptonic width measurements for the charmonium levels would be welcome in refining our knowledge of the $c\bar{c}$ interaction and testing flavor-independence to greater accuracy.

## 6.2    Elementary Power-Law Behavior

The similarity of $c\bar{c}$ and $b\bar{b}$ bound-state spectra suggests (45, 66) a phenomenological power-law potential $V(r) = Ar^{v}$ with $v \approx 0$. The $v = 0$ limit corresponds to $V(r) = C \ln(r)$, and would give spectra identical except for an overall shift. Confirming evidence is available from the level spacings

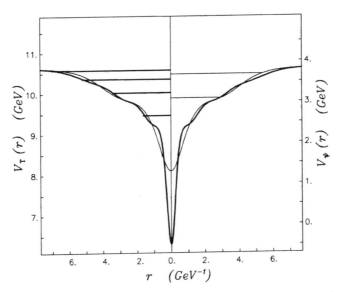

*Figure 11* Comparison of potentials constructed via the inverse-scattering method for $\Upsilon$ levels (*heavy curve*, levels on left) and charmonium levels (*light curve*, levels on right).

in the $\psi$ and $\Upsilon$ families separately, as summarized in Figure 12. It is notable that a simple power law interpolates between the anticipated Coulomb and linear regimes for all the known $c\bar{c}$ and $b\bar{b}$ states.

The potential $V(r) = C\ln(r)$ fits charmonium and upsilon data adequately for $C \approx 0.72$ GeV. In such a potential, all states have the same kinetic energy $T$. According to the virial theorem, $T = \langle(r/2)\mathrm{d}V/\mathrm{d}r\rangle = C/2 = 0.36$ GeV $= 2\langle\mathbf{p}^2/2m_Q\rangle$, so the average velocity squared of a quark is approximately $\langle\beta^2\rangle = 0.36$ GeV/$m_Q$. Thus we expect $\langle\beta^2\rangle$ to be approximately 0.2 for charmed quarks, and 0.07 for b quarks.

## 6.3 Role of Heavy Quarks in Probing Potential Near $r = 0$

In Figure 11 the $\Upsilon(1S)$ is more deeply bound (has a smaller mean radius) than the $\psi(1S)$. According to the Feynman-Hellmann theorem (45) ($\partial E/\partial m = -\langle T\rangle/m$), heavier quarks probe the potential more deeply. In

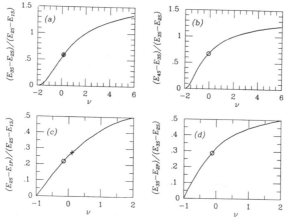

Figure 12    Ratios of level spacings implied by power-law potentials behaving as $r^\nu$. (a) $(E_{3S}-E_{2S})/(E_{2S}-E_{1S})$; (b) $(E_{4S}-E_{3S})/(E_{3S}-E_{2S})$; (c) $(E_{2S}-E_{1P})/(E_{2S}-E_{1S})$; and (d) $(E_{3S}-E_{2P})/(E_{3S}-E_{2S})$. [$\psi$: +; $\Upsilon$: O.]

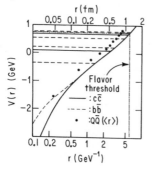

Figure 13    Comparison of predicted $c\bar{c}$, $b\bar{b}$, and $Q\bar{Q}$ $^3S_1$ levels (for $m_Q = 40$ GeV/$c^2$) in the potential (*solid curve*) of Ref. (87). For $Q\bar{Q}$ levels, both the energy and average radius are shown. The dashed curve describes $V(r) = (8\pi/27)(\lambda r - 1)^2/r\ln\lambda r$, $\lambda = 0.7325$ GeV (134).

Figure 13 (40) we compare predicted $c\bar{c}$, $b\bar{b}$, and $Q\bar{Q}$ ${}^3S_1$ levels, where Q is a hypothetical quark of mass 40 GeV/$c^2$. The unique information provided by the $Q\bar{Q}$ system comes from the lowest levels. The 1S level [especially its leptonic width, sensitive to $|\Psi_{1S}(0)|^2$] probes the potential at distances shorter than 0.05 fm, where $c\bar{c}$ and $b\bar{b}$ systems give little information. Potentials (87, 134) with very different behavior for $r < 0.1$ fm reproduce existing $c\bar{c}$ and $b\bar{b}$ data.

As the interquark separation decreases, one expects a trend toward more Coulomb-like behavior, which may be manifested in several ways (45). First, in the limit of a pure Coulomb force, the 2S and 1P levels become degenerate. Second, the values of $|\Psi(0)|^2$ for $n$S levels (and hence their leptonic widths) become proportional to $n^{-3}$, so that $|\Psi_{2S}(0)|^2/|\Psi_{1S}(0)|^2 \to$ 1/8. Third, the spacing between 1S and 2S levels should begin to grow with increasing quark mass, reflecting the proportionality of energy levels to reduced mass in a Coulomb potential.

The first two of these trends are visible as we pass from the charmonium to the upsilon family. The 1P level is closer to the 2S (relative to the 1S-2S spacing) in $b\bar{b}$ than in $c\bar{c}$, as we see in Figure 12. The ratio $|\Psi_{2S}(0)|^2/|\Psi_{1S}(0)|^2$ is $0.63 \pm 0.07$ for $c\bar{c}$ [characteristic of a power-law potential $V(r) \propto r^\nu$ with $\nu \approx 0.36$], and $0.50 \pm 0.03$ for $b\bar{b}$ (characteristic of $\nu \approx 0$). On the other hand, the 2S-1S spacing for $b\bar{b}$ is actually slightly less than for $c\bar{c}$. This has led to the use of a phenomenological power-law potential with a small positive power, in view of the expectation that level spacings in a potential behaving as $r^\nu$ should vary with quark mass $m_Q$ as $\Delta E \propto m_Q^{-\nu/(2+\nu)}$. If QCD provides a correct short-distance description of the quark-antiquark force, the success of this description should be transitory. Detailed predictions for the 2S-1S level spacing and leptonic widths in a variety of potentials compatible with $\psi$ and $\Upsilon$ data are given for heavy quarkonium systems in (40, 68, 135).

## 6.4 Flavor Thresholds: Counting Narrow Levels

The dissociation of a heavy quark-antiquark pair into a pair of flavored mesons should become less and less sensitive to the flavor of the heavy quark $Q$ as $m_Q \to \infty$, occurring simply when the heavy-quark pair has a given separation. This distance turns out to be about 1 fm, as shown in Figure 13. It then becomes possible to count the number $n$ of narrow s-wave quarkonium levels below flavor threshold (136), with the result

$$n \approx 1/4 + (1/\pi) \int dr \{m_Q[E_{TH} - V(r)]\}^{1/2}, \qquad 45.$$

where $E_{TH}$ is the threshold energy. The integral is taken from $r = 0$ to the classical turning point, where the integrand vanishes. The only quark-mass dependence arises from the explicit factor inside the square root, so that (scaling from the upsilon family) we expect $n \approx 4(m_Q/m_b)^{1/2}$. Thus, for a 40-GeV/$c^2$ quark, one would expect about 10 or 11 narrow quarkonium levels below flavor threshold, as indicated in Figure 13. As we shall see, discovery of much of this rich structure may be quite challenging.

## 7.    TOPONIUM

### 7.1    Present Experimental Situation Regarding the t Quark

The top (t) quark, as mentioned in Section 1.1.6, is an as yet hypothetical partner of the b quark, having charge $+2/3$. Its mass is known to be larger than about 23 GeV/$c^2$. Hints of its discovery in the mass range between 30 and 50 GeV/$c^2$ seem to have been premature, but there are not yet bounds that exclude its existence in this mass range. What new physics could be learned from a 40-GeV/$c^2$ top quark? Here we give only a brief sampling; much more complete discussions are presented in (38, 68, 137–139).

### 7.2    Prediction of Levels from Potentials and Vice Versa

As illustrated by the example of Figure 13, the lowest $t\bar{t}$ levels can be expected to probe the interquark potential at distances shorter than those for which it is known at present. Predicting the properties of these levels thus requires some extrapolation of our present phenomenological knowledge. If QCD is a reliable guide, we may expect a 2S-1S spacing that probably exceeds 700 MeV/$c^2$ (139) and a 1S leptonic width (for a $t\bar{t}$ 1S state at 80 GeV/$c^2$) ranging between about 4 and 8 keV (40, 140). These observables discriminate between a short-distance Coulomb interaction and extrapolation of the phenomenological $V(r) \sim \ln(r)$ form.

Higher $t\bar{t}$ levels should provide information about regions of the potential already probed by $b\bar{b}$ and $c\bar{c}$. Since the bound top quark will be quite nonrelativistic, the information coming from $t\bar{t}$ in these regions of $r$ may actually reflect the properties of the static potential more accurately.

### 7.3    Weak Decays of t

Toponium may decay by annihilation into gluons, mixing with the $Z^0$, strong or radiative cascades to other toponium levels, and weak decays of the t quark. The weak decays of t increase in importance as $m_t$ increases, as illustrated in Figure 14. For sufficiently high $m_t$, electric dipole transitions from excited S states and production of Higgs bosons via the decay

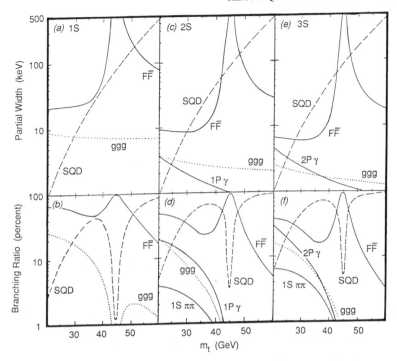

*Figure 14* Expected decay rates and branching ratios of 1S-3S levels of toponium. [From Kühn & Ono (138) potential "T."] FF̄: fermion pairs via virtual γ and Z; SQD: charged-current decay of a single t quark. The results for the Richardson potential (67) are very similar [see the erratum in (138)].

tt̄ → γ + H occur with too small a branching ratio to study effectively (for the electric dipole rates and branching ratios, see Figure 14c–f).

## 7.4   Toponium-Z Interference

The $n$S states of tt̄ can mix and interfere with the $Z^0$, depending on their masses (38, 141–143), and as a result may be more easily produced if they lie near the $Z^0$. On the other hand, since they decay quite readily via the $Z^0$, their transitions to other quarkonium states will occur with a smaller branching ratio.

## 7.5   Top-Flavored States

7.5.1   SPIN CORRELATIONS   As a result of the small predicted hyperfine splitting, the lightest $^3S_1$ state containing a single t quark is expected to undergo weak decay, carrying useful spin information (138), before it decays electromagnetically to the $^1S_0$ level.

7.5.2 CHARGED HIGGS BOSONS    If there is a charged Higgs boson $H^+$ lighter than the t quark, the semiweak decay $t \to H^+ + b$ will dominate over any conventional semileptonic decay of t. This will dramatically reduce the top-quark lifetime and the canonical leptonic branching ratio $\Gamma(t \to b\ell v)/\Gamma(t \to \text{all}) = 1/9$. For a discussion of rates, see (144).

## 7.6    Unusual Short-Range Forces

An enhanced coupling of neutral Higgs bosons to t quarks (145) can lead to distortions of the lowest $t\bar{t}$ levels from the expected patterns, including apparent violations of the level inequalities noted in (47).

# 8.    THE UNEXPECTED

## 8.1    A Fourth Generation (b')

Much of what we said in Section 6.3 applies to any heavy color-triplet quark, whether of charge 2/3 or $-1/3$. A fourth quark b' of charge $-1/3$ is not excluded by present $e^+e^-$ data above a mass of about 22.7 GeV/$c^2$ (146). A fourth generation would relax many constraints imposed by the three-generation Kobayashi-Maskawa (KM) model (52–55) on the phenomenology of $CP$ violation. Since the leptonic width of a $b'\bar{b}'$ state is expected to be 1/4 that of a corresponding $t\bar{t}$ state, such quarks will be more difficult (but not impossible) to study in $e^+e^-$ annihilations. One expects leptonic widths for the lowest $^3S_1$ state to be in the range of 1–2 keV for $m_{b'} = 40$ GeV/$c^2$. Scans for narrow resonances in $e^+e^-$ annihilations should be performed with at least this sensitivity.

## 8.2    Scalar Quarks: Level Structure, Bounds on Leptonic Widths

Scalar quarks, or "squarks," are expected partners of ordinary quarks on the basis of supersymmetry (147). The bound states of squarks with one another have properties that are easily calculated (112) but hard to study in the laboratory. As an example, the s-wave states are expected to be spinless, so they cannot be produced in $e^+e^-$ annihilations. The p-wave states are expected to have very small couplings to single virtual photons. The $e^+e^-$ cross section would grow very slowly, as $\beta^3$ (where $\beta$ is the center-of-mass squark velocity) as the energy crosses threshold for squark pair production, and would approach 1/4 the value for a corresponding spin-1/2 quark. Further suggestions for observing states of squarkonium have been given in (3, 112, 148).

## 8.3    Color Sextets

Quarks of higher color representations have been suggested in various contexts (149); in particular, color sextets should contribute twice as much to the $e^+e^-$ annihilation cross section as color triplets of the same charge and mass. The binding between color sextets should be stronger than between triplets, at least at short distances. It is interesting that modern superstring theories do not lead one to expect color sextet quarks, making the search for them particularly timely.

## 8.4   Exotic Quarks in Superstring Theories

Popular grand unified theories with an $E_6$ symmetry, loosely based on superstrings, contain electroweak singlet "h" quarks of charge $-1/3$. These couple purely vectorially to the $Z^0$, so the characteristic forward-backward asymmetry in $e^+e^-$ annihilations already observed for b quarks should be absent (150).

# 9.   SUMMARY AND OUTLOOK

In only a dozen years, the subject of heavy-quark systems has grown from its infancy to a rich spectroscopy, replete with insights into both the strong and electroweak interactions. The quarkonium ($Q\bar{Q}$) levels are described by a flavor-independent interaction with features of both perturbative and nonperturbative quantum chromodynamics. The number of $\Upsilon$ levels below flavor threshold was predicted in advance, as were numerous properties of members of the $\Upsilon$ family, on the basis of a nonrelativistic potential interpretation.

As one passes from $c\bar{c}$ ($\beta \approx 1/2$) to $b\bar{b}$ ($\beta \approx 1/4$) systems, a nonrelativistic description based on QCD acquires greater validity. Still heavier quarks will probe the interquark interaction almost completely freed from measurable relativistic effects.

Future electron-positron and antiproton-proton experiments will provide new data on $c\bar{c}$ and $b\bar{b}$ systems, of particular interest in the study of spin-dependent forces between quarks. Present data are converging on the notion of an effective scalar interaction at large distances, but it will be possible to test this idea in the next few years with far greater precision. The masses of $^1P_1$ candidates, for which far more conclusive evidence is needed, appear to lie near the spin-averaged $^3P_J$ values, a fact indicative of the short-range nature of the spin-spin interaction and its origin in a Coulomb-like gluon exchange.

We also look forward to more data on quarkonium systems with orbital angular momenta greater than one. At present, only one such state [the d-wave $\psi(3772)$] is known. Antiproton-proton collisions may be able to see two quasi-stable $J = 2$ charmonium states around $3.815$ GeV/$c^2$. Almost any flavor-independent central interquark potential describing the known quarkonium levels predicts the spin-averaged masses of the d-wave $b\bar{b}$ states to lie at $10.16 \pm 0.01$ (1D) and $10.44 \pm 0.01$ (2D) GeV/$c^2$.

Mesons and baryons containing single heavy quarks are yielding new information not only on the electroweak interaction, but also on hadron physics. The hyperfine splittings between D and D*, $D_s$ and $D_s^*$, and B and B* mesons reflect the expected interactions of quarks via their chromomagnetic moments. The first hints of p-wave excitations of D mesons have appeared; further information is eagerly awaited. Electromagnetic mass differences between charged and neutral D mesons, charged and neutral charmed baryons ($\Sigma_c$), and charged and neutral B mesons all appear to agree with expectations.

Bound states of heavier quarks (such as the anticipated top, or t) will provide a variety of information. The lowest states will probe new territory (below 0.05 fm) of the interquark force, as a result of the large mass of the t quark. The higher states are affected primarily by that range of the interquark force already studied in lighter systems, and will allow confirmation of the flavor independence of the interaction. Relativistic corrections, important for charmonium and still perceptible for the $\Upsilon$ family, ought to be negligible for most $t\bar{t}$ bound states.

The possibility of more quarks (beyond the t) or other colored objects means that it is crucial to understand the bound states of heavy quarks as well as we can. By so doing, we can infer properties of the fundamental constituents (such as their masses, electric charges, and any anomalous interactions they may possess) in terms of properties of the hadrons containing these constituents.

Experimental techniques for studying heavy-quark systems in the future include electron-positron colliders of modest energy (151), such as the SPEAR machine in operation at Stanford and the planned Beijing Electron-Positron Collider (BEPC); proton-antiproton collisions with precisely defined center-of-mass energy (77); and large electron-positron colliders operating at the $Z^0$ mass and above. Further specialized colliders, perhaps operating in the $\Upsilon$ energy range, also are under discussion (152). The Cornell (CESR) and Hamburg (DORIS) electron-positron colliders will continue to produce stimulating results under conditions of improved luminosity.

Heavy-quark spectroscopy is at once a mature field, and one with a rich future. We look forward in the next few years to enjoying its fruits as new means become available for its study.

ACKNOWLEDGMENTS

We wish to thank R. Cahn, Kuang-ta Chao, E. Eichten, P. Franzini, J.

Lee-Franzini, G. P. Lepage, P. Mackenzie, M. Oreglia, J. Prentice, L. Rosenberg, M. Sokoloff, and S. Stone for fruitful discussions. This work was supported in part by the United States Department of Energy under Contract No. DE-AC02-82ER40073. Fermilab is operated by Universities Research Association, Inc., under contract with the United States Department of Energy.

*Literature Cited*

1. Gell-Mann, M. *Phys. Lett.* 8: 214–15 (1964); Zweig, G. CERN reports Th.401 and 412 (1964) Unpublished; and in *Proc. Int. Sch. Phys. "Ettore Majorana,"* Erice, Italy, 1964, ed. A. Zichichi, pp. 192–234. New York/London: Academic (1965)
2. Lipkin, H. J. *Phys. Rep.* 8C: 173–286 (1973); Rosner, J. *Phys. Rep.* 11C: 189–326 (1974); Hendry, A. W., Lichtenberg, D. B. *Rep. Prog. Phys.* 41: 1707–80 (1978); Close, F. *An Introduction to Quarks and Partons.* New York: Academic (1979)
3. Rosner, J. In *Techniques and Concepts of High Energy Physics*, ed. T. Ferbel, pp. 1–141. New York: Plenum (1981); Gasiorowicz, S., Rosner, J. L. *Am. J. Phys.* 49: 954–84 (1981)
4. Quigg, C. In *Gauge Theories in High Energy Physics*, ed. M. K. Gaillard, R. Stora, pp. 645–822. Amsterdam: North-Holland (1982)
5. Greenberg, O. W. *Ann. Rev. Nucl. Sci.* 28: 327–86 (1978)
6. Greenberg, O. W. *Phys. Rev. Lett.* 13: 598–602 (1964)
7. Han, M., Nambu, Y. *Phys. Rev.* 139B: 1006–10 (1965); Nambu, Y. In *Preludes in Theoretical Physics*, ed. A. de Shalit, pp. 133–42. Amsterdam: North-Holland (1966)
8. Gross, D. J., Wilczek, F. *Phys. Rev. Lett.* 30: 1343–46 (1973); Politzer, H. D. *Phys. Rev. Lett.* 30: 1346–49 (1973)
9. Gross, D. J., Wilczek, F. *Phys. Rev.* D8: 3633–52 (1973); *Phys. Rev.* D9: 980–93 (1974); Politzer, H. D. *Phys. Rep.* 14C: 129–80 (1974)
10. Jaffe, R. L. *Nature* 268: 201–8 (1977); Bander, M. *Phys. Rep.* 75: 205–86 (1981)
11. Aubert, J. J., et al. *Phys. Rev. Lett.* 33: 1404–6 (1974)
12. Augustin, J. E., et al. *Phys. Rev. Lett.* 33: 1406–8 (1974)
13. Renard, F. M. *Basics of Electron-Positron Collisions*, p. 123. Gif-sur-Yvette, France: Editions Frontières (1981)
14. Glashow, S. L. In *Experimental Meson Spectroscopy—1974*, ed. D. A. Garelick, pp. 387–92. New York: AIP (1974)
15. Gaillard, M. K., Lee, B. W., Rosner, J. L. *Rev. Mod. Phys.* 47: 277–310 (1975)
16. Appelquist, T., Politzer, H. D. *Phys. Rev. Lett.* 34: 43–45 (1975)
17. De Rújula, A., Glashow, S. L. *Phys. Rev. Lett.* 34: 46–49 (1975); Appelquist, T. A., De Rújula, A., Politzer, H. D., Glashow, S. L. *Phys. Rev. Lett.* 34: 365–69 (1975)
18. Eichten, E., et al. *Phys. Rev. Lett.* 34: 369–72 (1975); Eichten, E., Gottfried, K., Kinoshita, T., Lane, K. D., Yan, T. M. *Phys. Rev.* 36: 500–4 (1976); *Phys. Rev.* D17: 3090–3117 (1978); *Phys. Rev.* D21: 313(E) (1980); *Phys. Rev.* D21: 203–33 (1980); Eichten, E. *Phys. Rev.* D22: 1819–23 (1980)
19. Goldhaber, G., et al. *Phys. Rev. Lett.* 37: 255–59 (1976); Peruzzi, I., et al. *Phys. Rev. Lett.* 37: 569–71 (1976)
20. Glashow, S. L., Iliopoulos, J., Maiani, L. *Phys. Rev.* D2: 1285–92 (1970)
21. De Rújula, A., Georgi, H., Glashow, S. L. *Phys. Rev.* D12: 147–62 (1976)
22. Niu, K., Mikumo, E., Maeda, Y. *Prog. Theor. Phys.* 46: 1644–46 (1971)
23. Herb, S. W., et al. *Phys. Rev. Lett.* 39: 252–55 (1977); Innes, W. R., et al. *Phys. Rev. Lett.* 39: 1240–42, 1640(E) (1977)
24. Brown, C. N., et al. *Fermilab Rep. No. FERMILAB-CONF-86/97-E*, May 1986. Presented at 2nd Aspen Winter Phys. Conf. Jan. 1986; Kaplan, D. M. In *Quarks, Strings, Dark Matter and All the Rest* (Proc. 7th Vanderbilt High Energy Phys. Conf., Nashville, Tenn., May 15–17, 1986), ed. R. Panvini, T. Weiler, p. 83. Singapore: World Scientific (1987)
25. Quigg, C., Thacker, H. B., Rosner, J. L. *Phys. Rev.* D21: 234–40 (1980); Quigg, C., Rosner, J. L. *Phys. Rev.* D23: 2625–37 (1981)
26. Andrews, D., et al. (CLEO). *Phys. Rev.*

*Lett.* 45: 219–21 (1980); Finocchiaro, G., et al. (CUSB). *Phys. Rev. Lett.* 45: 222–25 (1980); Behrends, S., et al. (CLEO). *Phys. Rev. Lett.* 50: 881–84 (1983)

27. Kane, G. L., Peskin, M. E. *Nucl. Phys.* B195: 29–38 (1982); Avery, P., et al (CLEO). *Phys. Rev. Lett.* 53: 1309 (1984)

28. Jackson, J. D. In *Weak Interactions at High Energy and the Production of New Particles* (Proc. SLAC Summer Inst. on Particle Phys., Aug. 2–13, 1976), ed. M. C. Zipf, SLAC Rep. No. 198, pp. 147–202. Stanford: SLAC (1976); *Proc. 1977 European Conf. on Particle Phys.*, Budapest, Hungary, ed. L. Jenik, I. Montvay, pp. 603–29. Budapest: Central Inst. Phys. (1977)

29. Chinowsky, W. *Ann. Rev. Nucl. Sci.* 27: 393–464 (1977)

30. Novikov, V. A., et al. *Phys. Rep.* 41C: 1–133 (1978)

31. Appelquist, T., Barnett, R. M., Lane, K. *Ann. Rev. Nucl. Sci.* 28: 387–499 (1978)

32. Goldhaber, G., Wiss, J. E. *Ann. Rev. Nucl. Sci.* 30: 337–81 (1980)

33. Martin, A. In *Proc. 21st Int. Conf. on High Energy Phys.* Paris, July 26–31, 1982 (*J. Phys.* 43: Colloque C-3, Suppl. 12), ed. P. Pétiau, M. Porneuf, pp. 96–101. Les Ulis, France: Les Editions de Physique (1982)

34. Martin, A. *Comments Nucl. Part. Phys.* 16: 249–66 (1986); *CERN Rep.* CERN-TH 4676/87, lecture presented at 26th Int. Universitätswochen für Kernphysik, Schladming (1987)

35. Franzini, P., Lee-Franzini, J. *Ann. Rev. Nucl. Part. Sci.* 33: 1–29 (1983)

36. Bloom, E., Peck, C. W. *Ann. Rev. Nucl. Part. Sci.* 33: 143–97 (1983)

37. Eichten, E. In *Proc. 11th SLAC Summer Inst. on Particle Phys.*, SLAC, Stanford, Calif., ed. P. McDonough, pp. 497–516. Stanford, Calif: SLAC (1983)

38. Eichten, E. In *The Sixth Quark* (12th SLAC Summer Inst. on Particle Physics, Stanford, Calif., Jul. 23–Aug. 3, 1984), ed. P. McDonough, SLAC Rep. No. 281, pp. 1–42. Stanford: SLAC (1984)

39. Peskin, M. E. See Ref. 37, pp. 151–90

40. Rosner, J. In *Proc. Int. Symp. on Lepton and Photon Interactions at High Energy*, Kyoto, Aug. 19–24, 1985, ed. M. Konuma, K. Takahashi, pp. 447–85. Kyoto: Kyoto Univ. (1986)

41. Berkelman, K. *Rep. Prog. Phys.* 49: 1–59 (1986)

42. Cooper, S. In *Proc. 23rd Int. Conf. on High Energy Phys.*, Berkeley, Calif., July 16–23, 1986, ed. S. Loken, pp. 67–104. Singapore: World Scientific (1987)

42a. Buchmüller, W., Cooper, S. *MIT Lab. Nucl. Sci. Rep. No. 159*, Mar. 1987

43. Lichtenberg, D. B. *Univ. Wash. Rep.* 40048-17-N7 (1987); submitted to *Int. J. Mod. Phys. A*

44. Cahn, R. N., ed. $e^+e^-$ *Annihilation: New Quarks and Leptons* (A volume in the Annual Reviews Special Collections Program). Menlo Park: Benjamin-Cummings (1985)

45. Quigg, C., Rosner, J. *Phys. Rep.* 56C: 167–235 (1979)

46. Grosse, H., Martin, A. *Phys. Rep.* 60C: 341–92 (1980)

47. Baumgartner, B., Grosse, H., Martin, A. *Phys. Lett.* 146B: 363–66 (1984); *Nucl. Phys.* B254: 528–42 (1985)

48. Duke, D. W., Roberts, R. G. *Phys. Rep.* 120: 277–368 (1985)

49. Shifman, M. A. *Ann. Rev. Nucl. Part. Sci.* 33: 199–233 (1983)

50. Reinders, L. J., Rubinstein, H., Yazaki, S. *Phys. Rep.* 127: 1–97 (1985)

51. Shifman, M. A. In *Proc. Int. Symp. on Production and Decay of Heavy Hadrons*, Heidelberg, May 20–23, 1986, ed. K. R. Schubert, R. Waldi, pp. 199–278. Hamburg: DESY (1986)

52. Chau, L. L. *Phys. Rep.* 95: 1–94 (1983)

53. Thorndike, E. H. *Ann. Rev. Nucl. Part. Sci.* 35: 195–243 (1985); See Ref. 40, pp. 405–45

54. Buras, A. J., Gérard, J.-M., Rückl, R. *Nucl. Phys.* B268: 16–48 (1986)

55. Caso, C., Touboul, M. C. *Riv. Nuovo Cimento* 9(12): 1 (1986); Gilchriese, M. See Ref. 42, pp. 140–68

56. Aguilar-Benitez, M., et al (Particle Data Group). *Phys. Lett.* 170B: 1–350 (1986)

57. Edwards, C., et al. *Phys. Rev. Lett.* 48: 70–73 (1982)

58. Lee-Franzini, J., et al (CUSB). See Ref. 42, pp. 669–76

59. Bowcock, T., et al (CLEO). *Phys. Rev. Lett.* 58: 307–10 (1987)

60. Macfarlane, D. B., et al (ARGUS). See Ref. 42, pp. 664–68

61. Schindler, R., et al (Mark III). See Ref. 42, pp. 745–51

62. Biagi, S. F., et al. *Phys. Lett.* 122B: 455–60 (1983); *Phys. Lett.* 150B: 230–34 (1985); *Z. Phys.* C28: 175–85 (1985); Coteus, P., et al. *Univ. Colo. Rep. COLO-HEP-140*, Dec. 1986. Presented at Div. Part. Fields Meet., Am. Phys. Soc., Salt Lake City, Utah, Jan. 14–17, 1987; *Salt Lake City Meeting: Proceedings*, ed. J. Ball, C. De Tar. Singapore: World Sci. (1987). In press

63. Bebek, C., et al (CLEO) (presented by A. Jawahery). See Ref. 42, pp. 773–77
64. Han, K., et al (CUSB). *Phys. Rev. Lett.* 55: 36–39 (1985)
65. Nambu, Y. *Phys. Rev.* D10: 4262–68 (1974)
65a. Creutz, M. *Quarks, Gluons, and Lattices.* Cambridge Univ. Press (1983); de Forcrand, P., Stack, J. *Phys. Rev. Lett.* 55: 1254–57 (1985); Michael, C. *Phys. Rev. Lett.* 56: 1219–21 (1986); Campostorini, M., Moriarty, K., Rebbi, C. *Phys. Rev. Lett.* 57: 44–47 (1986); Pisarski, R. D., Stack, J. D. *Fermilab Rep.* 86/122-T, Sept. 1986
66. Machacek, M., Tomozawa, Y. *Ann. Phys. (NY)* 110: 407–20 (1978); Quigg, C., Rosner, J. *Phys. Lett.* 71: 153–57 (1977); Martin, A. *Phys. Lett.* 93B: 338–42 (1980); Richard, J. M., Taxil, P. *Ann. Phys. (NY)* 150: 267–86 (1983)
67. Carlitz, R. D., Creamer, D. B. *Ann. Phys. (NY)* 118: 429–75 (1979); Richardson, J. L. *Phys. Lett.* 82B: 272–74 (1979); Buchmüller, W., Grunberg, G., Tye, S.-H. H. *Phys. Rev. Lett.* 45: 103–6 (1980)
68. Buchmüller, W., Tye, S.-H. H. *Phys. Rev.* D24: 132–56 (1981)
69. Landau, L. D., Lifshitz, E. M. *Quantum Mechanics,* Oxford: Pergamon. 3rd ed. (1965) p. 95
70. Eichten, E., Feinberg, F. *Phys. Rev.* D23: 2724–44 (1981); Gromes, D. *Z. Phys.* C26: 401–6 (1984)
71. Mackenzie, P. B., Lepage, G. P. *Phys. Rev. Lett.* 47: 1244–47 (1981); In *Perturbative Quantum Chromodynamics* (Proc. Conf. Tallahassee, Fla., 1981), ed. D. W. Duke, J. F. Owens, pp. 176–92. New York: AIP (1981); Lepage, G. P. In *Proc. 1983 Int. Symp. on Lepton and Photon Interactions at High Energies,* Cornell Univ., Aug. 4–9, 1983, ed. D. G. Cassel, D. L. Kreinick, pp. 565–92. Ithaca, NY: Newman Lab. Nucl. Studies, Cornell Univ. (1983)
72. Brodsky, S. J., Lepage, G. P., Mackenzie, P. *Phys. Rev.* D28: 228–35 (1983)
73. Barbieri, R., Gatto, R., Kögerler, R., Kunszt, Z. *Phys. Lett.* 57B: 455–59 (1975); Barbieri, R., Gatto, R., Kögerler, R. *Phys. Lett.* 60B: 183–88 (1976); Barbieri, R., Gatto, R., Remiddi, E. *Phys. Lett.* 61B: 145–68 (1976); Barbieri, R., Curci, G., d'Emilio, E., Remiddi, E. *Nucl. Phys.* B154: 535–46 (1979); Barbieri, R., Caffo, M., Gatto, R., Remiddi, E. *Phys. Lett.* 95B: 93–95 (1980); *Nucl. Phys.* B192: 61–65 (1981); Barbieri, R., Gatto, R., Remiddi, E. *Phys. Lett.*

74. Poggio, E. C., Schnitzer, H. J. *Phys. Rev.* D18: 1175–86 (1979); *Phys. Rev.* D21: 2034–35 (1980)
75. Photiadis, D. M. *Phys. Lett.* 164B: 160–66 (1985)
76. Baglin, C., et al. *Phys. Lett.* 171B: 135–41 (1986); 172B: 455–60 (1986); 187B: 191–97 (1987)
77. Menichetti, E. In *Proc. First Workshop on Antimatter Physics at Low Energy,* Fermilab, Apr. 10–12, 1986, ed. B. E. Bonner, L. S. Pinsky, pp. 95–118. Batavia, Ill: Fermilab (1986)
78. Thacker, H. B., Quigg, C., Rosner, J. L. *Phys. Rev.* D18: 274–86, 287–95 (1978); Schonfeld, J. F., Kwong, W., Rosner, J. L., Quigg, C., Thacker, H. B. *Ann. Phys. (NY)* 128: 1–28 (1981); Kwong, W., Rosner, J. L., Schonfeld, J. F., Quigg, C., Thacker, H. B. *Am. J. Phys.* 48: 926–30 (1980)
79. Kwong, W., Rosner, J. *Prog. Theor. Phys. (Suppl.)* 86: 366–76 (Festschrift volume in honor of Y. Nambu) (1986)
80. L. Köpke, et al (Mark III). See Ref. 42, pp. 692–99
81. Augustin, J. E., et al (DM2). *Orsay Preprint LAL/85-27,* July, 1985, submitted to Int. Symposium on Lepton and Photon Interactions at High Energy, Kyoto, Aug. 19–24, 1985, ed. M. Konuma, K. Takahashi. Kyoto: Kyoto University (1986); Jean-Marie, B. See Ref. 42, pp. 652–59
82. Scharre, D. L., et al (Mark II). *Phys. Rev.* D23: 43–55 (1981)
83. Baltrusaitis, R. M. et al (Mark III). *Phys. Rev. Lett.* 56: 107–10 (1986)
84. Gottfried, K. *Phys. Rev. Lett.* 40: 598–601 (1978); Yan, T.-M. *Phys. Rev.* D22: 1652–68 (1980); Kuang, Y.-P., Yan, T.-M. *Phys. Rev.* D24: 2874–85 (1981); Voloshin, M. B. *Yad. Fiz.* 43: 1571–75 (1986) [*Sov. J. Nucl. Phys.* 43: 1011–13 (1986)]; Ma, J., Kuang, Y.-P. *Commun. Theor. Phys.* 5: 67–78 (1986); Li, G.-Z., Kuang, Y.-P. *Commun. Theor. Phys.* 5: 79–88 (1986)
85. Franklin, M. E. B., et al (Mark II). *Phys. Rev. Lett.* 51: 963–66 (1983); Edwards, C., et al (Crystal Ball). *Caltech Rep.,* 1985 (unpublished)
86. Karl, G., Roberts, W. *Phys. Lett.* 144B: 263–65 (1984)
87. Moxhay, P., Rosner, J. L. *Phys. Rev.* D28: 1132–37 (1983)
88. McClary, R., Byers, N. *Phys. Rev.* D28: 1692–1705 (1983)
89. Gupta, S. N., Radford, S., Repko, W. *Phys. Rev.* D26: 3305–8 (1982); *Phys. Rev.* D30: 2424–25 (1984); *Phys. Rev. Lett.* 55: 3006 (1985); *Phys. Rev.* D31:

160–63 (1985); *Phys. Rev.* D34: 201–6 (1986)

90. Henriques, A. B., Kellett, B. H., Moorhouse, R. G. *Phys. Lett.* 64: 85–92 (1976)

91. Oreglia, M. J., et al (Crystal Ball). *Phys. Rev.* D25: 2259–77 (1982)

92. Oreglia, M. J. Thesis, Stanford Univ. (1980); *SLAC Rep. No. 236.* Stanford: SLAC (1980)

93. Gaiser, J. E., et al (Crystal Ball). *Phys. Rev.* D34: 711–21 (1986)

94. Olsson, M. G., Martin, A. D., Peacock, A. W. *Phys. Rev.* D31: 81–84 (1985); Olsson, M. G. See Ref. 77, pp. 119–30

95. Bloom, E. D. In *Proc. 1979 Int. Symp. on Lepton and Photon Interactions at High Energy,* Fermilab, 1979, ed. T. B. W. Kirk, H. D. I. Abarbanel, pp. 92–106. Batavia, Ill: Fermilab (1979)

96. Sucher, J. *Rep. Prog. Phys.* 41: 1781–1838 (1978); Hardekopf, G., Sucher, J. *Phys. Rev.* D25: 2938–43 (1982)

97. Zambetakis, V., Byers, N. *Phys. Rev.* D28: 2908–11 (1983); Zambetakis, V., Thesis, Univ. Calif. Los Angeles, 1985 (unpublished)

98. Grotch, H., Owen, D. A., Sebastian, K. J. *Phys. Rev.* D30: 1924–36 (1984)

99. Gaillard, M. K., Lee, B. W. *Phys. Rev. Lett.* 33: 108–11 (1974); Altarelli, G., Maiani, L. *Phys. Lett.* 52B: 351–54 (1974); The idea of "6* enhancement" for charm decays was explored in Altarelli, G., Cabibbo, N., Maiani, L. *Nucl. Phys.* B88: 285–88 (1975); *Phys. Lett.* 57B: 277–80 (1975); Kingsley, R. L., Treiman, S. B., Wilczek, F., Zee, A. *Phys. Rev.* D11: 1919–23 (1975); Einhorn, M. B., Quigg, C. *Phys. Rev.* D12: 2015–30 (1975); Ellis, J., Gaillard, M. K., Nanopoulos, D. V. *Nucl. Phys.* B100: 313–28 (1975); Quigg, C. *Z. Phys.* C4: 55–50 (1980)

100. Guberina, B., Nussinov, S., Peccei, R. D., Rückl, R. *Phys. Lett.* 89B: 111–15 (1979); Kobayashi, T., Yamazaki, N. *Prog. Theor. Phys.* 65: 775–78 (1981); Peccei, R. D., Rückl, R. In *Symp. Special Topics in Gauge Field Theories,* Ahrenshoop, E. Germany, Nov. 8–13, 1981, p. 1. Berlin: Akad. Wiss. (1981); Voloshin, M. B., Shifman, M. A. *Yad. Fiz.* 41: 187–98 (1985) [*Sov. J. Nucl. Phys.* 41: 120–26 (1985)]

101. Rosen, S. P. *Phys. Rev. Lett.* 44: 4–7 (1980); Bander, M., Silverman, D., Soni, A. *Phys. Rev. Lett.* 44: 7–9 (1980); Fritzsch, H., Minkowski, P. *Phys. Lett.* 90B: 455–59 (1980); Bernreuther, W., Nachtmann, O., Stech, B. *Z. Phys.* C4: 257–67 (1980)

102. Barger, V., Leveille, J. P., Stevenson,

P. M. *Phys. Rev. Lett.* 44: 226–29 (1980)

103. Cazzoli, E. G., et al. *Phys. Rev. Lett.* 34: 1125–28 (1975)

104. Le Yaouanc, A., Oliver, L., Pène, O., Raynal, J.-C. *Phys. Lett.* 72B: 53–56 (1977); Itoh, C., Minamikawa, T., Miura, K., Watanabe, T. *Prog. Theor. Phys.* 61: 548–58 (1979)

104a. Lane, K. D., Weinberg, S. *Phys. Rev. Lett.* 37: 717–19 (1976)

105. De Rújula, A., Georgi, H., Glashow, S. L. *Phys. Rev. Lett.* 37: 398–401 (1976); Thews, R. L., Kamal, A. N. *Phys. Rev.* D32: 810–12 (1985); Godfrey, S., Isgur, N. *Phys. Rev.* D32: 189–231 (1985) and Univ. Toronto preprint, 1986; Rosner, J. L. *Comments Nucl. Part. Phys.* 16: 109–30 (1986)

106. Lee, B. W., Quigg, C., Rosner, J. L. *Phys. Rev.* D15: 157–65 (1977)

107. Albrecht, H., et al (ARGUS). *Phys. Rev. Lett.* 56: 549–52 (1986)

108. Schnitzer, H. J. *Phys. Rev.* D18: 3482–3503 (1978)

109. Csorna, S., et al (CLEO). *Phys. Rev. Lett.* 56: 1222–25 (1986)

110. Schamberger, R. D., et al (CUSB). *Phys. Lett.* 138B: 225–29 (1984)

111. Wilczek, F. *Phys. Rev. Lett.* 39: 1304–6 (1977); Vysotsky, M. I. *Phys. Lett.* 97B: 159–62 (1980); Ellis, J., Enqvist, K., Nanopoulos, D. V., Ritz, S. *Phys. Lett.* 158B: 417–23 (1985); Nason, P. *Phys. Lett.* 175B: 223–26 (1986)

112. Krasemann, H. In Barbiellini, G., et al. DESY report 79/67 (Unpublished) (1979); Nappi, C. R. *Phys. Rev.* D25: 84–88 (1982); Tye, S.-H. H., Rosenfeld, C. *Phys. Rev. Lett.* 53: 2215–18 (1984); Moxhay, P., Ng, Y. J., Tye, S.-H. H. *Phys. Lett.* 158B: 170–74 (1985)

113. Peccei, R. D., Quinn, H. R. *Phys. Rev. Lett.* 38: 1440–43 (1977); *Phys. Rev.* D16: 1791–97 (1977); Weinberg, S. *Phys. Rev. Lett.* 40: 223–26 (1978); Wilczek, F. *Phys. Rev. Lett.* 40: 279–82 (1978)

114. Mageras, G., et al (CUSB). *Phys. Rev. Lett.* 56: 2672–75 (1986); Bowcock, T., et al (CLEO). *Phys. Rev. Lett.* 56: 2676–79 (1986)

115. Albrecht, H., et al (ARGUS). *Z. Phys.* C29: 167–73 (1985); *Phys. Lett.* 179B: 403–8 (1986); Keh, S., et al (Crystal Ball), Schmitt, P., et al (Crystal Ball), Irion, J., et al (Crystal Ball). (Presented by R. T. Van de Walle). See Ref. 42, pp. 677–84

116. Buchmüller, W., Ng, Y. J., Tye, S.-H. H. *Phys. Rev.* D24: 3003–6 (1981); see also (121); Igi, K., Ono, S. *Phys. Rev.*

D32: 232–36 (1985); Frank, M., O'Donnell, P. J. *Phys. Lett.* 159B: 174–76 (1985)

117. Rosner, J. L. In *Experimental Meson Spectroscopy—1983* (7th Int. Conf., Brookhaven), ed. S. J. Lindenbaum, pp. 461–78. New York: AIP (1984)

118. Peskin, M. Private communication; see also (39); Au, K. L., Morgan, D., Pennington, M. R. *Phys. Rev.* D35: 1626–64 (1987)

119. Walk, W., et al (Crystal Ball). *Phys. Rev.* D34: 2611–20 (1986); Skwarnicki, T., et al (Crystal Ball). *Phys. Rev. Lett.* 58: 972–75 (1987)

120. Chan, L. H. *Phys. Lett.* 71B: 422–24 (1977); Beavis, D., Chu, S.-Y., Desai, B. R., Kaus, P. *Phys. Rev.* D20: 743–47 (1979); *Phys. Rev.* D20: 2345–47 (1979); Bander, M., Silverman, D., Klima, B., Maor, U. *Phys. Rev.* D29: 2038–50 (1984); *Phys. Lett.* 134B: 258–62 (1984); Ito, H. *Prog. Theor. Phys.* 74: 1092–1104 (1985); 75: 1416–30 (1986); 76: 567–70 (1986), 77: 681–87 (1987); Yoshida-Habe, C., Iwata, K., Murota, T., Tsuruda, D. *Prog. Theor. Phys.* 73: 1274–77 (1985); *Prog. Theor. Phys.* 75: 333–39 (1986); 77: 917–25 (1987); Song, X., Lin, H. *Univ. Hangzhou Preprint*, 1986; Yoshida, C. *Prog. Theor. Phys.* 76: 474–80 (1986); Michael, C. *Phys. Rev. Lett.* 56: 1219–21 (1986); Olsson, M. G., Suchyta, C. J. *Phys. Rev.* D35: 1738–40 (1987)

122. Ng, Y. J., Pantaleone, J., Tye, S.-H. H. *Phys. Rev. Lett.* 55: 916–19 (1985); Pantaleone, J., Tye, S.-H. H., Ng, Y. J. *Phys. Rev.* D33: 777–800 (1986)

122. Ng, Y. J., Pantaleone, J., Tye, S.-H. H. *Phys. Rev. Lett.* 55: 916–19 (1985); Pantaleone, J., Tye, S.-H. H., Ng, Y. J. *Phys. Rev.* D33: 777–800 (1986)

123. Schnitzer, H. *Phys. Rev. Lett.* 35: 1540–42 (1975)

124. Pumplin, J., Repko, W., Sato, A. *Phys. Rev. Lett.* 35: 1538–40 (1975); Eichten, E., Feinberg, F. See Ref. 70

125. Mitra, A. N. *Z. Phys.* C8: 25–31 (1981); Mitra, A. N., Santhanam, I. *Z. Phys.* C8: 33–42 (1981); Faessler, A., Pfenninger, Th., Straub, U., Mitra, A. N. *Univ. Tubingen Preprint*, 1986

126. Adler, S. L., Davis, A. C. *Nucl. Phys.* B244: 469–91 (1984); Adler, S. L. *Prog. Theor. Phys. (Suppl.)* 86: 12–17 (Festschrift volume in honor of Y. Nambu) (1986); Byers, N., Zambetakis, V. In *AIP Conf. Proc. No. 150*, ed. D. F. Geesaman, pp. 170–72. New York: AIP (1986)

127. Kwong, W., Rosner, J. L. *D Wave*

*Quarkonium States*, 1987, in preparation

128. Voloshin, M. B. *Yad. Fiz.* 36: 247–55 (1982) [*Sov. J. Nucl. Phys.* 36: 143–48 (1982)]

129. Lovelock, D. M. J., et al (CUSB). *Phys. Rev. Lett.* 54: 377–80 (1985); Besson, D., et al (CLEO). *Phys. Rev. Lett.* 54: 381–84 (1985)

130. Ono, S., Törnqvist, N. A., Lee-Franzini, J., Sanda, A. I. *Phys. Rev. Lett.* 55: 2938–40 (1985)

131. Haas, P., et al (CLEO). *Phys. Rev. Lett.* 56: 2781–84 (1986); Alam, M. S., et al (CLEO). *Phys. Rev.* D34: 905–8 (1986); Albrecht, H., et al (ARGUS). *Phys. Lett.* 185B: 218–22 (1987)

132. Martin, A., Richard, J. M. *Phys. Lett.* 185B: 426–30 (1987)

133. Eggert, K., et al (UA1). In *Proc. 21st Rencontre de Moriond*, Les Arcs, France, Mar. 1986, ed. J. Tran Thanh Van, pp. 369–87. Gif-sur-Yvette: Ed. Front. (1986); Ellis, N., et al (UA1). See Ref. 42, pp. 801–5; Albajar, C., et al (UA1). *Phys. Lett.* 186B: 247–54 (1987)

133a. Albrecht, H., et al (ARGUS). *Phys. Lett.* 192B: 245–52 (1987)

134. Lichtenberg, D. B., Wills, J. G. *Lett. Nuovo Cimento* 32: 86–90 (1981)

135. Moxhay, P. J., Rosner, J. L., Quigg, C. *Phys. Rev.* D23: 2638–46 (1981)

136. Eichten, E., Gottfried, K. *Phys. Lett.* 66B: 286–90 (1977); Quigg, C., Rosner, J. L. *Phys. Lett.* 72B: 462–64 (1977)

137. Jackson, J. D., Olsen, S., Tye, S.-H. H. In *Proc. 1982 DPF Summer Study on Elementary Particles and Future Facilities*, Snowmass, Colo., ed. R. Donaldson, H. R. Gustafson, F. R. Paige, pp. 175–80. Batavia, Ill: Fermilab (1982)

138. Kühn, J. H., Ono, S. *Z. Phys.* C21: 395–402 (1984); (E) C24: 404 (1984); Kühn, J. H. *Acta Phys. Polon.* B16: 969–1006 (1985); Kühn, J. H. In *Heavy Quarks, Flavor Mixing, and CP Violation* (5th Moriond Workshop, La Plagne, France, Jan. 13–19, 1985), ed. J. Tran Thanh Van, pp. 91–107. Gif-sur-Yvette, France: Éditions Frontières (1985); Martin, A. D. *New Particles 1985: Proceedings* (Madison, Wis., May 8–11, 1985), ed. V. Barger, D. Cline, F. Halzen, p. 24. Singapore: World Scientific (1986); Igi, K., Ono, S. *Univ. Tokyo Rep. UT-486*, June, 1986, to appear in *Essays in Honor of the 60th Birthday of Professor Yoshio Yamaguchi*. Singapore: World Scientific (1987)

139. Lichtenberg, D. B., Clavelli, L., Wills, J. G. *Phys. Rev.* D33: 284–86 (1986)

140. Quigg, C. See Ref. 95, pp. 239–56

141. Güsken, S., Kühn, J. H., Zerwas, P. M. *Nucl. Phys.* B262: 393–438 (1985)
142. Franzini, P. J., Gilman, F. J. *Phys. Rev.* D32: 237–46 (1985)
143. Hall, L. J., King, S. F., Sharpe, S. R. *Nucl. Phys.* B260: 510–30 (1985)
144. Eichten, E., Hinchliffe, I., Lane, K. D., Quigg, C. *Phys. Rev.* D34: 1547–66 (1986)
145. Sher, M., Silverman, D. *Phys. Rev.* D31: 95–98 (1985); Abbott, L. F., Sikivie, P., Wise, M. B. *Phys. Rev.* D21: 1393–1403 (1980); Athanasiu, G. G., Franzini, P. J., Gilman, F. J. *Phys. Rev.* D32: 3010–19 (1985)
146. Komamiya, S. See Ref. 40, pp. 611–59; Cornet, F., Glover, E. W. N., Hagiwara, K., Martin, A. D., Zeppenfeld, D. *Phys. Lett.* 174B: 224–28 (1986)
147. Dawson, S., Eichten, E., Quigg, C. *Phys. Rev.* D31: 1581–1637 (1985); Haber, H., Kane, G. L. *Phys. Rep.* 117: 75–263 (1985)
148. Rosner, J. In *Proc. 6th Int. Symp. on High Energy Spin Phys.*, Marseille, France, Sept. 12–19, 1984 (*J. Phys.* 46: Colloque C2, Suppl. 2), ed. J. Soffer, pp. C2-77-93. Les Ulis, France:
Editions Physique (1985)
149. Ma, E. *Phys. Lett.* 58B: 442–44 (1975); Karl, G. *Phys. Rev.* D14: 2374 (1976); Gell-Mann, M. *Bull. Am. Phys. Soc.* 22: 541 (1977); Freedman, D. Z. In *Particles and Fields—1977* (Div. Part. Fields Ann. Meet., Argonne Natl. Lab.), ed. P. A. Shreiner, G. H. Thomas, A. B. Wicklund, pp. 65–75. New York: AIP (1978); Wilczek, F., Zee, A. *Phys. Rev.* D16: 860–68 (1977); Giles, R. C., Tye, S.-H. H. *Phys. Lett.* 73B: 30–32 (1978); Ng, Y. J., Tye, S.-H. H. *Phys. Rev. Lett.* 41: 6–9 (1978); Fritzsch, H. *Phys. Lett.* 78B: 611–14 (1978); Freund, P. G. O., Hill, C. T. *Phys. Rev.* D19: 2755–56 (1978); *Nature* 276: 250 (1978); Georgi, H., Glashow, S. L. *Nucl. Phys.* B159: 29–36 (1979); Dover, C. B., Gaisser, T. K., Steigman, G. *Phys. Rev. Lett.* 42: 1117–20 (1979)
150. Rosner, J. *Comments Nucl. Part. Phys.* 15: 195–221 (1986)
151. Rosner, J. *Comments Nucl. Part. Phys.* 14: 229–44 (1985)
152. Eichler, R., Nakada, T., Schubert, K. R., Weseler, S., Wille, K. *Swiss Inst. Nucl. Res. Preprint PR-86-13*, Nov., 1986

*Ann. Rev. Nucl. Part. Sci. 1987. 37: 383–409*
*Copyright © 1987 by Annual Reviews Inc. All rights reserved*

# THE THEOREMS OF PERTURBATIVE QCD

*John C. Collins*

Department of Physics, Illinois Institute of Technology, Chicago, Illinois 60616,[1] and Fermi National Accelerator Laboratory, Batavia, Illinois 60510

*Davison E. Soper*

Institute of Theoretical Science, University of Oregon, Eugene, Oregon 97403

CONTENTS

[1] Permanent address.

0163–8998/87/1201–0383$02.00

# 1. INTRODUCTION

Although on distance scales of the order of a hadronic size the strong interactions are indeed strong, they become weak at short distances. This property was first surmised from the (approximate) Bjorken scaling obeyed by deeply inelastic structure functions. The discovery of quantum chromodynamics (QCD) as the quantum field theory of strong interactions was based on the observation by Gross & Wilczek (1) and by Politzer (2) that QCD has the required property, called "asymptotic freedom": its effective coupling goes to zero as the distance of phenomena under consideration is taken to zero.

Asymptotic freedom implies that perturbative methods can be used to make reliable calculations of short-distance phenomena. This is very important for the phenomenology of strong interactions, since perturbation theory is the only systematic method we have at present for calculating scattering cross sections directly from the fundamental Lagrangian.

Short-distance phenomena are associated with large momentum transfers and thus with production of particles at large transverse momentum. On the other hand, all scattering processes involve at least some phenomena that happen on large distance scales. Thus applications of perturbative QCD rely on factorization theorems, whereby cross sections are written as the product of factors, each of which involves phenomena happening on a different distance scale. (The products are generally matrix or convolution products.)

In the early days of QCD, the only process that could be treated perturbatively were deeply inelastic scattering and the total cross section for $e^+e^-$ annihilation. Since then, perturbative QCD has been applied to many other processes—Drell-Yan, high $p_T$ particle production, etc. However, the reliability of the applications runs the gamut from complete statements of factorization, with more-or-less complete proofs, to approximate models merely inspired by QCD. It is not easy to tell from the literature which results are reliably established; proofs are often spread over several papers, with early work that claims to provide proofs often being seriously incomplete.

In this review we discuss a sampling of important results (particularly factorization theorems) for which there is a statement of the result at all orders of perturbation theory and a reasonable approximation to a proof. We consider as outside our scope results for which only the first term in a leading logarithm expansion is available. [For example, we do not discuss the algorithms used by perturbative Monte Carlo event generators (3).] We emphasize that some of the topics left out of this review will ultimately become the subject of theorems of the power of those that we do discuss.

We first summarize the kinds of theorems and possible theorems that are available. Then we cover the foundations of the subject, starting with renormalization. After that, we discuss the various factorization theorems and expose the key steps in their proof. Finally we indicate where we believe future work in the field should go, and also indicate our opinion of how well the theorems are proved.

We spend a substantial time explaining the theorems that provide the foundations of the subject rather than concentrating solely on the theorems of phenomenological interest. This is because, in our experience, progress in the field is made by having a clear formulation and understanding of the foundations, of the reasons why things work. Of course, many readers will regard the treatment of the basics as analogous to an experimentalist's lengthy explanation of apparatus; such readers should skip Sections 3 to 6 at a first reading.

## 2. WHAT ARE THE THEOREMS?

We may distinguish several classes of theorems (or potential theorems) in perturbative QCD:

1. The first we term "pre-theorems." These include the result, for example, that QCD is renormalizable. They also include the derivation of the renormalization group. Both of these are fundamental tools in any calculation or proof in perturbative QCD.

2. Then come the simplest of the theorems for cross sections, typified by $\sigma_{tot}$ in $e^+e^-$ annihilation. All infrared sensitivity cancels. Thus a simple renormalization-group transformation suffices to give a perturbative result that can be usefully compared to experiment, when only low-order calculations are done.

3. Next are what we term the "classical" factorization theorems. The oldest was the operator product expansion for the moments of deeply inelastic structure functions. Radyushkin (4) and Politzer (5) formulated the correct factorization equation for the Drell-Yan process. Similar factorization theorems for a wide class of similar processes were also written

down (6–8). No longer did one have to take moments. In each process there is a hard scattering. The long-distance effects no longer cancel but instead factor into parton distribution functions for incoming partons and decay functions for outgoing partons. Cancellation of soft gluon effects is needed to get the factorization. This soft gluon problem is particularly tricky when there are two incoming hadrons.

4. Finally, there are results for processes like Drell-Yan muon pair production at low measured transverse momentum for the muon pair. These are characterized by the noncancellation of soft gluon effects, the consequent presence of Sudakov factors in the result, and the need to use the concept of "intrinsic transverse momentum." These results came after a long history of leading logarithm approximations, starting with *the* Sudakov form factor in QED.

Other established results, for the effects of heavy quarks and for the electromagnetic form factor of the pion can be regarded as cases of the classical factorization theorems.

There are many other areas in which perturbative QCD has been applied, with and without reasonable justification. Among these are elastic scattering, the Monte Carlo event generators, small-$x$ calculations and the like. Typical here is the need to treat soft gluon effects in more detail than in the examples listed above. There is a good deal of knowledge of these matters, especially in the USSR, and there is much potential (and experimental need) for complete results.

## 3.    QCD IS RENORMALIZABLE

QCD is defined by quantizing the Lagrangian (e.g. 9)

$$L = -\frac{1}{4}F_{\mu\nu}F^{\mu\nu} + \sum_j \bar{\psi}_j(i\not{D} - M_j)\psi_j - \frac{1}{2\xi}(\partial \cdot A)^2 + \partial_\mu \bar{c}_a(\partial^\mu c_a + gf_{abc}c_b A_c^\mu).$$

1.

Here $A_{a\mu}$, $\psi_j$, and $c_a$ are the gluon, quark, and Faddeev-Popov ghost fields, $D_\mu$ is the covariant derivative $\partial_\mu + igA_{a\mu}t_a$, $F_{\mu\nu}$ is the field strength tensor $(1/ig)[D_\mu, D_\nu]$, $f_{abc}$ are the structure constants of SU(3), and $t_a$ are the generators of SU(3) in the fundamental representation. We have included a gauge-fixing term that allows perturbation theory to be defined, using a standard covariant gauge with $\xi$ as the gauge-fixing parameter. We let $N_{fl}$ be the number of quark flavors.

Because of ultraviolet divergences, renormalization is needed to make the theory well defined. Now, to show that the unphysical degrees of freedom (the longitudinal gluons and ghosts) decouple from physics and to show that the physical consequences of the theory are independent of the choice of gauge, one uses the Ward identities that follow from gauge invariance. Thus a key theorem for all studies in QCD is that renormalization can be achieved while preserving gauge invariance.

The most straightforward way of proving this involves regulating the divergences in a gauge-invariant manner (e.g. by the use of dimensional regularization). Then one provides renormalization factors $Z_i$ for the coupling, the masses, the gauge-fixing parameter $\xi$, and for the fields. Gauge invariance is preserved if no counterterms other than those given by these $Z_i$ are needed to make the Green functions finite as the regularization is removed.

The proofs that no extra counterterms are needed use either the Ward identities for the full Green functions (10, 11) or the Ward identities for the one-particle-irreducible Green functions (12). The latter method is mathematically more sophisticated, but is very compact. Some earlier proofs tended to be incomplete in their demonstration of renormalizability in the presence of fermions—see the comments in (11).

As part of the renormalization program, one specifies exactly how the factors $Z_i$ are to be fixed. One very useful method is to use dimensional regularization followed by the well-known "minimal subtraction" prescription MS (13) or the very similar $\overline{\text{MS}}$ prescription (14).

In much work on factorization theorems, the axial gauge $n \cdot A = 0$ is used, where $n^{\mu}$ is a fixed vector. One can thereby avoid many of the technical complications that appear in proofs using a covariant gauge, especially if $n^{\mu}$ is light-like. Unfortunately, this gauge condition introduces extra singularities into the gluon propagator. In the light-like case, the singularities are so severe that there is as yet no completely satisfactory definition of graphs with a gluon loop. Even for the cases of time-like or space-like $n^{\mu}$, there appear to be problems (e.g. 15) beginning at the two-loop level: the singular factors may be defined with a principal value prescription, but the product of such a factor with another singular function is not thereby defined. Although Konetschny & Kummer (16) have presented a proof of renormalizability in the space-like axial gauge, they appear not to have treated adequately the problem of existence of the axial gauge. We conclude that *complete* proofs of factorization must at present be made in a covariant gauge.

# 4. RENORMALIZATION GROUP

The renormalization coefficients $Z_i$ are somewhat arbitrary; one may add a finite part to the counterterm for a graph. As we explain in this section, one can use this flexibility to remove the large logarithms that otherwise spoil the usefulness of perturbation theory.

## 4.1  *Renormalization-Group Equation*

When we consider lines with high virtuality, we wish to neglect quark masses; but this is only legitimate provided that the $M_j \to 0$ limit of the renormalization counterterms exists. Since QCD is not a scale-invariant theory (1, 2), there must be some other mass parameter $\mu$ in the renormalization prescription in order for the limit to exist. The precise way in which $\mu$ enters depends on the renormalization prescription—it may be a renormalization point, or it may be the "unit of mass" in dimensional regularization (13). In any event we choose a prescription in which the $M_j \to 0$ limit exists.

Suppose one changes the value of $\mu$. Then to keep the physics of the theory fixed one must keep the unrenormalized parameters $g_0$ and $M_{j0}$ unchanged by adjusting the values of the renormalized parameters $g$ and $M_j$. This is an example of renormalization group invariance. The renormalized parameters are really functions of $\mu$, and one calls them the effective coupling and masses at scale $\mu$. They obey the renormalization-group equations (17)

$$\mu \frac{d\alpha_s(\mu)}{d\mu} = \beta(\alpha_s), \qquad \mu \frac{dM_j(\mu)}{d\mu} = \gamma_M(\alpha_s)M_j, \qquad \qquad 2.$$

where $\beta$ and $\gamma_M$ are perturbatively calculable renormalization-group coefficients. Here we use the notation $\alpha_s = g^2/4\pi$.

Since $\beta$ is negative (1, 2) when $\alpha_s$ is small, the effective coupling $\alpha_s(\mu)$ becomes small when $\mu$ is large. This is the celebrated property of asymptotic freedom. It is the key to the magic and utility of perburbative methods in strong interactions.

The $S$ matrix is renormalization-group invariant. That is, it obeys the equation $\mu DS/D\mu = 0$, where

$$\mu \frac{D}{D\mu} = \mu \frac{\partial}{\partial \mu} + \beta \frac{\partial}{\partial \alpha_s} + \gamma_M \sum_i M_i \frac{\partial}{\partial M_i}. \qquad \qquad 3.$$

Green functions $G$ obey renormalization-group equations with anomalous dimensions $\gamma_i$: $\mu DG/D\mu = \Sigma_i \gamma_i G$. Here the sum is over the external lines of $G$.

## 4.2  *Asymptotic Freedom*

From Equation 2, one can deduce the behavior of $\alpha_s(\mu)$ for large $\mu$. An expansion approximately in inverse powers of $\ln \mu$ is useful (17a):

$$\frac{\alpha_s(\mu)}{\pi} = \frac{12/(33-2N_{fl})}{\ln(\mu^2/\Lambda^2)} + \frac{72(153-19N_{fl})}{(33-2N_{fl})^3} \frac{\ln\ln(\mu^2/\Lambda^2)}{\ln^2(\mu^2/\Lambda^2)}$$

$$+ O\left\{ \frac{\ln^2[\ln(\mu^2/\Lambda^2)]}{\ln^3(\mu^2/\Lambda^2)} \right\}. \qquad \qquad 4.$$

The definition of the scale parameter $\Lambda$ is that the term proportional to $1/\ln^2(\mu^2/\Lambda^2)$ that could be in Equation 4 is absent. This scale may be considered the fundamental parameter that specifies the coupling in QCD. It depends, however, on the renormalization scheme being used. A one-loop calculation suffices to relate the values of $\Lambda$ in different schemes (18).

## 4.3  Application of the Renormalization Group

The simplest example of the use of perturbative methods is the prediction of the total cross section for $e^+e^-$ annihilation into hadrons (19, 20). Let $Q$ be the invariant mass of the final state, which we take to be large compared with an ordinary hadronic scale. In the next two sections, we find the important regions of momentum space for Feynman graphs for the cross section, and then we show that the infrared dependence of the cross section cancels. At that stage, we can neglect quark masses compared to $Q$, and we are ready to use the renormalization group.

One obtains the following for the ratio $R$ of the cross section for annihilation into hadrons to the cross section for annihilation into a muon pair (21–23):

$$R = 3\sum_i e_i^2 \left\{ 1 + \frac{\alpha_s(\mu)}{\pi} + \left[\frac{\alpha_s(\mu)}{\pi}\right]^2 [1.986 - 0.115N_{fl} \right.$$

$$\left. + \ln(\mu Q)\{5.50 - 0.333N_{fl}\}] + \ldots \right\}. \qquad \qquad 5.$$

Here we are using $\overline{\text{MS}}$ renormalization and the sum is over quark flavors. For each extra power of $\alpha_s$ one gets an extra factor of $\ln(Q/\mu)$.

Suppose that Equation 5 were to be used with large $Q$ and fixed $\mu \approx 1$ GeV. Then the perturbative expression of Equation 5 would not be useful because $\alpha_s(1\text{ GeV})$ is not small and, moreover, the logarithms $\ln(Q/\mu)$ in the coefficients are large.

However, $R$ is a renormalization-group invariant. Thus we can set $\mu$ to any value we like. A good choice is $\mu = Q$. (Any value for $\mu$ of order $Q$ will do.) Then there are no large logarithms and the coupling is small if $Q$ is large. We may expect, though it is by no means proved, that the difference between the exact value for $R$ and the approximate value given by Equation 5 with $\mu = Q$ is of the order of the first term omitted. Thus the difference tends to zero like a constant divided by $\ln^3(Q)$ as $Q \to \infty$. To prove this kind of result needs nonperturbative work of a kind that has not yet been done.

The factorization theorems of perturbative QCD express cross sections in terms of quantities such as $R$, which depend on a single scale. The factors with a large scale can be reliably estimated by low-order perturbation theory after a renormalization-group transformation. We discuss below how to handle the quantities whose scale is not large.

## 5.  LEADING INTEGRATION REGIONS

Consider a process in which there is a scale of virtuality $Q$ that is large compared to the 1-GeV scale typical of hadrons. A cut Feynman graph for the cross section is expressed as an integral over many momenta $k_1^\mu, k_2^\mu, \ldots, k_N^\mu$. The first step in deriving any factorization theorem in QCD is to characterize the important integration regions.

The necessary theorem is most clearly given by Sterman (24, 25). It relies, in turn, on a theorem of Coleman & Norton (26). The theorem can be adapted to cover many situations. Here, we consider the simplest case, the cross section $e^+e^-$ annihilation into hadrons at a high center-of-mass energy $Q$.

It is convenient to define scaled integration variables $\hat{k}_n^\mu = k_n^\mu/Q$. One important integration region for some of the $\hat{k}_n^\mu$ can be the region in which all of the components of these $\hat{k}_n^\mu$ are nonzero when $Q \to \infty$. One can neglect such quantities as quark masses in propagators that carry these "ultraviolet" momenta. There are often other important integration regions. These are regions near certain surfaces in the space of the $\hat{k}_n^\mu$ on which, in the $Q \to \infty$ limit, some of the propagator denominators vanish. The contributions from these regions are sensitive to the infrared cutoffs provided by such quantities as quark masses, hadronic binding energies, and transverse momenta of hadronic constituents.

Sterman therefore examines the integrands in the $Q \to \infty$ limit, in which all masses and other infrared cutoffs are set to zero. He looks for singular surfaces at which Feynman denominators of the limiting integrand vanish. The problem has therefore been changed from that of calculating the asymptotic behavior of graphs in a theory with all masses at their physical values to the problem of investigating the singularities of a massless theory with some external particles on shell.

However, it is crucial that one does not need to consider all such singularities. Only singularities at which the integration contour is pinched between two or more poles of the integrand are important. The reason is that if the integration contour is not pinched, then one may simply deform the contour away from the poles.

Following Sterman, we therefore ask, "What are the *pinch* singular surfaces for Feynman graphs for massless particles?" The answer, given by the Coleman-Norton theorem, is simple and intuitively appealing. It is that the momentum configuration represents a classical particle scattering.

One associates a reduced diagram with each pinch singular surface. Its vertices are formed by contracting to points all lines carrying "ultraviolet" momenta, i.e. those for which $\hat{k}_n^\mu$ is off shell by order 1. The lines of the reduced graph are formed by the remaining lines of the original graph, i.e. those whose scaled momenta are pinched on shell ($\hat{k}^2 = 0$). The reduced graph must depict a classically possible process that can be realized with momentum conservation at each reduced vertex and classical propagation of the massless particles between the vertices. In a zero-mass theory there are two kinds of line with pinched momentum:

1. Jet lines. These have nonzero (but light-like) scaled momenta $\hat{k}_n^\mu$ at the pinch singular surface.
2. Soft lines. These have $\hat{k}_n^\mu = 0$ at the pinch singular surface.

Kinematics constrains the form of the reduced graphs. In the case of $e^+e^-$ annihilation, several jet lines may emerge from the reduced vertex corresponding to the original annihilation. After that, any number of jet particles that are traveling in the same direction can interact at a reduced vertex and emerge as some other number of jet particles traveling in this same direction. Soft momenta, however, may be coupled anywhere.

After locating the pinch singular surfaces, Sterman asks which of them are sufficiently strong to give leading contributions to the cross section in the $Q \to \infty$ limit. This is answered by counting powers of the variables that parameterize the space orthogonal to the pinch singular surface. For example, near a soft singularity one counts $+1$ for each power of the soft momentum $\hat{k}^\mu$ in the numerator, $-1$ for each power in the denominator, and $+4$ for $d^4\hat{k}$. In $e^+e^-$ annihilation, surfaces with a net power $p$ give contributions to the dimensionless cross section, $Q^2\sigma_{\text{tot}}$ that are proportional to $(m/Q)^p$ times possible logarithms of $Q/m$. Thus $p = 0$ surfaces give the leading contributions, while the contributions from regions near surfaces with $p > 0$ vanish in the $Q \to \infty$ limit and can be ignored.

The result of this power counting is gauge dependent. We will state the result for a physical gauge, such as the space-like axial gauge. [The gauge used by Sterman (24, 25) is less conventional.] One finds first of all that no graphs have worse than logarithmic infrared divergence when the cutoffs are taken away (i.e. no graph has $p < 0$). The $p = 0$ reduced graphs may be characterized as follows:

1. Initial hard scattering subgraph. There is a reduced vertex in which the initial virtual photon enters and two or more jet-like quarks or gluons emerge, all of them going in different directions.

2. Other reduced vertices. Two-, three-, and four-gluon ultraviolet subgraphs and quark-antiquark and quark-antiquark-gluon ultraviolet
   subgraphs are allowed anywhere in the reduced diagram, but these are
   the only topologies for reduced vertices allowed outside of the initial
   hard scattering subgraphs. They just correspond to the ultraviolet divergences that are cancelled by counterterms in the QCD action.
3. Interactions of jet-like particles. Partons carrying parallel light-like
   momenta may interact at the allowed reduced vertices described above.
   For instance, an initial quark can divide into a quark and a gluon
   carrying parallel momentum.
4. Interactions of soft particles with each other. Soft quarks and gluons
   may interact with each other at any of the allowed reduced vertices.
5. Interactions of soft particles with jets. Soft gluon lines may connect to
   jets at three-point-reduced vertices. Soft quarks may not connect to
   jets, nor may soft gluons connect to jets at other than three-point-
   reduced vertices.

The resulting structure of a typical reduced diagram giving a leading
contribution to the $e^+e^-$ annihilation cross section is illustrated in Figure
1. We call the set of all jet lines going in a certain direction a jet subgraph.

The structure of the leading integration regions in Feynman gauge is
similar but a bit more complicated than that described above for a physical
gauge (24, 27). The chief difference is that extra longitudinally polarized
jet gluons may join the initial hard subgraph and any given jet subgraph.

There is an older power-counting theorem proved by Weinberg (28).
This theorem applies to Feynman graphs in which all particles carry a
mass and in which energies are rotated to the imaginary axis, so that the
metric is Euclidean. It shows, at a level of rigor not attempted in the
Minkowski-metric results described above, that the integral is bounded by

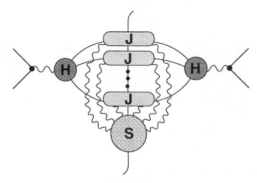

*Figure 1*    Leading integration regions for electron-positron annihilation.

the estimate provided by the power counting. Weinberg's theorem has often been used in proofs of renormalizability and of the operator product expansion.

# 6. CANCELLATION OF FINAL-STATE INTERACTIONS

For each cut graph for the total $e^+e^-$ annihilation cross section, there are a number of leading regions of integration, each of which can be characterized by a graph of the form of Figure 1. All of these regions involve some lines that have low virtuality, and it is not manifest that one can neglect masses, as was necessary to derive Equation 5. The proof that the infrared-sensitive regions indeed cancel in this case is an example of a standard procedure in perturbative QCD.

The intuitive reason for the cancellation is simple: the infrared-sensitive processes happen a long time (relative to the time scale $1/Q$) after the initial hard interaction. Because the evolution of the system after the hard interaction is unitary, these processes do not affect the probability that the hard interaction occurs.

In perturbation theory it is quite easy (8) to demonstrate the necessary theorem: all infrared divergences created when the quark masses are set to zero cancel upon summing over the various cut graphs corresponding to a given uncut Feynman graph. There are two approaches. The quickest is to realize that the graphs can be treated as the discontinuity of vacuum polarization graphs; we present this argument. The other approach uses time-ordered perturbation theory. This approach is rather easier to generalize to other processes (27), but is a little more cumbersome for our simple case.

It is a consequence of the spectral representation that the cross section ratio $R$ is proportional to the discontinuity of the hadronic part of the photon's vacuum polarization, $\Pi$. Let us take an average of $R$ with a smooth test function $t(Q^2)$:

$$\int R(Q^2)t(Q^2)\mathrm{d}Q^2 \propto \int \Pi(Q^2+i\varepsilon)t(Q^2)\mathrm{d}Q^2 - \int \Pi(Q^2-i\varepsilon)t(Q^2)\mathrm{d}Q^2. \qquad 6.$$

In each of the terms on the right-hand side, we may deform the contour of integration away from real positive $Q^2$, and then there are no longer any pinch singular surfaces in the massless limit.

The averaging over $Q^2$ is important (29, 30), even though it has no effect if the cross section is smooth. In general, nonperturbative phenomena may create bumps and jumps in the cross section, such as are seen even perturbatively near a heavy-quark threshold. The averaging smooths out such behavior and then provides a quantity amenable to low-order perturbative calculation. Generalizations of the above cancellation are common in perturbative QCD. Averaging over an appropriate kinematic variable is always implicit in such cancellations.

# 7.  FACTORIZATION FOR SIMPLE ONE-SCALE PROCESSES

In this section, we discuss the simplest inclusive processes in which there is a single large momentum scale. These are deeply inelastic lepton-hadron scattering, and particle and jet production in $e^+e^-$ annihilation. In such cases, QCD predictions depend on factoring the ultraviolet-dominated interaction from the infrared-sensitive pieces associated with the hadrons. Cases in which there are two hadrons in the initial state present more problems and are discussed in Section 8.

## 7.1  Deeply Inelastic Scattering

Consider scattering of a lepton by a hadron with the exchange of a single virtual boson (photon, W, Z, etc). The inclusive cross section summed over all hadronic final states can be expressed in terms of a current correlation function (e.g. 17):

$$W_{\mu\nu} = \int d^4 y e^{iq\cdot y} \langle p | j_\mu(y) j_\nu^\dagger(0) | p \rangle.    \qquad 7.$$

Then $W_{\mu\nu}$ is expressed in terms of several scalar dimensionless "structure functions" $F_i$. These are functions of two kinematic variables $Q = \sqrt{-q^2}$ and $x = Q^2/2p \cdot q$. Here $q^\mu$ is the momentum transfer from the lepton and $p^\mu$ is the momentum of the incoming hadron, which we take to be unpolarized.

The factorization theorem for the structure functions has the form (5, 31):

$$F_i(x, Q) = \sum_a \int_x^1 \frac{d\xi}{\xi} C_{ia}[x/\xi, Q/\mu, \alpha_s(\mu)] f_{a/H}(\xi, \mu) + O(1 \text{ GeV}/Q).    \qquad 8.$$

It is valid in the "Bjorken limit," in which $Q$ gets large with $x$ fixed. The theorem may be interpreted as saying that the cross section is dominated by scattering at short distances by constituents (partons) of the incoming hadron, with the variable $\xi$ representing the fraction of the hadron's momentum that is carried by the parton. The sum is over all species of parton (gluon and flavors of quark and antiquark), $f_{a/H}$ represents the distribution of partons of type $a$ in hadron $H$, and $C_{ia}$ is the hard scattering function (or Wilson coefficient)—effectively the short distance part of a structure function for scattering by partons of type $a$. Precise definitions of the parton distributions and of the Wilson coefficients are formulated as part of the proof of Equation 8.

The traditional treatment of deeply inelastic scattering (11, 17, 32, 33) involves the integer moments of the structure functions, $\int dx\, x^n F_i(x, Q)$. One uses the operator product expansion (34, 35) to establish the integer moment analog of Equation 8. Needed here is the result established by Joglekar (36) that only gauge-invariant operators have to be included in the operator product expansion. Current methods, which may be generalized to other processes, yield Equation 8—not just its integer moments.

Perhaps the best published treatment is by Efremov & Radyushkin (37, 38); it is in Feynman gauge. It is not clear that they have completely treated the problems due to soft gluons and final-state interactions that we allude to below. Much other work (e.g. 39), uses the light-like axial gauge, which is not yet well defined. Since an understanding of deeply inelastic scattering is useful in understanding other factorization theorems, we now outline how a full covariant-gauge proof of Equation 8 should go.

First, we choose light-cone coordinates in which $q^\mu$ and $p^\mu$ have components

$$q^\mu = (q^+, q^-, q^T) = (-xQ/\sqrt{2}, Q/x\sqrt{2}, 0^T),$$

$$p^\mu = (Q/\sqrt{2}, M^2/Q\sqrt{2}, 0^T),$$

9.

where for any four-vector $v^\mu$ we define $v^\pm = (v^0 \pm v^3)/\sqrt{2}$. (We have neglected a small correction proportional to $M^2/Q^2$ in $p \cdot q$.) The structure of the leading regions of momentum space can be derived by the methods of Section 5 (8, 24, 25). Just as in $e^+e^-$ annihilation, there is production of arbitrarily many jets in the final state, and these can be connected by soft interactions, as shown in Figure 2. But, just as in that case, a sum over cut graphs produces a cancellation of the final-state interactions. After this cancellation, there is a single hard scattering containing both vertices for the current. If we used axial gauge, then exactly one pair of lines would join the hard scattering subgraph to the beam-jet subgraph, but in a covariant gauge any number of additional longitudinal gluons are allowed.

A single graph will contribute in several regions. In each region we let $k_i^\mu$ be the momenta of the lines coming into the hard scattering from the jet. Within the hard scattering we neglect $k_i^-$ and $k_i^T$. Ward identities can be used to disentangle the longitudinal gluons from the hard scattering and associate them with the jet factors. The contribution to each region then has the form

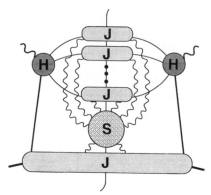

*Figure 2*    Leading integration regions for deeply inelastic scattering.

$$\left(\prod_i \int d\xi_i \, \text{Hard}(\xi, Q)\right)\left(\prod_j \int dk_j^- \, dk_j^T \, \text{Jet}\right). \qquad 10.$$

This is roughly like the factorization theorem, Equation 8. To obtain Equation 8 exactly, we must use a subtraction procedure to define the hard scattering coefficients $C_{ia}$, without double counting the contributions from the various regions. This is a standard, but quite technical, operation (see 11, 34). One makes use of the "forest" algorithm of Zimmermann (11, 40) to define a set of subtractions (38) within the hard part. After the subtraction procedure is completed, the Wilson coefficient $C_{ia}$ is exactly the $i$th structure function for scattering by parton $a$, minus infrared-dependent parts. It is a pure ultraviolet object to which the renormalization group may be applied, so that perturbation theory may be used for its calculation. It has a nontrivial generalized anomalous dimension, as discussed below.

### 7.2    *Parton Distribution Functions*

The subtraction procedure also generates a precise definition of the parton distributions $f_{a/H}$ (41, 42)

$$f_{q/h}(x, \mu) = (1/4\pi)\int dy^- \, e^{-ixp^+y^-} \langle p | \bar{\psi}_q(0, y^-, 0^T)\gamma^+ P\psi_q(0)|p\rangle_R, \qquad 11.$$

with corresponding definitions for antiquarks and gluons. The factor $P$ represents a path-ordered exponential of the gluon field, as is necessary to give gauge invariance. In light-cone gauge, $P = 1$, and then (43) the distribution function has a clear interpretation as the number density of quarks of fractional momentum $x$ in the hadron. Feynman rules for the parton distributions can be found in (42).

The subscript $R$ indicates that the ultraviolet divergences of the operator product are canceled by renormalization. These divergences arise from the integration over the transverse momentum through the external field vertices. The renormalization generates a nontrivial renormalization-group equation, the Altarelli-Parisi equation, which we discuss below.

7.2.1 SUM RULES   Certain moments of the parton distributions are matrix elements of local operators (31). For example,

$$\int_0^1 dx[f_{q/H}(x) - f_{\bar{q}/H}(x)] = (1/p^+)\langle H|\bar{q}\gamma^+q|H\rangle_R.$$   12.

This particular case gives the number of quarks of flavor $q$ minus the number of antiquarks of that flavor. Since this is a conserved quantum number, we can evaluate the right-hand side and thereby obtain a sum rule for the distributions. (Thus for $H$ = proton and $q$ = u the right-hand side is 2, whereas for $q$ = d it is 1.)

The second moments can be related to the energy momentum tensor:

$$\int_0^1 dx\, x \sum_a f_{a/H}(x) = (1/p^+)^2\langle H|\theta^{++}|H\rangle_R = 1.$$   13.

The need for renormalization somewhat complicates the argument, but the sum rules survive renormalization if one uses the MS or $\overline{\text{MS}}$ renormalization prescription for both the distribution functions and for the local operators.

7.2.2 ALTARELLI-PARISI EQUATION   It can be shown (41, 42, 44) that the renormalization-group equation for the parton distribution functions has the form

$$\mu \frac{df_{a/H}(x, \mu)}{d\mu} = \sum_b \int_x^1 \frac{d\xi}{\xi} \rho_{ab}[\xi/x, \alpha_s(\mu)] f_{b/H}(\xi, \mu).$$   14.

The kernel $\rho_{ab}$ plays the role of a generalized anomalous dimension, and can be deduced from the ultraviolet divergences of the parton distributions. This equation is known as the Altarelli-Parisi equation (41, 45–47). The original derivations were only given in the leading logarithm approximation, but Equation 14 is in fact true to all orders of logarithms.

## 7.3   Use of Factorization Theorem

The parton distributions are nonperturbative properties of the hadrons, but are process independent, while the Wilson coefficients are independent of the initial hadrons but depend on the precise process. The Wilson coefficients may be calculated by first calculating the structure functions for scattering by massless on-shell quarks and gluons, then calculating the distribution of partons inside these incoming particles. Finally, one uses the factorization theorem (Equation 8), applied to deeply inelastic scattering from quark and gluon targets, to deduce the Wilson coefficients.

One supposes that the parton distributions at some initial value $\mu_0$ of the renormalization scale $\mu$ are known from experimental measurements. The Altarelli-Parisi equation is used to determine the parton distributions at a scale $\mu$ of order $Q$. Finally, one convolutes them with the Wilson coefficients computed in low order. The setting of the renormalization scale $\mu$ to a value of order $Q$ ensures that there are no large logarithms in the perturbation expansion of the Wilson coefficients.

The Wilson coefficients in lowest order are constants times delta functions, $\delta(1 - x/\xi)$. Thus in leading order the quark distribution functions $f_{i/H}(x, Q)$ are simple linear combinations of structure functions $F_i(x, Q)$. However, this relation is not maintained in higher order, if one uses the $\overline{\text{MS}}$ definitions. Many authors define the quark distribution functions so that they remain the same linear combinations of structure functions $F_i(x, Q)$ to all orders in $\alpha_s$. (The gluon distribution is often left without a precise definition in this approach.) Whatever the phenomenological merits of this definition may be, we find it to be physically unnatural.

## 7.4  *Operator Product Expansion*

When one takes moments of the factorization formula in Equation 8, it turns into a simple matrix equation:

$$F_{i,N}(Q) = \sum_a C_{ia,N}[Q/\mu, \alpha_s(\mu)] f_{a/H,N}(\mu),\qquad\qquad 15.$$

where for example $F_{i,N}(Q) = \int_0^1 dx\, x^{N-1} F_i(x, Q)$. For integer $N$, this is equivalent to the final result of the operator product expansion approach (17); then the $f_{a/H,N}(\mu)$ are matrix elements of local operators. [More precisely, the combinations $f_{a/H,N}(\mu) + (-1)^N f_{\bar{a}/H,N}(\mu)$ of parton and antiparton densities are matrix elements of local operators, and these combinations are actually the ones that appear in Equation 8.] Mueller (48) has also considered a definition of corresponding operators for noninteger $N$ ("cut vertices"), at least in Abelian gauge theories. These may be considered simply as the Mellin transform of our parton distribution, Equation 11, with the local operators being particular special cases. If one has the moment Equation 15 valid for all $N$ with sufficiently large real part, then our first factorization formula, Equation 8, can easily be derived by an inverse Mellin transform.

## 7.5   Distribution of Partons in a Photon

The process $e^+e^- \to e^+e^-$ + hadrons may be used to probe deeply inelastic scattering by an almost real photon ($\gamma^* + \gamma \to$ hadrons). This may be treated the same way as deeply inelastic scattering by a hadron ($\gamma^* +$ H $\to$ hadrons). The only change is that one must include the photon as a possible parton (i.e. the sum over $a$ in Equation 8 is over quarks, antiquarks, gluon, and photon). There are corresponding changes in the Altarelli-Parisi equations. The original treatment was by Witten (49) and was given in the language of the operator product expansion.

## 7.6   Hadron Production in $e^+e^-$ Annihilation

Many of the features of the factorization theorem for deeply inelastic scattering generalize nicely to other processes. A good example is $N$-particle production in $e^+e^-$ annihilation: $e^+e^- \to \gamma^* \to H_1 + \cdots + H_N + X$. We suppose that the center-of-mass energies of the produced hadrons are of order $Q$ and that none of the angles between the hadrons is small.

We already know the form of the leading regions; they are symbolized in Figure 1. If there were no soft gluon attachments, then we could trivially generalize our work for deeply inelastic scattering.

The resulting theorem is in fact correct, but we must prove cancellation of the soft gluon effects. Since we detect final-state particles, we cannot perform an unrestricted sum over final states such as we did for the total cross section. Rather, we must examine the soft gluon couplings in more detail. At first sight the problem appears essentially identical to the infrared divergence problem in QED, where there is a well-established cancellation between real and virtual soft photon emission. However, there are two significant differences: First, the soft gluons can have arbitrary interactions among themselves and with quark and ghost loops. Secondly, the soft gluons can be emitted from internal jet lines of graphs, and it is this that makes the QCD problem so much harder than the QED problem. The problem was not properly recognized in the original papers on factorization (6–8, 37) except for that of Libby & Sterman (8), and even there it was not solved.

The necessary proof was given by Collins & Sterman (50). First one uses an approximation [generalizing the QED approximation of Grammer & Yennie (51)] for the attachment of a soft gluon to a jet line. Then the Ward identities allow one to sum the soft emissions of each jet, so that the soft gluon effects are gathered into an overall factor. Physically, the soft gluons couple to jets as a whole, but are insensitive to the internal structure of these jets. Finally, the standard cancellation theorem of Section 6 shows that this soft gluon factor equals 1.

After the cancellations we obtain the expected factorization theorem, in which the cross section is written as a hard scattering cross section for producing energetic partons convoluted with "parton decay distributions" $d_{H/a}(z)$. These functions give the distribution of hadrons in the jet created by an outgoing parton. The variable $z$ represents the ratio of the momentum of the produced hadron $H$ to that of the parton $a$. The parton decay functions are defined in a fashion (41, 42) analogous to the parton distribution functions. Mueller's cut vertex formulation (48) works with the Mellin transform of these decay functions. The decay functions obey sum rules and Altarelli-Parisi equations of the same form as the distribution functions. (However, the kernels of the Altarelli-Parisi equations are different.)

## 7.7    Jet Production in $e^+e^-$ Annihilation

Instead of measuring inclusive cross sections for production of particles of specific momenta, one may measure the cross sections for making the jets in which the hadrons are found. Intuitively, this amounts to measuring cross sections for making the parent partons of the jets. Sterman & Weinberg (52) suggested that the long-range final-state interactions cancel, so that jet cross sections in $e^+e^-$ annihilation are like the total cross section; they are purely short-distance quantities that may be calculated directly in renormalization-group-improved perturbation theory. The detailed proof of this assertion was given by Sterman (24, 25), and we explained the ideas behind this proof in previous sections.

One needs a suitable inclusive definition of the concept of a jet. One may ask for the jet to consist of a given range of energy in a given range of angle. This allows for a spreading in angle between the initial parton and the spray of hadrons that results, and for a loss of energy by soft radiation outside the angular range used to define the jet.

## 8.    TWO HADRONS IN THE INITIAL STATE

Single-scale processes with two hadrons in the initial state are of crucial importance in the era of hadron colliders. Examples of hard scattering processes in hadron collisions include the inclusive production of high-mass muon pairs and of W and Z bosons via the Drell-Yan mechanism, high $p_T$ jets, high $p_T$ photons, high $p_T$ single hadrons, a high-mass Higgs boson, etc. In this section, we discuss the Drell-Yan process, which has been extensively studied and can serve as a prototype. In the first work on factorization (4, 5), the Drell-Yan process was the process considered.

## 8.1   *Drell-Yan Muon Pair Production*

We consider the cross section $d\sigma/dQ^2\,dy$ for the process $H_A + H_B \rightarrow \gamma^* + X$, where the virtual photon $\gamma^*$ decays into a muon pair. The virtuality $Q^2$ of the virtual photon is taken to be large, of order $s$. Its rapidity $y$ is measured and may be varied freely, as long as we do not approach closely to the edge of phase space. We suppose here that the virtual photon transverse momentum $Q_T$ is not observed. The power-counting arguments discussed earlier show that a typical leading integration region for the Drell-Yan process is like that shown in Figure 3. Fast partons from the two hadrons collide to produce the observed virtual photon plus, possibly, a number of unobserved high $p_T$ jets. Spectator jets from the two hadrons are left behind. Soft gluons couple the jets.

One might suppose that the detailed proof of the factorization theorem for the Drell-Yan process would be a transcription, with minor modifications, of that for two-particle production in $e^+e^-$ annihilation. After all, $H_A + H_B \rightarrow \gamma^* + X$ differs kinematically from the process $\gamma^* \rightarrow H_A + H_B + X$ only by having the detected particles switched between the initial and final states. In a field theory without gauge fields, the proof is indeed essentially the same in both cases (6–8, 53), but in the presence of gauge fields, as in QCD, a more complicated argument is necessary.

The complications arise because (50, 54) the (generalized) Grammer-Yennie approximation for soft gluons coupling to jets that is used in the $e^+e^-$ argument (50) is not always correct. It breaks down in a certain part

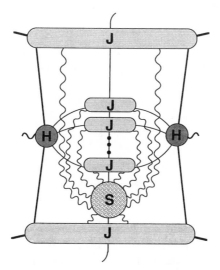

*Figure 3*   Leading integration regions for the Drell-Yan process.

of the soft region known as the "Glauber region." In $e^+e^-$ annihilation, because the low virtuality interactions are all in the final state, the momentum integrations are not pinched in this region. Thus one can deform the contours of integration to where the approximation is valid. The same situation does not hold in the Drell-Yan case, where there are both initial- and final-state interactions: the integrations in individual graphs are trapped in the Glauber region.

This reasoning may be put in more physical terms. It is quite intuitive that soft gluon interactions that happen long after the hard scattering in $e^+e^-$ annihilation cannot retroactively affect the probability for that hard scattering. However, in the Drell-Yan case, a soft gluon interaction that occurs long *before* the hard quark + antiquark $\rightarrow \gamma^*$ interaction can change the color of the quarks and thus apparently affect the probability that the subsequent interaction will occur.

The extra steps needed to establish factorization in the Drell-Yan case are given in (27, 55), with some corrections that are at present being worked out. The construction uses a contour deformation idea that was first given by DeTar, Ellis & Landshoff (56) and by Cardy & Winbow (57). Although the arguments are necessarily quite technical, a simple physical picture emerges. Soft gluons that couple spectator partons in one hadron to other spectator partons in the other hadron do not affect the cross section because these interactions occur when the hadrons collide, which is long after the spectators last interacted with the active partons that participate in the hard collision. Soft gluons that couple partons of hadron A to the active parton in hadron B do affect the cross section. However, their effect is absorbed into the definition of the parton distribution functions.

After this work, one obtains a factorization like Equation 8 except that it has two factors of parton distribution functions:

$$\frac{d\sigma_{DY}}{dy\,dQ^2} \propto \sum_{a,b} \int \frac{d\xi_A}{\xi_A}\frac{d\xi_B}{\xi_B}$$

$$\times C_{DY}^{ab}(\xi_A, \xi_B, s, Q^2, y; \alpha_s, \mu)\,f_{a/A}(\xi_A, \mu)\,f_{b/B}(\xi_B, \mu) + \text{higher twist.} \quad 16.$$

If one determines the parton distributions from data on deeply inelastic scattering, then this factorization formula provides an absolute prediction for the Drell-Yan cross section from perturbative calculations. We clearly have an important test of QCD.

Experimental measurements (57a) are a factor of 1.5 to 2 above the prediction obtained by using the lowest order values for the hard scattering coefficients. This factor is called the $K$ factor, and it is roughly independent of the kinematic variables. Perturbative calculations of higher order corrections to the hard scattering coefficients provide the prediction for the $K$ factor. The calculated corrections turn out to be large and positive, so that there is a reasonable agreement between theory and experiment (57b). Of course, when the higher order corrections are so large, one must at present regard the phenomenology as being somewhat delicate. Only at extremely high energy is the effective coupling $\alpha_s(Q)$ sufficiently small that $K$ is close to 1.

# 9.  HEAVY QUARKS AND OTHER COLORED OBJECTS

In this section we consider heavy quarks, like the bottom quark, whose mass is much larger than 1 GeV. Our remarks also apply to other heavy particles with strong interactions, such as appear in many extensions of the standard model.

## 9.1  *Decoupling Theorem*

Suppose we calculate Green functions on a momentum scale $P$ well below the mass $M_Q$ of some heavy quark Q. Then heavy-quark loops are suppressed by a power of $P/M_Q$ except when they are inside subgraphs that require renormalization counterterms. One finds that, up to power-suppressed corrections, one gets the same results by calculating in an "effective low-energy theory." This effective theory is obtained from the full theory by omitting the heavy quark and by making suitable adjustments to the renormalized couplings and light-quark masses. The details are explained by Symanzik (58) and by Appelquist & Carazzone (59); see also Witten (60). Their theorem applies to all renormalizable quantum theories, not just to QCD. Its proof uses techniques similar to those for the operator product expansion.

When one wants to calculate Green functions at a momentum scale $P$ that is comparable to or larger than $M_Q$, one must use the full theory. One then needs to know the relation between the renormalized parameters in the full theory and those in the effective low-energy theory. This relation may be calculated perturbatively in an expansion in powers of $\alpha_s(M_Q)$.

## 9.2  *Distribution of Heavy Quarks in a Hadron*

The distribution of heavy quarks in a hadron is just a hadron matrix element of a certain operator involving heavy-quark fields. Witten (60) showed that this distribution can be computed in a power series in $\alpha_s(M_Q)$, up to corrections suppressed by a power of (1 GeV$/M_Q$). The physical content of the calculation is that the creation of a heavy quark in a light hadron involves short-distance scales on the order of $\Delta x \approx 1/M_Q$. The heavy-quark distribution function is thus factorized as distributions of light partons convoluted with the perturbative short-distance graphs for creating heavy quarks from light partons (e.g. gluon → heavy-quark pair). Witten's argument used the operator product expansion and moments of distribution functions, but there appears to be no reason why his proof cannot be adapted to the more modern methods that directly use the distribution functions. A discussion may be found in (61, 62). Oliensis (63) applied similar ideas to the heavy-quark fragmentation function.

### 9.3 Reactions Involving Heavy Quarks

Heavy quarks can play a role in many processes. One simple example is deeply inelastic scattering, which is covered in the paper of Witten (60). If $Q^2 \gg M_Q^2$ then one should use distribution functions that include heavy quarks as constituents. These are convoluted with Wilson coefficients in which the heavy-quark mass is neglected compared to $Q$. On the other hand, if $Q^2 \approx M_Q^2$ then one should use light-parton distribution functions convoluted with Wilson coefficients involving heavy quarks, in which the heavy-quark mass is not neglected compared to $Q$.

The production cross section for heavy quarks and other colored objects is of great experimental interest. This process has not received a full treatment to all orders of perturbation theory. However, arguments given in (64) lead us to the conclusion that inclusive heavy-quark production is a short-distance process that can be expressed as a perturbative cross section for producing the heavy quarks from partons, convoluted with parton distribution functions. Thus the mechanism for heavy-quark production to leading order in $\alpha_s(M_Q)$ is parton-parton fusion.

## 10.  CROSS SECTIONS WITH SUDAKOV BEHAVIOR

In the previous sections, we indicated how factorization is proved for inclusive processes in which there is only one large momentum scale $Q$. Suppose that we now consider the Drell-Yan process in which the transverse momentum $Q_T$ of the muon pair (or produced W or Z boson) is measured. If $Q_T$ is of order $Q$, then there is but one scale and the factorization takes the same form as that obtained when one integrates over $Q_T$. However, if $Q_T \ll Q$, new features arise.

In the parton model [or in $(\phi^3)_6$ theory (65)], the $Q_T$ distribution of the muon pair would reflect only the "intrinsic" transverse momentum distribution of the annihilating partons. However, in QCD, the $Q_T$ distribution is broadened by the emission of relatively soft ($k_T \ll Q$) gluons from the parton lines.

Calculations indicate that something new is happening because there are two logarithms per loop, compared to one logarithm when one has integrated over $Q_T$. The extra logarithm arises because of an incomplete cancellation between real and virtual gluon graphs: gluons in virtual graphs can have any transverse momentum, while gluon emission into the final state is suppressed for transverse momentum larger than $Q_T$.

A detailed analysis (66) of the effect of soft gluon emissions has been given by us for the case of the production of nearly back-to-back hadrons in electron-positron annihilation, which is essentially the crossed version of the Drell-Yan process discussed here. This analysis covers all leading power contributions to all orders of perturbation theory, although it uses the space-like axial gauge, which we have criticized in Section 3. We formulated the corresponding results for the Drell-Yan process in (66a). Work at the leading logarithm level by Dokshitzer, Dyakonov & Troyan (67) and by Parisi & Petronzio (68) was important in the development of this subject.

One finds that one can write the cross section in a certain factorized form, which is rather more complicated than that obtained for the $Q_T$-integrated processes. Then the factors can be analyzed by using two differential equations. One is the renormalization-group equation, while the second reflects gauge invariance. (Roughly speaking, one needs two differential equations to eliminate two logarithms per loop.)

A characteristic result of this analysis is the emergence of a factor that, at the very crudest level of approximation, is

$$\exp\left[ -\frac{4}{3}\frac{\alpha_s}{\pi}\ln^2(Q^2/Q_T^2) \right]. \qquad 17.$$

Such factors in QCD are called Sudakov factors, after *the* Sudakov (69) form factor of the electron

$$F_{\text{QED}}(Q^2) = \exp\left[ -\frac{1}{4}\frac{\alpha}{\pi}\ln^2(Q^2) \right]. \qquad 18.$$

Many people contributed to the theory of the leading logarithm expansion of Sudakov form factors (e.g. 67–70), both in QED and QCD. A complete formulation in which the form factor is expressed in terms of perturbatively expandable quantities with no large logarithms was derived by Mueller (71) and Collins (72).

## 11. EXCLUSIVE PROCESSES

The statements of factorization theorems for inclusive processes can be generalized to many exclusive processes, e.g. electromagnetic form factors of hadrons at large $Q$ and elastic scattering of hadrons at large angle. The key ideas were explained by Brodsky & Farrar (72a). Then Brodsky & Lepage (73) formulated factorization formulae analogous to the inclusive process formulae, using light cone variables. [Efremov & Radushkin (73a) also developed the same ideas, though less generally.] In this formulation, exclusive wave functions appear that have definitions very close to those of the parton distributions in Equation 11.

However, many of these proposed factorization formulae are incorrect, at least when taken literally order by order in perturbation theory. When one attempts to imitate the proofs for inclusive processes, failure often occurs at the first stage: the power counting for the leading regions does not always correspond to a single hard scattering accompanied by soft gluons. In elastic hadron-hadron scattering, Landshoff (74) found that regions with more than one independent hard scattering are favored by a power of $Q$. Furthermore, in most exclusive processes, regions with soft quark exchange have the same power law as the basic hard scattering (72a, 78)—such regions were discussed in pre-QCD models for form factors (74a). As Brodsky & Farrar noted, there are nontrivial Sudakov form factor effects that suppress these extra regions. Unfortunately, the presence of Sudakov form factor effects shows that the hard scattering involves more than fixed order perturbation theory. There is at present no complete treatment that enables us to accord factorization formulae for most exclusive processes in QCD the same status as the formulae for the inclusive processes we have discussed. As we saw in Section 10, it is possible to give full treatments for Sudakov-dominated processes, so we conjecture that a solution can be found.

There are a few processes where the extra regions are power suppressed graph by graph and for which factorization theorems of the Brodsky-Lepage form appear to be valid. These are processes involving mesons and photons (for example, the pion form factor), and exclusive decays of quarkonia. See also the work of Duncan & Mueller (75). We do not think that the demonstrations of factorization for these processes treat soft gluon effects properly, but we believe that remedying this is a technicality.

## 12.    CONCLUSIONS

In the decade since its discovery, QCD has come to be accepted as the correct theory of strong interactions. The factorization theorems are now a standard tool used by both theorists and experimentalists to estimate cross sections. A new piece of physics such as the production of a new particle is regarded as another hard process to be added to the list of possible calculations. [See Eichten et al (76) and recent Snowmass proceedings (77) for examples.]

The most basic theorems are well-established by now, as we have seen. However, there are important gaps in what has been published. Notably, we regard as necessary proofs of factorization directly in a covariant gauge; this has not yet been done at the level we want even for deeply inelastic scattering. We would also like to see a correct formulation of the axial gauges, since the lack of such a formulation is the main defect of many current proofs. There are also areas in which theorems of the simplest type are presumably true, but for which no reasonably complete proof exists. A notable example is heavy-quark production.

We emphasize that careful analysis of such possible theorems is not an exercise that can safely be put off into the indefinite future. In some cases, a true theorem is not widely recognized. An example is provided by the various alternative mechanisms that have been proposed for heavy-quark production. In other cases, inadequacy in the proof of widely recognized theorems can allow claims that the theorems are false, as happened for the Drell-Yan case (54). On the other hand, in some cases an apparently attractive proposed theorem may prove upon careful examination to have fatal flaws, as in the case of many sorts of exclusive processes.

There are a number of examples of possible theorems that we have not even tried to cover. One is the $x \to 1$ limit of structure functions. [This is reviewed in (78); see also (79).] Others are the processes considered by Berger & Brodsky (80), such as the Drell-Yan process at $x_F \to 1$. These have contributions that are normally regarded as "higher twist," that is, suppressed by a power of the large variable $Q$. However, as $x_F \to 1$ they may have a smaller power of $(1 - x_F)$ than the leading twist term and therefore be the dominant contributions.

Much of the future development of perturbative QCD is dependent on an improved understanding of soft gluon physics. For instance, as experiments go to ever higher energies, one sees more and more small-$x$ minijets created by the emission of relatively soft gluons in hard collisions. Also, much of the event structure in Monte Carlo simulations is related to soft gluons. An improved understanding of soft gluon physics also bears on the important subject of exclusive scattering, where soft gluon graphs lead to Sudakov form factors. Finally, there is much overlap with the physics of the Pomeron (81), and there is the exciting prospect that perturbative methods may connect with the physics that dominates the bulk of hadronic cross sections.

ACKNOWLEDGMENTS

This work was supported in part by the Department of Energy. J.C.C. would like to thank the John Simon Guggenheim Memorial Foundation for a fellowship. We would like to thank many colleagues for extensive discussions on the subject matter treated, especially A. H. Mueller, G. Sterman, and J. C. Taylor.

*Literature Cited*

1. Gross, D. J., Wilczek, F. *Phys. Rev. Lett.* 30: 1343 (1973)
2. Politzer, H. D. *Phys. Rev. Lett.* 30: 1346 (1973)
3. Webber, B. *Ann. Rev. Nucl. Part. Sci.* 36: 253 (1986)
4. Radyushkin, A. V. *Phys. Lett.* 69B: 245 (1978)
5. Politzer, H. D. *Nucl. Phys.* B129: 301 (1977)
6. Amati, D., Petronzio, R., Veneziano, G. *Nucl. Phys.* B140: 54; B146: 29 (1978)
7. Ellis, R. K., Georgi, H., Machacek, M., Politzer, H. D., Ross, G. G. *Nucl. Phys.* B152: 285 (1979)
8. Libby, S., Sterman, G. *Phys. Rev.* D18: 3252, 4737 (1978)
9. Marciano, W., Pagels, H. *Phys. Rep.* 36: 137 (1978), and references therein
10. Brandt, R. *Nucl. Phys.* B116: 413 (1976)
11. Collins, J. C. *Renormalization: An Introduction to Renormalization, The Renormalization Group and the Operator Product Expansion.* Cambridge Univ Press (1984)
12. Becchi, C., Rouet, A., Stora, R. *Ann. Phys. (NY)* 98: 287 (1976)
13. 't Hooft, G. *Nucl. Phys.* B61: 455 (1973)
14. Bardeen, W. A., Buras, A. J., Duke, D. W., Muta, T. *Phys. Rev.* D18: 3998 (1978)
15. Carracciolo, S., Curci, G., Menotti, P. *Phys. Lett.* 113B: 311 (1982)
16. Konetschny, W., Kummer, W. *Nucl. Phys.* B124: 145 (1977)
17. Gross, D. J. In *Methods in Field Theory*, ed. R. Balian, J. Zinn-Justin. Amsterdam: North-Holland (1976)
17a. Buras, A. J., Floratos, E. G., Ross, D. A., Sachrajda, C. T. *Nucl. Phys.* B131: 308 (1977)
18. Celmaster, W., Gonsalves, R. J. *Phys. Rev.* D20: 1420 (1979)
19. Appelquist, T., Georgi, H. *Phys. Rev.* D8: 4000 (1973)
20. Zee, A. *Phys. Rev.* D8: 4038 (1973)
21. Chetyrkin, K. G., Kataev, A. L., Tkachov, F. V. *Phys. Lett.* 85B: 277 (1979)
22. Dine, M., Sapirstein, J. *Phys. Rev. Lett.* 43: 668 (1979)
23. Celmaster, W., Gonsalves, R. J. *Phys. Rev. Lett.* 44: 560 (1980); *Phys. Rev.*

D21: 3112 (1980)
24. Sterman, G. *Phys. Rev.* D17: 2773 (1978)
25. Sterman, G. *Phys. Rev.* D17: 2789 (1978)
26. Coleman, S., Norton, R. E. *Nuovo Cimento* 38: 438 (1965)
27. Collins, J. C., Soper, D. E., Sterman, G. *Nucl. Phys.* B261: 104 (1985)
28. Weinberg, S. *Phys. Rev.* 118: 838 (1960)
29. Poggio, E. C., Quinn, H. R., Weinberg, S. *Phys. Rev.* D13: 1958 (1976)
30. Barnett, R. M., Dine, M., McLaren, L. *Phys. Rev.* D22: 594 (1980)
31. Efremov, A. V., Radyushkin, A. V. *Phys. Lett.* 63B: 449 (1976), *Nuovo Cimento Lett.* 19: 83 (1977)
32. Gross, D. J., Wilczek, F. *Phys. Rev.* D8: 3633 (1973), D9: 980 (1974)
33. Georgi, H., Politzer, H. D. *Phys. Rev.* D9: 416 (1974)
34. Wilson, K. *Phys. Rev.* 179: 1499 (1969)
35. Christ, N., Hasslacher, B., Mueller, A. H. *Phys. Rev.* D6: 3543 (1972)
36. Joglekar, S. *Ann. Phys. (NY)* 108: 233, 109: 210 (1977)
37. Efremov, A. V., Radyushkin, A. V. *Teor. Mat. Fiz.* 44: 327 (1980) [Engl. transl: *Theor. Math. Phys.* 44: 774 (1981)]
38. Efremov, A. V., Radyushkin, A. V. *Teor. Mat. Fiz.* 44: 17 (1980) [Engl. transl: *Theor. Math. Phys.* 44: 573 (1981)]
39. Curci, G., Furmanski, W., Petronzio, R. *Nucl. Phys.* B175: 27 (1980)
40. Zimmermann, W. *Ann. Phys. (NY)* 77: 536, 570 (1973)
41. Lipatov, L. N. *Yad. Fiz.* 20: 181 (1974) [Engl. transl: *Sov. J. Nucl. Phys.* 20: 95 (1975)]
42. Collins, J. C., Soper, D. E. *Nucl. Phys.* B194: 445 (1982)
43. Soper, D. E. *Phys. Rev.* D15: 1141 (1977)
44. Baulieu, L., Floratos, E. G., Kounnas, C. *Nucl. Phys.* B166: 321 (1981)
45. Gribov, V. N., Lipatov, L. N. *Yad. Fiz.* 15: 781 (1972) [Engl. transl: *Sov. J. Nucl. Phys.* 46: 438 (1972)]
46. Altarelli, G., Parisi, G. *Nucl. Phys.* B126: 298 (1977)
47. Johnson, P. W., Tung, W.-T. *Phys. Rev.* D16: 1769 (1977)
48. Mueller, A. H. *Phys. Rev.* D18: 3705 (1978)

49. Witten, E. *Nucl. Phys.* B120: 189 (1976)
50. Collins, J. C., Sterman, G. *Nucl. Phys.* B185: 172 (1981)
51. Grammer, G., Yennie, D. *Phys. Rev.* D8: 4332 (1973)
52. Sterman, G., Weinberg, S. *Phys. Rev. Lett.* 39: 1436 (1977)
53. Efremov, A. V., Radyushkin, A. V. *Teor. Mat. Fiz.* 44: 157 (1980) [Engl. transl: *Theor. Math. Phys.* 44: 664 (1981)]
54. Bodwin, G., Brodsky, S. J., Lepage, G. P. *Phys. Rev. Lett.* 47: 1799 (1981)
55. Bodwin, G. T. *Phys. Rev.* D31: 2616 (1985); D34: 3932 (1986)
56. DeTar, C., Ellis, S. D., Landshoff, P. V. *Nucl. Phys.* B87: 176 (1975)
57. Cardy, J., Winbow, G. *Phys. Lett.* 52B: 95 (1974)
57a. Stroynowski, R. *Phys. Rep.* 71: 1 (1981)
57b. Altarelli, G., Ellis, R. K., Martinelli, G. *Nucl. Phys.* B143: 521; B146: 544; B157: 461 (1978); Kubar-Andre, J., Paige, F. *Phys. Rev.* D19: 221 (1979)
58. Symanzik, K. *Comments Math. Phys.* 34: 7 (1973)
59. Appelquist, T., Carazzone, J. *Phys. Rev.* D11: 2856 (1975)
60. Witten, E. *Nucl. Phys.* B104: 445 (1976)
61. Collins, J. C., Tung, W.-K. *Nucl. Phys.* B278: 934 (1986)
62. Haber, H. E., Soper, D. E., Barnett, R. M. In *Proc. Workshop on Physics Simulations at High Energy, Madison, 1986*, ed. V. Barger, T. Gottschalk, F. Halzen. Singapore: World Scientific (1987)
63. Oliensis, J. *Phys. Rev.* D23: 1420 (1981)
64. Collins, J. C., Soper, D. E., Sterman, G. *Nucl. Phys.* B263: 37 (1986)
65. Collins, J. C. *Phys. Rev.* D21: 2962 (1980)
66. Collins, J. C., Soper, D. E. *Nucl. Phys.* B193: 381 (1981)
66a. Collins, J. C., Soper, D. E., Sterman, G. *Nucl. Phys.* B250: 199 (1985)
67. Dokshitzer, Yu. L., Dyakonov, D. I., Troyan, S. I. *Phys. Rep.* 58: 269 (1980)
68. Parisi, G., Petronzio, R. *Nucl. Phys.* B154: 427 (1979)
69. Sudakov, V. *Zh. Eksp. Teor. Fiz.* 30: 87 (1956) [Engl. transl: *Sov. Phys.-JETP* 3: 65 (1956)]
70. Jackiw, R. *Ann. Phys. (NY)* 48: 292 (1968)
71. Mueller, A. H. *Phys. Rev.* D20: 2037 (1979)
72. Collins, J. C. *Phys. Rev.* D22: 1478 (1980)
72a. Brodsky, S. J., Farrar, G. R. *Phys. Rev. Lett.* 31: 1153 (1973); *Phys. Rev.* D11: 1309 (1975)
73. Brodsky, S. J., Lepage, P. *Phys. Rev.* D22: 2157 (1980)
73a. Efremov, A. V., Radushkin, A. V. *Phys. Lett.* 94B: 245 (1980)
74. Landshoff, P. V. *Z. Phys.* C6: 69 (1980); Donnachie, A., Landshoff, P. V. *Z. Phys.* C2: 55 (1979), erratum *Z. Phys.* C2: 372 (1979)
74a. Drell, S. D., Yan, T.-M. *Phys. Rev. Lett.* 24: 181 (1970); West, G. B. *Phys. Rev. Lett.* 24: 1206 (1970)
75. Duncan, H. A., Mueller, A. H. *Phys. Rev.* D21: 1636 (1980)
76. Eichten, E. J., Hinchliffe, I., Lane, K. D., Quigg, C. *Rev. Mod. Phys.* 56: 579 (1984)
77. Donaldson, R., Morfin, J., eds. *Proc. 1984 DPF Summer Study of the Design and Utilization of the Superconducting Super Collider.* New York: Am. Phys. Soc. (1985)
78. Mueller, A. H. *Phys. Rep.* 73: 239 (1981)
79. Sterman, G. *Nucl. Phys.* B281: 310 (1987)
80. Berger, E. L., Brodsky, S. J. *Phys. Rev. Lett.* 42: 940 (1979); *Phys. Rev.* D24: 2428 (1981)
81. Gribov, L. V., Levin, E. M., Ryskin, M. G. *Phys. Rep.* 100: 1 (1983)

*Ann. Rev. Nucl. Part. Sci. 1987. 37:411–61*

# PION-NUCLEUS INTERACTIONS[1]

## W. R. Gibbs and B. F. Gibson

Theoretical Division, Los Alamos National Laboratory, Los Alamos, New Mexico 87545

CONTENTS

## INTRODUCTION

The pion is an hadronic probe with many facets. The multiple research aspects contain complex interconnections. In an attempt to present some of the developments in pion-nucleus physics, we have separated the field into areas closely associated with specific goals. The inherent danger in such an approach is clear: the reader's attention may be too narrowly focused and important collateral observations missed. This danger is largely offset by the advantage of giving the reader a thread to follow.

---

[1] The US Government has the right to retain a nonexclusive, royalty-free license in and to any copyright covering this paper.

Thus, we present a series of minireviews, each one largely independent of the others, along with the basic physics background needed by a nonexpert. Each topic is worthy of a separate review, but we seek to gather here an overview of the field.

There are several topics that, although interesting, have not matured to the point where definite results can be simply stated. These include inclusive double charge exchange (1), $\pi^+$ and $\pi^-$ scattering from $^3$He and $^3$H (2), nonanalog exclusive double charge exchange (3), and eta production with the $\pi\eta$ reaction on nuclei (4). No doubt these research areas will form the basis for a more complete look at pion-nucleus physics within another year or two. We omit the topic of deltas in nuclei; the reader can consult the proceedings (5) of the recent conference on this subject for information. In addition, there exist several partial reviews related to the present subject (6–8) and one excellent book (9).

Contributions to our understanding of pion-nucleus physics appear to fall most naturally into the three areas: fundamental properties of hadronic interactions, the structure of nuclei, and hadronic reactions in a many-body environment. It is along these lines that we have sorted the achievements we selected to review. Others would no doubt have chosen a different organization or different examples. We invite comments and suggested additions to this look at the accomplishments of the past decade.

## 2.   FUNDAMENTAL HADRONIC INTERACTIONS

Any attempt to comprehend the many-body hadronic system from a "first principles" viewpoint would prove fruitless without a substantial understanding of the fundamental hadronic interaction. In which circumstances may one consider mesonic degrees of freedom to be appropriate? When, if ever, must quark-gluon degrees of freedom be accounted for explicitly? Clearly the answers to these questions must involve a synthesis of information obtained in the two-body and multibody environments. In Section 2 we review our understanding of the hadron-hadron interaction with an emphasis on pion beams. We recognize that many other fundamental measurements have been made with pion beams [$\pi^+\pi^0$ mass difference (10), $\Delta$ magnetic moment (11), pion beta decay (12), etc] but we bypass them here to concentrate on those phenomena that have more direct implications for many-body hadronic physics.

We have selected three examples in which pion experiments have played a significant role in our effort to understand the strong force in nuclear physics: (a) pion-nucleon scattering provides our basis for modeling the behavior of the pion within the nucleus and the window through which we may view the excitations of the three-valence-quark nucleon; (b) pion-deuteron scattering has yielded evidence of charge-symmetry breaking that has been interpreted in terms of quark mass differences in the delta isobars; and (c) radiative pion capture in deuterium has provided our most definitive measurement of the low-energy scattering parameters for the neutron-neutron system and direct evidence for charge-symmetry breaking in the nucleon-nucleon interaction. They illustrate the direct role of pion physics in investigating the meson-nucleon strong interaction as well as the use of the pion as a tool to study the nucleon-nucleon interaction and properties of the nucleon.

## 2.1  The Pion-Nucleon Interaction

The availability of intense pion beams has permitted the determination of pion-nucleon scattering observables in a manner, and with an accuracy, not previously possible. The importance of this scattering system, as our direct window on the excited states of three quark systems, is simply illustrated. Since the pion consists (in lowest order) of a light quark-antiquark pair, the full set of combinations of the three light quark systems can be reached in the pion-nucleon system. The excitation energy of the state is directly determined by the pion beam energy.

A review of recent data at low energies shows that the elastic scattering, especially with $\pi^-$ beams, remains problematic, although considerable improvements in the data have been made (13). The use of charge exchange (14, 15) and photoabsorption (15) measurements have filled the gap. Measurements of pion-nucleon charge exchange have now been pushed down in energy to 10 MeV (D. H. Fitzgerald, private communication), and the Panofsky ratio has been measured to an accuracy of 0.5% (17). Analyses, including isospin-breaking effects, have now placed (18) all of the threshold data in a unified framework.

The pion-nucleon scattering lengths remain among the most fundamental of all quantities in hadron physics. Along with the $\pi\pi$ scattering lengths, they form much of the basic input for modern theories embodying our concepts of low-energy quark-gluon physics. Since QCD has not yet provided unambiguous guidance concerning how to describe hadrons and their interactions, we have been led to develop many different models, motivated by QCD, in attempting to answer these questions. Among the most active endeavors are calculations within the cloudy bag model (19) and those based on chiral QCD (20). While the value of the $\sigma$ term [a quantity derived from pion-nucleon scattering data (21) and reflecting the degree to which chiral symmetry is broken. The current value is $\sim 64$ MeV] is in substantial disagreement with the chiral QCD models, which predict 35 MeV, this discrepancy can be removed if the $\bar{q}q$ sea of the nucleon is included (22). Using the $\Xi$-$\Lambda$ mass difference and the measured value of the $\sigma$ term, the sea population of the nucleon has been extracted (23). Surprisingly, in this work the $\bar{s}s$ content is found to be 21–29% of the $\bar{q}q$ sea.

Since the confinement range is still debated, useful comparisons can be made between the energy dependence of the low-energy phase shifts and that predicted by these models because the size of the $\pi N$ system and the energy dependence of low-energy scattering phase shifts are closely related. For example, the isospin-1/2 s-wave $\pi$-nucleon interaction has a much longer range than the isospin 3/2 (18).

A recent measurement of the level shift of the pionic hydrogen atom (24) yields a value of the $\pi^- p$ scattering length in disagreement with the more classical methods (21). This measurement has caused considerable discussion since it would yield a value of the $\sigma$ term in agreement with the chiral QCD calculations (20). If this discrepancy persists as the atomic measurement is refined, it will be puzzling indeed.

While the development of QCD has presumably given us the fundamental physics of hadronic interaction, our evident inability to convert this information to practical understanding on a nuclear physics level is nowhere clearer than in the lowest $P_{11}$ pion-nucleon resonance. This first $(T = 1/2, J = 1/2)$ excited state of the nucleon (the Roper resonance) clearly illustrates the question of the appropriate degrees of freedom to be used to describe hadronic systems. Modern attempts have been made using pure quark systems (25), gluon degrees of freedom (26), meson-baryon degrees of freedom (27), and lately, $\pi\pi$ interactions (28). Several of these models predict, or are able to accommodate, a doublet Roper resonance.

However, the experimental situation has been very unsettled. From the beginning there have been hints of two resonances in the data (29). While several analyses see no evidence for such splitting (30) the most recent analysis claims two distinct pole-zero pairs (31).

It is at this point that intense pion beams and modern polarized targets are making a decisive contribution. Measurements of the left-right asymmetry in $\pi^- p$ scattering (32) failed to distinguish among the various partial wave solutions, but recent charge exchange asymmetry results (33) greatly favor the phase-shift solutions that support the doublet hypothesis.

The impact of these measurements on our fundamental understanding of hadronic physics cannot be overemphasized. This data base is the source of a large fraction of our knowledge of hadronic structure. Its relevance for interpreting nuclear data is of no less importance. The description of the behavior of the pion in a many-body environment requires not only a knowledge of the basic pion-nucleon interaction but, perhaps even more important, a deep understanding of its origin, so that hadronic effects in complex hadronic systems can be realistically modeled.

## 2.2  Pion-Deuteron Scattering and Quark Masses

Analysis of $\pi^\pm p$ scattering produced evidence for a slight violation of charge independence in the vicinity of the spin 3/2, isospin 3/2 (3,3) resonance (34). Parameterized in terms of $\Delta$ masses, the $\Delta^0$ resonance was found to be heavier than the $\Delta^{++}$. This is consistent with the experimental observation that the baryon having the larger negative charge is always the heavier: $m_n - m_p = 1.3$ MeV, $m_{\Sigma^-} - m_{\Sigma^0} = 4.9$ MeV, $m_{\Sigma^-} - m_{\Sigma^+} = 8.0$ MeV, etc. Thus, it was natural to test charge symmetry in a pion-nucleus system by investigating $\pi^\pm$ scattering from the isoscalar deuteron (35). After correcting for the direct effects of the Coulomb interaction, a charge-symmetry violation was found that can be parameterized in terms of mass and width differences among the various $\Delta$ isobars. These differences appear to be in agreement with simple quark model predictions.

Comparison of $\pi^+$ and $\pi^-$ total cross sections for scattering from deuterium is a direct test of charge-symmetry conservation, if electromagnetic effects are properly taken into account. Myhrer & Pilkuhn (36) first estimated the electromagnetic corrections and predicted dynamical effects in the cross-section difference

$$\delta\sigma_t = \sigma_t(\pi^- d) - \sigma_t(\pi^+ d)$$

of up to a few percent. Pedroni et al (35) then measured the $\pi^\pm p$ total cross sections in a standard transmission geometry, obtaining essential agreement with the results from Carter et al (34), and went on to make the first precision $\pi^\pm d$ measurements in the energy range of 70–370 MeV. The average $\pi d$ cross section was found to differ from impulse approximation by only 20% at the peak of $\sim 200$ mb around 180 MeV and by less in the wings where it falls to 100 mb near 120 and 280 MeV. Most of the difference could be accounted for by multiple scattering and by Fermi motion corrections. It was shown (34) that the only significant Coulomb distortion in the $\pi p$ total cross section came from the $J = 3/2$ partial wave. Thus, Coulomb corrections were made only in that amplitude in the analysis. Eschewing the introduction of the Coulomb potential into the Faddeev equations and noting that Fermi motion and multiple scattering corrections were only some 20%, Pedroni et al (35) calculated the Coulomb distortion corrections to the $\pi^\pm d$ cross-section difference $\delta\sigma_t$ in impulse approximation. These corrections are essentially negligible above 190 MeV but do affect the extracted parameters, as discussed below.

The resulting $\delta\sigma_t$ is shown in Figure 1. The origin of the asymmetry ($< 3\%$) is not completely agreed upon. The observed violation of charge symmetry cannot be due to Coulomb-induced isospin mixing in the deuteron, because it is ruled out by parity conservation and the Pauli principle. It could be due to dynamical effects such as the neutron-proton mass difference, $\rho\omega$ mixing in the $\pi$-nucleon interaction, etc. Regardless, it was parameterized (36) in terms of $\Delta$ mass and width differences (of a Breit-Wigner form) as quoted in Table 1. Only the combinations

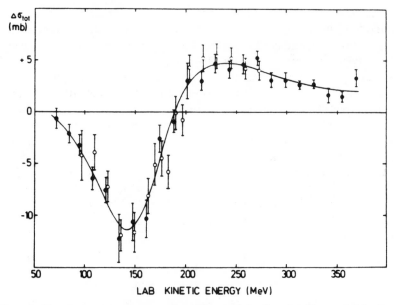

*Figure 1*    Pion-deuteron total cross-section difference $[\sigma(\pi^- d) - \sigma(\pi^+ d)]$ corrected for Coulomb distortion. The solid curve is a Breit-Wigner form fitted to the data.

$$\delta M = M^- - M^{++} + \frac{1}{3}(M^0 - M^+)$$

and

$$\delta\Gamma = \Gamma^- - \Gamma^{++} + \frac{1}{3}(\Gamma^0 - \Gamma^+)$$

can be determined from the data.

There exist many models for estimating baryon mass differences. Simply assuming that u and d quarks differ slightly in mass but have isospin-

**Table 1**    Mass and width differences (in MeV) from a Breit-Wigner fit to $\delta\sigma_T$ with and without Coulomb distortion corrections

| Coulomb distortion | $\delta M$ | $\delta\Gamma$ |
|---|---|---|
| Corrected | $4.6 \pm 0.2$ | $3.6 \pm 0.3$ |
| Uncorrected | $3.2 \pm 0.2$ | $7.1 \pm 0.3$ |

invariant interactions leads one to a simple quark-bond counting estimate of

$$M^0 - M^+ = m_n - m_p = 1.3 \text{ MeV}$$

and

$$M^- - M^{++} = 3(M^0 - M^+) = 3.9 \text{ MeV}.$$

Bag model estimates give similar results. If one associates these mass estimates with the mass values in the Breit-Wigner formula (the real part of the resonance pole position in the complex energy plane), then the theoretical estimate of $\delta M = 4.3$ MeV agrees reasonably well with the value of 4.6 MeV in Table 1. Thus, the breaking of charge symmetry in the difference of $\pi^{\pm}d$ scattering appears to follow quark model expectations.

## 2.3  Neutron-Neutron Scattering Length

The interaction of two physical nucleons is not invariant under rotations in isotopic-spin space (37). This is most evident at very low energy where the scattering lengths magnify differences in the nucleon-nucleon (NN) interactions. The most obvious manifestation is in the Coulomb force acting between two protons, but this is by no means the only one. The mass difference between charged and neutral pions that mediate the nuclear force (38, 39), the mixed isospin character of certain mesons such as the $\rho$ and $\omega$ or the $\pi^0$ and $\eta$ (40–44), and the differing coupling constant normalizations that arise from the breaking of SU(3) symmetry (45, 46) all contribute to a lack of isospin invariance in the NN interaction. If we use charge-independence breaking and charge-symmetry breaking to refer to differences in the NN interactions after removal of the Coulomb effect in proton-proton scattering, then a striking example of the lack of charge independence is afforded by the difference between the Coulomb-corrected proton-proton scattering length (38, 47, 48) $a_{pp}^C = -17$ fm and the $^1S_0$ neutron-proton scattering length (47, 48) $a_{np}^s = -23.7$ fm. The case for charge-symmetry breaking is much less clear cut, because the consensus value (49) of $a_{nn} = -16$ to $-17$ fm from kinematically complete nd → nnp measurements was close to $a_{pp}^C$. It was not until the measurements of $a_{nn}$ in radiative pion capture by deuterium, leading to a final state in which there are only two strongly interacting particles, was pushed to its ultimate precision that charge-symmetry breaking was directly established in the strong NN interaction.

Several three-body reactions have been used in the effort to determine $a_{nn}$. The third particle has been a photon ($\pi^-$d), a proton (nd), a deuteron (nt), a $^3$He nucleus (dt), and an alpha particle (tt). In each case the final-state nn interaction produces a significant enhancement in the spectrum near the maximum energy of the third particle (i.e. at low nn relative energy). Unfortunately, except for the first case cited, the interactions of the third particle with the nn pair affect the resulting spectrum. A proper theoretical treatment is complicated. On the other hand, the $\pi^- d \to \gamma nn$ reaction is an ideal way to determine the low-energy nn scattering parameters, because it is free of such final-state complications. The analysis is actually simplified because the final state contains one zero-mass photon and two equal-mass neutrons, and the photon energy is uniquely related to the nn relative momentum (50, 51).

Because of resolution difficulties, three early measurements of the photon spectrum (52–54) each yielded sizeable uncertainties in the extracted value of $a_{nn}$. The kinematically complete measurement of Haddock et al (55, 56) led to an estimate of $a_{nn} = -16.7 \pm 1.3$ fm, where the quoted uncertainty is essentially statistical. The innovative Lausanne experiment (57–59) used the radiative nucleon capture reaction with a monoenergetic line at 129.404 MeV to provide the required stable energy scale and response function of the spectrometer. This permitted extraction of a precision value of $a_{nn}$ from a comparison of the $\pi^- d \to \gamma nn$ and $\pi^- p \to \gamma n$. The resolution in the experiment of 720 keV at a peak energy of 130 MeV corresponds to an uncertainty of only $\pm 5$ MeV/$c$ in the relative momentum of the two neutrons at 38 MeV/$c$ (see Figure 2).

The final estimate of the nn scattering length is (59)

$$a_{nn} = -18.5 \pm 0.4 \text{ fm}.$$

The uncertainty is a one-standard-deviation estimate. The value is unambiguous, independent of the energy range used in the analysis, and independent of the model used to generate the theoretical comparison spectrum (50, 59). It is incompatible with values obtained from hadronic three-body final-state reactions and falls outside of the range of the generally accepted value of $a_{pp}^C = -17.1 \pm 0.3$ fm. Thus, there appears to be direct evidence for charge-symmetry breaking in the strong NN force, although the sign differs from what one might infer from an analysis of the $^3$H-$^3$He binding energy difference (60–62).

This same radiative pion capture experiment of Gabioud et al (57–59) also yields a relatively precise value for the effective range $r_{nn} = 2.80 \pm 0.11$ fm. The result is compatible with the value extracted from pp scattering (38, 39, 47, 48) of 2.84 fm. It is also consistent with more recent results from the analysis of quasi-free nd $\to$ pnn scattering (63–65).

The uncertainty in the radiative pion capture technique is much smaller than that obtained by means of hadronic reactions involving more than two strongly interacting particles in the final state. The measurement

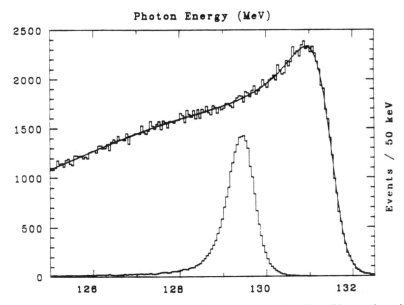

*Figure 2*  Photon spectra from capture on hydrogen and deuterium. The solid curve through the deuterium spectrum was obtained by convoluting the measured hydrogen spectrum (shown) with the theoretical spectrum calculated using $a_{nn} = -18.5$ fm and $r_{nn} = 2.80$ fm. This procedure corrects for the lack of perfect resolution of the spectrometer.

appears to have provided definitive values for the low-energy neutron-neutron scattering parameters.

## 3.  NUCLEAR STRUCTURE

The study of classical nuclear structure has a long history. Detailed comparison of many individual states with the best (debated) theoretical models for each of these states is involved. Needless to say, we do not attempt such a study here but instead illustrate the unique properties of the pion that make it a valuable probe for nuclear structure investigations. While an improved understanding of reaction mechanisms is leading to a more profound knowledge of the hadronic environment, as discussed in Section 4, the traditional structure of nuclear states has been investigated using features so obvious that little or no sophisticated theoretical treatment is necessary.

Pion nuclear structure investigations have been, to a large extent, pursued in the classical mode, i.e. delineating aspects of nucleonic degrees of freedom. While recent attempts have been made to go beyond this traditional picture, the investigation of nuclear structure in terms of its recognized principal degrees of freedom remains a vital and intellectually challenging endeavor.

The study of the nucleus with pions relies largely on specific characteristics of the probe:

1. The isospin properties of the pion are most useful in this regard. The $\pi^+$ couples strongly to the proton and only weakly to the neutron, at least in the region of the delta resonance. Since the $\pi^-$ behaves in the reverse manner, a strong distinction between neutron and proton properties is possible.

2. The existence of three charge states of the pion permits the double-charge-exchange reactions, which ensure that a minimum of two nucleons participate in the process.

3. Because $\ell = 0$ and $\ell = 1$ pion-nucleon partial waves dominate in the energy region considered, and no spin flip is possible for the $\ell = 0$ amplitude, the very simple properties of the p-wave spin-flip amplitude are easily identified experimentally.

We have selected eight examples illustrating how pion-nucleus interactions have expanded our knowledge of nuclear structure. Neutron radius determinations have contributed to our knowledge of ground-state properties. Excited-state investigations have uncovered new information about giant resonances, collective states, spin-flip transitions, and non-spin-flip excitations. Double charge exchange has probed short-range correlations and has provided an effective means of producing rare isotopes. It is in the investigation of the structure of nuclei that one finds the largest number of simple examples of the application of pion-nucleus interactions to enhance our understanding of nuclear physics.

## 3.1  Neutron Radii of Nuclear Ground States

To date, pion-nucleus scattering theories have not been sufficiently well developed to allow extraction of absolute neutron radii. However, in those cases in which an isospin-zero nucleus lies nearby, a relative measurement can be made.

This has been done for the pairs $^{12}$C-$^{13}$C and $^{16}$O-$^{18}$O with elastic scattering of low-energy pions (66). Using the electron scattering measurements to define the proton body radius, and assuming the neutron radius to be the same as the proton for the isospin-zero member of the pair, one obtains values of $r_n^{12} = r_p^{12} = 2.31$ fm; $r_n^{13} = 2.35 \pm 0.03$ fm for carbon and $r_n^{16} = r_p^{16} = 2.60$ fm; $r_n^{18} = 2.81 \pm 0.03$ fm for oxygen.

Similar experiments and analyses with the low-energy comparisons mentioned above were performed (67) to measure the proton difference $^{12}$C-$^{11}$B. The value obtained is $0.07 \pm 0.021$ fm in agreement with, but more accurate than, a value obtained from pionic x-ray studies (68) of $0.09 \pm$ 0.04 fm. This experiment was carried out primarily to verify the accuracy of the method for measuring neutron radii, but in this case it actually gives a value of the proton radius difference that is more accurate than electromagnetic studies.

The oxygen isotopes have also been studied by means of total cross-section measurements (69). There the neutron-proton radius difference in $^{18}$O is seen to be $0.19 \pm 0.02$ fm, in agreement with the above result. This agreement occurs in spite of the fact that the elastic scattering measurement was made at 30-MeV pion kinetic energy and the total cross-section measurement sampled the energy region around the (3,3) resonance and above.

A relatively complete study of the $^{18}$O-$^{16}$O pair was completed (70) using all of the data in the higher energy region available at that time. The authors concluded that the results are best expressed as a set of inequalities:

$$r_p^{18} \lesssim r_p^{16} + 0.05 \text{ fm}; \; r_n^{18} \lesssim r_n^{16} + 0.15 \text{ fm}; \; r_n^{18} \gtrsim r_p^{16} - 0.06 \text{ fm}.$$

It is necessary to include in the analysis the effects of energy variation due to the Coulomb potential [i.e. the fact that upon entering the nucleus the $\pi^+$ ($\pi^-$) has lost (gained) the energy that came from the Coulomb barrier, and the strong amplitudes are modified accordingly]. The authors make this correction with a simple constant shift. Figure 3 demonstrates the importance of this effect.

Comparisons with several specific density models have been made. Harmonic oscillator densities (71) ($r_n^{16} = r_p^{16} = 2.529$; $r_p^{18} = r_p^{16}$; $r_n^{18} = 2.670$ fm) are successful in reproducing the data. While Hartree-Fock densities based on a Skyrme III force (72) lead to the same conclusion in terms of relative radii, the absolute radii appear to be 2–3% too large.

The calcium isotopes have also been studied extensively at energies around the delta resonance. These analyses must cope with the problem that, with the stronger interaction and the larger nucleus, only the surface region is sampled. Hence, a neutron radius measurement in a pure sense is not possible. Nonetheless, pion-nucleus data can be analyzed (69) in these terms, and total cross-section comparisons give a small $^{48}$Ca-$^{40}$Ca neutron radius difference ($\sim 0.05$ fm) in agreement with alpha-particle (73) and proton (74) scattering as well as Coulomb energy differences (75). All of these extracted radii disagree with state-of-the-art Hartree-Fock calculations (76).

Angular distributions of pion elastic scattering from the calcium isotopes have also been measured (77). A difference in "black disk" $\pi^+/\pi^-$ radii for $^{48}$Ca of 0.30 fm is obtained. However, since the measurement is only

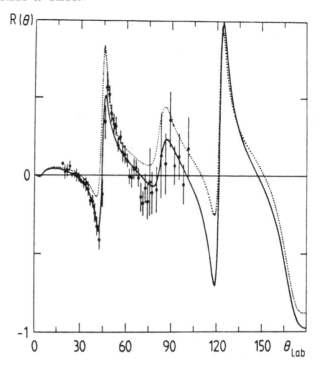

*Figure 3*  The charge asymmetry $R(\theta)$ $\{\equiv [\sigma^-(\theta) - \sigma^+(\theta)]/[\sigma^-(\theta) + \sigma^+(\theta)]\}$ for $^{18}O$, where $\sigma^\mp(\theta)$ is the differential cross section for the elastic scattering of $\pi^\mp$ from $^{18}O$. The solid curve corresponds to the use of a Coulomb energy shift in the evaluation of the amplitudes, and the dotted curve corresponds to its neglect.

sensitive to the tail of the density, it is not clear that the difference comes from a variation in radius rather than diffuseness (78, 79).

Sensitivity to the neutron density only in the region beyond the half-density radius can be an asset, since differences in density are magnified in this region. However, to analyze the measurements, one must deal with other quantities, such as the density at some fixed radius, and make the comparison directly with the corresponding theoretical quantity. The calcium isotopes were recently analyzed with this method (80).

A number of neutron radii have been measured (relative to proton radii) and a general disagreement with Hartree-Fock calculations is suggested in several cases.

## 3.2  *Giant Isovector Monopole Resonance*

The subject of collective degrees of freedom in the nuclear medium has been of intense interest to nuclear physicists since the discovery of the giant dipole (or Goldhaber-Teller) resonance (81), the collective vibration of protons against neutrons depicted schematically in Figure 4a. Predictions of other giant resonances, such as the isoscalar and isovector monopole and quadrupole resonances, followed. Giant resonances are characterized by (a) the concentration of a large transition strength having a specific multipolarity ($L = 0$ for monopole, $L = 1$ for dipole, $L = 2$ for quadrupole, etc) in a relatively narrow region of excitation energy; (b) the existence of such structure for many nuclei; and (c) a smooth variation of the excitation energy and width of the resonance with mass number $A$. It was not until the $(\pi^-, \pi^0)$ single-charge-exchange reaction was elevated to the status of a nuclear structure probe (because of the construction of a special $\pi^0$ spectrometer) that the giant isovector monopole resonance (82) was identified experimentally (83) (see Figure 5). This collective excitation of the nucleus, in which the protons and neutrons oscillate out of phase along the radius of the nucleus in a compression mode (the protons breathe in while the neutrons breathe out), is illustrated in Figure 4b.

Why had this nuclear collective degree of freedom been missed by other probes? The alpha particle is, of course, an isospin-zero particle and hence excites only the giant isoscalar monopole resonances (84). The (n,p) charge exchange reaction (85) excites numerous spin-flip transitions, which severely hinder the data analysis. The failure of electron scattering (86) to see the isovector monopole is more subtle. Because the electron is a weak

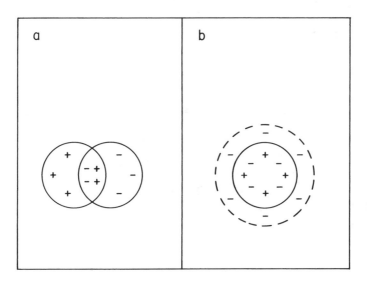

*Figure 4*    The motion of a dipole isovector type (a) and a monopole isovector type (b). Here "+" represents proton matter and "−" represents neutron matter.

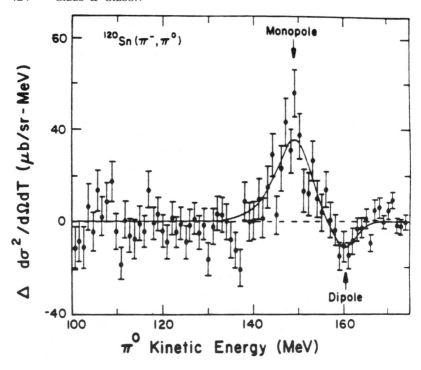

*Figure 5* Differences in the 4.5° and 11.0° double differential cross section for the reaction $^{120}$Sn$(\pi^-, \pi^0)$ at $T_{\pi^-} = 165$ MeV as a function of the kinetic energy of the outgoing $\pi^0$. The positive excursion is the isovector monopole resonance and the dip is the well-known giant dipole resonance.

probe and therefore not strongly absorbed, it interacts uniformly with the entire volume of the nucleus. Because of this feature, the electron's differing interaction with protons and neutrons easily excites the giant dipole motion shown in Figure 4a. However, the electron's interaction with the nucleus averages to zero when one considers exciting the isovector monopole motion shown in Figure 4b. This is due to the fact that the volume integral of the isovector monopole transition density (and hence the cross section near $q = 0$) vanishes since excited monopole states are orthogonal to the ground state.

On the other hand, the $(\pi^-, \pi^0)$ single-charge-exchange reaction is ideally suited for exciting the giant isovector monopole resonance (87). First, the $(\pi^-, \pi^0)$ reaction eliminates the possibility of $T = 0$ isoscalar excitations, which are dominant when the projectile interacts strongly with all nucleons in the target as is true in $(\pi, \pi')$ reactions (Figure 6). Thus, $T = 1$ isovector

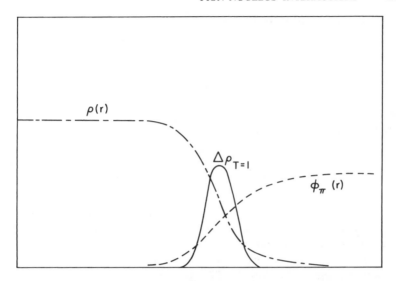

*Figure 6* Graph showing how the overlap of the nuclear density and the strongly attenuated pion wave function cause the reaction to take place in the surface of the nucleus.

transitions are brought to the fore. (Any $T = 2$ transitions are an order of magnitude smaller.) Second, the $(\pi^-,\pi^0)$ reaction lowers the energy of the excited state by the Coulomb displacement energy, as can be seen in Figure 7, and therefore narrows the isovector monopole resonance compared to that seen via $T = 0$ transitions [e.g. in the $(\pi,\pi')$ reaction] in the target nucleus. At forward angles, where the isovector monopole cross section peaks, the pion single-charge-exchange reaction excites primarily non-spin-flip transitions. This avoids the analysis difficulty associated with the (n,p) reaction. Last, the strong absorption of pions in the nuclear interior

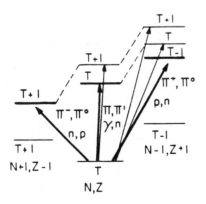

*Figure 7* Level scheme displaying the various charge exchange reactions.

for pion energies near the (3,3) resonance insures that the surface nature of the pion-nucleus interaction does excite the isovector monopole mode, that the interaction does not average to zero as in the case of electron scattering. The surface-peaked $T = 1$ transition density accounting for the excitation of the motion illustrated in Figure 4b meshes perfectly with the overlap of the nuclear density (which, for purposes of this discussion, may be thought of as a Fermi function) and the pion wave function (which dies rapidly as the pion penetrates the nucleus). This is illustrated in Figure 6.

It is interesting to note that the properties of the pion-nucleus interaction are sufficiently well understood that the consequences discussed above were appreciated *before* (88) the actual discovery of the isovector monopole state (89).

Pion single-charge-exchange excitation of the giant isovector monopole resonance is another example of the utility of the complementary nature of this versatile probe. Experiments on targets from $^{40}$Ca to $^{208}$Pb established that this structure exhibits the general features required of giant resonance phenomena. Angular distributions follow the expected $L = 0$ diffraction pattern. The cross section exhausts a substantial fraction of the sum rule prediction (90), which confirms the collective nature of the resonance. The excitation energies and widths decrease with increasing mass, in agreement with RPA calculations. The cross sections for the $T+1$ components of the resonance decrease with increasing isospin (as does the giant dipole resonance), consistent with the expected influence of Pauli blocking in the final $(N+1, Z-1)$ final state.

Of more than passing interest is the fact that the isovector quadrupole resonance, predicted by RPA calculations to be easily observable with the same technique, was not found at the level of an order of magnitude below that expected. Recent work (91) suggests that the fault may lie with the inadequacy of the NN interaction model used.

These results are expected to influence our modelling of the isovector part of the effective nucleon-nucleon interaction in nuclei and, therefore, our understanding of the symmetry energy in nuclei. They should also improve our understanding of isospin mixing, Coulomb displacement energies (92), and widths of analog states.

## 3.3   *Proton/Neutron Selectivity in $\pi^+/\pi^-$ Scattering*

An outstanding feature of the pion-nucleon interaction is the ratio $R = \sigma(\pi^+ p)/\sigma(\pi^- p)$ [or $\sigma(\pi^- n)/\sigma(\pi^+ n)$] in the scattering of $\pi^+$ and $\pi^-$ from free protons [neutrons] at pion energies near the $(J = 3/2, T = 3/2)$ or (3,3) pion-nucleon resonance at 180 MeV. Because of this feature, comparisons of $\pi^+$ and $\pi^-$ inelastic scattering should provide a sensitive method of separating proton and neutron components of inelastic transitions. Striking examples of this include the discovery of the essentially pure proton and pure neutron transitions in $^{12}C$ and $^{13}C$ (93, 94).

Isospin mixing had been observed by other means in systems such as $^8Be$, where the $2^+$ states at 16 MeV are strongly mixed ($T = 0$ and $T = 1$) states leading to isospin-forbidden $\alpha$-decay widths (95, 96). However, it came as a surprise when essentially pure proton and pure neutron state transitions were discovered in $^{12}C$. The spectra for $\pi^+$ and $\pi^-$ inelastic scattering from $^{12}C$ at an incident pion kinetic energy of 162 MeV and at a laboratory scattering angle of 70° are depicted in Figure 8, along with the difference spectrum (93) showing a strong bipolar shape near 19.5 MeV, exactly as one would expect from strong isospin mixing. The proton state, most strongly excited in $\pi^+$ scattering, can be seen at 19.25 MeV. The neutron state, most strongly excited in $\pi^-$ scattering, can be seen at 19.65 MeV. These data clearly established the role to be played by $\pi^+$ and $\pi^-$ scattering in the determination of isospin mixing among inelastic transitions in nuclei.

Similar striking results were observed in the case of the non-self-conjugate nucleus $^{13}C$ (94). The spectra for $\pi^-$ and $\pi^+$ scattering from that nucleus at an incident pion kinetic energy of 162 MeV and at laboratory scattering angles of 62°–86° are shown in Figure 9, along with the difference spectrum. In this example, the $\pi^+$ excitation of the $9/2^+$ state at 9.5 MeV is so weak that it can barely be discriminated from the background, whereas the $\pi^-$ excitation of this state is quite strong. The identification of this state as a pure neutron state seems clear. Furthermore, it is to be expected from a weak coupling picture, in which a $d_{5/2}$ neutron is coupled to a $^{12}C$ ($2^+$) $T = 0$ core. Shell model calculations yield states having a large overlap with such a weak coupling state (97).

The $\pi^+/\pi^-$ comparisons have also been exploited to examine isospin structure in other p-shell nuclei. A triplet of isospin-mixed $4^-$ particle-hole states was discovered in $^{16}O$ lying between 17.8 and 19.8 MeV (98). The isospin structure of the $^{13}C$ spectrum ranging from single states at 3.68 MeV to groups of states at 21.6 MeV was investigated, with comparisons made to distorted wave impulse approximation (DWIA) predictions based upon shell model wave functions (99). Evidence for strong isospin mixing to the states near 21 MeV was found (94, 99). In $^{14}C$, ratios of $\pi^+$ and $\pi^-$ cross sections were found (100) to exceed the free nucleon ratio of 9 for the first time; that is, the pion-nucleus ratios were larger than for pure proton and pure neutron states, as one can seen in Figure 10. Strong cancellations of neutron and proton transition amplitudes connecting to the $2^+$ states at 7.01 and 8.32 MeV are the cause of this phenomenon. One observes constructive interference between the proton and neutron

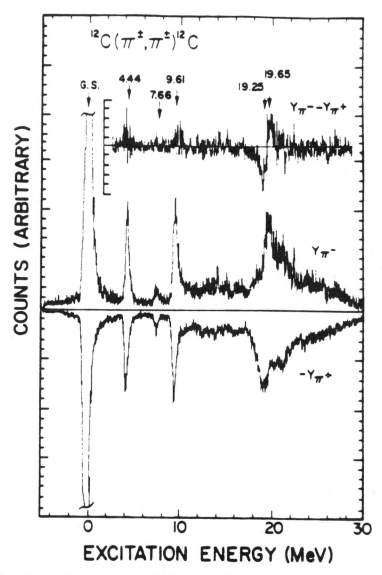

*Figure 8*   Inelastic scattering spectra for $\pi^+$, $\pi^-$, and the difference spectrum for $^{12}$C showing how the isospin-mixed states can be seen directly in the data.

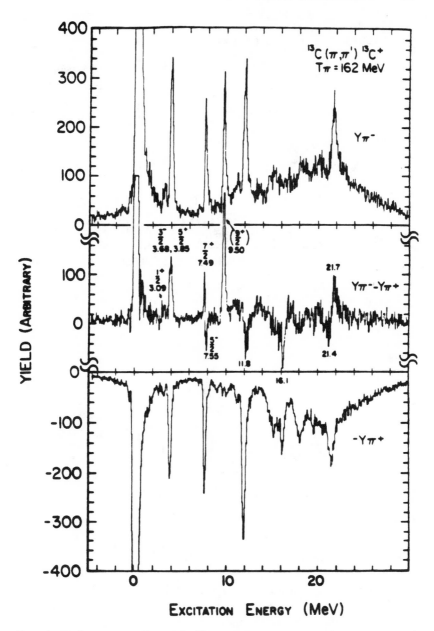

*Figure 9*   Similar spectra to Figure 8 for $^{13}$C. Note the strong neutron character of the $9/2^+$ state at 9.5 MeV.

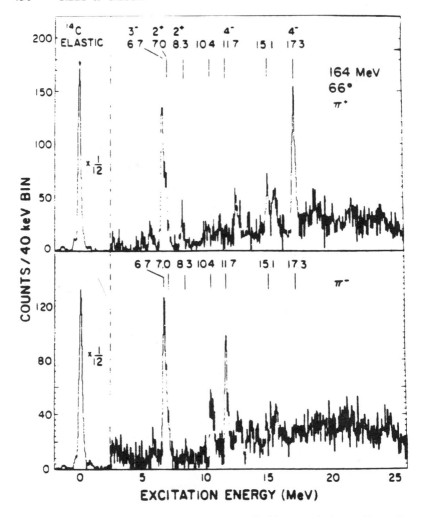

*Figure 10*    The $\pi^+$ and $\pi^-$ inelastic scattering spectra for $^{14}$C. Note the large $\pi^+/\pi^-$ ratio at 17.3 MeV.

components in the transition from the ground state to the $2_1^+$ excited state and destructive interference for the transition to the $2_2^+$ excited state.

Such $\pi^+$ and $\pi^-$ scattering studies indeed proved to be a superb tool for investigating the isospin structure of p-shell transition amplitudes. Furthermore, it was just such investigations of isospin structure of the excited-state transitions in the p shell that led to the realization that inelastic pion scattering was an excellent tool for identifying high-spin stretched states in light nuclei (see Section 3.5).

## 3.4   Collective States in Nuclei

In inelastic $\pi^+$ and $\pi^-$ scattering to the first $2^+$ state in $^{18}$O it was observed (101–103) that the $\pi^-$ angular distribution is essentially the same as the $\pi^+$ except for a scaling factor. Measurements showed this factor to be $1.58 \pm 0.13$, $1.58 \pm 0.15$, and $1.86 \pm 0.16$ for 230, 180, and 164 MeV respectively. At the same time, the ratio for the first $2^+$ state in the isospin-zero nucleus $^{28}$Si was (102) $0.85 \pm 0.06$, so that evidently one would expect the enhancement observed in the $^{18}$O to be little affected by distortion differences between the $\pi^+$ and $\pi^-$.

One can approach the interpretation of the result from two points of view. If the low-lying states in $^{18}$O are considered to be made up purely of rearrangements of the valence neutrons, then the ratio should be the square of the $\pi^+$n to $\pi^-$n amplitude $\sim 3^2 = 9$. Since the ratio is considerably less than this, the protons and neutrons in the $^{16}$O core must be participating in the transition as well. These can be brought into the picture by including core excitation, and it was found (104) that it was possible to reduce the ratio to $\sim 2.5$ in this manner.

A second approach is to consider the state to be completely collective. If *all* nucleons participate *equally*, then the ratio should be $\sim (3N+Z/3Z+N)^2$ or 1.25. From this viewpoint the measured ratio is too large, so that only some fraction of the nucleons must participate. This is the usual problem in the interpretation of collective nuclear states, and one common method of treatment is to assume that each valence neutron (proton) looks partly like a proton (neutron) because of its coupling to the nucleons in the core. This leads to the concept of an "effective charge" (the idea originated with electromagnetic probes) and has a thorough theoretical foundation (105). The same physical quantities (neutron and proton transition matrix elements) can be obtained using combinations of other probes as well, and the comparison of these various methods using several pion measurements (106, 107) finds general consistency (108).

A distinct advantage of using pion beams is that the full neutron-proton comparison is made using only pions, and the probe dependence (109), arising from a variation (among different probes) of the region of the nucleus being sampled, is avoided. It is, of course, recognized that the pion, like other probes, measures the neutron or proton transition strength over some specific region of the nucleus.

These experiments were done in the resonance region and hence are sensitive to only the surface of the nucleus. One might expect that measurements made at lower pion energies would sample the entire nuclear volume.

In fact $\pi^+/\pi^-$ comparisons have been made at 50 MeV for $^{18}$O (110), $^{26}$Mg (110), $^{30}$Si (111), and $^{34}$S (112) and the ratios of neutron to proton transition matrix elements compared with the lifetime measurements on mirror nuclei (111). The results agree reasonably well (except possibly for $^{30}$Si) given the spread in the latter (113).

For heavier nuclei, the structure is naturally expressed in terms of different types of collective motion of the entire nucleus, such as vibration or rotation of deformed shapes. Recent progress in understanding the relationship between this very collective picture and the microscopic nucleon picture has been largely within the framework of the interacting boson model (IBM). In this approximation the neutron-neutron and proton-proton pairing force is used to reexpress the structure of the valence nucleons as $J = 0$ and $J = 2$ bosons, which then interact with each other. The core nucleons participate in the structure through effective charges induced by the valence particles. The IBM is able to represent many types of collective motion as well as the evolution of the character of nuclear structure that typically occurs with the variation of the number of valence nucleons outside a closed shell.

The transition strengths to the first $2^+$ state in Pd isotopes were recently measured using both $\pi^+$ and $\pi^-$ beams (114). Analysis (115) of this data has shown the IBM to give the correct dependence on the number of valence nucleons and, perhaps more significantly, the effective charges were extracted. With both $\pi^+$ and $\pi^-$ measurements available, the ratio of the neutron to proton effective charge can be determined independently of the character of the collective state. (Using only the $\pi^+$ results, the ratio depends strongly on the assumed character of the collectivity.) These studies are limited by the experimental resolution presently available.

For giant resonances, resolution is not a problem. The majority of the studies on giant resonances have been made on the giant quadrupole (GQR) and, while the others provide interesting comparisons, we restrict our discussion to this case. The GQR was first observed with pions, using $\pi^+$ only, in $^{40}$Ca (116) and $^{89}$Y (117). Since that time studies have been carried out with both charges on $^{40}$Ca and $^{118}$Sn at 130 MeV (118), $^{90}$Zr and $^{118}$Sn at 164 MeV (119), and $^{208}$Pb and $^{238}$U at 162 MeV (120, 121). These results are summarized in Table 2.

Again these ratios are larger than can be explained by collective models (122) but much smaller than would be expected if the valence neutrons alone participated, i.e. $9:1$.

While it is often said that direct comparison of $\pi^-/\pi^+$ inelastic scattering is the best way to separate neutron and proton degrees of freedom, this statement does not fully convey the importance of such investigations to the total understanding of states of high collectivity. As seen by the ex-

**Table 2**   Ratio of $\pi^+$ to $\pi^-$ cross sections for excitation of the giant quadrupole resonance in several nuclei

| Isotope | $\sigma(\pi^-)/\sigma(\pi^+)$ | Theory (122) |
|---|---|---|
| $^{40}$Ca | 1.05 | 1.19 |
| $^{90}$Zr | 1.65 | 1.22 |
| $^{118}$Sn | 2.20 | 1.61 |
| $^{208}$Pb | 2.70 | 1.70 |
| $^{238}$U | 3.00 | — |

amples presented here, the principal question revolves around the relative degree of participation of "valence" vs "core" particles. In fact, in some sense, it is to define the active shell model space for each transition. Since the least bound nucleons are usually the neutrons, their participation is most important and a definitive technique for directly delineating neutron degrees of freedom is crucial.

## 3.5   Spin Excitations in $(\pi,\pi')$ Reactions

An unanticipated feature of pion-nucleus scattering, which became evident as the effort to utilize $\pi^+$ and $\pi^-$ inelastic scattering to explore isospin mixing in nuclei increased, is the strong excitation of high-spin stretched states. (Here stretched states encompass those whose total angular momentum is the maximum that can be attained by means of a one-particle–one-hole excitation.) Such transitions have been observed in $(\pi,\pi')$ experiments in almost all p-shell nuclei and in $^{28}$Si (97, 99, 100, 123–127).

The large asymmetry in $\pi^+/\pi^-$ scattering discussed in Section 3.3 is due to the nearly maximal isospin mixing between $T = 0$ and $T = 1$ states to form essentially pure proton and pure neutron states. Many of these states were later identified to be M4 excitations, the angular distributions being well described by transition densities having $\Delta J = 4$ (orbital angular momentum transfer $\Delta L = 3$ and spin transfer $\Delta S = 1$) (128). It was, in fact, observed that cross sections for natural-parity transitions rise dramatically as the pion kinetic energy is increased, in contrast to a decrease for unnatural-parity transitions (129). This results from the fact that only s- and p-wave pion-nucleon interactions are important and the p wave is dominant. Angular distributions are $\cos^2 \theta$ and increasing for non-spin-flip transitions; they are $\sin^2 \theta$ and decreasing for spin-flip transitions (unnatural-parity states) (130), as is illustrated for $^{12}$C in Figure 11. While the shape of the form factor and projectile distortion effects can sig-

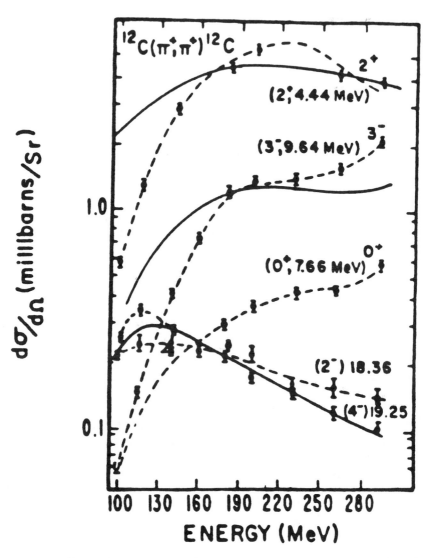

*Figure 11* Energy dependence of inelastic pion scattering at fixed-momentum transfer showing the discrimination of natural- from unnatural-parity states. The natural-parity states show a monotonic rise with energy while the unnatural-parity state cross sections decrease with energy, for the reasons given in the text. The dotted curves are to guide the eye.

nificantly alter the angular distribution, making measurements at a fixed momentum transfer nullifies the form factor effect and reduces significantly the distortion effect. The $\sin^2 \theta$ dependence of the $2^-$ (18.36 MeV) and $4^-$ (19.25 MeV) states of $^{12}C$ in the $(\pi, \pi')$ spectra confirmed their stretched nature.

A similar analysis of $^{13}C$ shows the 9.5-MeV $9/2^+$ state to be a $4^-$ excitation (99, 131). The $\Delta S = 1$ nature of the excitation function confirmed the assignment deduced from angular distribution data. (The stretched nature of the transition limits it to only one allowed orbital angular momentum transfer, which is not otherwise generally true.) Several $4^-$ excitations in $^{13}C$ were identified and thereby proved the value of the $(\pi, \pi')$ reaction to be complementary to $180°$ electron scattering. Electron scattering is sensitive to $\Delta T = 1$ transitions, whereas the pion excites both $\Delta T = 1$ and $\Delta T = 0$ transitions. More importantly, the $\pi^+$ and $\pi^-$ comparisons are sensitive to relative proton and neutron contributions. For example, the first $9/2^+$ state (at 9.5 MeV) is a pure neutron excitation, while the second (at 16.1 MeV) is mostly proton. Both states are excited in $180°$ (e,e'). A third $4^-$ state, about equally excited by $\pi^+$ and $\pi^-$ and therefore probably a pure isospin state, is not seen in (e,e'), which indicates that it is likely at $\Delta T = 0$ transition.

Two well-identified $4^-$ states are seen in $^{14}C$ (100, 132). The lower state (at 11.7 MeV) is $\pi^-$ enhanced; the upper state (at 17.3 MeV) is $\pi^+$ enhanced. However, model calculations indicate that the 11.7-MeV state as well as a third $4^-$ state (at 15.2 MeV) may contain appreciable three-particle–three-hole components (124). An M4 transition to a $5^-$ state in $^{14}N$ has been identified (123). In this case, approximately equal $\pi^+$ and $\pi^-$ excitations indicate that it is a relatively pure isospin state. M4 transitions have also been identified in $^{15}N$ (126) and $^{16}O$ (98). Stretched states exist in heavier systems. Pion excitation of these states has been investigated thoroughly only for the $^{28}Si$ nucleus (125). Angular distributions for the $6^-$ $T = 0$ and $T = 1$ states as well as for a $T = 0$ $5^-$ state are shown in Figure 12. The $6^-$ states represent spin-flip and spin-isospin-flip excitations in the s-d shell, respectively. The extracted spectroscopic factors for the $6^-$ states are consistent with those from (p,p') data analysis (132).

The $(\pi, \pi')$ reaction has been found to be a selective probe of high-spin, unnatural-parity states. The high-momentum-transfer excitations mesh well with the strong backward scattering of the pion. The unique signature of the $\Delta S = 1$ transitions is of great assistance in establishing their identity. Because the pion excites both $\Delta T = 0$ and $\Delta T = 1$ transitions, the $(\pi, \pi')$ data complement data from (e,e'), and a combined analysis can be used to disentangle the isospin character of the transitions. Stretched states have been identified throughout the p shell and in $^{28}Si$. Similar inves-

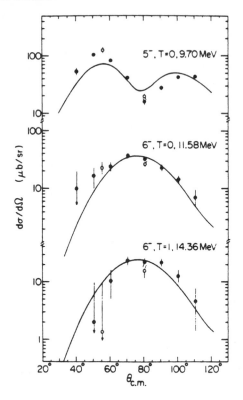

*Figure 12*   Angular distribution for the excitation of three states in $^{28}$Si by $\pi^+$ (*full dots*) and $\pi^-$ (*open circles*).

tigations may well prove useful in attempts to identify spin-mode giant resonances.

## 3.6   *Analog Double-Charge-Exchange Reactions as a Probe of Nucleon-Nucleon Correlations*

Unlike the other sections under nuclear structure, which describe the use of the pion to probe one-nucleon properties of the nucleus, pion double charge exchange requires that (at least) two nucleons participate in the reaction. In the most direct description within the nucleon picture, that of two subsequent scatterings, the pion wave propagating between the two nucleons disperses, hence the reaction should be sensitive to the relative spacing of the nucleon pair.

The most straightforward baseline calculations are those in which one assumes totally uncorrelated nucleon wave functions. These are most often implemented by the use of spherically symmetric, optical model, coupling potentials that are proportional to the difference between neutron and proton densities (133). It was found (134) that these models are qualitatively successful for the highest energies measured to date ($\sim 300$ MeV) but the agreement becomes progressively worse as the energy is decreased until, in the resonance region, there are discrepancies of an order of magnitude, as illustrated in Figure 13.

One can understand that the high-energy double-charge-exchange (DCX) process is not strongly dependent on the relative position of the two nucleons because, at these energies, the single-charge-exchange cross section is very forward peaked. After the first transfer of charge, the pion proceeds along the beam direction, and this (almost parallel) interior neutral pion beam spreads very little with distance. As the energy is decreased, the pion-nucleon charge-exchange (CEX) cross section becomes more nearly isotropic in angle, and the $1/r$ spread of the pion scattering wave function gives a preference for scatterings in which the second nucleon is close to the first.

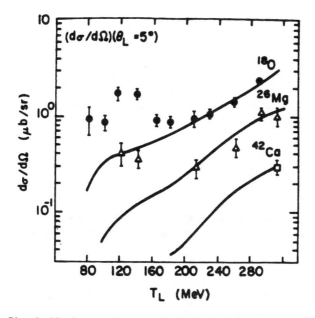

*Figure 13*  Pion double charge exchange to the double isobaric analog for the reactions $^{18}O(\pi^+,\pi^-)^{18}$Ne (*solid circles*), $^{26}$Mg$(\pi^+,\pi^-)^{26}$Si (*open triangles*), and $^{42}$Ca$(\pi^+,\pi^-)^{42}$Ti (*open square*). The solid lines are calculations from (134). Note that the baseline calculation (without nucleon-nucleon correlations) is an order of magnitude below the data at the lower energies.

The features of the reaction can also be discussed profitably in terms of the intermediate states involved and the multipolarity required to reach these states. The relationship, for two-body operators, between ground-state correlations and intermediate excited states has been known for some time (135) but perhaps it is worthwhile reviewing since the discussion of mechanisms in the literature often uses the two languages interchangeably. Consider a pure single-particle model with the two nucleons described by the wave functions $\phi_n(\bar{r}_1)\phi_{n'}(\bar{r}_2)$. Thus we start from $\phi_0(\bar{r}_1)\phi_0(\bar{r}_2)$ and after the first interaction, with particle 1, the intermediate state is $\phi_n(\bar{r}_1)\phi_0(\bar{r}_2)$. After the second interaction, the wave function would be $\phi_n(\bar{r}_1)\phi_{n'}(\bar{r}_2)$ but, since *only* two interactions are being considered and the final state is the ground (double isobaric analog) state, the only possibility is $n = n' = 0$. Thus, the only allowed intermediate state, without correlations and with only two-body interactions, is the single analog state. It then follows that the excitation of an intermediate state, other than the analog, requires some kind of correlation in the description of the nuclear system. Generally speaking, the correlations appear in the intermediate system as well as the initial and final "ground" states. Note that the intermediate analog state route proceeds via a monopole operator, while the excited states usually involve higher multipoles.

If one assumes that the excitation energies of the intermediate states are negligible in comparison with the pion energy, then a closure approximation can be used to eliminate the specific dependence on the intermediate states, and the correlation effect appears only in the initial and final states. Early calculations using closure (136) and those explicitly including several intermediate states (137) demonstrated the approximate equivalence of these two methods.

Note also that in neglecting the nuclear Hamiltonian (the energy of the intermediate excited states) the motion of the nucleons during the time the pion passed from the first to the second nucleon is also neglected. For this reason the closure methods are sometimes referred to as "fixed-nucleon" techniques. For a local pion-nucleon interaction, fixed-nucleon and closure techniques are equivalent, while for a separable interaction the correspondence is only approximate.

Aside from the obvious discrepancy in magnitude shown in Figure 13, there are two other very strong indications that the simple (uncorrelated) picture requires major modification. The first is the variation of the cross section with the number of excess neutrons in a semimagic nucleus. If all neutron pairs are to be treated similarly, then the cross section should be proportional to $n(n-1)/2$, where $n$ is the number of excess neutrons ($n = N - Z$). This was shown not to be the case in measurements at 164 MeV for the calcium isotopes, where the $^{48}Ca/^{42}Ca$ ratio was found (138) to be about 6 instead of the 28 expected from the uncorrelated estimate.

A second indication of the deviation from the uncorrelated model comes from the angular distribution in the resonance region (139–142). The first minimum occurs at an angle much smaller than that of the uncorrelated theory, which gives a diffraction-like pattern with a minimum determined in a strong absorption picture by the nuclear radius (Figure 14). Such clear deviations from the predictions of the simple (uncorrelated nucleon) model have made the study of DCX currently the most active field of research in pion-nucleus physics.

In the resonance region, where these observations were made, the analysis is complicated by the very strong distortion effects and the possibility of multinucleon excitations. Nonetheless it has been suggested (143) that the anomalous forward position of the minimum results from an inter-

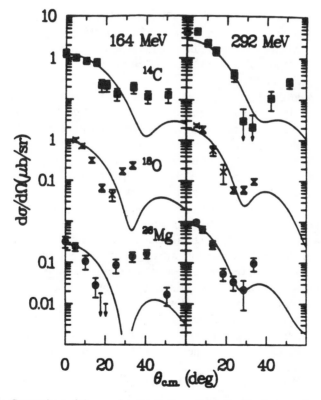

*Figure 14* Comparison of the angular distribution of pion double charge exchange to the isobaric analog state with the baseline calculations showing the anomalous position of the first minimum at 164 MeV.

ference of the "intermediate analog" amplitude with that of another mechanism. This second amplitude has often been associated with DCX on the core nucleus. The relationship between these two amplitudes has been the subject of some study (144).

Since many of the complications that cloud the correlation interpretation of the cross section are associated with the reaction in the (3,3) resonance energy region and since the difference from the baseline comparison seemed to be growing with decreasing energy, it was natural to extend the measurements to lower energy. Indeed the single-charge-exchange cross section to the analog state has a zero-degree minimum at $\sim 46$ MeV, so that the baseline calculation gives a small value and hence maximizes our chances of seeing other terms.

Measurements made near 50 MeV, first at TRIUMF (145) and then at Los Alamos (146), showed that the $^{14}$C DCX cross section was much larger (about two orders of magnitude) than the baseline expectation. Further measurements on $^{18}$O (147) and $^{26}$Mg (148) showed this to be a general phenomenon. Calculations addressing these cross sections have included summing over the intermediate-state wave functions with closure and using (a) pure shell model configurations (149, 150), (b) mixed shell configurations (149), and (c) three-body (the core plus two neutrons) model wave functions with "realistic" nucleon-nucleon potentials (149). Expressing the calculation directly in terms of selected nucleon intermediate states required, results have also been obtained (151, 152). Semiphenomenological calculations using a mixed configuration shell (due to core polarization) (153) and including an isotensor term in the coupled channel formalism (154) have also been performed. While conclusions among these authors vary, it clearly emerges that the correlation coming from the shell model (the requirement that the two valence neutrons on which the reaction takes place are coupled to total spin zero) provides a large fraction of the needed enhancement. How much short-range modification to the nucleon-nucleon wave function is required to match the cross section has not yet been determined. One should note, however, that, at the time of the calculations cited, the free pion-nucleon single-charge-exchange cross sections were not well known. Those values now exist (13, 18) so that the calculations can now be performed with greater certainty.

One way to focus on the role of these shell model correlations is to examine the dependence of the cross section on the number of neutrons outside of a closed shell core. The semimagic isotopes of calcium ($^{42}$Ca, $^{44}$Ca, and $^{48}$Ca) provide an ideal testing ground. Measurements just performed at 35 MeV on $^{48}$Ca (155) and $^{42,44}$Ca (156) show a deviation from the baseline calculation that is even larger than that seen at the resonance. The $^{44}$Ca/$^{42}$Ca ratio is about 1/2, which deviates by a factor of 12 from the "uncorrelated nucleon" prediction of 6. This ratio can be understood provided that the correlation effect is assumed to be important (157).

The present discussion has been totally in terms of pion-nucleon degrees of freedom. Since a large part of the reaction takes place when the two-nucleon spacing is small (closure approximation estimates state that a substantial fraction of the reaction takes place when the two-nucleon spacing is less than 1 fm) (149, 150), one must take into account the interaction with two nucleons simultaneously. This is implemented in the nucleon picture through the use of a pion-nucleon off-shell $t$-matrix. The parameters of these $t$-matrices (there are actually six of them, corresponding to two s and four p waves) are fixed from phenomenological considerations. This phenomenology is at the pion-nucleon, not the pion-nucleus, level. One may choose to look more closely at the origin of these functions, to the meson or quark level. In fact, it might be more efficient to represent the $\pi NN$ interaction directly in terms of these variables. These $\pi NN$ vertices have been treated directly in the language of meson exchange currents (158) and six-quark bags (159).

Evidence for the direct observation of the effect of two-nucleon correlations is provided by deviations of more than an order of magnitude from the uncorrelated model calculations in both absolute value and ratio measurements. A strong sensitivity to correlation effects has been established, and three methods for studying their properties (absolute magnitude, angular distribution, and isotopic ratios) have been unearthed.

## 3.7  Rare Isotope Production in Pion Double Charge Exchange

High-intensity pion beams at the meson factories made possible the utilization of the pion double-charge-exchange reaction as a means of determining nuclear masses for isotopes off the line of stability. The mass of $^{16}$Ne was first determined by means of the $^{16}O(\pi^+,\pi^-)$ reaction (160). Heavy-ion reactions can, of course, be used to reach the same rare isotopes as pion double charge exchange and, in fact, the cross sections are of the same order of magnitude. However, because of the clean pion identification that is possible, pion double charge exchange appeals to many as an efficient method.

The pion experiments have measured masses by comparing the magnetic rigidities of pions from comparable double-charge-exchange reactions. For the calibration reaction, a well-known mass is produced and the final pion energy compared with the reaction of interest, in which an unknown mass is observed. By selecting a calibration reaction having a $Q$ value similar to that of the test reaction, the two can be examined under the same conditions (spectrometer and pion channel settings) to minimize errors resulting from absolute calibration uncertainties.

Two examples of early precision mass measurements using pion double charge exchange are the mass determinations of $^{18}$C and $^{26}$Ne, which used the $(\pi^-,\pi^+)$ reaction on $^{18}$O and $^{26}$Mg, respectively (161, 162). The spectra for $^{18}$O$(\pi^-,\pi^+)^{18}$C and for the calibration reaction $^{12}$C$(\pi^-,\pi^+)^{12}$Be are shown in Figure 15. The ground-state transition is clearly seen with almost no background contamination. The resulting mass excesses of 24.91(15) MeV and 0.44(7) MeV provided stringent tests for model mass predictions of (isospin) $T_z = -3$ nuclei.

The $T_z = -2$ isotopes $^{12}$O, $^{16}$Ne, $^{24}$Si, and $^{32}$Ar were also investigated, using the $(\pi^+,\pi^-)$ reaction on targets of $^{12}$C, $^{16}$O, $^{24}$Mg, and $^{32}$Si, respectively (163). The following mass excesses were obtained: 32.059(48) MeV for $^{12}$O, 24.051(45) MeV for $^{16}$Ne, 10.682(52) MeV for $^{24}$Si, and $-2.181(50)$ MeV for $^{32}$Ar. These may be compared with the results of the $(^4$He, $^8$He) heavy-ion reaction of 32.10(12) MeV for $^{12}$O and 23.92(8) MeV for $^{16}$Ne done at a comparable time (164). The mass excesses were important additions to the data base used for testing the mass formulae

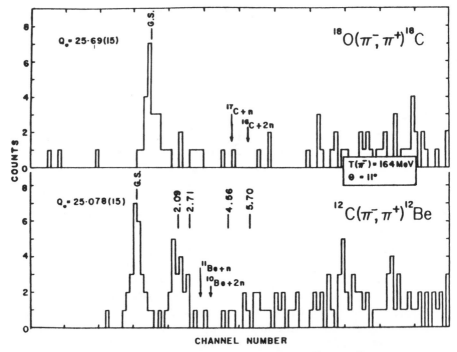

*Figure 15*   Comparison of DCX spectra leading to $^{12}$Be and $^{18}$C.

for isotope quintets. The masses were also used to determine the coefficients of the isobaric mass multiplet equation for the $T = 2$ systems.

The $(\pi^+,\pi^-)$ double-charge-exchange reaction was also employed to obtain a value for the mass of $^{58}$Zn, the heaviest $T_z = -1$ nucleus ever measured (165). The ($^4$He, $^8$He) heavy-ion reaction is not possible here, because the $^{62}$Zn target does not exist. Pion double charge exchange provided the only means of reaching $^{58}$Zn. [The calibration was made against the known $^9$Be$(\pi^+,\pi^-)^9$C reaction.] A mass excess of 42.295(50) MeV was found. This mass value was used in the transverse Garvey-Kelson mass excess formula to estimate the mass excess of $^{57}$Cu, a value that agrees well with the most recent measurement (166).

The clean discrimination of the pion double-charge-exchange reaction against other particles makes the reaction an excellent tool for investigating light nuclei near the neutron drip line. This is especially true for unbound systems, where background peaks from excited states of detected particles, etc are very undesirable. An example is the pion double-charge-exchange study confirming that the ground state of $^9$He is unstable with respect to decay into $^8$He + n. The spectrum for the $^9$Be$(\pi^-,\pi^+)^9$He reaction at an incident pion kinetic energy of 194 MeV is shown in Figure 16 (167). The ground state is unstable by more than 1 MeV. Similarly, the search for the tetraneutron was led by the $(\pi^-,\pi^+)$ reaction studies (168, 169). The four-neutron ground state was also found to be unbound.

High-intensity pion beams proved their worth in the determination of nuclear masses off the line of stability by means of pion double charge

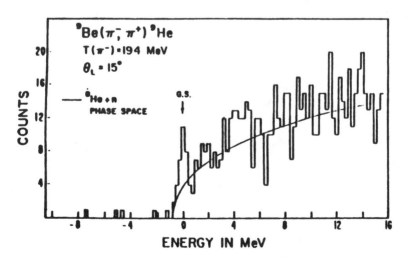

*Figure 16*   Spectrum of pion DCX leading to $^9$He and showing the ground state.

exchange. In addition, the clean discrimination against other particles makes the pion double-charge-exchange reaction an excellent tool for probing light systems near the neutron drip line.

## 3.8    *Discrimination Between Spin-Flip and Non-Spin-Flip Excitations*

The dominance of the p-wave pion-nucleon interaction near and below the (3,3) resonance in $(\pi,\pi')$ scattering, as discussed in Section 3.5, results in the noted signatures for spin-flip and non-spin-flip excitations: angular distributions are $\cos^2\theta$ and rising with incident pion kinetic energy for non-spin-flip (unnatural-parity) transitions, while they are $\sin^2\theta$ and falling for spin-flip (natural-parity) transitions (130). Although the primary use of $(\pi,\pi')$ studies to date has been in the investigation of the properties of stretched states (Section 3.5) and isospin mixing (Section 3.3), a study of the 4.45-MeV state in $^{18}$O provides an illustration of the utility of the $(\pi,\pi')$ reaction for other purposes (127, 170).

The angular distribution for $\pi^+$ and $\pi^-$ scattering from the 4.45-MeV $1^-$ state in $^{18}$O is shown in Figure 17 (127). The curves are DWIA calculations from the code by T. S. H. Lee (unpublished) and show a dramatic difference between $1^-$, $\Delta S = 0$ excitations and $1^-$, $\Delta S = 1$ excitations. The model predictions show the two spin possibilities to be completely out of phase. Only the $\Delta S = 1$ calculation appears to represent the data reasonably. Thus, the natural-parity (non-spin-flip) excitation appears to be dominated by a $\Delta S = 1$ transition, and this provides another example of the usefulness of pion scattering as a probe of the structure of the nucleus.

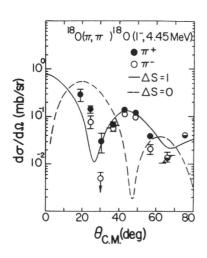

*Figure 17* Angular distribution for a $\Delta S = 1$ vs a $\Delta S = 0$ transition.

A striking example of a discrepancy in our present understanding is the comparison of the $1^+$, $T = 0$ (12.71 MeV) and $1^+$, $T = 1$ (15.11 MeV) states in $^{12}C$. The structure of these states is generally believed to be very similar. However, while the isospin-zero state shows the expected "spin-flip" behavior, the $T = 1$ state behaves much more like the "non-spin-flip" type of transition (172, 173). Explanations advanced to explain this result include multistep processes (172) and delta admixtures into the nuclear state (173). Recent measurements (B. Ritchie, private communication) at low energy show that the discrepancy persists in this weakly interacting region as well, so that a nuclear structure effect is made more likely, but as of this writing the question is unresolved.

These examples illustrate that the simple, established properties of the spin-independent and spin-dependent components of the pion-nucleon interaction provide a powerful tool in our effort to classify nuclear states according to their structure.

# 4.   PROPERTIES OF HADRONIC SYSTEMS

While Sections 2 and 3 had neatly defined aims ("What are the interaction strengths and ranges of the fundamental hadron-hadron systems?" and "What are the values of the parameters that delineate the structure of the nucleus?"), the present section has no such restricted goal but rather directly addresses the study of the hadronic interaction. The type of questions addressed in this section are "How is the Pauli principle to be applied to nucleons when we now know that it should be valid on the quark—not on the nucleon—level?" or "How is hadronic substructure rearranged when it is not restricted by baryon conservation, as is the case in pion absorption?" As the reader will realize, these questions have not been definitively answered. However, considerable progress has been made and our understanding of strongly interacting hadronic systems is becoming clearer.

To this end we look at three examples. In pion-deuteron scattering we see how the pion-nucleon interaction becomes coupled to the properties of the nucleon, when the $\pi N$ interaction is embedded in the nuclear medium. In pion absorption we examine a fundamental difference between pion propagation in nuclear matter and nucleon propagation: the pion can disappear. Finally, the ways in which the Pauli principle and the Lorentz-Lorenz effect modify pion propagation in the nuclear medium are explored. In each of these areas considerable progress has been achieved, but there remain questions to be answered. It is our hope that the reader will be motivated to investigate the possibilities.

## 4.1    *Pion-Deuteron Polarizations*

Investigation of the $\pi NN$ system is of fundamental importance in intermediate-energy physics. The few-body nature of the system makes it amenable to precise microscopic description. Accurate data have been generated (total and differential cross sections, $iT_{11}$ vector polarization, and $T_{20}$ tensor polarization), motivated in part by a search for dibaryon resonances. Much of the data can be qualitatively understood in terms of the strong $\Delta$-isobar formation ($\pi N$ $P_{33}$ rescattering) in intermediate states. Nonetheless, there do exist large and intriguing discrepancies between the data and the unitary model calculations that attempt to take into account absorption of the pion.

Three form factors are required to describe the deuteron: monopole (or charge), quadrupole, and magnetic. The quadrupole and charge form factors enter in the cross section incoherently, so that it is difficult to measure them separately. No minimum is seen in the non-spin-flip cross section because the quadrupole term fills in the zero in the charge form factor. However, the tensor polarization $T_{20}$ (a measure of the $m = 0$ magnetic substate population relative to the average of the $m = \pm 1$ substates) is very sensitive to the ratio of the quadrupole to monopole form factors. It was suggested (175) that pion-deuteron scattering at $180°$ (where the magnetic form factor does not enter since the spin-flip pion-nucleon amplitude is zero) would be a good way of studying the properties of the deuteron d state. It had already been pointed out that a similar measurement with electrons would be sensitive, not only to deuteron properties (176), but to meson exchange currents as well (177).

The elastic scattering differential cross section has been measured at laboratory energies ranging from 47 to 325 MeV (178). Except for the back angle measurements near 256 MeV (e.g. 179), the unitary model approaches existing in the literature reproduce the features of the data reasonably well (180–185). (See Ref. 186 for a discussion of the equivalence of the various models.) Similarly, the vector polarization $iT_{11}$ data (187) show reasonable qualitative agreement with the unitary model approach, although the minimum near $70°$ that develops with increasing pion energy is still not quantitatively understood. The presence of pion absorption in the models does not appear to play a significant role in the vector polarization calculations (179).

It is in the tensor polarization $T_{20}$(Lab) measurements that the discrepancy between theory and data stands out. The early measurements at LAMPF (188), in which recoil deuterons were analyzed by means of a second scattering in a $^3$He cell polarimeter utilizing the $^3$He(d,p)$^4$He reaction, were confirmed by a similar experiment at TRIUMF (189). More recently, agreement with these data has been obtained by a second group working at TRIUMF and using an entirely different experimental technique. In this case a tensor-polarized deuteron target was employed, so that $T_{20}$(cm) was obtained in a single scattering experiment (190). [Because $T_{20}$(Lab) is a linear combination of $T_{2m}$(cm), calculated values of $T_{21}$(cm) and $T_{22}$(cm) from Garcilazo (191) were used to make the comparison between data sets.] Results from all three measurements are shown in Figure 18.

The nature of the discrepancy between the data and existing unitary model calculations is illustrated in Figure 19. While the single scattering impulse approximation gives a very reasonable representation of the $\pi d$ tensor polarization data, it was found that three-body treatments did much worse (180–185). The problem was traced fairly quickly to the manner in which pion absorption was included (192). It is in the isospin-1/2–spin-1/2 pion-nucleon channel that the absorption takes place, and this partial wave amplitude, normally very weak, now exerts a large influence. Conceptually the $P_{11}$ amplitude may be viewed as being composed of two parts, a nucleon pole piece and a nonpole piece, when coupling to the $\pi d \leftrightarrow NN$ channel is included in the model (180, 182, 193–196). The phase shift of

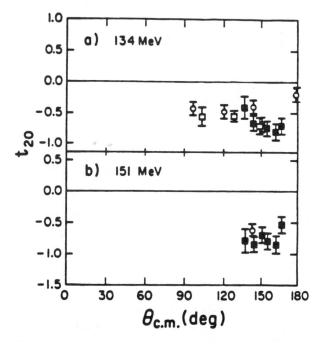

*Figure 18*  $T_{20}$ tensor polarization data at 134 and 151 MeV. Three different data sets are consistent.

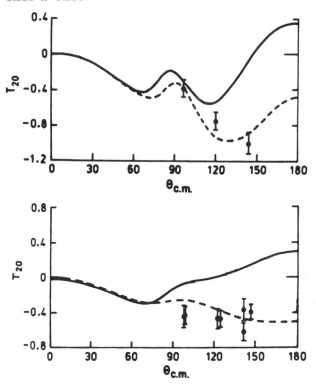

*Figure 19*  Comparison of theory and experiment showing the effect of including the $P_{11}$ channel. The solid curve includes absorption and the dotted one does not.

this amplitude is small, and if it is neglected entirely a satisfactory fit to the data is obtained (179). However, if a two-potential formalism is used or if a Chew-Low or quark bag model (197) of the amplitude is applied, then the two parts are not individually small—only their sum is. The pole in the $P_{11}$ amplitude is needed to account for the nucleon; however, it is argued that this pole part of the $P_{11}$ amplitude must be removed from intermediate states in which the nucleon is excluded due to the action of the Pauli principle. It is this Pauli exclusion (blocking) of the pole part of the $P_{11}$ amplitude that gives rise to the disagreement with the data (Figure 19), because the remaining (nonpole) piece of the amplitude is large (179). Clearly, the burning question is "What is the missing physics?" The spin observables measured in elastic pion-deuteron scattering, plus the corresponding observables in the absorption and breakup channels, will certainly provide the data base needed to answer this question. They will also thoroughly test our ability to produce a valid three-body model calculation of this fundamental system.

It was recently pointed out (198) that the quark scaling laws (199) predict that the form factor ratio, $x$, will become unity for high momentum transfer and that this feature may serve as a signature for the onset of perturbative QCD. In this limit $T_{20}$ takes its maximal negative value of $-\sqrt{2}$. Figure 20 shows values of $T_{20}$ measured for pion scattering near $180°$ where the contribution of the magnetic form factor is very small. As can be seen, the agreement between theory (200) and experiment is quite good. The negative extremum in the curve occurs when $x = 1$, near the zero of the charge form factor. While the theoretical curve appears to follow the data, the values are still too near $-\sqrt{2}$ for us to be certain that the QCD result is *not* correct. However, it is certainly not unique, since recent calculations (201) suggest, for the case of electron scattering, that the meson exchange currents and/or delta contributions can give a result similar to the QCD limit. While present theoretical treatments of extranucleon degrees of freedom are probably inadequate, they do suggest the utility of the $T_{20}$ measurements as a tool to use in separating the relevant deuteron form factors with both pion and electron beams and they help develop our picture of hadronic structure.

## 4.2   Pion Absorption

Pion absorption, one of the most fundamental aspects of hadronic inter-actions, has received a good deal of attention. While the two-body nuclear reaction $\pi + A \rightarrow (A-1) + N$ has been studied (202), the most complete investigations of this type have been made of the inverse reaction with proton beams and lie outside the scope of this review.

The simplest absorption reaction, $\pi + d \rightarrow p + p$, has received its share

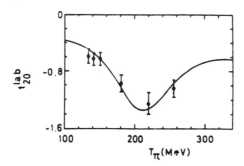

*Figure 20*   Comparison of experimental values of $T_{20}$ for pion-deuteron scattering near $180°$ with the results of Garcilazo (200).

of attention (see 203 for an analytical representation of the low-energy data). While a number of studies have focused on the mechanism and fundamental hadronic properties, the general tendency has been to use the reaction for other purposes, such as the search for dibaryons. While the large separation of two nucleons in the deuteron is less than ideal for the study of a strongly interacting multiquark system, suggestions have been made for nonpionic, intermediate-range, color singlet exchange, primarily the $\rho$ (204).

Nuclear systems provide higher nucleon densities and an extra variable (the mass number) along with the additional complication in interpretation. Several facets of this problem have been well delineated experimentally.

The one-nucleon spectrum has been measured (205) in several cases. Hints of possible multinucleon mechanisms are present in these results, but the interpretation leading to these conclusions is not yet considered compelling since scattering of the hadrons before and/or after a "two-nucleon absorption" can simulate a multinucleon mechanism (206).

By combining measurements of the inelastic cross sections with estimates for the small, unobserved, single-charge-exchange cross sections, direct measurements of the cross section for pion disappearance have been made (207). Using a less direct technique, one involving the observation of gamma rays from nuclei only excited if the mass of the pion is available, these same absorption cross sections have also been extracted (208). These two sets of observations provide a fundamental data base for examining the pion absorption problem.

However, these measurements alone do not directly address the questions of hadronic interaction, which revolve around the mechanism of absorption in a dense system. The next step has been to attempt to specify the fraction of absorption occurring on only two nucleons. The precise concept of "two-nucleon absorption" is not self-evident. For an operational definition here, we make the (admittedly theoretical) distinction that in the absorption process the energy corresponding to the mass of the pion is given up to only two nucleons. Thus any truly multinucleon mechanism is assumed to distribute the pion energy among three or more nucleons at the instant of disappearance, so that it is extremely unlikely that any pair of nucleons would carry off all of the energy. The experiments (209, 210) then aim at determining, by means of coincidence measurements, the number of events in which a single pair of nucleons do indeed carry the total energy of absorption. The remaining events seen in the total absorption cross section must then be attributed to some other process involving more than two nucleons. Substantial corrections must be made for nuclear properties in relating the correlated pairs observed to the original pair absorption. While the exact amount of non-two-body absorption is still debated, some finite fraction seems to be indicated.

As a step between the deuteron and large nuclear systems, the absorption on the three-nucleon system has been thoroughly studied (211). Here the ability to define completely the final-state kinematics is very useful. A remarkably detailed picture has been achieved for stopped $\pi^-$ absorption in $^3$He. Here the two-nucleon absorption can be clearly seen, along with the form of three-nucleon absorption that is still two-body in nature, i.e. where one nucleon recoiling against a correlated pair of nucleons forms the final system (Figure 21). Of particular interest is the failure to observe a three-nucleon absorption mechanism with an approximate equipartition of energy, even at the level of 10%. This result directly addresses such questions as whether absorption occurs on nine-quark clusters in the three-nucleon system.

The form of three-nucleon absorption in which the energy arising from the rest mass of the pion is carried by two clusters was also observed (212) in stopped pion absorption in $^6$Li, where composite final particles make up 20% of the reaction products. While no claim is made of understanding the process (in fact neither a sequential, nor a cluster mechanism seems to work) it has been established that this is an important absorption mode, at least for pions at rest.

## 4.3   *Nuclear Transparency and Low-Energy Pion Reactions*

There are two classic many-body phenomena that clearly manifest themselves in the propagation of pions through nuclear material. The study of these characteristics in many-body systems in which the number of constituents can be varied from 2 to 200 permits us to explore the approach to the true many-body limit. The medium-to-short-range properties of the interactions among constituents are often coupled into this study in a manner analogous to van der Waals forces in molecular physics in the sense of an effective interaction.

The first of these two phenomena, the Lorentz-Lorenz effect (213), is now just over 100 years old in the realm of classical many-body theory. It arises from the multiple scattering in a granular system when the fundamental interaction is p-wave in character. The fluid picture, used as the first-order approximation for either the dielectric constant in the original electromagnetic calculation, or for the optical potential in the pionic case, results from averaging over all positions of the scattering centers using the free space radiation function (very singular for small distances) to represent the scattering between any two centers. For the case of p-wave scattering, the modification of the averaging over this strong singularity (due to a nonuniform distribution of relative distances between the pairs) has

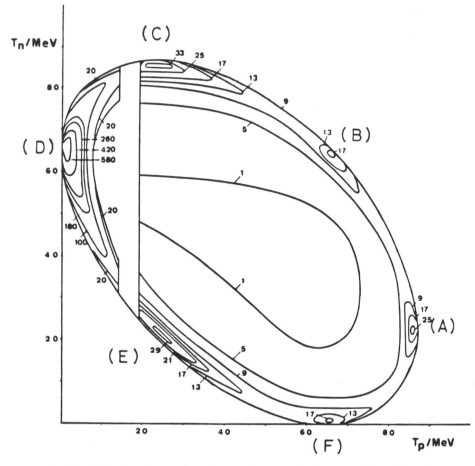

*Figure 21* Dalitz plot of events from the $\pi^-$ $^3$He $\rightarrow$ n + p + X for stopped pions. The point D corresponds to absorption on an np pair with one of the energetic neutrons and the spectator being detected. B and F correspond to absorption on a pp pair with either the two energetic particles being detected or the energetic neutron and spectator proton being detected. A, C, and E correspond to absorption on three nucleons with one nucleon recoiling against a pair with equal energy.

observable consequences. Thus, if the relative separation of a pair of scattering centers is required to be larger than a given distance, then the total scattering will be different than the continuum approximation (214). Clearly this correction itself depends on the average distance between scattering centers and so leads to a nonlinear dependence on the density. Furthermore, the near-zone radiation function itself is modified if the scattering from the second member of a pair occurs while the projectile is still under the influence of its interaction with the first member when the second scattering occurs. Thus, it is the interplay of the short-range correlation wave function between nucleons and the range of the pion-nucleon interaction that controls the Lorentz-Lorenz effect in the nuclear medium (215).

The second phenomenon arises from the fact that the Pauli principle severely restricts the struck particles' phase space for recoil (216). In a very large system this is expressed in terms of a Fermi sea. However, in a finite system additional quantum numbers (such as angular momentum) must be respected, so that wave functions describing finite systems must be used. For the largest nuclei, such as $^{208}$Pb, enough nuclear shells exist so that a continuum approximation is reasonable. Again, by studying multibody systems for 2 to 200 particles, one can investigate the approach to condensed matter.

Both of these effects lead to greater transparency for pion propagation through the nucleus. While they persist into the strongly interacting (3,3) resonance energy region, they are best studied at energies below 100 MeV where the high-density regions of the nucleus are sampled.

Analysis of elastic scattering yields indications of the nonlinear dependence of pion propagation on the nuclear density, since the varying interaction strength with increasing density implies that the pion-nucleus potential does not become as strong in the nuclear interior as would be expected by extrapolating from the surface region using the known nucleon density. The result is to give a larger apparent nuclear radius than that obtained from electron scattering (217).

The cancellation between pion-nucleon s and p partial waves, which leads to an extremely sharp minimum in the zero-degree single-charge-exchange cross section, has been exploited to determine the extent to which the nucleus is transparent to pions at energies below 80 MeV. This distinct feature in the cross section has many of the same aspects as a resonance: it is altered by the distortion of the pion waves, and modified by various medium effects. The experimental observation was made [first for $^{15}$N (218) and later for $^{7}$Li (219) and $^{14}$C (220)] that the position and depth of the minimum are almost unchanged from the free value in the pion-nucleus single charge exchange (SCX) reactions (Figure 22). Most of this transparency is accounted for in terms of the naturally long mean-free-path at these low energies (the total cross sections for pions on protons and neutrons are approximately equal at pion kinetic energy of 50 MeV and lead to a mean free path of $\sim 10$ fm in nuclear matter) coupled with substantial Pauli blocking (216) and Lorentz-Lorenz effects (221). In heavier nuclei the minimum persists (222), although it is broadened and

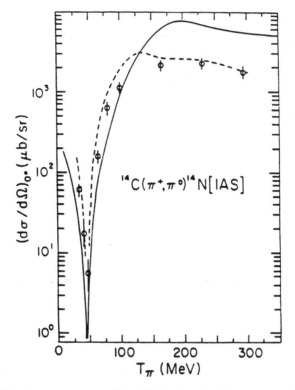

*Figure 22*    Pion charge exchange at 0° on $^{14}$C showing that the strong cancellation of the free case is not strongly modified in the nucleus. The solid curve is twice the free pion-nucleon charge exchange cross section (which is the nuclear result for no distortion of the initial and final pion waves). The dotted curve includes distorted waves using the techniques of Kaufmann & Gibbs (216).

shifted to higher pion energies. A large part of the observed shifts in the position of the minimum can be understood in terms of the Coulomb potential (the incident $\pi^+$ is slowed down by the repulsive potential) and the larger $Q$ values (the analog state is now an excited state in the final nucleus so the outgoing $\pi^0$ has lower energy) for the heavier nuclei.

Another case in which the degree of transparency is indicated is the population of $0^+$ states excited by inelastic pion scattering in a $0^+$ target nucleus. Because of the orthogonality of the states, the cross section must vanish at zero degrees for a single-step process in which the $Q$ value is negligible if the nucleus is totally transparent (Figure 23). Analysis of such a reaction on $^{12}$C shows a high degree of transparency (223).

The existence of many-body effects analogous to those of classical phys-

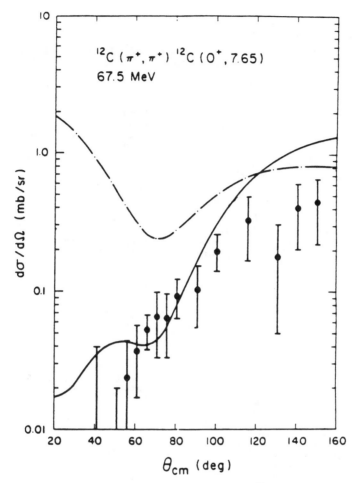

*Figure 23*   Inelastic scattering of $\pi^+$ to the excited $0^+$ state in $^{12}$C at 7.65 MeV. The dash-dot curve shows the result of a standard phenomenological Kisslinger optical potential. The solid curve includes only Coulomb distortion of the initial and final waves.

ics has been observed in systems as small as $^{12}$C. The relative importance of the Pauli and Lorentz-Lorenz effects is still under study and depends on the pion-nucleon and nucleon-nucleon interaction ranges.

ACKNOWLEDGMENTS

We would like to thank each of the many physicists in North America, Europe, Israel, and Australia who read and commented on various aspects of this review and who provided material for us to use. Unfortunately, such a list would be far too long to be included. The work of the authors was performed under the auspices of the US Department of Energy.

*Literature Cited*

1. Kinney, E. R., Matthews, J. L., Gram, P. A. M., MacArthur, D. W., Piasetzky, E., et al. *Phys. Rev. Lett.* 57: 3152 (1986)
2. Nefkens, B. M. K., Briscoe, W. J., Eichon, A. D., Fitzgerald, D. H., Holt, J. A., et al. *Phys. Rev. Lett.* 52: 735 (1984)
3. Gilman, R., Fortune, H. T., Johnson, M. B., Siciliano, E. R., Toki, H., Wirzba, A. *Phys. Rev.* C32: 349 (1985)
4. Peng, J.-C. In *Proc. Hadronic Probes and Nuclear Interactions*, Tempe, AZ, March 11–14, 1985, AIP 133. New York: AIP (1985)
5. Gessaman, D., ed. *Proc. Symp. on Delta-Nucleus Dynamics*, May 2–4, 1983, ANL-PHY-83-1; Conf. 830588. Argonne, Ill: ANL (1983)
6. Johnson, M. B. *From Nuclei to Stars*, XCI Corso Soc. Italiana di Fisica, Bologna (1985)
7. Vogt, E. *Prog. Theor. Phys. Suppl.* 85: 190 (1985)
8. Rosen, L. *LAMPF—Its History and Accomplishments*, LA-UR-85-3437. Los Alamos: LAMPF (1985)
9. Eisenberg, J. M., Koltun, D. *Theory of Meson Interactions with Nuclei.* New York: Wiley (1980)
10. Crawford, J. F., Daum, M., Frosch, R., Jost, B., Kettle, P.-R. *Phys. Rev. Lett.* 56: 1043 (1986)
11. Heller, L., Kumano, S., Martinez, J. C., Moniz, E. J. *Phys. Rev.* C35: 718 (1987); Nefkens, B. M. K., Arman, M., Ballagh, H. C., Glodis, P. F., Haddock, R. P., et al. *Phys. Rev.* D18: 3911 (1978)
12. McFarlane, W. K., Auerbach, L. B., Gaille, F. C., Highland, V. L., Jastrzembski, E., et al. *Phys. Rev.* D32: 547 (1985)
13. Frank, J. S., Browman, A. A., Gram, P. A. M., Heffner, R. H., Klare, K. A., et al. *Phys. Rev.* D28: 1569 (1983); Brack, J. T., Kraushaar, J. J., Mitchell, J. H., Peterson, R. J., Ristinen, R. H., et al. *Phys. Rev.* C34: 1771 (1986); Göring, R., Klein, U., Kluge, W., Matthäu, H., Metzler, M., et al. In *Proc. Intersections Between Particle and Nuclear Physics*, Lake Louise, Canada, AIP 150: 646 (1986)
14. Fitzgerald, D. H., Baer, H. W., Bowman, J. D., Cooper, M. D., Irom, F., et al. *Phys. Rev.* C34: 619 (1986)
15. Salomon, M., Measday, D. F., Poutissou, J.-M., Robertson, B. C. *Nucl. Phys.* A414: 493 (1984)
16. Deleted in proof
17. Spuller, J., Berghofer, D., Hasinoff, M. D., MacDonald, R., Measday, D. F., et al. *Phys. Lett.* 67B: 479 (1977)
18. Siegel, P. B., Gibbs, W. R. *Phys. Rev.* C33: 1407 (1986)
19. Veit, E. A., Jennings, B. K., Thomas, A. W. *Phys. Rev.* D33: 1859 (1986)
20. Gasser, J. *Nucl. Phys.* B279: 65 (1987); Gasser, J. *Ann. Phys.* 136: 62 (1981); Gasser, J., Leutwyler, H. *Ann. Phys.* 158: 142 (1984)
21. Koch, R. *Nucl. Phys.* A448: 707 (1986)
22. Cheng, T. P. *Phys. Rev.* D13: 2161 (1976)
23. Donoghue, J. F., Nappi, C. R. *Phys. Lett.* 168B: 105 (1986)
24. Bovet, E., Antonuk, L., Egger, J.-P., Fiorucci, G., Gabathuler, K., et al. *Phys. Lett.* B153: 231 (1985)
25. DeGrand, T., Rebbi, C. *Phys. Rev.* D17: 2358 (1978)
26. Umland, E., Duck, I., von Witsch, W. *Phys. Rev.* D27: 2678 (1983)
27. Blankleider, B., Walker, G. E. *Phys. Lett.* 152B: 291 (1985)
28. Johnstone, J., Lee, T.-S. H. *Phys. Rev.* C34: 243 (1986)
29. Ayed, R. *Rev. Mod. Phys.* 52: S190 (1980); Ayed, R., Bareyre, P. In *Proc. 2nd Int. Conf. on Elementary Particles*, Aix-en-Provence, 1973 [*J. Phys. (Paris) Suppl.* 34: C1 (1973)]
30. Höhler, G., Kaiser, F., Koch, R., Pietarinen, E. *Handbook of Pion-Nucleon Scattering* (Physics Data No. 12-1) (Fachsinformationzentrum, Karlsruhe, 1979); Koch, R., Pietarinen, E. *Nucl. Phys.* A336: 331 (1980); Cutkosky, R. E., Hendrick, R. E., Alcock, J. W., Chao, Y. A., Lipes, R. G., et al. *Phys. Rev.* D20: 2804 (1979): 2839 (1979)
31. Arndt, R., Ford, J. M., Roper, L. D. *Phys. Rev.* D32: 1085 (1985)
32. Mokhtari, A., Briscoe, W. J., Eichon, A. D., Fitzgerald, D. H., Kim, G. J., et al. *Phys. Rev.* D33: 296 (1986)

33. Mokhtari, A., Eichon, A. D., Kim, G. J., Nefkens, B. M. K., Wightman, J. A., et al. *Phys. Rev.* D35: 810 (1987)
34. Carter, J. R., Bugg, D. V., Carter, A. A. *Nucl. Phys.* B58: 378 (1973)
35. Pedroni, E., Gabathuler, K., Domingo, J. J., Hirt, W., Schwaller, P., et al. *Nucl. Phys.* A300: 321 (1978)
36. Myhrer, F., Pilkuhn, H. *Z. Phys.* A276: 29 (1976)
37. Henley, E. M. In *Isospin in Nuclear Physics*, ed. D. H. Wilkinson. Amsterdam: North Holland (1969)
38. Heller, L., Signell, P., Yoder, N. R. *Phys. Rev. Lett.* 13: 577 (1964)
39. Sher, M. S., Signell, P., Heller, L. *Ann. Phys. (NY)* 58: 1 (1970)
40. McNamee, P. C., Scadron, M. D., Coon, S. A. *Nucl. Phys.* A249: 483 (1975)
41. McNamee, P. C., Scadron, M. D., Coon, S. A. *Nucl. Phys.* A287: 381 (1978)
42. Friar, J. L., Gibson, B. F. *Phys. Rev.* C17: 1752 (1978)
43. Brandenburg, R. A., Coon, S. A., Sauer, P. U. *Nucl. Phys.* A294: 305 (1978)
44. Coon, S. A., Scadron, M. D. *Phys. Rev.* C26: 562, 2402 (1978)
45. Meiere, F. T., Fischbach, E., McDonald, A., Nieto, M. M., Scott, C. K. *Phys. Rev.* D8: 4209 (1973)
46. Nagels, M. M., Rijken, T. A., deSwart, J. J. *Phys. Rev.* D12: 744 (1975)
47. Noyes, H. P. *Ann. Rev. Nucl. Sci.* 22: 465 (1972)
48. Noyes, H. P., Lipinski, H. M. *Phys. Rev.* C4: 995 (1972)
49. Kuhn, B. 1975. *Fiz. Elem. Chastits At. Vadra* 6: 347 [*Sov. J. Part. Nucl.* 6: 139 (1976)]
50. Gibbs, W. R., Gibson, B. F., Stephenson, G. J. Jr. *Phys. Rev.* C11: 90; 12: 2130 (1975)
51. Gibbs, W. R., Gibson, B. F., Stephenson, G. J. Jr. *Phys. Rev.* C16:327; 17: 856 (1977)
52. Phillips, R. H., Crowe, K. M. *Phys. Rev.* 96: 484 (1954)
53. Ryan, J. W. *Phys. Rev. Lett.* 12: 564 (1964)
54. Butler, P. G., Cohen, N., James, A. N., Nicholson, J. P. *Phys. Rev. Lett.* 21: 470 (1968)
55. Haddock, R. P., Salter, R. M., Zeller, M., Czirr, J. B., Nygren, D. R. *Phys. Rev. Lett.* 14: 318 (1965)
56. Salter, R. M., Haddock, R. P., Zeller, M., Nygren, D. R., Czirr, J. B. *Nucl. Phys.* A254: 241 (1975)
57. Gabioud, B., Alder, J. C., Joseph, C., Loude, J. F., Morel, N., et al. *Phys. Rev. Lett.* 42: 1508 (1975)
58. Gabioud, B., Alder, J. C., Joseph, C., Loude, J. F., Morel, N., et al. *Phys. Lett.* 103B: 9 (1981)
59. Gabioud, B., Alder, J. C., Joseph, C., Loude, J. F., Morel, N., et al. *Nucl. Phys.* A420: 496 (1984)
60. Gibson, B. F., Stephenson, G. J. Jr. *Phys. Rev.* C8: 1222 (1973)
61. Gibson, B. F., Stephenson, G. J. Jr. *Phys. Rev.* C11: 1448 (1975)
62. Friar, J. L., Gibson, B. F. *Phys. Rev.* C17: 1752 (1978)
63. Soukup, J., Cameron, J. M., Fielding, H. W., Hussein, A. H., Lam, S. T., et al. *Nucl. Phys.* A322: 109 (1979)
64. Guratzsch, H., Kühn, B., Kumpf, H., Mösner, J., Neubert, W., et al. *Nucl. Phys.* A342: 239 (1980)
65. von Witsch, W., Moreno, B. G., Rosenstock, W., Franke, R., Steinheuer, B. *Phys. Lett.* 91B: 342 (1980)
66. Johnson, R. R., Masterson, T., Bassalleck, B., Gyles, W., Marks, T., et al. *Phys. Rev. Lett.* 43: 844 (1979)
67. Barnett, B. M., Gyles, W., Johnson, R. R., Erdman, K. L., Johnstone, J., et al. *Phys. Lett.* 97B: 45 (1980)
68. Beer, G. A., et al. Contribution 1B27 to 7th Int. Conf. on High Energy Physics and Nuclear Structure (Vancouver, Canada, August, 1979)
69. Cooper, M. D. In *Meson-Nuclear Physics—1976*, 33: 237. New York: AIP (1976)
70. Dedonder, J.-P., Maillet, J.-P., Schmit, C. *Ann. Phys.* 127: 1 (1980)
71. Lawson, R. L., Serduke, F. J. D., Fortune, H. T. *Phys. Rev.* C14: 1245 (1976)
72. Beiner, M., Flocard, H., Giai, N. V., Quentin, P. *Nucl. Phys.* A238: 29 (1975)
73. Lerner, G. M., Hiebert, J. C., Rutledge, L. L., Papanicolas, C., Bernstein, A. M., et al. *Phys. Rev.* C12: 778 (1975)
74. Alkhazor, G. D., Belostotsky, S. L., Domchenkov, O. A., Dotsenko, Yu. V., Kuropatkin, N. P., et al. *Phys. Lett.* 57B: 47 (1975)
75. Nolan, J. A., Schiffer, J. P. *Ann. Rev. Nucl. Sci.* 19: 471 (1969)
76. Negele, J. W. *Phys. Rev.* C1: 1260 (1970)
77. Egger, J.-P., Corfu, R., Gretillat, P., Lunke, C., Piffaretti, J., et al. *Phys. Rev. Lett.* 39: 1608 (1977)
78. Gibbs, W. R. In *Common Problems in Low- and Medium Energy Nuclear Physics, NATO Adv. Study Inst. Ser.* B: *Physics*, 45: 595. New York: Plenum (1978)
79. Maillet, J.-P., Dedonder, J.-P., Schmit, C. *Nucl. Phys.* A316: 267 (1979)
80. Germond, J.-F., Johnson, M. B., John-

stone, J. A. *Phys. Rev.* C32: 983 (1983); Johnson, M. B., Bethe, H. A. *Comments Nucl. Part. Phys.* 8: 75 (1978)
81. Baldwin, G. C., Klaiber, G. S. *Phys. Rev.* 73: 1156 (1948); Goldhaber, M., Teller, E. *Phys. Rev.* 74: 1046 (1948)
82. Bohr, A., Mottelson, B. R. *Nuclear Structure* New York: Benjamin (1975)
83. Erell, A., Alster, J., Lichtenstadt, J., Moinester, M. A., Bowman, J. D., Cooper, M. D., et al. *Phys. Rev. Lett.* 52: 2134 (1984)
84. Youngblood, D. H. In *Giant Multipole Resonances*, ed. F. E. Bertrand. Harwood (1979)
85. Sterrenburg, W. A., Austin, S. M., deVito, R. P., Galonsky, A. *Phys. Rev. Lett.* 45: 1839 (1980)
86. Shoda, K. *Phys. Rep.* 53: 341 (1979)
87. Erell, A. *Study of Isovector Giant Resonances with Pion Charge Exchange*, Thesis, Tel Aviv Univ. (1984)
88. Bowman, J. D., Johnson, M. B., Negele, J. W. *Phys. Rev. Lett.* 46: 1614 (1981)
89. Bowman, J. D., Baer, H. W., Bolton, R., Cooper, M. D., Cverna, F. H., et al. *Phys. Rev. Lett.* 50: 1195 (1983); Erell, A., Alster, J., Lichtenstadt, J., Moinester, M., Bowman, J. D., et al. *Phys. Rev.* C34: 1822 (1986); Irom, F., et al. *Phys. Rev.* C34: 2231 (1986)
90. Auerbach, N., Klein, A. *Phys. Rev.* C28: 2075 (1983)
91. Leonardi, R., Lipparini, F., Stringari, S. *Phys. Rev.* C35: 1439 (1987)
92. Auerbach, N. *Phys. Rep.* 98: 274 (1983)
93. Morris, C. L., Piffaretto, J., Thiessen, H. A., Cottingame, W. B., Braithwaite, W. J., et al. *Phys. Rev. Lett.* 86B: 31 (1979)
94. Dehnhard, D., Tripp, S. J., Franey, M. A., Kyle, G. S., Morris, C. L. *Phys. Rev. Lett.* 43: 1091 (1979)
95. Oothoudt, M. A., Garvey, G. T. *Nucl. Phys.* A284: 41 (1977)
96. Marion, J. B., Nettles, P. H., Cocke, C. L., Stephenson, G. J. Jr. *Phys. Rev.* 157: 847 (1967)
97. Millener, D. J., Kurath, D. *Nucl. Phys.* A255: 315 (1975)
98. Holtkamp, D. B., Braithwaite, W. J., Cottingame, W. B., Green, S. J., Joseph, R. J., et al. *Phys. Rev. Lett.* 45: 420 (1980)
99. Seestrom-Morris, S. J., Dehnhard, D., Franey, M. A., Kyle, G. S., Morris, C. L., et al. *Phys. Rev.* C26: 594 (1982)
100. Holtkamp, D. B., Seestrom-Morris, S. J., Chakravarti, S., Dehnhard, D., Baer, H. W., et al. *Phys. Rev. Lett.* 47: 216 (1981)
101. Iversen, S., Obst, A., Seth, K. K., Thiessen, H. A., Morris, C. L., et al.

*Phys. Rev. Lett.* 40: 17 (1978)
102. Lunke, C., Corfu, R., Egger, J.-P., Gretillat, P., Piffaretti, J., et al. *Phys. Lett.* 78B: 201 (1978)
103. Iversen, S., Nann, H., Obst, A., Seth, K. K., Tanaka, N., et al. *Phys. Lett.* 82B: 51 (1979)
104. Oset, E., Strottman, D. *Phys. Lett.* 84B: 396 (1979)
105. Brown, V. R., Madsen, V. A. *Phys. Rev.* C11: 1298 (1975)
106. Olmer, C., Geesaman, D. F., Zeidman, B., Chakravarti, S., Lee, T.-S. H., et al. *Phys. Rev.* C21: 254 (1980)
107. Boyer, K. G., Cottingame, W. B., Smith, L. E., Greene, S. J., Moore, C. F., et al. *Phys. Rev.* C24: 598 (1981)
108. Bernstein, A. M., Brown, V. R., Madsen, V. A. *Phys. Lett.* 103B: 255 (1981)
109. Bernstein, A. M., Brown, V. R., Madsen, V. A. *Phys. Lett.* 106B: 259 (1981)
110. Tacik, R., Erdman, K. L., Johnson, R. R., Roser, H. W., Gill, D. R., et al. *Phys. Rev. Lett.* 52: 1276 (1984)
111. Wienands, U., Hessey, N., Barnett, B. M., Rozon, F. M., Roser, H. W., et al. *Phys. Rev.* C35: 708 (1987)
112. Sobie, R. J., Drake, T. E., Erdman, K. L., Johnson, R. R., Roser, H. W., et al. *Phys. Rev.* C30: 1612 (1984)
113. Ball, G. C., Alexander, T. K., Davies, W. G., Forster, J. S., Mitchell, I. V. *Nucl. Phys.* A377: 268 (1982)
114. Saha, A., Seth, K. K., Casey, L., Godman, D., Kielczewska, D., et al. *Phys. Lett.* 132B: 51 (1983)
115. Ginocchio, J. N., Van Isacker, P. *Phys. Rev.* C33: 365 (1986)
116. Arvieux, J., Albanese, J. P., Buenerd, M., Lebrun, D., Boschitz, E., et al. *Phys. Rev. Lett.* 42: 753 (1979)
117. Arvieux, J., Albanese, J. P., Buenerd, M., Bolger, J., Boschitz, E., et al. *Phys. Lett.* 90B: 371 (1980)
118. Ullmann, J. L., Kraushaar, J. J., Masterson, T. G., Peterson, R. J., Raymond, R. S., et al. *Phys. Rev.* C31: 177 (1985)
119. Ullmann, J. L., et al. Unpublished. Cited in Ref. 121
120. Seestrom-Morris, S. J., Morris, C. L., Moss, J. M., Carey, T. A., Drake, D., et al. *Phys. Rev.* C33: 1847 (1986)
121. Seestrom-Morris, S. J. In *Proc. Nuclear Structure at High Spin, Excitation, and Momentum Transfer*, Indiana Univ., 142: 54. New York: AIP (1985)
122. Auerbach, N., Klein, A., Siciliano, E. R. *Phys. Rev.* C31: 682 (1985)
123. Geesaman, D. F., Kurath, D., Morri-

son, G. C., Olmer, C., Zeidman, B., et al. *Phys. Rev.* C27: 1134 (1983)
124. Holtkamp, D. B., Seestrom-Morris, S. J., Dehnhard, D., Baer, H. W., Morris, C. L., et al. *Phys. Rev.* C31: 957 (1985)
125. Olmer, C., Zeidman, B., Geesaman, D. F., Lee, T.-S. H., Segel, R. E., et al. *Phys. Rev. Lett.* 43: 612 (1979)
126. Seestrom-Morris, S. J., Dehnhard, D., Morris, C. L., Bland, L. C., Gilman, R., et al. *Phys. Rev.* C31: 923 (1985)
127. Seestrom-Morris, S. J., Holtkamp, D. B., Cottingame, W. B. *Proc. Int. Conf. Spin Excitations in Nuclei*, Telluride, p. 291. New York: Plenum (1984)
128. Carr, J. A., Petrovich, F., Halderson, D., Holtkamp, D. B., Cottingame, W. B. *Phys. Rev.* C27: 1636 (1983)
129. Moore, C. F., Cottingame, W., Boyer, K. G., Smith, L. E., Harvey, C., et al. *Phys. Lett.* 80B: 38 (1978)
130. Siciliano, E. R., Walker, G. E. *Phys. Rev.* C23: 2661 (1981)
131. Seestrom-Morris, S. J., Dehnhard, D., Holtkamp, D. B., Morris, C. L. *Phys. Rev. Lett.* 46: 1447 (1981)
132. Petrovich, F., Love, W. G., Picklesimer, A., Walker, G. E., Siciliano, E. *Phys. Lett.* 95B: 166 (1980)
133. Miller, G. A., Spencer, J. E. *Ann. Phys. (NY)* 100: 562 (1976)
134. Miller, G. A. *Phys. Rev.* C24: 221 (1981)
135. Kerman, A. K., McManus, H., Thaler, R. M. *Ann. Phys.* 8: 551 (1959)
136. Kaufmann, W. B., Jackson, J. C., Gibbs, W. R. *Phys. Rev.* C9: 1340 (1974)
137. Sparrow, D. A., Rosenthal, A. S. *Phys. Rev.* C18: 1753 (1978)
138. Seth, K. K., Kaletka, M., Barlow, D., Kielczewska, D., Saha, A., et al. *Phys. Lett.* B155: 339 (1985)
139. Seidl, P. A., Brown, M. D., Kiziah, R. R., Moore, C. F., Baer, H., et al. *Phys. Rev.* C30: 973 (1984)
140. Greene, S. J., Braithwaite, W. J., Holtkamp, D. B., Cottingame, W. B., Moore, C. F., et al. *Phys. Rev.* C25: 927 (1982)
141. Seth, K. K. *Nucl. Phys.* A434: 287c (1985)
142. Gilman, R., Fortune, H. T., Zumbro, J. D., Seidl, P. A., Moore, C. F., et al. *Phys. Rev.* C33: 1082 (1986)
143. Seth, K. K., Iversen, S., Nann, H., Kaletka, M., Hird, J., Theissen, H. A. *Phys. Rev. Lett.* 43: 1574 (1979); Green, S. J., Holtkamp, D. B., Cottingame, W. B., Moore, C. F., Burleson, G. R., et al. *Phys. Rev.* C25: 924 (1982)
144. Lee, T.-S. H., Kurath, D., Zeidman, B. *Phys. Rev. Lett.* 39: 1307 (1977)
145. Navon, I., Leitch, M. J., Bryman, D. A., Numao, T., Schlatter, P., et al. *Phys. Rev. Lett.* 52: 105 (1984)
146. Leitch, M. J., Piasetzky, E., Baer, H. W., Bowman, J. D., Burman, R. L., et al. *Phys. Rev. Lett.* 54: 1482 (1985)
147. Altman, A., Johnson, R. R., Wienands, U., Hessey, N., Barnett, B. M., et al. *Phys. Rev. Lett.* 55: 1273 (1985)
148. Barnett, B. M., Grion, N., Hessey, N., Mills, D., Rozon, F. M., et al. *Bull. Am. Phys. Soc.* 31: 802 (1986)
149. Gibbs, W. R., Kaufmann, W. B., Siegel, P. B. *Proc. Los Alamos Workshop on DCX*, ed. H. Baer. Los Alamos. Rep. Los Alamos: LAMPF (1985)
150. Bleszynski, M., Glauber, R. See Ref. 13, AIP 150: 644
151. Karapiperis, T., Kobayashi, M. *Phys. Rev. Lett.* 54: 1230 (1985)
152. Gerace, W. J., Leonard, W. J., Sparrow, D. *Phys. Rev.* C34: 353 (1986)
153. Liu, L. C. *Phys. Rev.* C27: 1611 (1983)
154. Johnson, M. B., Siciliano, E. R. *Phys. Rev.* C27: 1647 (1983)
155. Baer, H., Leitch, M. J., Burman, R. L., Cooper, M. D., Cue, A., et al. *Phys. Rev.* C35: 1382 (1987)
156. Weinfeld, Z., Piasetzky, E., Baer, H. W., Burman, R. L., Leitch, M. J., et al. *Phys. Rev. C.* In press; and contribution to PANIC 87, Kyoto
157. Auerbach, N., Gibbs, W. R., Piasetzky, E., *Phys. Rev. Lett.* In press (1987)
158. Oset, E., Strottman, D., Vicente-Vacas, M. J., Wei-Hsing, M. *Nucl. Phys.* A408: 461 (1983)
159. Miller, G. A. *Phys. Rev. Lett.* 53: 2008 (1984)
160. Burman, R. L., Baker, M. P., Cooper, M. D., Heffner, R. H., Lee, D. M., et al. *Phys. Rev.* C17: 1774 (1978)
161. Seth, K. K., Nann, K., Iversen, S., Kaletka, M., Hird, J. *Phys. Rev. Lett.* 41: 1589 (1978)
162. Nann, H., Seth, K. K., Iversen, S. G., Kaletka, M. O., Barlow, D. B., et al. *Phys. Lett.* 96B: 261 (1980)
163. Burleson, G. R., Blanpied, G. S., Daw, G. H., Viescas, A. J., Morris, C. L., et al. *Phys. Rev.* C22: 1180 (1980)
164. KeKelis, G. J., Zisman, M. S., Scott, D. K., Jahn, R., Vieira, D. J., et al. *Phys. Rev.* C17: 1929 (1978)
165. Seth, K. K., Iversen, S., Kaletka, M., Barlow, D., Saha, A., et al. *Phys. Lett.* 173B: 397 (1986)
166. Sherrill, B., Beard, K., Benson, W., Bloch, C., Brown, B. A., et al. *Phys. Rev.* C31: 875 (1985)
167. Nann, H. In *VI Int. Conf. on Atomic Masses, 1979*, East Lansing; Seth, K. K., Artuso, M., Barlow, D., Iversen, S.,

Kaletka, M., et al. *Phys. Rev. Lett.* 58: 1930 (1987)

168. Gilly, L., Jean, M., Meunier, R., Spighel, M., Stroot, J. P., et al. *Phys. Lett.* 19: 335 (1965)

169. Kaufman, L., Perez-Mendez, V., Sperinde, J. *Phys. Rev.* 175: 1358 (1968)

170. Seestrom-Morris, S. J., Morris, C. L., Holtkamp, D. B., Dehnhard, D., Blilie, D., et al. *Phys. Rev.* C35: 2210 (1987)

171. Deleted in proof

172. Peterson, R. J., Boudrie, R. L., Kraushaar, J. J., Ristinen, R. A., Shepard, J. R., et al. *Phys. Rev.* C21: 1030 (1980)

173. Morris, C. L., Cottingame, W. B., Greene, S. J., Harvey, C. J., Moore, C. F., et al. *Phys. Rev. Lett.* 108B: 172 (1982)

174. Deleted in proof

175. Gibbs, W. R. *Phys. Rev.* C3: 1127 (1971)

176. Brady, T. J., Tomusiak, E. L., Levinger, J. L. *Can. J. Phys.* 52: 1322 (1974)

177. Blankenbecler, R., Gunion, J. F. *Phys. Rev.* D4: 718 (1971)

178. Axen, D., Duesdieker, G., Felawka, L., Ingram, Q., Johnson, R., et al. *Nucl. Phys.* A256: 387 (1976); Gabathuler, K., Domingo, J., Gram, P., Hirt, W., Jones, G., et al. *Nucl. Phys.* A350: 253 (1980); Stanovnik, A., Kernel, G., Tanner, N. W., Bressani, T., Chiavassa, E., et al. *Phys. Lett.* 94B: 323 (1980); Norem, J. H. *Nucl. Phys.* B33: 512 (1971); Cole, R. H., McCarthy, J. S., Minehart, R. C., Wadlinger, E. A. *Phys. Rev.* C17: 681 (1978)

179. Afnan, I. R., McLeod, R. J. *Phys. Rev.* C31: 1821 (1985)

180. Avashai, Y., Mizutani, T. *Nucl. Phys.* A326: 352 (1979); A338: 377 (1980); *Phys. Rev.* C27: 312 (1983)

181. Thomas, A. W., Rinat, A. S. *Phys. Rev.* C20: 216 (1971)

182. Afnan, I. R., Blankleider, B. *Phys. Rev.* C22: 1638 (1980)

183. Mizutani, T., Fayard, C., Lamot, G. H., Nahabetian, R. S. *Phys. Lett.* 107B: 177 (1981)

184. Fayard, C., Lamot, G. H., Mizutani, T. *Phys. Rev. Lett.* 45: 524 (1980)

185. Rinat, A. S., Starkand, Y. *Nucl. Phys.* A397: 381 (1983)

186. Afnan, I. R., Stelbovics, A. T. *Phys. Rev.* C23: 1384 (1981)

187. Bolger, J., Boschitz, E., Proebstle, G., Smith, G. R., Mango, S., et al. *Phys. Rev. Lett.* 46: 167 (1981); 48: 1667 (1982); Smith, G. R., Mathie, E. L., Boschitz, E. T., Ottermann, G. R., Mango, S., et al. *Phys. Rev.* C29: 2206 (1984)

188. Holt, R. J., Specht, J. R., Stephenson, E. J., Zeidman, B., Burman, R. L., et al. *Phys. Rev. Lett.* 43: 1229 (1979); Holt, R. J., Specht, J. R., Stephenson, K., Zeidman, B., Frank, J. S., et al. *Phys. Rev. Lett.* 47: 472 (1981); Ungricht, E., Freeman, W. S., Geesaman, D. F., Holt, R. J., Specht, J. R., et al. *Phys. Rev.* C31: 934 (1985)

189. Shin, Y. M., Itoh, K., Stevenson, N. R., Gill, D. R., Ottewell, D. F., et al. *Phys. Rev. Lett.* 55: 2672 (1985)

190. Smith, G. R. See Ref. 13, AIP 150: 1219 (1986)

191. Garcilazo, H. *Phys. Rev. Lett.* 53: 652 (1984)

192. Rinat, A. S. See Ref. 78, 45: 621 (1978)

193. Blankleider, B., Afnan, I. R. *Phys. Rev.* C31: 1380 (1985)

194. Mizutani, T., Fayard, C., Lamot, G. H., Nahabetian, S. *Phys. Rev.* C24: 2633 (1981)

195. Rinat, A. S. *Nucl. Phys.* A377: 341 (1982)

196. Morioka, S., Afnan, I. R. *Phys. Rev.* C26: 1148 (1982)

197. Pearce, B. C., Afnan, I. R. *Phys. Rev.* C34: 991 (1986)

198. Carlson, C. E., Gross, F. *Phys. Rev. Lett.* 53: 127 (1984)

199. Brodsky, S. J., Farrar, G. R. *Phys. Rev. Lett.* 31: 1153 (1973)

200. Garcilazo, H. *Phys. Rev. Lett.* 53: 652 (1984)

201. Dymarz, R., Khanna, F. C. *Phys. Rev. Lett.* 56: 1448 (1986)

202. Blanpied, G. S., Mishra, C. S., Adams, G. S., Preedom, B., Whisnant, C. S., et al. *Phys. Rev.* C35: 1567 (1987)

203. Ritchie, B. G. *Phys. Rev.* C28: 926 (1983)

204. Brack, W. M., Riska, D. O., Weise, W. *Nucl. Phys.* A287: 425 (1977)

205. McKeown, R. D., Sanders, S. J., Schiffer, J. P., Jackson, H. E., Paul, M. et al. *Phys. Rev.* C24: 211 (1981)

206. Girija, V., Koltun, D. S. *Phys. Rev. Lett.* 52: 1397 (1984); *Phys. Rev.* C31: 2147 (1985)

207. Ashery, D., Navon, I., Azuelos, G., Walter, H. K., Pfeiffer, H. J., et al. *Phys. Rev.* C23: 2173 (1981)

208. Nakai, K., Kobayashi, T., Numao, T., Shibata, T. A., Chiba, J., et al. *Phys. Rev. Lett.* 44: 1446 (1980)

209. Altman, A., Piasetzky, E., Lichtenstadt, J., Yavin, A. I., Ashery, D., et al. *Phys. Rev. Lett.* 50: 1187 (1983); Altman, A., Ashery, D., Piasetzky, E., Lichtenstadt, J., Yavin, A. I., et al. *Phys. Rev.* C34: 1757 (1986)

210. Burger, W. J., Beise, E., Gilad, S., Redwine, R. P., Roos, P. G., et al. *Phys. Rev. Lett.* 57: 58 (1986)

211. Gotta, D., Dörr, M., Fetscher, W., Schmidt, G., Ullrich, H., et al. *Phys. Lett.* 112B: 129 (1982)
212. Dörr, M., Fetscher, W., Gotta, D., Reich, J., Ullrich, H., et al. *Nucl. Phys.* A445: 557 (1985)
213. Lorentz, H. A. *Ann. Phys.* 9: 64 (1880); Lorenz, L. *Ann. Phys.* 11: 70 (1881)
214. Ericson, M., Ericson, T. E. O. *Ann. Phys.* 36: 323 (1966)
215. Eisenberg, J. M., Hufner, J., Moniz, E. J. *Phys. Lett.* B47: 381 (1973)
216. Kaufmann, W. B., Gibbs, W. R. *Phys. Rev.* C28: 1286 (1983)
217. Gibbs, W. R., Gibson, B. F., Stephenson, G. J. *Phys. Rev. Lett.* 39: 1316 (1977)
218. Cooper, M. D., Baer, H. W., Bolton, R., Bowman, J. D., Cverna, F., et al. *Phys. Rev. Lett.* 52: 1100 (1984)
219. Irom, F., Baer, H. W., Bowman, J. D., Cooper, M. D., Piasetzky, E., et al. *Phys. Rev.* C31: 1464 (1985)
220. Ullmann, J. L., Alons, P. W. F., Kraushaar, J. J., Mitchell, J. H., Peterson, R. J., et al. *Phys. Rev.* C33: 2092 (1986)
221. Siciliano, E. R., Cooper, M. D., Johnson, M. B., Leitch, M. J. *Phys. Rev.* C34: 267 (1986)
222. Irom, F., Leitch, M. J., Baer, H. W., Bowman, J. D., Cooper, M. D., et al. *Phys. Rev. Lett.* 55: 1862 (1985)
223. Jennings, B. K., de Takacsy, N. *Phys. Lett.* 124B: 302 (1983)

*Ann. Rev. Nucl. Part. Sci.* 1987. 37: 463–91

# NUCLEAR EFFECTS IN DEEP INELASTIC LEPTON SCATTERING[1]

*E. L. Berger and F. Coester*

Argonne National Laboratory, Argonne, Illinois 60439

CONTENTS

## 1. INTRODUCTION

Fractionally charged quark constituents of spin $1/2$ were postulated early in the 1960s to explain significant patterns in the spectroscopy of mesons and baryons. Toward the end of that decade, regularities observed in data from scattering experiments provided complementary evidence for constituent substructure in nucleons. In these investigations, high-energy electrons were scattered inelastically from nucleons. The cross section was observed to remain large at large momentum transfer, suggestive of the presence of point-like scattering centers within nucleons, called "partons." Studies, which continue to this day with high-energy electron, muon, and neutrino beams, have established that these parton constituents have the properties expected of quarks: spin, fractional charge, baryon number.

[1] Work supported by the US Department of Energy, Division of Nuclear Physics and Division of High Energy Physics, Contract W-31-109-ENG-38. The US Government has the right to retain a nonexclusive royalty-free license in and to any copyright covering this paper.

Data from lepton scattering experiments measure directly the momentum distributions of quarks and antiquarks within the target hadron (for reviews, see 1).

In parallel with significant experimental advances, the 1970s saw the development of a non-Abelian gauge theory of interacting quarks and gluons, quantum chromodynamics (QCD). This theory provides a basis for understanding both the confinement of quarks and their scattering dynamics. One of its heralded quantitative successes is the prediction of the detailed momentum transfer dependence of inelastic lepton scattering in the very large momentum transfer or "deep inelastic" limit.

In this review, we discuss deep inelastic lepton scattering as a probe of the parton structure of nuclei. [For previous reviews of both experiment and theory, as well as an extensive list of references, see (2).] A common prejudice before 1983 held that, in deep inelastic scattering, a nucleus acts as an incoherent collection of free nucleons. The cross section per nucleon of a nucleus was therefore expected to be the same for all nuclei, except for kinematic effects of the Fermi motion of the nucleons. This implied that the quark momentum distribution in a nucleus was given simply by the appropriate average of the quark momentum distributions of the neutrons and protons in the nucleus. Some excitement resulted therefore from the discovery by the European Muon Collaboration (EMC) (3) that the ratio of cross sections from iron and deuterium targets differs from unity in a manner that cannot be attributed to Fermi motion of free nucleons. This observation, known as the "the EMC effect," has been confirmed in subsequent experiments with various nuclei (4–8).

The data show directly that the quark momentum distribution of a nucleus differs significantly from that of a free nucleon. It does not, however, follow logically that quark effects in nuclear structure are being observed. At minimum, the data call for a quantitative examination of the consequences of the conventional assumption that a nucleus is a bound system of nucleons or of nucleons, mesons, and isobars. The deviations of the cross section obtained in this manner from the cross section of a collection of free nucleons are called nuclear effects. Successful interpretations of all available data have been obtained in this manner (9–15). This success does not exclude the possible relevance of explicit quark effects, which may be mandated by theoretical beliefs about confinement scales and quark deconfinement in nuclei. Indeed, in one approach, known as "rescaling," the EMC effect is believed to show that the typical length associated with quark confinement in the nuclear ground state is longer than the corresponding length in a free nucleon (16–21).

In this paper, we review the data and interpretations of the EMC effect, concluding with a summary of desirable future investigations. We begin with a summary of the relevant kinematic relations and the definition of the invariant structure functions. Data are summarized in Section 3 along with general statements regarding their interpretation. In Section 4, we review the spectrum of models (2) proposed to explain the data, concentrating specifically on the rescaling approach (16–21) and on explanations in terms of nuclear binding (9–15). One of our conclusions, expressed in Section 5, is that current data are fully compatible with an interpretation in terms of conventional nuclear binding. There is no hard evidence in these data for new QCD effects such as a change in the confinement size of nucleons. However, more precise data, especially at small values of the Bjorken scaling variable $x$ ($x \lesssim 0.3$) are desirable for future quantitative tests of all interpretations.

## 2.   DEFINITIONS AND KINEMATICS

We consider inclusive scattering of a lepton $\ell$ from either a nucleon N or a nuclear target A. This process is usually denoted $\ell A \to \ell'X$, where the symbol X represents an inclusive sum over all final states. The initial four-momenta of the lepton and the target are denoted by $k$ and $p$. The four-vector $q$ is the momentum transfer from the initial lepton to the target; that is, it is the difference between the four-momenta of the initial and final leptons. The incident lepton may be an electron, a muon, or a neutrino. The laboratory energies of the initial and final leptons are $E$ and $E'$; $\nu$ is the energy transfer, $E-E'$, in the laboratory frame. It is conventional to define $Q^2 = \mathbf{q}^2 - \nu^2 = -q^2 > 0$, and two dimensionless variables $x$ and $y$, $x = Q^2/2M_N\nu$, $y = \nu/E$, where $M_N$ is the mass of the nucleon. (In this review, boldface symbols represent Euclidean three-vectors.)

The deep inelastic domain is that in which the energy transfer is large compared to the four-momentum transfer,

$$\nu^2/Q^2 \equiv Q^2/(2M_N x)^2 \gg 1. \qquad\qquad 1.$$

Light-front components $p^{\pm}$ of any four-vector $p$ are defined by

$$p^{\pm} \equiv p^0 \pm \mathbf{n \cdot p}, \qquad\qquad 2.$$

where $\mathbf{n}$ is a unit vector, which we choose in the direction of the momentum transfer, $\mathbf{n} = -\mathbf{q}/|\mathbf{q}|$. For deep inelastic scattering, $q^- \approx 2\nu$, $q^+ \approx 0$, $p \cdot q \approx \frac{1}{2}p^+q^- \approx p^+\nu$. Light-front momentum fractions are defined as ratios of plus-components of momenta and are thus invariant under longitudinal boosts.

In the one-photon-exchange approximation, the differential cross section for inclusive inelastic scattering of a charged lepton (eA $\to$ e'X or $\mu$A $\to$ $\mu$'X) by any target is proportional to tensors $f_{\lambda\rho}(q,k)$ and $F^{\lambda\rho}(q,p)$

that depend respectively on the properties of the lepton and the target only:

$$\frac{d^2\sigma}{dx\,dy} = \frac{4\pi\alpha^2 2M_N y}{Q^4} F^{\lambda\rho} f_{\lambda\rho}.$$    3.

The structure tensor of the target, $F^{\lambda\rho}$, is a bilinear functional of the matrix elements of the electromagnetic current operator, $j^\lambda$:

$$F^{\lambda\rho} \equiv \int d^4 p' \int d\alpha 2\delta[p'^2 + M(\alpha)^2]\delta(p' - p - q)$$

$$\times \langle p|j^\lambda(0)|\alpha, p'\rangle \langle p', \alpha|j^\rho(0)|p\rangle.$$    4.

Here $M(\alpha)$ labels the mass of the system X in $\ell A \to \ell' X$, and $\alpha$ designates the set of quantum numbers of X. The tensor $F^{\lambda\rho}$ is always a linear combination of "structure functions," invariant functions of $Q^2$ and $p\cdot q$, multiplied by universal covariant functions of $p$ and $q$. In the case of the scattering of unpolarized particles, there are two structure functions $F_i(x', Q^2)$, $x' \equiv Q^2/(2p\cdot q) = x(M_N \nu/p\cdot q)$, for the conserved electromagnetic current.

In the deep inelastic approximation we have

$$F_N^{\lambda\rho} f_{\lambda\rho} = E\{(1-y)F_2^N(x', Q^2) + 2x' F_1^N(x', Q^2)(y^2/2)\}/y,$$    5.

for scattering by a nucleon, and

$$F_A^{\lambda\rho} f_{\lambda\rho}/A = E\{(1-y)F_2^A(x'_A, Q^2) + 2x'_A F_1^A(x'_A, Q^2)(y^2/2)\}/y,$$    6.

for a nucleus. We use the nucleon number $A$ also as a label identifying the nucleus. In Equation 6 for the nucleus, we have introduced a new variable $x'_A$, which is a multiple of $x'$; $x'_A \equiv Q^2 A/(2P_A \cdot q)$. If the nucleus is at rest, $x'_A$ is approximately equal to $x$: $x'_A = xAM_N/M_A \approx x$. The ranges allowed kinematically for $x'$ of the nucleon and for $x'_A$ are $(0,1)$ and $(0,A)$ respectively. The structure functions $F_1^A$ and $F_2^A$ are structure functions *per nucleon*.

We note in Equations 3 and 5 that the momentum of the target enters only through the variable $x'$. This means that for a collection of incoherent free nucleons the momentum-averaged cross section is related to momentum-averaged structure functions in the same manner as the cross section per nucleon of the nucleus is related to the structure functions of the nucleus per nucleon. The momentum average is called Fermi smearing (22, 23).

For neutrino and antineutrino charged current processes ($\nu A \to \mu^- X$, $\bar\nu A \to \mu^+ X$), the cross section has again the general form of Equation 3 with the electromagnetic current replaced by the weak current, and

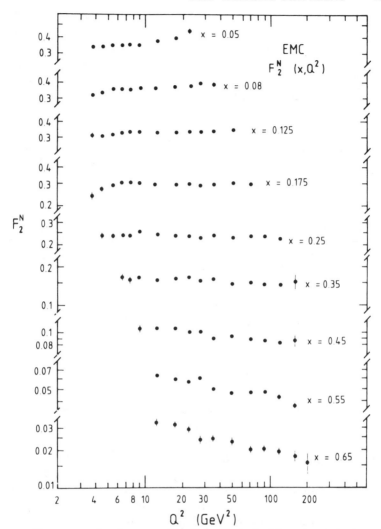

*Figure 1*    The structure function per nucleon $F_2(x, Q^2)$ obtained from deep inelastic muon scattering on iron is shown as a function of $Q^2$ at different values of $x$. These data were published by the European Muon Collaboration (25).

$4\pi\alpha^2/Q^4$ replaced by $g^2/2\pi$. In that case there are three invariant structure functions

$$\frac{\mathrm{d}^2\sigma}{\mathrm{d}x\,\mathrm{d}y} = \frac{g^2 M_N E}{\pi}\{(1-y)F_2^{\nu A}(x'_A, Q^2) + 2x'_A F_1^{\nu A}(x'_A, Q^2)(y^2/2)$$

$$\pm x'_A F_3^{\nu A}(x'_A, Q^2)y(1-y/2)\}. \qquad\qquad 7.$$

The set $(F_1, F_2)$ may be replaced by the set $(F_2, R)$, where $R \equiv \sigma_L/\sigma_T \equiv (F_2 - 2xF_1)/2xF_1$; $\sigma_T$ and $\sigma_L$ are cross sections for absorption of transversely and longitudinally polarized virtual photons. If $R$ is independent of $A$, the ratio of cross sections, Equation 3, for two different values of $A$ yields directly the ratio of the values of the functions $F_2^A(x, Q^2)$. Data (24) show that $R \approx 0$, except for small $x$. If $R \equiv 0$ is inserted in the equations, one may use data on $\mu A \to \mu' X$ to determine $F_2^{\mu A}$ directly, and use data on $\nu A \to \mu^- X$ and $\bar{\nu}A \to \mu^+ X$ to isolate $F_2^{\nu A}$ and $F_3^{\nu A}$. Values of $F_2^A(x, Q^2)$ obtained from experiments with muon beams are shown in Figure 1; in the current data sample $Q^2$ extends up to about 200 GeV$^2$.

In the parton model (26), as extended by quantum chromodynamics (e.g. 27), probabilities $q_f^A(z, Q^2)$, $\bar{q}_f^A(z, Q^2)$, and $G^A(z, Q^2)$ are defined that represent the quark, antiquark, and gluon number *densities in a nucleus*, $A$. These are densities per nucleon, just as are the structure functions $F_i^A(x, Q^2)$, meaning that a factor of $A$ has been divided out. Subscript f on $q_f$ and $\bar{q}_f$ labels the flavor of the quark or antiquark. Variable $z$ is the light-front fraction of the momentum of the target $A$ carried by a parton of a given type.

In the parton model (26), $z$, $q_f^A(z, Q^2)$, $\bar{q}_f^A(z, Q^2)$, and $G^A(z, Q^2)$ are measurable quantities. Indeed, $z \equiv x$, with $x$ determined from the lepton kinematics, as defined above. Furthermore,

$$F_2^{\mu A}(x, Q^2) = \sum_f e_f^2 x[q_f^A(x, Q^2) + \bar{q}_f^A(x, Q^2)]. \qquad 8.$$

Thus, the observable $F_2^{\mu A}$ measures the fraction of momentum of the target $A$, per nucleon, carried by quarks and antiquarks, weighted by the squares of the fractional quark charges $e_f$. Constraints on $G^A(x, Q^2)$ are obtained from detailed studies of the $Q^2$ dependence of $F_2(x, Q^2)$ as well as from data on other hard scattering processes mentioned in Section 5.

For neutrino and antineutrino charged current processes ($\nu A \to \mu^- X$, $\bar{\nu}A \to \mu^+ X$) on an "isoscalar" target A containing an equal number of protons and neutrons, the structure functions per nucleon are

$$F_2^{\nu A} = x[u + d + \bar{u} + \bar{d} + 2s];$$

$$xF_3^{\nu A} = x[u + d - \bar{u} - \bar{d} + 2s];$$

$$xF_3^{\bar{\nu} A} = x[u + d - \bar{u} - \bar{d} - 2s]. \qquad 9.$$

For the up and down flavors, $u(x) = u_p(x) = d_n(x)$, and $d(x) = d_p(x) = u_n(x)$; subscripts p and n denote proton and neutron. For the strange quark ocean, $s_p(x) = s_n(x) = s(x) = \bar{s}(x)$. Note that structure function

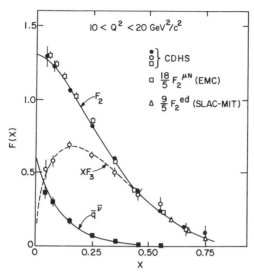

*Figure 2*  The structure functions $F_2(x, Q^2)$ and $xF_3(x, Q^2)$ and the antiquark density $\bar{q}^\nu(x, Q^2)$ measured by the CERN-Dortmund-Heidelberg-Saclay collaboration (28) are displayed as a function of $x$; $Q^2$ is restricted to the range $10 < Q^2 < 20$ GeV$^2$. The data on $F_2(x, Q^2)$ are compared with measurements of $F_2^{\mu N}$ by the EMC group (47) and of $F_2^{ed}$ by the SLAC-MIT collaboration (48). The $\mu N$ and ed data are multiplied by factors specified by the parton model.

$F_2$ provides information on the sum $q_f(x) + \bar{q}_f(x)$, whereas $F_3$ yields the difference. Together, the two may be used to isolate the ocean $\bar{q}_f(x)$ from $q_f(x)$. An example of this decomposition is presented in Figure 2.

For each quark flavor, a decomposition may be made of the form $q(x) = q_V(x) + q_O(x)$, where subscripts V and O denote "valence" and "ocean," with $u_V(x) = u(x) - \bar{u}(x)$ and $d_V(x) = d(x) - \bar{d}(x)$. As indicated in Figure 2, data (28, 29) show that at $Q^2 \approx 20$ GeV$^2$, the valence components are dominant for $x \gtrsim 0.1$, with the ocean vanishing for $x \gtrsim 0.3$. Below $x \approx 0.1$, the ocean quark densities are substantial.

## 3.  NUCLEAR DEPENDENCE

Turning to the principal subject of this review, we wish to compare the structure functions $F_i^A(x, Q^2)$ with $F_i^N(x, Q^2)$, and, by inference, $q^A(x, Q^2)$ with $q^N(x, Q^2)$, and $\bar{q}^A(x, Q^2)$ with $\bar{q}^N(x, Q^2)$. Functions with superscript N refer to those of free nucleons.

A standard expectation before 1983 was that a nucleus behaves as an incoherent collection of free nucleons at large $Q^2$, at least for

0.05 $< x <$ 0.7. This means that the structure functions of the nucleus are simply momentum averages of the nucleon structure functions:

$$xF_i^A(x, Q^2) = \int_{z>x} \mathrm{d}z(x/z)F_i^N(x/z, Q^2)f_N(z), \quad \text{for } i = 1 \text{ or } 3,$$

$$F_2^A(x, Q^2) = \int_{z>x} \mathrm{d}z\, F_2^N(x/z, Q^2)f_N(z). \qquad\qquad 10.$$

The function $f_N(z)$ is determined by the momentum distribution $\rho(\mathbf{p})$ of the nucleons, normalized to unity, $\int \mathrm{d}^3\mathbf{p}\,\rho(\mathbf{p}) = 1$:

$$f_N(z) = \int \mathrm{d}^3\mathbf{p}\,\rho(\mathbf{p})\delta[z - p\cdot q/M_N\nu] \approx \int \mathrm{d}^3\mathbf{p}\,\rho(\mathbf{p})\delta[z - p^+/M_N], \qquad 11.$$

with $p^+ \approx \mathbf{n}\cdot\mathbf{p} + M_N$.

At small enough $x$ or $Q^2$, shadowing (30) is expected to invalidate Equation 10. Because the virtual photon does not resolve individual nucleons within the nucleus, the cross section cannot grow linearly with $A$. Nonvanishing values of the structure functions $F_i^A(x, Q^2)$ for $x > 1$, which must be expected in principle, are associated with the behavior of $f_N(z)$ for large $z$ (23, 31).

Except for shadowing at very small $x$ and Fermi smearing at large $x$, it was generally believed (see 32 for an exception) that Equation 10 with $f_N(z) = \delta(z-1)$ would hold to a good approximation. For quark densities, this statement may be expressed in the form

$$Aq_i^A(x, Q^2) = Zq_i^p(x, Q^2) + (A - Z)q_i^n(x, Q^2). \qquad 12.$$

Moreover, prior to 1983, experiments were not of sufficient statistical or systematic precision to discern more than gross deviations from Equations 10 or 12. The first "test" was published by the European Muon Collaboration (3). Since it is known that the $x$ dependences of up and down quark densities differ, it would be inappropriate to compare heavy-target data with hydrogen data. The EMC group compared data from iron with data from deuterium. The fact that there is a slight neutron excess in Fe leads to only a small correction. The data were published in the form of a ratio:

$$R_{EMC}(x) \equiv \frac{F_2^{Fe}(x)}{F_2^D(x)}. \qquad\qquad 13.$$

A compilation is presented in Figure 3 of data published prior to 1986. The data of the CERN-EMC (3) and Bologna-CERN-Dubna-Munich-Saclay (BCDMS) (5) collaborations are taken at "large $Q^2$" in the sense

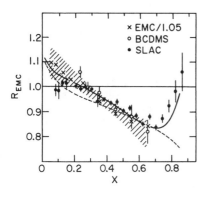

*Figure 3* A compilation of data published prior to 1986 on the ratio of structure functions $R_{EMC}(x, Q^2) \equiv F_2^{Fe}(x, Q^2)/F_2^D(x, Q^2)$ for deep inelastic electron and muon scattering. Shown are published results from the EMC Collaboration (3), divided by 1.05, as well as data from the BCDMS Collaboration (5), and from SLAC experiments (4). The shaded band indicates the EMC group's estimate of experimental systematic uncertainties. The solid curve is calculated from the pion exchange model of Ref. 12. The dashed curve shows the expectation of $Q^2$ rescaling (17) with $Q^2 = (200x + 10)\,\text{GeV}^2$ appropriate for the kinematics of the EMC data.

that $Q^2 \gtrsim 10\,\text{GeV}^2$ for all $x$. On the other hand, in the SLAC-E139 data sample (4), $Q^2$ descends to rather small values at small $x$ (i.e. $Q^2 \approx 1$ GeV$^2$). The shaded band in Figure 3 indicates the systematic uncertainty in the EMC measurement. In addition, there is a $\pm 7\%$ normalization uncertainty in the ratio (3). No marked $Q^2$ dependence of $R_{EMC}(x, Q^2)$ is observed within either the CERN or SLAC data samples. Nevertheless, $Q^2$ dependence of $R_{EMC}(x, Q^2)$ might be responsible for the discrepancy between the CERN and SLAC data in Figure 3. Another potential resolution lies in a possible dependence of $\sigma_L/\sigma_T$ on $A$ at low $Q^2$ (33). These possibilities will be examined in future experiments.

The incoherence assumption expressed in Equations 10 and 12 implies that $R_{EMC}(x, Q^2) \approx 1$, for $0.05 \lesssim x \lesssim 0.7$. The data in Figure 3 show a $\pm 15\%$ deviation from this expectation. The "EMC effect" is precisely this deviation, especially the excursion above $R_{EMC} = 1$ at small $x$.

Although it is necessary to try to understand the $x$ dependence in Figure 3 for all $x$ in a unified fashion, it is useful to begin with an examination of the data in three distinct regions of $x$:

$x \lesssim 0.2$ *("small $x$")*. In this region the antiquark density is important, and shadowing presumably influences results in a $Q^2$-dependent fashion. Neutrino data (6), shown in Figure 4, indicate that the antiquark density is increased at small $x$ by at most a modest amount. While of limited precision, the neutrino data contradict early proposed interpretations (34) in terms of a 40 to 60% increase in the ocean.

The greatest source of confusion about the small-$x$ region has been whether one should focus on the original EMC results, which show a clear rise of $R_{EMC}(x)$ above unity, or on the SLAC data, which show no rise. This problem is now resolved as a result of new experiments carried out

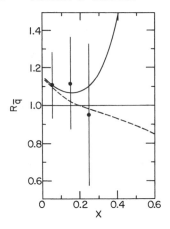

*Figure 4* Data of Abramowicz et al (6) on the ratio $R_{\bar{q}}(x) = \bar{q}^{Fe}(x)/\bar{q}^{P}(x)$ from neutrino scattering. Here $\bar{q} = (\bar{u} + \bar{d} + 2\bar{s})$. The solid curve illustrates predictions of the pion exchange model (12), whereas the dashed curve shows expectations of the rescaling model (17).

with particular attention devoted to systematic and relative normalization uncertainties at small $x$. New data (7, 8) from both the EMC and BCDMS collaborations are shown in Figure 5. The new BCDMS results in Figure 5 show clearly that $F_2^{Fe}(x)/F_2^{D}(x)$ exceeds unity for $x < 0.3$. The excess is in the neighborhood of 5%, less than the $\sim 15\%$ effect suggested in the original EMC data. In Figure 6, the new BCDMS results are compared with the original EMC data. The two sets of measurements are seen to be compatible as long as one takes into account the estimated systematic and normalization uncertainties. The new EMC data shown in Figure 5 also confirm that $R_{EMC}^{Cu}(x)$ exceeds unity for $0.05 < x < 0.3$. Bearing in mind that $A_{Cu} = 63$ but $A_{Fe} = 56$, one would expect a slightly larger "EMC effect" in Cu than in Fe at the same $Q^2$. However, the magnitude of the deviation of $R_{EMC}^{Cu}(x)$ from 1 at small $x$ appears to be 3 to 5% smaller than

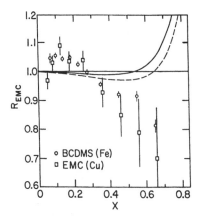

*Figure 5* Recent data on the ratio $F_2^{A}(x, Q^2)/F_2^{D}(x, Q^2)$ from the EMC (7) and BCDMS (8) collaborations; $A = $ Cu in the EMC case, whereas $A = $ Fe for BCDMS. The normalization uncertainty is quoted as $\pm 1.5\%$ by BCDMS. The curves illustrate the expectation based on Fermi smearing, which we have calculated with a function $f_N(z)$ obtained from a Fermi gas momentum distribution folded with the local densities of nucleons in Fe, illustrated in Figure 3 of Ref. 12. The solid curve is calculated with $M_A = AM_N$; the dashed curve is calculated with $(M_A/A) = M_N - 8$ MeV.

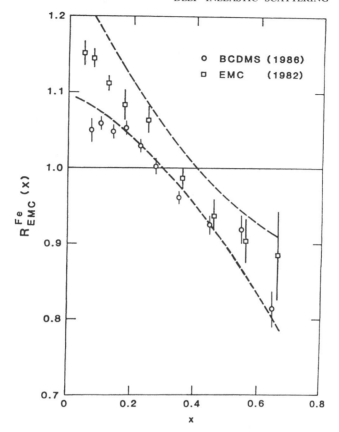

*Figure 6*  Comparison of the original EMC data (3) with recent BCDMS data (8). Shown is the band indicative of estimated systematic uncertainties for the EMC data. For BCDMS, systematic uncertainties (not shown) are comparable to the statistical errors for all points.

that of the original data on $R^{Fe}_{EMC}(x)$. The new BCDMS and EMC data lend further justification to the 5% renormalization of the original EMC data made in Figure 3.

There is some evidence that the ratio in Figure 5 decreases below unity as $x$ becomes very small, perhaps indicative of the shadowing limit, $R_{EMC} = A^{-1/3}$. However, caution is in order because radiative corrections become significant at small $x$ and because the decrease sets in at values of $x$ larger than those normally associated with shadowing (30, 35).

Our conclusion is that there is indeed an excursion of $R_{EMC}(x)$ above unity for $0.05 < x < 0.2$ and $Q^2 \gtrsim 10$ GeV$^2$. The magnitude of the excursion is in the range $R_{EMC}(x) \approx 1.05$. We surmise that the absence of an excess in the SLAC data is to be attributed to the fact that the SLAC data at small $x$ are taken at very small $Q^2$.

*0.2 ≲ x ≲ 0.7.* In this intermediate region of $x$ all experiments are in agreement that $R_{EMC}(x) < 1$. The ratio $R_{EMC}(x, Q^2)$ shows no pronounced $Q^2$ dependence. Since the valence quarks are dominant in this region, the data show directly that the momentum distribution of the valence quarks is softened.

In order to maintain momentum balance, softening of the valence quark component means that, at a given $Q^2$, the ocean quark and/or glue components must carry more momentum per nucleon in a nucleus than in a free nucleon. The net change in the charge-weighted momentum fraction carried by the quarks and antiquarks is

$$\Delta \equiv \int [F_2^{Fe}(x, Q^2) - F_2^D(x, Q^2)] \, dx. \qquad 14.$$

The new data show that $\Delta = 0$ within errors, which implies that the net momentum fraction per nucleon carried by the glue is the same in iron and deuterium.

*x ≳ 0.7.* In this region of $x$, 70% of the momentum of the nucleon is carried by a single parton. The structure functions $F_2^A(x, Q^2)$ and $F_2^N(x, Q^2)$ are both very small, and $R_{EMC}(x)$ is the ratio of two small quantities. The effects of Fermi motion are large in this region, as shown in Figure 5.

To recapitulate, the original ECM effect was defined in terms of

1. a strong, $\sim 15\%$, enhancement of $R_{EMC}^{Fe}(x)$ above unity for $x \lesssim 0.2$ and $Q^2 \gtrsim 10 \text{ GeV}^2$;
2. a depletion, $R_{EMC}(x) < 1$, for $0.2 \lesssim x \lesssim 0.7$; and
3. a possibly large, $\sim 40$ to $60\%$, enhancement of the antiquark ocean, $\bar{q}^A(x)$.

The data now present a different picture. The effect consists of

1. a confirmed, approximately 5% enhancement of $R_{EMC}^{Fe}(x)$ above unity for $x \lesssim 0.2$ and $Q^2 \gtrsim 10 \text{ GeV}^2$, but no enhancement for $Q^2 \approx 1 \text{ GeV}^2$;
2. a depletion, $R_{EMC}(x) < 1$, for $0.2 \lesssim x \lesssim 0.7$, independent of $Q^2$ (for $Q^2 \gtrsim 5 \text{ GeV}^2$);
3. a modest enhancement of the antiquarks $\bar{q}^A(x, Q^2)$ of order 10%: $\langle R_{\bar{q}} \rangle = 1.10 \pm 0.10 \pm 0.07$; and
4. no change in the net momentum fraction carried by the glue, $\int xG^A(x, Q^2) \, dx$.

In attempts to model the EMC effect, various authors have emphasized phenomena in different regions of $x$. Some have focussed on the enhancement in $R_{EMC}(x)$ at small $x$, others on the depletion at intermediate $x$, and others on the region $x \gtrsim 0.7$. In our view, the essential aspects of the "effect" are the simultaneous enhancement of $R_{EMC}(x)$ above unity for $x \lesssim 0.2$ and the decrease at larger $x$.

We emphasize that measurements with nuclear targets determine the structure functions of nuclei, not the structure functions of "nucleons within nuclei." The data discussed in this section show that Equation 12 is violated for all $x$. It is clear that $q^A(x, Q^2) \neq q^N(x, Q^2)$, and that, as shown in Figure 5, this difference cannot be explained by Fermi motion and the small binding effect, $M_A < A M_N$. A bias-free approach to interpretations begins with the realization that $q^A(x, Q^2)$, $\bar{q}^A(x, Q^2)$, and $G^A(x, Q^2)$ are also parton densities of a nucleus, not necessarily parton densities of nucleons within a nucleus.

## 4.   INTERPRETATIONS

Since deep inelastic lepton scattering experiments provided the empirical basis for the parton model of nucleons, it was natural for many to regard the EMC effect as directly indicative of a nontrivial role for parton degrees of freedom in nuclear structure. At present, it is not possible to be categorical on this point. As we describe below, quantitative interpretations of the data can be obtained in terms of conventional nuclear models. These approaches are based, however, on plausible assumptions whose validity may eventually be challenged by a more fundamental description of the nucleus in terms of QCD. Descriptions of the EMC effect in the language of QCD, such as color conductivity (16) and the rescaling model (17–21), also require model assumptions about the solution to the confinement problem in QCD. In either case, we face squarely the fact that while QCD has been tested in the perturbative large momentum transfer domain, it has not yet been solved to yield hadron structure. We must settle for simplified models of constituent confinement, motivated by or consistent with QCD.

Intepetations of nuclear effects in deep inelastic scattering cover the full spectrum of perspectives of the structure of nuclei. In the paragraphs to follow, we first describe the range of possibilities and then discuss two interpretations in detail.

The most fundamental approach is based on general properties of quantum field theory. It starts from the observation that the Fourier transform of the structure tensor $F^{\lambda\rho}(p, q)$ is given by

$$\int d^4q \, e^{iq\cdot\xi} F^{\lambda\rho}(p,q) = \langle p|j^\lambda(\xi)j^\rho(0)|p\rangle, \qquad\qquad 15.$$

and that only the neighborhood of the light cone $\xi^2 = 0$ matters in the deep inelastic domain. The property of asymptotic freedom of QCD permits the derivation of the main features of the parton model, including the logarithmic $Q^2$ dependence of the structure functions. Since quantum field theory does not yield explicit representations of the state vectors of bound states, this approach does not lead to complete predictions of the structure functions of nucleons, much less of nuclei. It does lead to parameterizations that respect the requirements of field theories in general and QCD in particular. Model features enter through the particular functional forms and the choice of parameters. The same basic approach applies to all targets, nucleons and nuclei. The interpretation of the nuclear dependence correlates the parameters with known features of nuclear structure (16, 19, 21). The justifications are qualitative at best. In this approach, intuitive interpretations of the space-time properties of the Fourier transform of the structure tensor play an important role in discussions of the physical significance (21, 36) of the EMC data.

Instead of starting from Equation 15, the nuclear physics approach begins with the definition Equation 4 and attempts to obtain the structure tensor from explicit representations of the target states and compatible matrix representations of the current operators. The essence of conventional nuclear models is that they are quantum dynamical models constructed on the tensor product of one-nucleon Hilbert spaces. They can be extended to include constituent mesons and excited states of the nucleons, i.e. isobars. Requirements of relativity can be satisfied by assuring that the state vectors and the current operators transform properly under a unitary representation of the Poincaré group. As long as the one-nucleon problem is not solved, it is futile to attempt to derive such models as mathematical approximations to QCD (18, 21). The validity of nuclear dynamics as a model representation of physical reality rests on established empirical success with a wide range of phenomena. Past successes in other areas do not guarantee the applicability of such models to description of deep inelastic scattering, but they motivate serious efforts to test their validity empirically. By implication, it is assumed that even for deep inelastic scattering the relevant constituents of nuclei are hadronic color singlets. Thus success or failure should ultimately serve to illuminate the role of quark confinement in nuclear structure.

Within the class of nuclear models, mean-field independent-particle models have the virtue of intuitive simplicity and empirical success in many aspects of nuclear spectroscopy and nuclear dynamics. This picture of the nucleus is closest to the naive notion of an incoherent collection of free nucleons. In such models the effects of the interactions among nucleons are described by a mean field and assigned to individual nucleons as their properties. For instance, bound nucleons may be assumed to act like free nucleons except for a medium-dependent mass (14, 15). It is a natural extension of this picture to ascribe to bound nucleons medium-dependent structure functions attributable to a medium dependence of the nucleon's pion cloud (9, 37), or of the quark confinement scale (20, 34, 38, 39). The size of a nucleon is defined experimentally by the $Q^2$ dependence of the elastic form factor at low $Q^2$, which measures its charge radius. A medium dependence of the form factor of independent nucleons at low $Q^2$ has been postulated to account for quasi-elastic scattering data (38). However, the data do not require this interpretation (40, 41). Soliton models of the nucleon suggest that the nucleon size may change in the nuclear medium (39). The inelastic structure functions $F_i(x, Q^2)$ are dimensionless functions of a dimensionless variable $x$ and of $Q^2$. It has been suggested (17) that they also depend on an infrared confinement cutoff parameter whose postulated medium dependence explains the difference between $F_2^A(x, Q^2)$ and $F_2^N(x, Q^2)$.

Mean-field models are justified in low-energy nuclear physics by the fact that they can, in principle, be derived as self-consistent mean-field approximations to microscopic many-body dynamics of nucleons. In mean-field models of deep inelastic scattering, the active nucleon and the nucleons in the medium are not treated self-consistently.

In microscopic nuclear many-body dynamics the target nucleus is treated as a bound state of mutually interacting nucleons (or baryons and mesons) (11–13). In this approach, one assumes that the confinement of quarks to hadrons is not disturbed by nuclear binding. The role of QCD is restricted to determining the structure functions of the hadrons, and all nuclear effects in deep inelastic lepton scattering must be due to the hadronic composition and dynamics of the nuclear ground state. This approach is reviewed in detail in Section 4.2.

Quark deconfinement is the prominent feature of models based on color conductivity (16, 42, 43), multiquark clustering (44), or the rescattering of quarks from neighboring nucleons (45) as the mechanisms responsible for nuclear dependence of deep inelastic scattering. In models based on multiquark components in the nucleus, constituent counting rules are used to specify the form of the parton densities. There are no independent data to support the specific choices made, and, moreover, the parameterizations are used in regions of $x$ far from the limits where constituent counting rules apply. The presence of components with more than three quarks (e.g. dibaryons) can lead to a rise in $R_{EMC}(x)$ as $x \to 1$. It is difficult to distinguish this dynamical effect from the kinematic rise associated with Fermi smearing.

Implementations of all the approaches sketched above yield fits to $F_2^A(x, Q^2)$ of roughly similar quality. Two curves are shown in Figure 7. Models differ, however, in their predictions for the antiquark density $\bar{q}_A(x)$, as shown in Figure 4. Experiments designed to measure $\bar{q}_A(x)$ with sufficient precision could provide substantial new insight into the short distance structure of nuclei.

In the paragraphs below we review in more detail the QCD-based rescaling models (16–21) and conventional nuclear structure models (9–15). The models yield the $A$ dependence of the structure functions per nucleon of nuclei.

## 4.1   $Q^2$ Rescaling Models

The structure function $F_2(x, Q^2)$ is known (1) to vary slowly with $Q^2$, as is shown in Figure 1. For $x \lesssim 0.1$, $F_2(x, Q^2)$ grows with increasing $Q^2$, whereas, for $x \gtrsim 0.1$, $F_2(x, Q^2)$ decreases with increasing $Q^2$. These deviations from scaling are understood in perturbative QCD in terms of gluonic radiative corrections (e.g. 46).

The scaling deviations imply that the ratio $F_2^N(x, Q_2^2)/F_2^N(x, Q_1^2)$, $Q_2^2 > Q_1^2$, is a decreasing function of $x$. At typical values of $Q_1^2$ and $Q_2^2$, the ratio crosses unity at $x \approx 0.1$. Focussing on $x$ dependence, one observes a similarity:

$$\frac{F_2^N(x, Q_2^2)}{F_2^N(x, Q_1^2)} \quad \text{resembles} \quad \frac{F_2^A(x, Q_1^2)}{F_2^N(x, Q_1^2)}. \qquad \qquad 16.$$

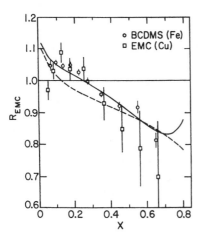

*Figure 7*   Calculations of $R_{EMC}(x)$ compared with the 1986 data. The solid curve is calculated from the pion exchange model of Ref. 12. The dashed curve shows the expectation of $Q^2$ rescaling (17) with $Q^2 = (200x + 10) \, \text{GeV}^2$ appropriate for the kinematics of the EMC data.

On the left-hand side, the target N is fixed, but $Q^2$ is changed. On the right-hand side, $Q^2$ is fixed, but the target is changed.

The resemblance in $x$ dependences of the two ratios suggests that the effective value of $Q^2$, the "scale" used to measure $Q^2$, may not be the same for a nucleus and for a nucleon. This observation is part of the motivation for the $Q^2$ rescaling model (17) and for the proposal of color conductivity (16).

To proceed toward a possible dynamical understanding of the resemblance, one may begin by examining the source of $Q^2$ dependence in QCD. For the valence quark density in a hadron h, $q_v^h(x, Q^2)$, the theory specifies that (46)

$$\frac{dq_v^h(x, t)}{dt} = \frac{\alpha_s(Q^2)}{\pi} \int_x^1 \frac{dy}{y} \, q_v^h(y, t) P_{qq}\left(\frac{x}{y}\right), \qquad \text{17a.}$$

where

$$t = \ln (Q^2/\mu_h^2). \qquad \text{17b.}$$

Equation 17 expresses the fact that the valence quark distribution at $x$ is determined by its value in the range $x \le y \le 1$, and by the probability $P_{qq}(x/y)$ for the transition $q(y) \to q(x)$ via gluon radiation. In Equation 17, $\alpha_s(Q^2)$ is the strong coupling strength describing the qqg vertex.

Because the integral over the transverse momentum of the radiated gluon in q $\to$ qg is formally divergent, $\propto \int^{Q^2} dp_T^2/p_T^2$, a scale $\mu_h^2$ is introduced as a lower momentum "confinement" cutoff. This scale $\mu_h^2$ appears in Equation 17b; its value is unspecified in QCD perturbation theory.

Within the context of Equation 17, one may propose at least two reasons for the EMC effect, $q^A(x, Q^2) \ne q^N(x, Q^2)$: either $\mu_A^2 \ne \mu_N^2$, or $q_v^A(x, \mu^2) \ne q_v^N(x, \mu^2)$. In other words, the lower momentum confinement cutoff $\mu_h^2$ may be different for a quark in a nucleus, or the initial (nonperturbative) distributions from which one begins the QCD evolution may be different.

In a series of papers (17–21), Close, Jaffe, Roberts, and Ross begin with the assumption that all dependence of $q_v^h(x, t)$ on the hadron label h resides in $\mu_h^2$. This length scale is a property of the hadron h and is distinct from the universal QCD scale $\Lambda_{QCD}$. They then argue that the empirical difference between $q^A(x, Q^2)$ and $q^N(x, Q^2)$ implies that $\mu_A^2 < \mu_N^2$. They state that the softening of the quark distributions in nuclei is due to the radiation of gluons between $Q^2 = \mu_A^2$ and $Q^2 = \mu_N^2$. Implicit in this view is the assumption that $q^N(x, Q^2 = \mu_N^2) = q^A(x, Q^2 = \mu_A^2)$. While not inconsistent with what is known in QCD, these important assumptions are not necessarily justified either.

Development of a definite model based on these assumptions requires that contact be made with known features of nuclear structure. The ratio $\mu_N/\mu_A$ is related (19) to a ratio $\lambda_A/\lambda_N$ of length scales associated with the two-nucleon distribution function. The ratio $\lambda_A/\lambda_N$ is assumed to vary between 1 and $2^{1/3}$ depending upon the "crowding" of nucleons in a nucleus. The value for a specific nucleus A is obtained by a linear interpolation:

$$\frac{\lambda_A}{\lambda_N} = 1 + (2^{1/3} - 1)V_A, \qquad\qquad 18.$$

where $V_A$ is the probability for overlap of spheres the size of which is given by the charge radius of a nucleon. This ratio of scales is a property of the nucleus that need not be interpreted as a property of nucleons in nuclear matter (19, 21). However, acknowledging that the "details of quark confinement in free or bound nucleons are not yet solved, [Close et al] believe that when a free nucleon is placed in nuclear matter, the modification to confinement is characterized by a change of a length scale" (20), the confinement size. They use $\lambda_A > \lambda_N$ to conclude that the physical size of the nucleon is greater in a nucleus. Within this model, the magnitude of the EMC effect implies $\lambda_A/\lambda_N \approx 1.15$, a 15% "swelling" of nucleons in iron. Although swelling of nucleons in nuclei has been discussed (38–41) in nuclear physics for reasons other than the EMC effect, it is not obvious that the large physical size ($\sim 1$ fermi) of a nucleon is relevant for determining a short-distance quantity such as $\mu$. The decrease of $\mu$ may also be viewed as a manifestation of partial deconfinement in the sense that colored constituents may be interchanged when two nucleons are sufficiently close.

The specific rescaling relationship proposed is

$$q^A(x, Q^2) = q^N[x, \xi(Q^2)Q^2] \qquad\qquad 19.$$

with

$$\xi(Q^2) = \xi(Q_0^2)^{[\alpha_s(Q_0^2)/\alpha_s(Q^2)]}. \qquad\qquad 20.$$

At $Q_0^2 = 20$ GeV$^2$, $\xi_{Fe} = 2.02$.

In the $Q^2$ rescaling model, momentum lost by valence quarks is taken up by an increase in the ocean q and $\bar{q}$ pairs and by the glue. A calculation based on the rescaling approach is compared with data in Figure 7. The rescaling curve crosses $R_{EMC} = 1$ at $x \approx 0.1$, as is expected in the model since scaling deviations in $F_2(x, Q^2)$ vanish at $x \approx 0.1$. However, the crossing point is closer to $x \approx 0.3$ in the CERN data shown in Figures 3 and 5. There does not seem to be any simple way to resolve this inconsistency within the context of the model. The model is not intended to be applicable for $x \gtrsim 0.7$. In QCD, evolution of $\bar{q}(x, Q^2)$ with $Q^2$ leads to a narrowing of the $x$ dependence of the antiquark distribution in a nucleus relative to that in the nucleon. The expected antiquark ratio is shown in Figure 4.

Rescaling is also a byproduct of the hypothesized color conductivity (16). According to this proposal, the color confinement of partons into color-singlet degrees of freedom changes with $Q^2$, allowing color conductivity throughout the nucleus for sufficiently small values of $1/Q^2$. In this approach, the cutoff parameter $\mu_A^2$ of Equation 17b is related to the radius of nucleus, and the rescaling parameter $\xi$ of Equation 19 becomes the ratio of the squares of the nuclear radii (16): that is, $F_2^A(x, Q^2) = F_2^N(x, Q^2 R_A^2/R_N^2)$. Two options are explored in the color conductivity approach; one is unusual in that it implies a different QCD scale parameter $\Lambda_{QCD}$ for different targets. The second option is compatible with a universal scale parameter. In the color conductivity model, the $A$ dependence of $R_{EMC}(x)$ varies as $\ln A$; it does not saturate for large $A$.

As we mentioned above, another way to maintain Equation 17 along with $q^A(x, Q^2) \neq q^N(x, Q^2)$ is to assert that boundary conditions $q^A(x, Q_0^2)$ and $q^N(x, Q_0^2)$ are different. In the conventional nuclear physics approach to which we now turn, the initial boundary conditions are indeed different.

## 4.2 Nuclear Structure

We discuss mean-field models and nuclear many-body dynamics in this section. The underlying assumptions are quite different in the two cases.

In considering mean-field models of nuclear structure, it is useful to think of the nuclear medium as a very heavy inert particle of mass $M$ to which the active nucleon is bound to form a nucleus of mass $M_A$. The active nucleon in the target moves with nonrelativistic velocity. The masses $M_A$ and $M$ of the nucleus and the spectator can be arbitrarily large while the "removal energy" of the nucleon, $\varepsilon \equiv M_N + M - M_A$, remains fixed. The electromagnetic current operator is effectively a medium-dependent, single-nucleon current operator. It is assumed that final-state interactions between the residue of the struck nucleon and the medium do not affect the inclusive cross section. These model assumptions lead to Equation 10 with

$$f_N(z) = \int d^3\mathbf{p} \rho(\mathbf{p}) \delta[z - (\mathbf{n} \cdot \mathbf{p} + M_A - M)/M_N] \qquad 21.$$

replacing Equation 11. It follows that the average light-front momentum fraction $\langle z \rangle$ carried by a nucleon is

$$\langle z \rangle \equiv \int dz \, z f_N(z) = 1 - \varepsilon/M_N < 1. \qquad 22.$$

This decrease in the momentum fraction carried by the nucleons is of the right size to account quantitatively for the observed decrease of $R_{EMC}(x)$ at intermediate values of $x$ (14), but not the rise at small $x$.

As mentioned above, mean-field models of nuclear structure easily accommodate a medium dependence of the single-nucleon current, which implies medium modifications of the nucleon's structure functions. The role of pion exchange in the long-range part of the nuclear force suggests a medium dependence of the nucleon's pion cloud and hence of the nucleon's structure function. The small-$x$ enhancement of $R_{EMC}$ has been described in this manner (9, 36). Rescaling arguments may also be used to assign a medium dependence to structure functions of the nucleon in nuclear matter (20). In that case it is necessary to consider simultaneously the binding effects (15) implied by Equations 10 and 21.

In microscopic nuclear many-body dynamics, the nucleus is treated as a bound state of mutually interacting nucleons (or baryons and mesons) (12). Convolution formulae of the form of Equation 10 are not an automatic feature of the model but require a derivation from appropriate assumptions. The Lorentz covariance of the tensor $F_A^{\lambda\rho}$ must be assured by consistent covariance properties of the current operators and of the bound-state wave function. The covariant structure tensor $F_A^{\mu\nu}$ of the nucleus, and hence the invariant structure functions are defined as matrix elements of current operators. Nuclear many-body dynamics is a model in which states are represented by vectors in a multihadron Hilbert space, and currents are specified as operators acting on that space. The current operators are, in general, sums of one- and multibody operators. Lorentz transformations and space-time translations of operators and states are in principle given by a unitary representation of the Poincaré group, which also specifies the dynamics. Essential assumptions in the treatment of deep inelastic lepton scattering in this fashion are (13) as follows:

1. Only one-body currents contribute to the structure tensor $F_A^{\mu\nu}$. This formal assumption implies physically the absence of a "medium dependence" of the nucleon structure functions. Effects appearing in mean-field models as such a medium dependence are due in this model to the effects of constituent pions, excitations of the nucleon (i.e. isobar effects), and/or multibody currents.
2. The contribution of interference terms $j_i^\mu(0)j_k^\nu(0)$, $i \neq k$, to the tensor $F_A^{\mu\nu}$ can be neglected. Intuitively this means that hadrons in a nucleus contribute incoherently in deep inelastic scattering.
3. In the sum over final states, final-state interactions between the residue of the struck hadron and the spectator nucleus do not affect the inclusive cross section.

The bound state of the nucleus must be an eigenstate of the four-momentum operator $P_{op}$. Ordinary bound-state wave functions, defined as functions of relative momenta in a nucleus at rest, are eigenfunctions of the rest energy. They are then augmented to eigenfunctions of the four-momentum operator $P_{op}$ by the imposition of purely kinematic three-momentum conservation. The choice of the components of $P_{op}$ that are kinematically conserved selects a kinematic subgroup of the Poincaré group. In order that the current operator remain a sum of one-body operators under longitudinal Lorentz transformations, assumption 1 above requires that the nuclear bound state have light-front symmetry as a kinematic symmetry (49). This implies that the light-front three-momentum $\{P^+, \mathbf{P}_T \equiv \mathbf{P} - \mathbf{n} \cdot \mathbf{Pn}\}$ is kinematically conserved.

The light-front momentum density of the nucleons, $f_N(z)$, is

$$f_N(z) = \int dp^+ \int d^2p_T \delta[z - Ap^+/P_A^+]\rho_N(p^+, \mathbf{p}_T). \qquad 23.$$

Our normalization condition, $\int dp^+ \int d^2p_T \rho_N(p^+, p_T) = 1$, implies

$$\int dz f_N(z) = 1. \qquad 24.$$

If nucleons are the only constituent particles of the nucleus, kinematic conservation of the light-front momenta,

$$\sum_i^A p_i^+ = P_A^+, \qquad 25.$$

implies the conservation law

$$\int dz \, z f_N(z) = 1. \qquad 26.$$

The success of the mean-field model embodied in Equation 22 rests on the tacit choice of kinematic conservation of the spatial three-momentum, which is inconsistent in the context of many-body dynamics. Only the presence of other constituent hadrons can reduce the average light-front momentum carried by the nucleons. In the presence of constituent pions, the ground-state wave function determines number densities per nucleon,

$$\rho_N(p^+, \mathbf{p}_T) = \langle a^\dagger(p^+, \mathbf{p}_T)a(p^+, \mathbf{p}_T)\rangle/A, \qquad 27.$$

and

$$\rho_\pi(p^+, \mathbf{p}_T) = \langle c^\dagger(p^+, \mathbf{p}_T)c(p^+, \mathbf{p}_T)\rangle/A. \qquad 28.$$

where $a$ and $c$ are respectively nucleon and pion annihilation operators. The light-front momentum density of pions per nucleon is then

$$f_\pi(z) = \int dp^+ \int d^2p_T \delta[z - Ap^+/P_A^+]\rho_\pi(p^+, \mathbf{p}_T).$$ 29.

Kinematic conservation of the light-front momenta in this case implies

$$\int dz\, zf_N(z) + \int dz\, zf_\pi(z) = 1.$$ 30.

The average number of pions per nucleon is $\int f_\pi(z)\,dz = \langle n_\pi \rangle$. Although not indicated explicitly, the functions $f_N(z)$ and $f_\pi(z)$ both depend on $A$. Use of light-front dynamics guarantees that the number densities $f_N(z)$ and $f_\pi(z)$ are invariant under longitudinal Lorentz boosts. The presence of pions degrades the momentum distribution of the nucleons in the nucleus just as the gluons degrade the momentum distribution of the quarks and antiquarks in the nucleon.

Under the three assumptions listed above, one may derive (13) convolution formulae of the form of Equation 10 for the structure functions,

$$xF_i^A(x, Q^2) = \int_{z>x} dz(x/z)F_i^N(x/z, Q^2)f_N(z) + \int_{z>x} dz(x/z)F_i^\pi(x/z, Q^2)f_\pi(z),$$

for $i = 1$ or 3, and 31.

$$F_2^A(x, Q^2) = \int_{z>x} dz\, F_2^N(x/z, Q^2)f_N(z) + \int_{z>x} dz\, F_2^\pi(x/z, Q^2)f_\pi(z).$$

The structure functions of a nucleus, per nucleon, $F_i^A(x, Q^2)$ are expressed in terms of the structure functions $F_i^N(x, Q^2)$ and $F_i^\pi(x, Q^2)$ measured on unbound nucleons and pions. Measurements of massive lepton pair production (the Drell-Yan process) with pion beams (see 50 for a recent review), $\pi^- N \to \mu\bar\mu X$ can be used to construct $F_i^\pi(x, Q^2)$.

By implication, expressions analogous to Equation 31 may be derived for the quark and antiquark densities per nucleon, $q^A(x, Q^2)$ and $\bar q^A(x, Q^2)$, as well as for the gluon density $G^A(x, Q^2)$. For example,

$$q^A(x, Q^2) = \int_x \frac{1}{z} f_\pi(z) q^\pi\left(\frac{x}{z}, Q^2\right) dz + \int_x \frac{1}{z} f_N(z) q^N\left(\frac{x}{z}, q^2\right) dz.$$ 32.

If the difference in the structure functions of nucleons and isobars is sufficiently important, then appropriate isobar terms should be added to the right-hand sides of Equations 31 and 32.

The number densities $f_N(z)$ and $f_\pi(z)$ are determined by the ground-state wave function of the nucleus. In practice, a variety of model assumptions and approximations are involved in obtaining numerical results.

In conventional nuclear dynamics, an $NN\pi$ vertex interaction is assumed as the dominant mechanism for the two-nucleon interaction at large and medium distances (51). The presence of constituent exchange pions in nuclei is a necessary consequence of such an interaction. The requirement that bound-state wave functions be eigenfunctions of the total spin as well as the mass operator $\sqrt{P_{op}^2}$ is nontrivial in light-front dynamics. It is possible, but not necessary, to construct the spin operator in such a manner that it is independent of interactions. This choice was used in Ref. 12. It affects the relation between the light-front momentum fractions $z_i$ of the constituents and the longitudinal components of intrinsic relative momenta used in the construction of the spin operator. The corresponding model ambiguity in $f_N(z)$ is unimportant for nonrelativistic nucleons but could be significant in $f_\pi(z)$.

In the pion exchange model of Ref. 12, Fermi gas momentum distributions averaged over local densities were used to determine $f_N(z)$. The momentum density of the exchange pions was obtained as the ground-state expectation value of a two-body operator $N_\pi(p)$, which is the variational derivative of the one- and two-pion-exchange potentials with respect to the pion frequency (12):

$$\rho_\pi(p) = \langle N_\pi(p) \rangle = \text{Tr} \left\{ \rho_{2N} N_\pi(p) \right\}. \qquad 33.$$

The same $\pi NN$ vertex form factor must appear in the operator $N_\pi(p)$ and in the two-nucleon density matrix $\rho_{2N}$ (12). This constraint is not used in the corresponding mean-field models, where different form factors appear in the active nucleon and in the medium (9). In the approximations used in Ref. 12, only the two-nucleon density $\rho_{2N}(\mathbf{r}_1, \mathbf{r}_2)$ is needed. In the local density approximation, this two-nucleon density is equal to the product of single-nucleon densities multiplied by the pair distribution function $g(\mathbf{r}_1 - \mathbf{r}_2)$,

$$\rho_{2N}(\mathbf{r}_1, \mathbf{r}_2) = \rho_{1N}(\mathbf{r}_1) \rho_{1N}(\mathbf{r}_2) g(\mathbf{r}_1 - \mathbf{r}_2). \qquad 34.$$

The pair distribution functions have been obtained by variational calculations (52). They can also be obtained from the response functions (53) of Fermi liquid models. In the one-pion-exchange approximation, only the spin-longitudinal component of the pair distribution function contributes to the pion density. Contrary to earlier expectations, the relevant spin-longitudinal response functions cannot be extracted from the polarization transfer observables measured for inclusive 500-MeV proton scattering (54, 55).

In principle, the momentum distributions $f_\pi(z_\pi)$ and $f_N(z_N)$ could be measured directly in semi-inclusive deep inelastic scattering experiments. For example, the cross section for the process $eA \to e'\pi X$ is expected theoretically to include a contribution in which the virtual photon scatters coherently from a constituent pion in the nucleus. At large $Q^2$, this specific contribution takes the form

$$\frac{d\sigma^A}{dx\, dy\, d\varsigma} \propto (1-y)\delta(\varsigma-1)xf_\pi(x)F_\pi^2(Q^2). \qquad 35.$$

Here $F_\pi(Q^2)$ is the electromagnetic form factor of the pion. The variable $y$ in Equation 35 is identical to that defined in Section 2, and $\varsigma = p_\pi \cdot p_A/q \cdot p_A$ specifies the fraction of the energy of the virtual photon that is carried by the observed final pion.

It is beyond the scope of this review to discuss semi-inclusive processes in greater detail (55a). We caution that the expected signal is small, since it is proportional to $F_\pi^2(Q^2)$, and that backgrounds from competing mechanisms may be significant.

The pion exchange model described above provides a unified description of $R_{EMC}(x)$ for all $x$. Scattering by the pions provides the small-$x$ enhancement (9, 12). The momentum carried by the pions requires the momentum loss of the nucleons, which explains the decrease of $R_{EMC}(x)$ for larger $x$ (12). The average momentum carried by the pions corresponds roughly to the "binding effect" of the potential produced by exchange of these mesons (14). Comparisons with data are shown in Figures 3 and 7. In this calculation $\langle n_\pi^{Fe} \rangle = 0.095$, meaning that in an iron nucleus there are on the average 5 to 6 pions from which deep inelastic scattering occurs. The average momentum fraction carried by those pions is $\langle z_\pi^{Fe} \rangle = 0.05$. The corresponding average nucleon momentum fraction is $\langle z_N \rangle = 1 - \langle z_\pi \rangle = 0.95$. The average number of pions per nucleon $\langle n_\pi \rangle$ controls the size of the enhancement of $R_{EMC}(x)$ above unity at small $x$, while $\langle z_\pi \rangle$ controls the size and shape of the depression below unity at intermediate $x$. There are uncertainties in the prediction of $R_{EMC}(x)$ for large $x$ associated with the approximations made in treating the high momentum tail of the nucleon momentum distribution. The changes of $\langle n_\pi \rangle$ and $\langle z_\pi \rangle$ with $A$ are modest, in agreement with data (12): for Al, Fe, and Au, $(\langle n_\pi \rangle, \langle z_\pi \rangle) = (0.089, 0.049)$, $(0.095, 0.052)$, and $(0.114, 0.061)$. The $A$ dependence of these quantities is determined by the $A$ dependence of the two-nucleon density $\rho_{2N}(\mathbf{r}_1, \mathbf{r}_2)$, which also determines the ratio $\lambda_A/\lambda_N$ in the rescaling model (19). Since the central density in nuclei is essentially independent of $A$, we expect that the change of $R_{EMC}(x)$ with $A$ will be proportional to $A^{-1/3}$ for large $A$. By considering variations in the pion exchange potential and the effects of other approximations made, we estimate an overall uncertainty of $\pm 30\%$ in $\langle n_\pi \rangle$.

Comparison with the data in Figures 3 and 7 shows that the pion exchange model accounts successfully for the data at large $Q^2$ for all $x$. It does not reproduce the SLAC data at low $Q^2$. Shadowing has not been incorporated. The prediction for the antiquark ratio is shown in Figure 4. The presence of exchange pions, with their valence antiquarks, leads to an enhancement of the antiquark density in nuclei. While predictions for $R_{EMC}(x)$ are relatively insensitive to model ambiguities of the pion momentum distribution $f_\pi(z)$, the shape of $R_{\bar{q}}(x)$, but not the sign, is subject to substantial model ambiguities.

As discussed earlier, the net change in the momentum fraction carried by gluons is proportional to the integral over $x$ of the difference $F_2^A(x) - F_2^N(x)$, Equation 14. In the pion exchange model, the integral can be carried out explicitly. We find

$$\Delta = \langle z_\pi \rangle \int [F_2^\pi(x, Q^2) - F_2^N(x, Q^2)]\,\mathrm{d}x. \qquad 36.$$

We note that $\int F_2^N(x, Q^2)\,\mathrm{d}x \approx 0.5$. It is generally expected that $\int F_2^\pi(x, Q^2)\,\mathrm{d}x$ will also be in the neighborhood of 0.5. Since $\langle z_\pi \rangle \approx 0.05$, no reasonable difference between $\int F_2^\pi(x)\,\mathrm{d}x$ and $\int F_2^N(x)\,\mathrm{d}x$ could result in appreciable values for $\Delta$. We conclude, therefore, that the pion exchange model is in accord with the observation that $\Delta$ is consistent with zero.

# 5.  SUMMARY AND FUTURE INVESTIGATIONS

The original EMC effect stimulated broad interest in quark and antiquark distributions of nuclei. New data from the BCDMS and EMC collaborations confirm that the ratio $R_{EMC}^{Fe}(x, Q^2)$ exceeds unity for $x \lesssim 0.3$ and $Q^2 \gtrsim 10$ GeV$^2$. The excess is in the neighborhood of 5%. The data provide tests of models of nuclear binding and, perhaps, information on properties of constituent confinement. Either rescaling of the nucleons quark structure, reviewed in Section 4.1, or the effects of conventional nuclear dynamics, reviewed in Section 4.2, can account at least qualitatively for the observed nuclear dependence in charged lepton scattering. The pion exchange model yields a unified quantitative description of all features of the present data sample at large $Q^2$. It reproduces the magnitude and shape of $R_{EMC}(x)$, for all $x$, and the weak enhancement of the antiquark distribution $\bar{q}^A(x)$ shown by neutrino experiments. Since this conventional nuclear physics approach suffices to explain the data, one cannot claim that the EMC effect is evidence for new QCD effects such as a proposed change of the color confinement size of nucleons within nuclei.

The above conclusions are based on an analysis of the current data sample. New data, especially in the small $x$ region, could have a significant impact. The absolute normalization of $R_{EMC}(x)$ at small $x$ is crucial. If a large enhancement of $R_{EMC}(x)$ above unity were to be established in future experiments—for example, the approximately 15% effect suggested in the original EMC paper—then none of the current models would survive. Neither the $Q^2$ rescaling nor the pion exchange model yields an enhancement above unity that would be sufficiently large. Shadowing is another important feature of the small-$x$ region. The $Q^2$ rescaling and the pion exchange models do not incorporate this physics as yet. Better data should stimulate the required effort.

As shown in Figure 4 the two approaches provide quite different expectations for antiquark distributions. The presence of the exchange pions implies an enhancement of the antiquarks as $x$ increases, whereas the rescaling model predicts a decrease. Unfortunately, the differences are significant only for $x > 0.35$, where there is insufficient information on $\bar{q}(x)$ from neutrino experiments. Even for $x < 0.3$, precise neutrino data would be valuable for establishing whether the antiquark ratio is indeed greater than unity, as shown in both models in Figure 4. Another source of information on $R_{\bar{q}}$ is massive lepton-pair production, the Drell-Yan process. Measurements of the $A$ dependence in massive lepton-pair production are important for at least two reasons (56): ($a$) to establish whether the $A$ dependence of $q^A(x, Q^2)$ observed in deep inelastic lepton scattering is a process-independent property of parton densities, and ($b$) to determine the $A$ dependence of the ocean, $\bar{q}^A(x, Q^2)$.

An experiment is underway (57) at SLAC to measure the $A$ dependence of the ratio $\sigma_L/\sigma_T$. Two important new experiments will take data on $\mu A \to \mu' X$ at large $Q^2$: CERN NA37 and Fermilab E-665. These experiments should provide further confirmation of the magnitude of $R_{EMC}(x)$ for values of $x$ down to $\sim 0.005$, as well as determine the $x$ and $Q^2$ characteristics of shadowing. Regrettably, we know of no new experiment planned for $\nu A \to \mu X$ that would be sufficiently precise to reduce substantially the size of the error flags on $\bar{q}^A(x)$ in Figure 4. On the other hand, a study (58) of the Drell-Yan process $pA \to \mu\bar{\mu}X$ should begin at Fermilab in 1987. The proponents intend to accumulate enough data in the relevant kinematic region to determine $\bar{q}^A(x)$ precisely at small $x$.

We commented briefly in Section 4.2 on semi-inclusive processes (55a): $eA \to e'\pi X$ and $eA \to e'pX$. Data from these reactions at large $Q^2$ would provide independent information on the momentum distribution of pions and nucleons in nuclei, $f_\pi(z)$ and $f_N(z)$. A careful theoretical estimate should be made of the expected backgrounds to the desired coherent signals.

We have said little about the gluon distribution $G^A(x, Q^2)$ other than to remark that the data on Fe and D are consistent with no change in the momentum integral $\int x G^A(x) \, dx$. There is no simple process in which $G^A(x, Q^2)$ is probed in the precise way $\mu A \to \mu' X$ or $hA \to \mu\mu X$ probe the quark and antiquark densities. However, inclusive prompt photon production (59) at large $p_T$, $pA \to \gamma X$; inclusive chi production, $pA \to \chi X$; and inelastic photoproduction (60) of the $J/\psi$, $\gamma^* A \to J/\psi X$ are all good candidates. Data on $\mu A \to \mu J/\psi X$ were interpreted (61) in terms of a significant nuclear enhancement of $G^A(x, Q^2)$ for $x \approx 0.03$. However, after a careful separation of the contribution from coherent diffractive production, the Fermilab Tagged Photon Collaboration (62) finds no indication of such an enhancement in $\gamma A \to J/\psi X$. We expect that further studies of the processes listed here will yield additional interesting constraints on $G^A(x, Q^2)$.

*Literature Cited*

1. Eisele, F. *Proc. XXI Int. Conf. on High Energy Physics, Paris (1982)*, ed. P. Petiau, M. Porneuf. *J. Phys.* 43: C3-337; Dydak, F. *Proc. 1983 Int. Symp. on Lepton and Photon Interactions at High Energies, Cornell*, ed. D. G. Cassel, D. L. Kreinick, p. 634. Ithaca: Cornell Univ. (1983); Drees, J., Montgomery, H. E. *Ann. Rev. Nucl. Part. Sci.* 33: 383 (1983); Hughes, V. W., Kuti, J. *Ann. Rev. Nucl. Part. Sci.* 33: 611 (1983); Sciulli, F. *Proc. 1985 Int. Symp. on Lepton and Photon Interactions at High Energies, Kyoto*, ed. M. Konuma, K. Takahashi, p. 8. Kyoto: Kyoto Univ. (1986); Berger, E. L. *Proc. Workshop on Fundamental Muon Physics, Los Alamos (1986)*, ed. C. M. Hoffman, V. W. Hughes, M. Leon, p. 109. Los Alamos Natl. Lab. (1986)
2. Berger, E. L. *XXIII Int. Conf. on High Energy Physics, Berkeley, July, 1986*, ed. S. C. Loken, p. 1433. Singapore: World Scientific (1987); Berger, E. L. *Proc. Semin. on Few and Many Quark Systems, San Miniato, 1985*, ed. F. Navarria, Y. Onel, P. Pelfer, A. Penzo, p. 153. Trieste: INFN; Nachtmann, O. *Proc. 11th Int. Conf. on Neutrino Physics and Astrophysics, Nordkirchen near Dortmund, 1984*, ed. K. Kleinknecht, E. A. Paschos, p. 405. Singapore: World Scientific (1987); Llewellyn Smith, C. H. Proc. PANIC (Heidelberg) 1984, ed. P. Povh, G. zu Putlitz. *Nucl. Phys.* A434: 35c (1985); Krzywicki, A. Proc. 11th Europhysics Conf. "Nuclear Physics with Electromagnetic Probes" Paris 1985, ed. A. Gerard, C. Samour. *Nucl. Phys.* A446: 135c (1985)
3. Aubert, J. J., et al. *Phys. Lett.* 123B: 275 (1983)
4. Bodek, A., et al. *Phys. Rev. Lett.* 50: 1431, 51: 534 (1983); Arnold, R. G., et al. *Phys. Rev. Lett.* 52: 727 (1984)
5. Bari, G., et al. *Phys. Lett.* 163B: 282 (1985)
6. Abramowicz, H., et al. *Z. Phys.* C25: 29 (1984); see also Parker, M. A., et al. *Nucl. Phys.* B232: 1 (1984); Cooper, A. M., et al. *Phys. Lett.* 141B: 133 (1984); Hanlon, J., et al. *Phys. Rev.* D32: 2441 (1985); Ammosov, V. V., et al. *JETP Lett.* 39: 393 (1984); Asratian, A. E., et al. *Sov. J. Nucl. Phys.* 43: 380 (1986); WA25 and WA59 Collaborations, Guy, J., et al. *CERN Rep.* CERN/EP 86-217 (Dec. 1986). Submitted to *Z. Phys. C*
7. EMC Collaboration, Norton, P. *XXIII Int. Conf. on High Energy Physics, Berkeley, July, 1986*, ed. S. C. Loken, p. 1399. Singapore: World Scientific (1987)
8. BCDMS Collaboration, Benvenuti, A. C., et al. *Phys. Lett.* 189B: 483 (1987)
9. Llewellyn Smith, C. H. *Phys. Lett.* 128B: 107 (1983); Ericson, M., Thomas, A. W. *Phys. Lett.* 128B: 112 (1983); Thomas, A. W. *Phys. Lett.* 126B: 97 (1983)
10. Szwed, J. *Phys. Lett.* 128B: 245 (1983)
11. Berger, E. L., Coester, F., Wiringa, R. B. *Phys. Rev.* D29: 398 (1984)
12. Berger, E. L., Coester, F. *Phys. Rev.* D32: 1071 (1985)
13. Berger, E. L., Coester, F. In *Quarks and Gluons in Particles and Nuclei*, ed. S. Brodsky, E. Moniz, p. 255. Singapore: World Scientific (1986)
14. Akulinichev, S. V., et al. *Phys. Lett.* 158B: 485 (1985); Akulinichev, S. V., et al. *Phys. Rev. Lett.* 55: 2239 (1985);

Birbrair, B. L., et al. *Phys. Lett.* 166B: 119 (1986); Staszel, M., et al. *Phys. Rev.* D29: 2638 (1984)

15. Dunne, G. V., Thomas, A. W. *Phys. Rev.* D32: 2061 (1986); *Nucl. Phys.* A455: 701 (1986)

16. Nachtmann, O., Pirner, H. J. *Z. Phys.* C21: 277 (1984); *Ann. Phys. (Leipzig)* 44: 13 (1987); Chanfray, G., et al. *Phys. Lett.* 147B: 249 (1984); Pirner, J. *Comments Nucl. Part. Phys.* 21: 199 (1984)

17. Close, F. E., Roberts, R. G., Ross, G. C. *Phys. Lett.* 129B: 346 (1983)

18. Jaffe, R. L., et al. *Phys. Lett.* 134B: 449 (1984)

19. Close, F. E., et al. *Phys. Rev.* D31: 1004 (1985)

20. Close, F. E., Roberts, R. G., Ross, G. C. *Phys. Lett.* 168B: 400 (1986); Bickerstaff, R. P., Miller, G. A. *Phys. Lett.* 168B: 409 (1986)

21. Jaffe, R. L. In *Relativistic Dynamics and Quark-Nucleon Physics*, ed. M. B. Johnson, A. Picklesimer, p. 573. New York: Wiley (1986)

22. Bodek, A., Ritchie, J. L. *Phys. Rev.* D23: 1070, D24: 1400 (1981)

23. Frankfurt, L. L., Strickman, M. I. *Phys. Rep.* 76: 215 (1981)

24. CHARM, Bergsma, F., et al. *Phys. Lett.* 141B: 129 (1984); EMC, Aubert, J. J., et al. *Nucl. Phys.* B259: 189 (1985); SLAC, Mestayer, M. D., et al. *Phys. Rev.* D27: 285 (1983)

25. Aubert, J. J., et al. *Nucl. Phys.* B272: 158 (1986)

26. Bjorken, J. D., Paschos, E. A. *Phys. Rev.* 185B: 1975 (1969); Feynman, R. P. *Phys. Rev. Lett.* 23: 1415 (1969)

27. Buras, A. *Rev. Mod. Phys.* 52: 199 (1980); Collins, J. C., Soper, D. E. *Nucl. Phys.* B194: 445 (1982)

28. Abramowicz, H. *Z. Phys.* C17: 283 (1983)

29. Allasia, D. *Phys. Lett.* 135B: 231 (1984); Abramowicz, H., et al. *Z. Phys.* C25: 29 (1984)

30. Gramer, G., Sullivan, J. In *Electromagnetic Interactions of Hadrons*, ed. A. Donnachie, G. Shaw, 2: 195. New York: Plenum (1978); Mueller, A. In *Quarks, Leptons, and Supersymmetry*, Vol. 1 of *Proc. XVIIth Recontre de Moriond, Les Arcs, France 1982*, ed. J. Tran Thanh Van, Gif-sur-Yvette: Editions Frontieres (1982), and references therein; Gribov, L. V., Levin, E. M., Ryskin, M. G. *Phys. Rep.* 100: 1 (1983)

31. Schmidt, I. A., Blankenbecler, R. *Phys. Rev.* D15: 3321, D16: 1318 (1977); Efremov, A. V. *Phys. Lett.* 174B: 219 (1986); Akulinichev, S. V., Shlomo, S. *Phys. Rev.* C33: 1551 (1986)

32. Nikolaev, N. N., Zakharov, V. I. *Phys. Lett.* 55B: 397 (1975); Krzywicki, A. *Phys. Rev.* D14: 152 (1976)

33. Sacquin, Y. In New Particle Production, *Proc. XIX Recontre de Moriond*, ed. J. Tran Thanh Van, 2: 659. Gif-sur-Yvette: Editions Frontières (1984); Rock, S. See Ref. 7, p. 1109

34. Jaffe, R. L. *Phys. Rev. Lett.* 50: 228 (1983)

35. Nikolaev, N. N., Quoted in Ref. 32; Frankfurt, L. L., Strickman, M. I. *Sov. J. Nucl. Phys.* 41: 308 (1985)

36. Llewellyn Smith, C. H. Quoted in Ref. 2

37. Sullivan, J. D. *Phys. Rev.* D5: 1732 (1972)

38. Noble, J. V. *Phys. Rev. Lett.* 46: 412 (1981); Celenza, L. S., Rosenthal, A., Shakin, C. M. *Phys. Rev.* C31: 232 (1985)

39. Jändel, M., Peters, G. *Phys. Rev.* D30: 1117 (1984); Celenza, L. S., et al. *Phys. Rev. Lett.* 53: 892 (1984); Rho, M. *Phys. Rev. Lett.* 54: 767 (1985); Oka, M., Amado, R. D. *Phys. Rev.* C35: 1586 (1987)

40. Sick, I. *Nucl. Phys.* A434: 677c (1985); *Phys. Lett.* B157: 13 (1985)

41. Pandharipande, V. R. *Univ. Ill. preprint ILL-(NU)-87-#7* (Jan. 1987)

42. Furmanski, W., Krzywicki, A. *Z. Phys.* C22: 391 (1984); Krzywicki, A. *Phys. Rev.* D14: 152 (1976); Angelini, C., Pazzi, R. *Phys. Lett.* 154B: 328 (1985)

43. Goldman, T., Stephenson, G. J. Jr. *Phys. Rev. Lett.* 146B: 143 (1984); Hoodbhoy, P., Jaffe, R. L. *Phys. Rev.* D35: 113 (1987)

44. Baldin, A. M. *Nucl. Phys.* A434: 695C (1985); Faissner, H., Kim, B. R. *Phys. Lett.* 130B: 321 (1983); Faissner, H., et al. *Phys. Rev.* D30: 900 (1984); Berlad, G., Dar, A., Eilam, G. *Phys. Rev.* D22: 1547 (1980), *Nucl. Phys.* B181: 22 (1981); Pirner, H. J., Vary, J. P. *Phys. Rev. Lett.* 46: 1376 (1981); Carlson, C. E., Havens, T. J. *Phys. Rev. Lett.* 51: 261 (1983); Titov, A. I. *Sov. J. Nucl. Phys.* 38: 964 (1983); Chemtob, M., Peshanski, R. J. *Phys.* G10: 599 (1984); Dias de Deus, J., et al. *Phys. Rev.* D30: 697 (1984), *Z. Phys.* C26: 109 (1984); Clark, B. C., et al. *Phys. Rev.* D31: 617 (1985)

45. Levin, E. M., Ryskin, M. G. *Sov. J. Nucl. Phys.* 41: 1027 (1985)

46. Buras, A. Quoted in Ref. 27

47. Aubert, J. J., et al. *Phys. Lett.* 105B: 322 (1981)

48. Bodek, A., et al. *Phys. Rev.* D20: 1471 (1979)

49. Dirac, P. A. M. *Rev. Mod. Phys.* 21: 392 (1949); Leutwyler, H., Stern, J. *Ann.*

*Phys. (NY)* 112: 94 (1978); Bakker, B. L. G., et al. *Nucl. Phys.* B158: 497 (1979)

50. Grosso-Pilcher, C., Shochet, M. *Ann. Rev. Nucl. Part. Sci.* 36: 1 (1986)
51. Machleidt, R., Hollinde, K., Elster, C. *Phys. Rep.* 149: 1 (1987)
52. Schiavilla, R., et al. *Univ. Ill. Rep.* ILL-(NU)-86 # 60 (Nov. 1986)
53. Fetter, A. L., Walecka, D. *Quantum Theory of Many Particle Systems.* New York: McGraw-Hill (1971); Fantoni, S., Pandharipande, V. R. *Univ. Ill. Rep.* ILL-(NU)-86 # 52 (Oct. 1986)
54. Carey, T. A., et al. *Phys. Rev. Lett.* 53: 184 (1984); Rees, L. B., et al. *Phys. Rev.* C34: 627 (1986)
55. Alberico, W. M., et al. *Phys. Lett.* 183B: 135 (1987)
55a. Berger, E. L. *Argonne Rep.* ANL-HEP-87-45 (Apr. 1987); *Proc. Workshop on Electronuclear Physics with Internal Targets.* Stanford, Calif: SLAC

56. Berger, E. L. *Nucl. Phys.* B267: 231 (1986); Berger, E. L. In *Weak and Electromagnetic Interactions in Nuclei,* ed. H. V. Klapdor, p. 374. Berlin: Springer-Verlag (1986); Bickerstaff, R. P., Birse, M. C., Miller, G. A. *Phys. Rev. Lett.* 53: 2532 (1984); Ericson, M., Thomas, A. W. *Phys. Lett.* 148B: 191 (1984); Gabellini, Y., Meunier, J. L., Plaut, G. Z. *Phys.* C28: 123 (1985)
57. Dasu, S., et al. *Univ. Rochester Rep.* UR-991 (1987)
58. Moss, J., et al. Fermilab experiment E-772 (1986)
59. Berger, E. L., Braaten, E., Field, R. D. *Nucl. Phys.* B239: 52 (1984)
60. Berger, E. L., Jones, D. *Phys. Rev.* D23: 1521 (1981)
61. Aubert, J. J., et al. *Phys. Lett.* 152B: 433 (1985)
62. Sokoloff, M., et al. *Phys. Rev. Lett.* 57: 3003 (1986)

*Ann. Rev. Nucl. Part. Sci. 1987. 37 : 493–535*
*Copyright © 1987 by Annual Reviews Inc. All rights reserved*

# NUCLEAR FRAGMENTATION IN PROTON- AND HEAVY-ION-INDUCED REACTIONS

## W. G. Lynch

National Superconducting Cyclotron Laboratory,
Michigan State University, East Lansing, Michigan 48824

CONTENTS

## 1. INTRODUCTION

When an energetic, strongly interacting projectile collides with a heavy target nucleus, low energy complex nuclei (fragments) with $A \geq 6$ are produced in a process called nuclear fragmentation.[1] The cross sections for most fragments are very small at low bombarding energies. They

---

[1] The term "fragmentation product" is used here to describe all fragments in the mass distribution. In this definition, projectile and target fragmentation are the same process and differ only by the choice of rest frame.

0163–8998/87/1201–0493$02.00

increase with bombarding energy for energies up to about 10 GeV, and remain essentially constant thereafter. Experimental evidence indicates that the lighter fragments with $A < 40$ are preferentially produced in the most violent collisions. These light fragments therefore provide information about the decay of very highly excited nuclear systems, information that could help determine the thermodynamic properties of nuclear systems, perhaps even nuclear matter, at high temperatures.

Over the past 35 years, radiochemical measurements and, more recently, counter telescope measurements have provided a wealth of inclusive fragment cross-section data. The fragment mass spectra (1, 2), shown in Figure 1, are commonly separated into three domains. Heavy fragments with $A \gtrsim \frac{2}{3} \cdot A_{\text{target}}$ are associated with a process called spallation. Lighter fragments with $6 \leq A \lesssim 40$, seen most clearly in Figure 1a, are attributed to a process termed intermediate mass fragment emission or, alternatively, multifragmentation. Since multiplicities of the lighter fragments are small ($< 1$) at low incident energies, the term "intermediate mass fragment emission" is used throughout the review. The third domain, corresponding to fragments with $A \approx \frac{1}{2} A_{\text{target}}$, occurs for heavy targets such as gold (see Figure 1b); it lies between the intermediate mass fragment and spallation domains, and is sometimes identified with fission or a process called deep spallation.

Spallation fragments are the residues of the target following a rapid disintegration in which somewhat less than half of the target nucleons are emitted, mainly in the form of nucleons or light composite nuclei (3–5). The origins of fragments in the fission–deep spallation domain are less certain. Some fragments may arise as the residues of even more violent collisions, while others may originate in a binary splitting of the target similar to low energy fission (6, 7). In collisions at low incident energy, some of the intermediate mass fragments are emitted from an equilibrated compound nucleus (8), while others are emitted before the compound nucleus comes to thermal equilibrium (9–11). The corresponding mechanisms for high energy collisions are poorly understood and intensely debated, much of the debate being stimulated by the hypothesis that intermediate mass fragment emission may be a signature of statistical clustering near the critical point in the liquid-gas phase diagram (12).

The energy dependence of the cross sections of some fragments produced in proton-induced reactions is shown in Figure 2 (13). Cross sections for intermediate mass fragments and for many spallation products increase strongly with incident energy. Increasing intermediate mass fragment cross sections are also observed for heavy-ion-induced reactions (14). Intuitively, this trend suggests that the cross sections are determined by the excitation energy of the fragmenting system. However, one should recognize that

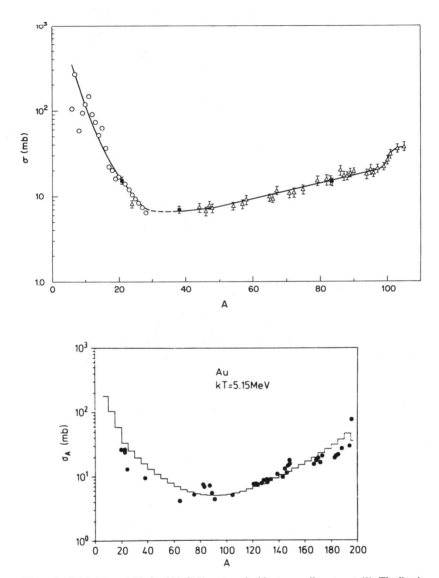

*Figure 1* (*Top*) Mass yields for 300-GeV protons incident on a silver target (1). The line is included to guide the eye. Cross sections for $A < 30$ were interpolated from measurements on xenon and krypton targets. (*Bottom*) Mass yields for 11.5-GeV protons incident on a [197]Au target. The solid line is a calculation by Gross et al (2).

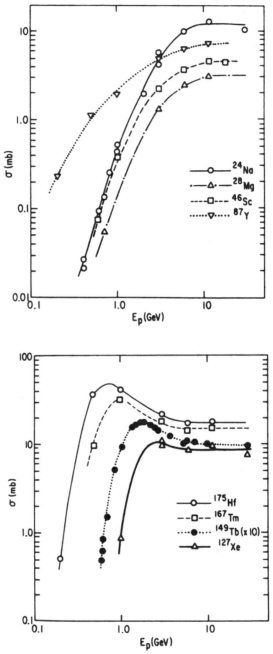

*Figure 2*   Energy dependence of fragment cross sections for protons incident on $^{197}$Au (13).

there is very little direct experimental information correlating the fragment cross sections with the energy deposition. Although the fragment cross sections are reduced at low energies, they are still large enough to permit detailed investigations at energies of a few hundred MeV (8–11). At these energies, the energy deposition can be assessed and therefore a more quantitative understanding of the production mechanism can be achieved. Such understanding could help to unravel the production mechanism at higher energies.

In this review, the traditional discussion of fragmentation is broadened to include recent results obtained with energetic but slow-moving heavy projectiles where the two-step approximation (defined below) is invalid. Present understanding does not allow the interpretation of fragmentation in terms of a standard model. Instead, the various fragmentation models are introduced and later, when the experimental results are discussed, some applications of models to the data are presented. In the interest of brevity, some aspects of high energy fragmentation reactions, such as the momentum distributions extracted from projectile fragmentation, are not fully explored. Discussions of these topics can be found in the recent reviews of Hüfner (15) and Friedlander & Heckman (16). More detailed discussions of theoretical work on fragment production can be found in the review by Csernai & Kapusta (17)

## 2.  FRAGMENTATION MODELS

### 2.1  *Reaction Dynamics*

For simplicity, consider the collision of a small projectile with a large target. If the energy is sufficiently low, there is a nonnegligible probability that the projectile may fuse with the target nucleus. In this case, fragments are emitted with relatively small cross sections by a partially or completely equilibrated compound nucleus. In reactions at higher energies, where fragment cross sections are much larger, much of the energy and momentum of the projectile are carried away by past particles emitted before the compound nucleus can reach thermal equilibrium (18, 19). With increasing incident energy, the trajectories of fast nucleons become increasingly straight; the region of geometrical overlap between projectile and target for a straight-line trajectory will be most violently disrupted. Projectile and target nucleons in this region are frequently called "participants" and the rest of the nucleons "spectators."

2.1.1  THE TWO-STEP APPROXIMATION  Fragmentation reactions have been most intensively investigated at high energies, $E > 1$ GeV. There, models of the fragmentation process are frequently formulated in terms of the two-step approximation originally proposed by Serber (20). In the first step, the participant nucleons interact violently. Some of these nucleons scatter out of the participant region and into the comparatively cold spectator matter, heating it up and perhaps knocking out an appreciable number of spectator nucleons. Then, in the second step, the fragments are generated by the breakup of an excited and equilibrated prefragment consisting primarily of spectator matter.

Extensive calculations have been performed in the two-step approximation. The intranuclear cascade model has frequently been used for the first step, sequential decay of an equilibrated compound nucleus for the second step. Cascade calculations for proton-induced reactions have been performed by Metropolis et al (21), Chen et al (22), and Bertini (23). Calculations for heavy-ion projectiles have been performed by Yariv & Fraenkel (24), Gudima et al (25), and Cugnon et al (26).

The energy deposition predicted by cascade calculations varies strongly with the target mass and the incident energy. Figure 3 shows the predicted relationship between the multiplicity of fast cascade nucleons and the energy deposition into the excited prefragment (21). Fast nucleon multiplicity is strongly correlated with energy deposition. For protons with incident energies of 1.84 GeV, energy depositions in excess of 700 MeV are predicted for heavy targets (uranium) when a large number ($> 15$) of fast nucleons are ejected during the cascade. If, however, for a given target and bombarding energy, one averages over collisions with different nucleon multiplicities, the mean energy deposition is predicted to be somewhat lower, around 450 MeV. Part of the variation in energy deposition is due to impact parameter, with larger impact parameters having smaller mean energy depositions. Part of the variation is due to fluctuations because of the stochastic nature of the energy loss mechanism. These fluctuations are relatively small for heavy-ion projectiles (24). Analytic expressions for the energy deposition based on the Glauber multiple scattering theory have been developed (4, 5, 27). These analytic expressions are in qualitative agreement with some of the results of cascade calculations (5).

For collisions with 1-GeV protons, Cugnon (28) calculated, with the intranuclear cascade model, that nucleons would be emitted by a fast nonthermal process for an elapsed time of about 20–30 fm/$c$ ($\approx 7$–$10^{-23}$ sec) after the projectile entered the target. If the cessation of fast nucleon emission signals the onset of thermal equilibrium, a breakup occurring at a later time would, for this reaction, be consistent with the two-step approximation. The experimental spectra and angular distributions of fragmentation products in relativistic collisions with $E > 10$ GeV (see Section 3) do suggest the validity of the two-step approximation.

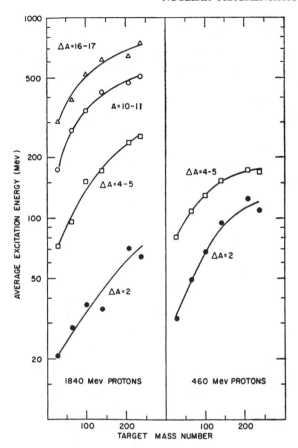

*Figure 3* Intranuclear cascade predictions (*open* and *closed points*) for the prefragment excitation energy as a function of target mass and the multiplicity of ejected fast nucleons (21). The lines are drawn to guide the eye.

Fragmentation cross sections are also appreciable in reactions with energetic but slow-moving heavy projectiles (10, 14, 29–31). There, the experimental evidence indicates that the two-step approximation breaks down, and fragments are emitted from both participant and spectator zones. Indeed, in this case, many intermediate mass fragments have been identified as originating in the participant zone in the collision (29–31). These collisions have been modeled with intranuclear cascade (24, 26), nuclear hydrodynamics (32, 33; additional references can be found in 34), molecular dynamics (35, 36), and with approaches based on the Boltzmann or Vlasov equation (37–39). Such calculations indicate that the interior density increases as the projectile and target begin to overlap. After reaching a maximum density, the system begins to expand isentropically (40, 41). (It proceeds without the dissipation of collective motion into internal excitation, leaving the entropy of the system unchanged.) For incident energies of $E/A \approx 400$ MeV, where interior densities twice normal nuclear matter density can be achieved, expansion can be sufficiently rapid that the participant zone of the prefragment breaks up explosively into a multi-fragment final state (29, 31). The breakup is often modeled by multi-particle phase space calculations (17, 31, 42, 43). Models for calculating the breakup of the prefragment are discussed in Section 2.2. At lower incident energies, the collision proceeds more slowly and the variations in the density are less pronounced. The prefragment may not explode but instead equilibrate to form a compound nucleus. For particles emitted from the compound nucleus, the two-step approximation would again be appropriate.

2.1.2   EQUILIBRATION AND FRAGMENT FORMATION   If there is sufficient time before the prefragment breaks up, it will equilibrate. Fragments will form and will be accelerated away from the prefragment by the Coulomb interaction. As the fragments move away, they may continue to interact with each other. These interactions will continue until the density becomes low enough that the probability for additional interactions becomes significantly less than unity. This density is called the freeze-out density. Many of the fragments at freeze-out will be excited and will decay sequentially afterwards.

Much depends on the time sequence of these events. One issue concerns the relative time sequence of equilibration and fragment formation or breakup. For some reactions, more than 100 fm/$c$ may be required before one can assume global thermal equilibrium. (Local thermal equilibrium might be achieved on shorter time scales.) The properties of fragments formed on shorter time scales may reflect the nonequilibrium properties of the prefragment through anisotropies in the fragment angular distributions. It is also important to know the elapsed time between breakup and freeze-out. If there is little or no time between the two, the experimental observables will be defined by the properties of the prefragment at breakup and the breakup process itself. If not, many of the experimental observables will be determined by the properties of the more dilute system at freeze-out.

## 2.2   *Fragmentation Models*

At present, there are many fragmentation models. To an extent, the differences between these models reflect differences in assumptions concerning the dynamics of the fragmentation process. Since the dynamics of the reaction evolve with bombarding energy and with projectile mass, more than one model may be required to describe the full range of observations. In liquid-gas transition (Section 2.2.1) and multiparticle phase space models (Section 2.2.2), it is assumed that the excited prefragment can be approximated by a static, spatially homogeneous, and thermally equilibrated system at a specific excitation energy and density. Cold shattering models (Section 2.2.3) assume that thermalization does not occur. Rate equation approaches (Section 2.2.4) incorporate into statistical models a time dependence based on statistical reaction rates. There is, however, no assurance that statistical reaction rates govern the time dependence of the fragmenting system. Other time-dependent features of the prefragment breakup (e.g. the time development of collective motion) may require the development of fully dynamical calculations (Section 2.2.5).

2.2.1 LIQUID-GAS PHASE TRANSITION    Many calculations have demonstrated that nuclear matter has a liquid-gas phase transition (44–47). Since the interaction between nucleons, attractive at small distances and repulsive at even smaller distances, is similar to that between constituents in a Van der Waals gas, the phase diagram for nuclear matter, shown in Figure 4a (48), is similar to that of a Van der Waals gas, having regions of liquid, gas, and mixed liquid and gas phases. In the liquid-gas phase transition model, the system is assumed to break up in the phase coexistence domain (grey shaded region) in the nuclear matter phase diagram. In this domain, fragments are considered to be droplets of liquid embedded in a gas of nucleons. The multiplicity of droplets of mass $A$ for a prefragment at temperature $T$ is given by (17, 49)

$$P_A = \text{const} \cdot \exp\left[ -\frac{(\mu_L - \mu_G)}{T} A - \frac{4\pi r_0^2 \sigma A^{2/3}}{T} - \tau \ln A \right]. \qquad 1.$$

Here, $\mu_L$ and $\mu_G$ are the nucleonic chemical potentials in the liquid and gas phases, $r_0$ is the radius parameter and $\sigma$ is the surface tension of the liquid drop, and $\tau$ is the critical exponent (50). When the system is in phase equilibrium in the mixed phase region, the chemical potentials for the liquid and gas phases become equal. At the critical point (shown as the open point in Figure 4a), $\sigma$, the surface tension coefficient, vanishes and $P_A \propto A^{-\tau}$, with $2.0 < \tau < 2.5$ (50). Elsewhere in the phase coexistence region, $\sigma > 0$; therefore the mass distribution predicted by Equation 1 falls off more rapidly than at the critical point. According to this argument, one could determine when a reaction produces systems at the critical point

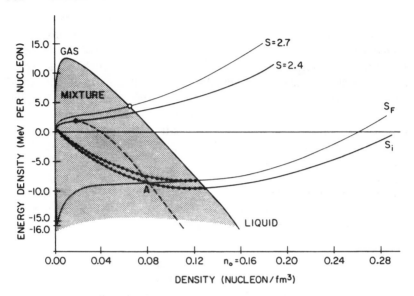

*Figure 4a* Phase boundaries for nuclear matter are shown as a function of density and energy density (48). Also shown are isentropes corresponding to the critical point (*open circle*, $S = 2.7$) and the endpoint (*solid circle*, $S = 2.4$). Inside the mixed phase region are shown the spinodal (*dashed line*) and the Maxwell construction of the proper mixed phase entropy curve (*solid line*) and the mechanically unstable isentropes $S_i$ and $S_f$ (*dotted lines*).

by finding the energy at which the mass distribution falls off most gradually (49, 52, 53).

Intermediate mass fragment cross sections at very high incident energies, shown as the solid points in Figure 4*b*, are not in disagreement with a power law (solid curve) dependence of the mass yield (12, 51, 52). Finn et al (12) first proposed that this power law behavior, characterized by a critical exponent of 2.64, is a signature for statistical clustering near the critical point of the phase diagram. Because the shape of the mass distribution is practically energy independent for $80 < E_{\text{proton}} < 400$ GeV and because no selection on impact parameter or energy deposition was made, this interpretation requires that the excited system pass close to the critical point irrespective of the initial energy deposition in the collision. In addition, the expanding system must be in thermal and phase equilibrium at densities close to the critical density but must lose thermal contact very quickly afterwards.

Lopez & Siemens argue the latter can occur if fragmentation takes place in the neighborhood of a mechanical instability (48). Nuclear matter is

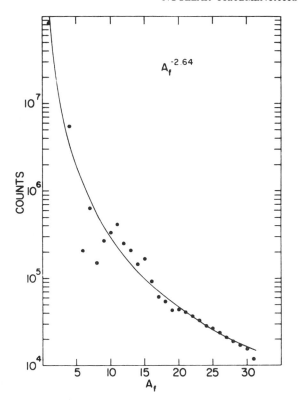

*Figure 4b* Intermediate mass fragment yields for 80–400-GeV protons incident on xenon (12). The solid curve corresponds to the power law $Y(A) = \text{const} \cdot A^{-2.64}$.

mechanically unstable with respect to breakup into fragments when both the pressure

$$P = \rho^2(\partial E/\partial \rho)|_s \qquad \qquad 2.$$

and the compressibility

$$k = \rho(\partial P/\partial \rho)|_s \qquad \qquad 3.$$

are negative (54). Here, $\rho$ is the density and $s$ is entropy per nucleon of nuclear matter. In the model of Lopez & Siemens (48), the excited prefragment expands isentropically (for example along the isentrope $S_i$ in Figure 4a) as a uniform liquid until it encounters the region to the left of the adiabatic spinodal (the dashed line in Figure 4a), where the compressibility changes sign. There it is assumed that a chaotic breakup occurs and phase equilibrium is achieved nearly instantaneously (48). [Alternative expansions in which entropy could increase continuously due to a partial phase equilibrium have been explored by Csernai (55).] Lopez & Siemens estimate that the system will expand outward after breakup within a characteristic time of about 10–40 fm/$c$, while expansion times of at least 400 fm/$c$ would be required for the system to maintain phase equilibrium on its way out. Similar expansion times were obtained by Schulz et al (56) in a fluid dynamical framework assuming a dynamical phase transition. Grant (57) and Boal (58), however, calculate that the systems could approach phase equilibrium on a much shorter time scale, $\leq 50$ fm/$c$. If the prefragment can maintain phase equilibrium during expansion, the fragment observables will reflect the properties of the prefragment at much lower densities and temperatures, where multiparticle phase space models would be applicable, and details of phase equilibrium at higher densities will be very difficult to extract.

2.2.2 MULTIPARTICLE PHASE SPACE CALCULATIONS    Here, the multiparticle phase space of nucleons and bound fragments is taken to be uniformly populated, consistent with specific energy and nucleon density at freeze-out. Unlike the liquid-gas phase transition model, interactions between nucleons not in the same fragment are neglected. This approximation clearly improves with decreasing freeze-out density. Calculations have been performed in both the grand canonical (2, 59, 60) and approximate microcanonical formalisms (42, 61). These calculations usually predict the average number of fragments observed in specific ground and excited states. They do not usually predict other observables, such as energy spectra. The expressions can be simple. For example, in the grand canonical approach of Randrup & Koonin (59), the yield of a given fragment is:

$$Y(Z, A) = \text{const} \cdot (AT)^{3/2} \zeta(A, Z, T) \exp \left\{ -[V_{AZ} - \mu_{\mathrm{p}}Z + \mu_{\mathrm{n}}(A - Z)]/T \right\},$$
4.

where $A$ and $Z$ are the mass and charge of the fragment, $\zeta(A, Z, T)$ is the partition function over the internal states of the fragment, $T$ is the temperature of the ensemble, $V_{AZ}$ is the ground-state mass excess of the fragment, and $\mu_{\mathrm{n}}$ and $\mu_{\mathrm{p}}$ are the neutron and proton chemical potentials.

While all phase space calculations are similar in principle, they differ in their technical details and in some of their predictions (17, 42). In some calculations (42, 60), particle-unstable states are included in the internal partition functions of the fragments, but these states are not included in others (2, 59, 61). The influence of the long-ranged Coulomb interaction is included approximately in some (2, 61), but not in the other calculations.

Further discussions of the differences between these models can be found in (17).

The descriptive power of the phase space calculations can be quite good. For example, Gross et al (2) can describe the U-shaped fragmentation yield for proton-induced reactions on gold, shown in Figure 1b. The calculation is shown in the figure as the solid line. Other examples can be found in (31, 43).

Fragment multiplicities provided by phase space models are dependent on the temperature and density at freeze-out. If one chooses instead, density and entropy per nucleon as the intensive parameters, calculations (31) have shown that the multiplicities provided by multiparticle phase space calculations are much more sensitive to entropy than to density. This makes entropy, in principle, a fairly robust observable (61a). A review of the effforts to extract the entropies produced in nuclear collisions from measurements of the fragment multiplicities can be found in (17).

2.2.3 COLD SHATTERING MODELS    Several models have been developed to describe a nonthermal disassembly of the prefragment in the mechanical instability domain (62, 63). One model developed by Aichelin et al (63) assumes that all distinct partitions of the $Z_0$ protons in the prefragment are equally probable. Here, a partition is defined to be a decomposition of the number $Z_0$ into integers $z_1, z_2, z_3, \ldots$, with $Z_0 = \Sigma z_i$. They obtain an elemental yield of the form

$$\sigma(Z) = \sigma_0 \cdot [\exp (1.28Z/\sqrt{Z_0}) - 1]^{-1}. \qquad 5.$$

Here, $\sigma_0$ is a free parameter. The prefragment charge, $Z_0$, determines the shape of the calculated elemental distribution. In postulating a nonthermal breakup, it is assumed that there is insufficient time between breakup and freeze-out for thermalization to occur. This is an assumption that could be tested in a multifragment rate equation approach (57, 58). Calculations with the cold shattering model (63) are in qualitative agreement with a wide range of experimental data, including the fragment yield in Figure 4b for $Z > 8$.

2.2.4 RATE EQUATION APPROACHES    Some time dependence can be explicitly included in a statistical calculation by casting it in the form of a rate equation. Rate equation approaches to multifragmentation final states have been attempted by Mekjian (64), Boal (58), and Grant (57). The most frequently used rate equation, however, involves the sequential binary decay of an equilibrated compound nucleus. This approximation has been used for calculating cross sections of spallation and deep spallation fragments where the fragments are produced as residues at the end of the evaporation calculation (3–5). Fission fragments in the fission–deep spallation domain have also been calculated in this approximation (65).

Intermediate mass fragment cross sections have been calculated by assuming that intermediate mass fragments are evaporated particles (66, 67). The multiplicity of a fragment in a specific internal state can be expressed most simply in terms of the Weisskopf formula (68)

$$\frac{d^2 N}{dE_{rel}\, dt} = \frac{2J+1}{\pi^2 \hbar^3} M E_{rel} \sigma_{inv} \exp(\Delta S), \qquad\qquad 6.$$

where $E_{rel}$ is the kinetic energy released by the emission of the fragment; $\sigma_{inv}$ is the cross section for capture of the fragment by the daughter nucleus to form the parent nucleus; and $N$, $J$, and $M$ are the multiplicity, spin, and mass of the fragment in the specific internal state. The factor $\exp(\Delta S)$ is frequently computed as the ratio of the level density of the daughter nucleus to that of the parent nucleus. In an evaporative approach, Friedman & Lynch (67) have described the mass distribution of Figure 4b as well as the associated (12, 52) energy spectra and isotopic distributions.

Moretto argues instead that intermediate mass fragment emission is better regarded as a form of asymmetric fission (69). The predicted energy-integrated multiplicities depend on similar phase space factors in the two approaches. Fluctuations in the potential barrier in the fission description, however, can broaden the maximum in the fragment energy spectrum located near the Coulomb barrier (69).

2.2.5  DYNAMICAL CALCULATIONS    Certain dynamical or time-dependent features of the prefragment breakup (e.g. the mechanical stability of the prefragment or time development of collective motion) may require the development of fully dynamical calculations. Dynamical calculations have been performed in the approximation of nuclear hydrodynamics (56, 70), in time-dependent Hartree-Fock (71, 72), by solving the Boltzmann equation (73, 74, 74a), and by solving the classical $A$-body equations of motion (75, 76). Both Sagawa & Bertsch (71) and Schulz et al (56) find a maximum temperature of 8 MeV above which a compound nucleus no longer exists. Pandharipande and coworkers (76) find that fragmentation occurs within the region of mechanical instability bounded by the adiabatic spinodal. Most dynamical calculations, however, have yet to be developed to the point at which direct comparisons with experimental data can be made.

2.2.6  DISCUSSION    Even though there are significant differences between the various fragmentation models, there is an important area of agreement. Most statistical models predict small intermediate mass fragment multiplicities for prefragments at either very low or very high excitation (2, 49,

67). Multiplicities are low at very low excitation energy because of the $Q$-value and Coulomb barrier energies required to separate the fragment from the prefragment matter. Multiplicities are low at very high excitation energy because of the very low probability of finding a cluster of $A$ nucleons with low internal excitation energy in a very hot prefragment.

Certain approaches have been neglected in this discussion. Bauer et al (77), Biro et al (78), and Campi (78a) have calculated fragmentation within the context of percolation theory. Bondorf and coworkers (79) have developed approaches incorporating features of both the liquid-gas phase transition and multiparticle phase space calculations.

Most of the calculations discussed in this section are moderately successful in reproducing the fragment elemental and mass distributions. Thus, it has been very difficult to rule out models by comparison with experimental elemental and mass distributions. It would help if more effort were expended in calculating other observables such as energy spectra and excited state populations (discussed in Section 4.3.2). It is also relevant to consider whether the dynamical assumptions of a given model are met by the system under investigation. Both liquid-gas and multiparticle phase space models are better suited to rapid multifragment breakup mechanisms. Compound nuclear decay models are more appropriate for a slower sequential binary disassembly mechanism. Some clues about the time sequence of the disintegration process can be provided by the experimental data. In addition, dynamical calculations can be used to examine the validity of the breakup assumptions of fragmentation models presently considered.

# 3.   FRAGMENTATION AT RELATIVISTIC ENERGIES

## 3.1   *Coincidence Measurements: Reaction Dynamics*

Some features of fragmentation reactions are clearly illustrated by coincident measurements. Fragmentation products and the multiplicity of coincident light particles were measured by Warwick et al (6, 80) for relativistic light- and heavy-ion projectiles incident on a gold target. Intranuclear cascade calculations suggest that the light-particle multiplicity is proportional to energy deposition (24, 80). Contour plots of the measured fragment yields as a function of fragment energy and "observed" light-particle multiplicity are shown in Figure 5a. (The "observed" multiplicity is defined to be the number of associated light particles actually detected in the multiplicity array. The "observed" multiplicities are then corrected for detection efficiency to obtain the multiplicities shown in Figure 5b.)

Consider first, fragments in the fission–deep spallation domain with

$80 \leq A \leq 89$. Two distinct processes are observed. At high fragment energy and low multiplicity, fission fragments are observed for peripheral collisions characterized by low energy deposition. At low fragment energy and high multiplicity, deep spallation fragments are produced in reactions

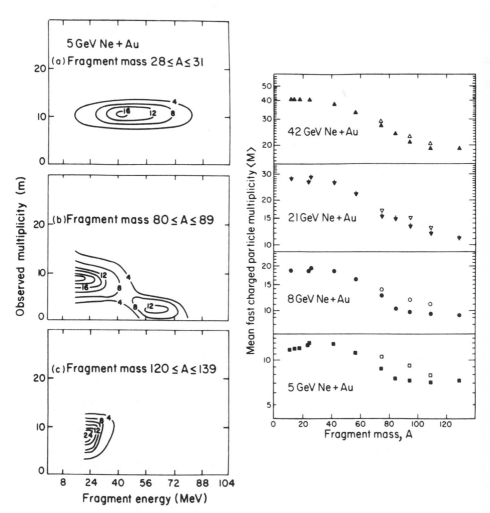

*Figure 5a*  Fragment yield vs observed multiplicity and fragment energy for neon-induced reactions on $^{197}$Au (80).

*Figure 5b*  Mean associated light-particle multiplicity as a function of fragment mass and bombarding energy. The solid points correspond to all fragments. For the open points, low multiplicity fission was excluded (80).

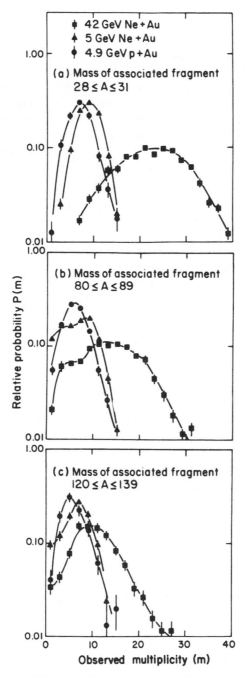

*Figure 5c* Fragment multiplicity distributions for a variety of reactions and fragment masses (80).

characterized by large energy deposition. The presence of both processes was previously deduced from measurements of the energy dependence of the isotopic and isobaric yields (81–83a) and mean fragment velocities (81, 83a). For low energy projectiles, one observes mainly neutron-rich fission fragments. With increasing energy, deep spallation fragments closer to the valley of stability grow in importance (81–83a). For lighter fragment masses, violent fission-like processes may also contribute. For 11.5-GeV protons incident on $^{238}$U, Wilkins et al (7) have observed fission-like decay processes leading to fragments with combined masses as light as 140 and with a total kinetic energy release significantly higher than expected from low energy fission systematics.

The dependence of light-particle multiplicity upon fragment mass is shown in Figure 5b for different bombarding energies (80). The solid points are the average multiplicities with all events included. For the open points, the low multiplicity fission events have been excluded. For fragments with $A < 40$ the multiplicities are roughly constant. For $A > 40$, the multiplicities for spallation and deep spallation fragments decrease monotonically with fragment mass. According to the cascade calculations (24, 80), this trend implies a descreasing residue excitation energy. Indeed, one should not expect to reproduce the entire fragmentation mass yield by a statistical calculation using a single temperature or excitation energy [as was done in the statistical calculations shown in Figure 1b (2)]. Rather, one should take the dynamics of energy deposition explicitly into account.

By comparisons with the cascade calculations, the average prefragment excitation energies for reactions yielding intermediate mass fragments were estimated to be 0.8, 1.0, and 1.2 GeV at incident neon energies of 5.0, 8.0, and 21.0 GeV respectively (80). The multiplicity distributions shown in Figure 5c become increasingly broad with incident energy. This suggests that the prefragment excitation energy in inclusive measurements is very poorly defined.

## 3.2  Mass Yields and Isobaric and Isotopic Distributions

Factorizability and limiting fragmentation are important concepts for the description of total and differential cross sections of fragments produced by relativistic projectiles. Target (projectile) fragmentation cross sections are said to be factorizable if the cross sections for different projectiles (targets) differ only by a multiplicative cross section term. A cross section is consistent with the hypothesis of limiting fragmentation, first discussed in elementary particle physics (84, 85), if it has attained a limiting value that remains unchanged as the energy is increased. This condition appears to be satisfied by the cross sections shown in Figure 2 for $E_p > 10$ GeV.

For the energy dependence of other fragment cross sections, see Kaufman & Steinberg (13). Further discussions can be found in (86–88).

3.2.1 SPALLATION YIELDS For incident energies greater than 10 GeV, spallation cross sections are consistent with the hypothesis of limiting fragmentation (86–88). To considerable accuracy, the spallation cross sections shown in Figure 6a, for different projectiles on a copper target, are proportional to one another and therefore factorizable. The constant of proportionality is nearly the same as the ratio of the corresponding reaction cross sections (86). Factorizability also apears to be reasonably well satisfied in relativistic collisions with other projectiles and targets (87–89). Empirical expressions, developed to permit extrapolation from measured to unmeasured cross sections (52, 88–91), have considerable

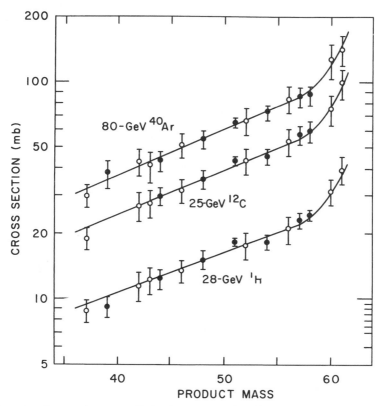

*Figure 6a* Cross sections for spallation fragments produced in reactions on a copper target with a variety of different projectiles (86). The solid curves were used in the integration of the spallation cross sections.

predictive power if the cross sections are factorizable and satisfy the hypothesis of limiting fragmentation.

By assuming that exactly one spallation fragment is produced per collision and integrating the spallation cross sections shown in Figure 6a, Cumming et al (86) obtained total spallation cross sections that are about 70% of the measured total reaction cross sections. Spallation takes up similar fractions of the total reaction cross section for other systems (1, 92).

The decay of the prefragment and subsequent production of the spallation residue, which occurs during the second step in the two-step approximation, has been calculated both by multiparticle phase space calculations (2) and by assuming the sequential binary (compound nuclear) decay of an equilibrated prefragment (3-5). Many Monte Carlo evaporation calculations (3, 13, 65, 92-96) use prefragment masses and excitation energies supplied by intranuclear cascade calculations. [Campi & Hüfner (4) and Abul-Magd et al (5) have simplified the evaporation calculation so that a Monte Carlo calculation can be avoided.] Unfortunately, energy deposition has been accorded only a cursory treatment in currently available multiparticle phase space calculations of the spallation cross sections (2) (see also Section 2.1). This precludes a quantitative comparison of multiparticle phase space calculations with experimental data.

In the simplified (4, 5) and Monte Carlo (3, 13, 65, 92-96) evaporation calculations, the mass $A$ of a spallation residue in the calculation can be obtained approximately from the mass $A_{pf}$ and excitation energy $E^*$ of the prefragment via the relation

$$A = A_{pf} - \frac{E^*}{\varepsilon},    7.$$

where $\varepsilon$ is the average excitation lost per evaporated nucleon. Typically, $\varepsilon = 10-15$ MeV and depends on the mass and excitation energy of the prefragment (4, 5, 13). Thus the spallation mass distributions predicted by evaporation calculations reflect directly the assumed distribution of prefragment mass and excitation energy.[2] The histograms in Figure 6b correspond to the spallation yield calculations for 8.0-GeV $^{20}$Ne ions incident on $^{181}$Ta (92) in which the excitation energies and prefragment masses were supplied by the cascade calculations (24). The solid line corresponds to similar calculations in which excitation energies and prefragment masses were taken from an abrasion-ablation model (65). The

---

[2] Cumming has pointed out that factorizability, in this sequential decay picture, implies very similar prefragment mass and excitation energy distributions for light and heavy projectiles (86). This surprising conclusion is difficult to explain.

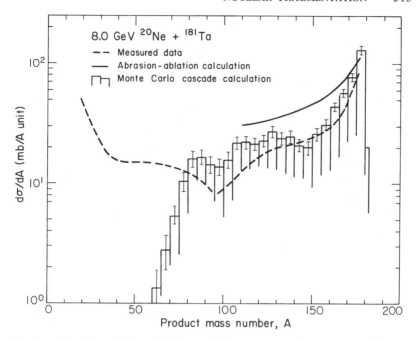

*Figure 6b* Experimentally determined mass yield curve (*dashed line*) and evaporative spallation calculations for 8.0-GeV $^{20}$Ne ions incident on $^{197}$Au (92). The histograms are calculations in which energy deposition was calculated with an intranuclear cascade calculation. The solid line depicts calculations in which energy deposition was calculated from an abrasion-ablation model.

agreement between the experimentally determined mass yield curve and the calculations shown in the figure is typical of this type of calculation.

3.2.2 INTERMEDIATE MASS FRAGMENT YIELDS Most of the fragmentation models, discussed in Section 2.2, predict monotonically decreasing mass yields roughly consistent with the yields shown in Figure 4b. Thus it is difficult to eliminate models by comparisons with the measured mass yields. Irregularities that are evident in the mass distributions of Figure 4b (e.g. minima at mass 6 and 8) appear to be caused by nuclear structure effects that influence the decay of heavier, particle-unstable, intermediate mass fragments that are populated at thermal freeze-out. These can be reproduced if one uses the tabulated spectroscopic information (97) to describe the excited states (98). Using a primary distribution of fragments in ground and excited states given by Equation 6, with $\Delta S = -(E+B)/T$ and $T = 5$ MeV and letting the particle-unstable states decay sequentially, Fields et al (98) reproduced the deviations of the fragment yields from a smooth monotonic decrease with fragment mass. (Here, $E$ and $B$ are the kinetic energy released and the fragment separation energy, respectively.) The results are shown in Figure 7a. It is probable that these features would also be reproduced by many other models discussed in Section 2.2 if sequential decay were similarly taken into account.

Intermediate mass fragment cross sections appear to attain limiting fragmentation values at incident energies greater than 10 GeV. Elemental multiplicities of intermediate mass fragments gated by the detection of another intermediate mass fragment at 90° are shown in Figure 7b for 5- and 42-GeV $^{20}$Ne ions incident on $^{197}$Au (80). Like the elemental distributions of Hirsch et al (52), these distributions fall off gradually with fragment charge and have likewise been characterized by a power law distribution $\sigma(Z) \propto Z - \tau$, with $\tau \approx 2.6$ (80). This could suggest that the intermediate mass fragment cross sections are factorizable. Unlike spallation cross sections, however, the intermediate mass fragment cross sections do not factorize simply. Instead, the measurements suggest that relatively more intermediate mass fragments are produced in heavy-ion-induced reactions than in proton-induced reactions (87, 88, 98a). Similar elemental yields are also observed in low energy, heavy-ion reactions. This implies that a power law dependence of the fragment yield may not be a characteristic signature of very high energy deposition process. More experimental data would help to clarify the issue.

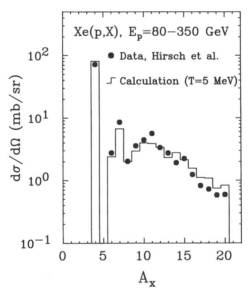

*Figure 7a*   Intermediate mass fragment yields for 80–350-GeV protons incident on xenon (12). The histogram corresponds to the results of sequential decay calculations (98).

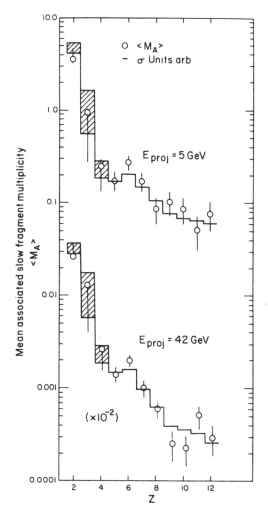

*Figure 7b*  Intermediate mass fragment multiplicities for 5- and 42-GeV $^{20}$Ne ions incident on $^{197}$Au (80). The open circles indicate the fragment multiplicities. The solid histogram corresponds to the intermediate mass fragment, single-particle, inclusive data arbitrarily normalized.

The distributions in Figure 7b can be integrated over $3 \leq Z \leq 12$ to provide an associate multiplicity of roughly 2.0–2.5 intermediate mass fragments, comprised of about 30 nucleons. Thus, intermediate mass fragment emission is clearly a multifragmentation process at these energies. Unfortunately, the cross sections for heavy fragment emission are not known accurately enough to constrain the impact parameter distribution and therefore the cross section of collisions that yield intermediate mass fragments. Additional information obtained from emulsion data can be found in (99).

3.2.3    ISOBARIC AND ISOTOPIC DISTRIBUTIONS    There have been extensive measurements of inclusive isotopic and isobaric distributions of fragmentation products (e.g. 51, 81–83, 86, 88, 89, 100–102). Entrance-channel-dependent variations in the isotopic and isobaric distributions have been observed for fragments in the fission–deep spallation domain (81–83, 100, 101). These variations reflect the competition between fission and deep spallation, which depends upon the target mass and charge (in part because the fissility depends on the mass and charge of the target) and the bombarding energy (because the deep spallation cross sections are strongly energy dependent). Outside of the fission–deep spallation domain, however, isotopic and isobaric distributions are relatively insensitive to the entrance channel (86, 88, 89, 102, 103). This insensitivity is illustrated by the isotopic distributions for sodium shown in Figure 8 measured for

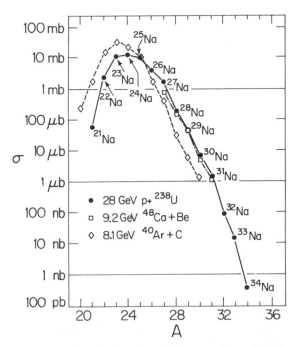

*Figure 8*    Isotopic distributions for sodium isotopes produced in a variety of reactions (103). The lines are drawn to guide the eye.

three very different entrance channels (103). For entrance channels that are more similar, the isotopic distributions are often indistinguishable (89, 102).

This insensitivity has a simple origin. The observed fragments are formed in the sequential decay of heavier particle-unstable nuclei. Since sequential decay is a stochastic process, information about the prefragment is lost in each step of the sequential decay chain. If the excitation energy of the prefragment is a hundred MeV or more, calculations (4, 94) indicate that details of the collision dynamics and energy deposition influence the isotopic and isobaric distributions very little. The isotopic and isobaric distributions are most strongly influenced by phase space considerations such as the fragment ground-state binding energies and the number of particle-stable excited states (4). Memory of the primary distribution appears to be reflected primarily in a dependence of the centroids of the isobaric and isotopic distributions upon the isospin of the prefragment (4, 103, 104).

## 3.3    *Energy and Angular Distributions*

3.3.1    VELOCITY AND ANGULAR DISTRIBUTIONS    Information about fragment velocity distributions, obtained with recoil range techniques, has been frequently analyzed with the *two-vector* model. It is assumed in this model that the fragment velocity, $\mathbf{v}$, can be written as $v_f = \mathbf{v} + \mathbf{V}$, where $\mathbf{v}$ is a velocity imparted to the prefragment during the first step of the two-step approximation ($\mathbf{v}$ is usually approximated by its component $v_\parallel$ parallel to the beam axis) and $\mathbf{V}$ is a velocity imparted during the decay of the prefragment and is assumed to be isotropically distributed. Both $\langle \beta_\parallel \rangle = \langle v_\parallel \rangle / c$ and $\langle |\mathbf{V}| \rangle$ can be obtained by measuring the fractions of fragments that recoil out of the target into the forward and backward directions (105).

Mean fragment velocities, $\langle \beta_\parallel \rangle$, are shown in Figure 9a, for projectiles incident on heavy targets (16). Two trends are apparent. First, for a given reaction, the mean fragment velocity $\langle \beta_\parallel \rangle$ decreases with increasing fragment mass. Second, for a given projectile and target, $\langle \beta_\parallel \rangle$ decreases with increasing projectile energy.

In the two-step approximation, one should be able to calculate $\langle \beta_\parallel \rangle$ from an intranuclear cascade calculation. Porile (106) examined the results of intranuclear cascade calculations for incident protons with energies of $E_p \leq 2$ GeV and obtained the relation

$$\langle \beta_\parallel \rangle / \beta_{CN} = 1.25 E^* / E^*_{CN}, \qquad\qquad 8.$$

where $E^*$ is the excitation energy of the prefragment; $\beta_{CN}$ and $E^*_{CN}$ are the velocity and excitation energy of a hypothetical compound nucleus formed

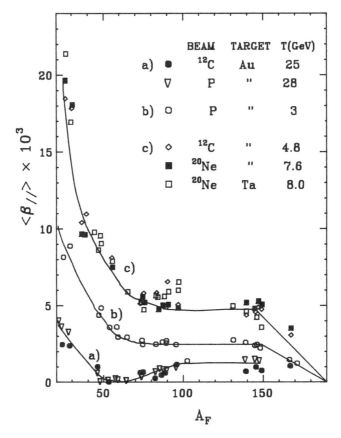

*Figure 9a*  Measured mean fragment velocities are shown as a function of fragment mass $A_f$ for a variety of reactions (16). The solid lines are drawn to guide the eye.

by fusing the projectile with the target. Using Equation 7 to relate $E^*$ to the mass loss $\Delta A$, one can obtain

$$\langle \beta_\parallel \rangle = 1.25 \frac{\varepsilon \Delta A \beta_{CN}}{E^*_{CN}} \qquad \qquad 9.$$

This formula predicts that $\langle \beta_\parallel \rangle$ should decrease linearly with $A$ and decrease monotonically with incident projectile energy for fragments in the spallation domain ($A > 140$). There is some indication that both trends are followed for the reactions in Figure 9a. Kaufman et al (93) report $\langle \beta_\parallel \rangle$ to be more consistent with Equation 9 for spallation fragments closer in mass to the target ($\Delta A \leq 20$). Unfortunately, for large mass losses ($\Delta A \approx 40$) Equation 9 overpredicts $\langle \beta_\parallel \rangle$ by a factor of almost two. Similar conclusions are reached by other studies (93, 95, 96). It is important to know whether this discrepancy portends deeper problems that could also affect estimates of the energy deposition provided by intranuclear cascade calculations.

Better agreement with the data is obtained with the empirical expression

$$\langle p_{\parallel} \rangle = \frac{E^*}{\beta_{p}} \cdot (1 + k\sqrt{1 - \beta_{p}^2}), \qquad\qquad 10.$$

where $\langle p_{\parallel} \rangle$ is the average momentum supplied to the prefragment, $\beta_{p}$ is the projectile velocity, and $k$ is a constant that could assume values ranging from 0 to 3 (6, 107).

Since $\beta_{p}$ is approximately one for $E_{p} \geq 3$ GeV, Equation 10 does not predict the further decrease in $\langle \beta_{\parallel} \rangle$ shown in Figure 9a for spallation fragments produced in reactions at higher energies. Indications that the reaction dynamics may be different at higher energies are provided by the angular distributions for fragments with masses near the boundary of the intermediate mass fragment and deep spallation domains, shown in Figure 9b (108, 109). For heavy-ion- and proton-induced reactions at low energies, the angular distributions are forward peaked in the laboratory frame, consistent with isotropic emission in a frame moving with velocity $\langle \beta_{\parallel} \rangle$ as expected in the two-vector model. At higher energies, the angular distribution peaks at 90°, which indicates that a significant amount of transverse momentum was supplied to the prefragment during the early stages of the reaction. Similar effects were observed by others (110, 111). Sidewards peaking for proton-induced reactions has been predicted by the cleavage model (62), which assumes that a high energy proton splits the target into two fragments that are preferentially emitted to 90°. At present, however, there is no calculation that predicts the observed sidewards peaking for both proton- and heavy-ion-induced reactions at $E_{p} \leq 25$ GeV.

The dramatic increase in $\langle \beta_{\parallel} \rangle$ for $A < 40$ indicates that collisions producing intermediate mass fragments are more violent, on average, than the collisions producing fragments in the spallation and deep spallation domains. Detailed analysis reveals, however, that the energy spectra of intermediate mass fragments produced in reactions at energies less than 10 GeV are not consistent with isotropic emission in any rest frame (112–114). In the discussions in Section 4 about intermediate mass fragment emission at low energies, it is shown that the large values of $\langle \beta_{\parallel} \rangle$ at low energies are accompanied by the breakdown of both the two-vector model and the two-step approximation.

The energy $\langle E \rangle = \langle |\mathbf{V}|^2 \rangle / 2m_{0}A$ of the fragment ($m_{0}$ is the mass of the nucleon), in the rest frame defined by $\langle \beta_{\parallel} \rangle$, has been analyzed by Winsberg (115) with the parameterization

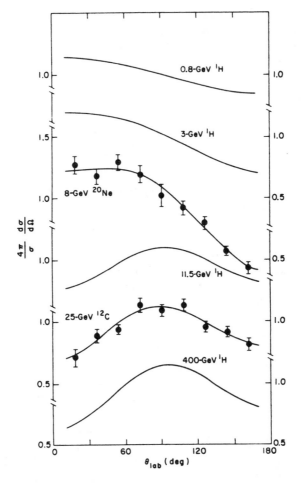

*Figure 9b* Angular distributions for $^{37}$Ar fragments produced in heavy-ion-induced reactions on $^{197}$Au and angular distributions for scandium isotopes produced in proton-induced reactions on $^{238}$U (108). Data for $^{37}$Ar are obtained from (108). The curves without data are fits based on the data of Fortney & Porile (109).

$$\langle E \rangle = \varepsilon_{s}(A_{T} - A)/A_{T}, \qquad\qquad 11.$$

where $A$ is the mass of the fragment, $A_{T}$ is the mass of the target, and $\varepsilon_{s}$ is a parameter. He finds that $\langle E \rangle$ is accurately described by Equation 11 for $0.4A_{T} \leq A \leq A_{T}$ and $\varepsilon_{s} \approx 17$ MeV. For $A \leq 0.4A_{T}$, $\varepsilon_{s}$ increases with decreasing fragment mass, which suggests the occurrence of a change in the fragment production mechanism in the intermediate mass fragment domain.

If the fragment is created suddenly, the momentum of the fragment could reflect the Fermi momentum of its constituent nucleons. If this were the dominant contribution to the final fragment momentum, then $\varepsilon_s = \frac{3}{5}\varepsilon_f$, where $\varepsilon_f$ is the Fermi energy (116, 117). This hypothesis has been explored extensively in studies of projectile fragmentation for fragments with masses close to that of the projectile (118, 119 and references there). If, instead, the fragment is the residue of a long evaporative decay chain, then $\varepsilon_s$ will reflect the incoherent superposition of the recoil boosts from successive stages in the evaporation chain. The most important boosts are generated by the evaporation of $\alpha$ particles and heavier complex fragments. Crespo et al (95) compared the spectra of $^{149}$Tb fragments produced by 2.2-GeV protons incident on $^{197}$Au with a two-step intranuclear cascade and evaporation model calculation. They found that the calculated spectra are extremely sensitive to uncertainties concerning the momentum transferred to the prefragment during the cascade and to the possible contributions from heavy fragment emission during the evaporation of the prefragment. When these effects were included in the calculations, Kaufman et al (93) were able to reproduce values for $\langle E \rangle$ measured in collisions on $^{197}$Au at a variety of energies.

3.3.2 ENERGY SPECTRA OF INTERMEDIATE MASS FRAGMENTS    Intermediate mass fragment energy spectra measured at 90° are shown in Figure 10 for 4.9-GeV protons incident on uranium (120). Each spectrum decreases exponentially with energy from a peak that occurs, for example, for fluorine at about 60 MeV. Westfall et al fit the energy spectra with a thermal moving-source parametrization. Isotropic emission of a fragment with mass $M$ and energy $E^*$ is assumed in a rest frame moving with velocity $\beta_\parallel$. In the laboratory frame the cross section is given by

$$\sigma(E,\Omega) = N(E/E^*)^{1/2} \int_{k=\langle k \rangle - \Delta}^{k=\langle k \rangle + \Delta} dk(vE^* - kV_{\mathrm{Coul}})^{1/2}$$

$$\times \exp\left[-(vE^* - kV_{\mathrm{Coul}})/T\right], \qquad\qquad 12.$$

where

$$E^* = E + \tfrac{1}{2}M\beta_\parallel^2 - 2(\tfrac{1}{2}E \cdot M\beta_\parallel^2)^{1/2}\cos\theta. \qquad\qquad 13.$$

$N$ is a normalization factor, and $V_{\mathrm{Coul}}$ is the Coulomb barrier between the fragments and a hypothetical residual nucleus considered as two tangent spheres

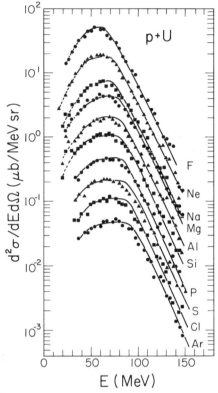

*Figure 10*    Intermediate mass fragment energy spectra for 4.9-GeV protons incident on uranium (120). The solid lines correspond to moving-source fits described in the text.

$$V_{Coul} = \frac{Z(Z_R - Z)e^2}{1.44[A^{1/3} + (A_R - A)^{1/3}]};$$    14.

$v = A_R/(A_R - A)$ is a factor that takes the recoil of the residual nucleus into consideration, and $A_R$ and $Z_R$ are the mass and charge of the residual nucleus. Better agreement with the measured spectra can be obtained by assuming a distribution of Coulomb barriers centered at about half of the value provided by Equation 14 (112). This requirement is accommodated in Equation 12 by the parameter $k$, which is distributed about a mean value $\langle k \rangle$ with a width $2\Delta$; $\langle k \rangle$ is about 0.5 for the spectra in Figure 10 (120).

The energy spectra of Figure 10 were consistent with "temperatures" $T$ and source velocities $\beta_{\parallel}$ of 14 MeV and 0.006. Nearly identical "temperatures" and similar source velocities were obtained by Gaidos et al

(121) for 80–400-GeV protons incident on xenon ($T = 15$ MeV, $\beta = 0.002$) and krypton ($T = 14.5$ MeV, $\beta = 0.006$) targets. This suggests that the energy spectra of intermediate mass fragments are consistent with limiting fragmentation. In comparison, "temperatures" ranging from 22 to 30 MeV were estimated for 5- to 42-GeV $^{20}$Ne ions incident on $^{197}$Au, but no clear relationship between "temperature" and bombarding energy was established (80). The interpretation (53) of $T$ as the temperature of an equilibrated prefragment may be somewhat problematic, however, because the slopes of energy spectra can be strongly affected by Fermi motion (63), collective motion (122), and by the temporal evolution of the emitting system (10, 67, 123).

Aichelin & Hüfner (63) calculated these energy spectra using the assumption of a sudden bulk disintegration of the prefragment. In this calculation, fragments are assumed initially to have energies given principally by the Fermi momenta of their nucleon constituents. Then, fragments are accelerated by the Coulomb field; fragments that originate in the interior of the prefragment are accelerated least and give the major contributions to the low energy portions of the energy spectra. Similar assumptions were invoked by Hirsch et al (52). In a different model (67), the prefragment is taken to be a compound nucleus from which intermediate mass fragments were assumed to be emitted sequentially. Most emission occurs after the prefragment has cooled below 10 MeV. Low energy fragments can occur when (a) the prefragment charge is reduced at the time of emission because of prior particle emission and (b) the prefragment is recoiling from the accumulative recoil boosts from prior particle emission; thus the fragment energy spectrum is smeared out and extends to lower energies.

Both sudden and sequential decay calculations have internal consistency problems, however. Calculations (56, 71) suggest that the compound nucleus is hydrodynamically unstable and may not exist for temperatures in excess of 8 MeV. The sequential decay calculation would be tenable only if much of the initial mass and excitation energy were dissipated through nonequilibrium particle emission before the prefragment equilibrates. The sudden model has problems because is assumes that the prefragment expands quite slowly after breakup. Indeed, the model provides for an expansion velocity ($< 0.05c$ for $Z = 8$) dictated by the Coulomb interaction, which is not large compared to the random velocities of the nucleons ($\approx 0.2c$) and fragments ($\approx 0.05c$ for $Z = 8$) assumed to arise from Fermi motion. Thus, it appears likely that many interactions will occur during expansion. Since in the liquid-gas model calculations (52) and the cold shattering calculations (63) the fragment observables are assumed to be defined at breakup, it is relevant to estimate how much these interactions alter the mass, charge, and velocity distributions after breakup.

## 4.  FRAGMENTATION WITH NONRELATIVISTIC PROJECTILES

### 4.1  *Characteristics of Nonrelativistic Collisions*

Failure of the two-step approximation and the two-vector model for non-relativistic projectiles is illustrated by the invariant cross sections shown in Figure 11a for carbon fragments produced by 200-MeV $^3$He ions incident on silver (11). For $\theta_{lab} \geq 80°$, contours of constant cross section follow the semicircles of constant velocity $v_{rel} = |\mathbf{v} - \mathbf{V}_{CN}|$ in the center of mass, consistent with evaporation of the fragment from an equilibrated compound nucleus. For $\theta_{lab} \leq 80°$, contrary to the two-vector model, the cross sections increase with decreasing scattering angle and the velocity spectra extend to larger fragment velocities. Many of the fragments at forward angles are emitted quickly, long before the compound nucleus can equilibrate.

The two-step approximation is generally inaccurate for the intermediate mass fragment spectra and angular distributions measured in light- and heavy-ion induced reactions at $E/A < 1$ GeV (8–11, 14, 29–31, 124, 125). Analysis of the intermediate mass fragment energy spectra with moving-source parametrizations indicates that lighter fragments are described by

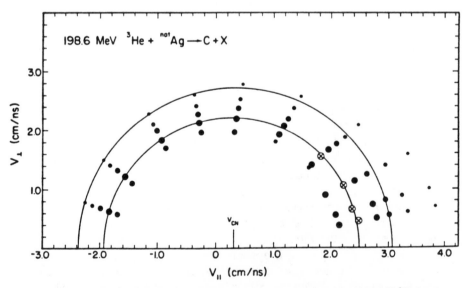

*Figure 11a*  Invariant cross sections for carbon fragments produced by 200-MeV $^3$He ions incident on silver (11). The diameter of the dots is proportional to the cross section; adjacent dots differ in cross section by a factor of two.

moving sources that have higher "temperatures" and faster source velocities than the sources that fit the energy spectra of heavier fragments (10, 124, 125). For 720-MeV $^{32}$S ions incident on silver, at least half of the lighter fragments with $Z < 8$ are emitted by a fast nonequilibrium process, while the spectra of heavier fragments with $Z > 16$ are not that different from emission by an equilibrated residual nucleus (125). Similar observations were also reported by others (9–11, 14).

The spectra of nonequilibrium intermediate mass fragments produced by 360-MeV $^{12}$C ions incident on $^{197}$Au are shown in Figure 11$b$ (10). These spectra have been calculated with two different models (10), shown

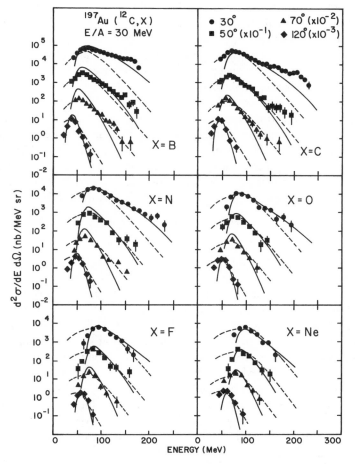

*Figure 11b*   Spectra of intermediate mass fragments produced by 360-MeV $^{12}$C incident on $^{197}$Au (10). The curves are discussed in the text.

as the dashed and solid lines in the figure. In the model described by the solid lines, fragments are emitted by a sequential binary decay mechanism, according to Equation 6, from a hot thermalized subset of nucleons that is in the process of equilibrating with the remaining cold target spectator matter (10). The dashed lines correspond to predictions of the cold shattering model (10, 63). Both models suggest that the slopes of the energy spectra do not provide a temperature. For the cold shattering model, the "temperature" actually results from the fragment Fermi momentum. In the equilibrating source model, emission from different stages of the reaction with different temperatures contributes to the energy spectra in amounts that depend on the scattering angle. The two models predict different multiplicities, however. The cold shattering model predicts a multifragment final state, while the equilibrating source model predicts intermediate mass fragment multiplicities less than 0.5.

Low intermediate mass fragment multiplicities are observed in non-relativistic heavy-ion-induced reactions. References (125–127) report the dominance of binary final states, characterized by incomplete momentum transfer, for heavy-ion-induced reactions at $E/A < 50$ MeV. Direct measurements of fragment multiplicities confirm this picture (125). Pelte et al (128) observe ternary decay channels but find that the binary decay channel is much more important for reactions at $E/A = 15$ MeV. Since sequential binary decay models predict some multiple heavy fragment production at sufficiently high excitation energies (67), it is not clear that the latter measurement necessitates a multifragmentation theory.

## 4.2   Emission from the Equilibrated Residual Nucleus

The two-step approximation can be valid in collisions at nonrelativistic energies for intermediate mass fragments emitted from an equilibrated reaction residue. In contrast to the situation in high energy collisions where the prefragment excitation energy is frequently poorly known, one can often accurately estimate the excitation energy of a reaction residue. Angular distributions measured by Sobotka et al (8) for 90-MeV $^3$He ions incident on silver are nearly angle independent at backward angles, consistent with emission from an equilibrated compound nucleus with relatively low spin. The yields of fragments are in agreement with simple compound nucleus calculations (8).

Estimating the cross section for emission from the compound nucleus by extrapolating the backward angle measurements over $4\pi$, McMahan et al (129) obtained the intermediate mass fragment cross sections shown in Figure 12. The intermediate mass fragment cross sections increase by several orders of magnitude as the center-of-mass energy is increased from 60 to 140 MeV. The curves correspond to two parameter fits to the

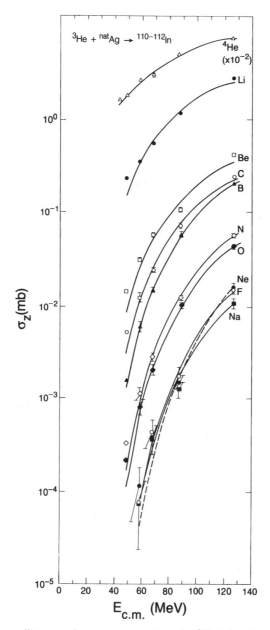

*Figure 12*  Intermediate mass fragment cross sections for $^3$He-induced reactions on silver (129). The curves are explained in the text.

excitation function in which intermediate mass fragment emission was regarded as a form of asymmetric fission. (The dashed curve is the fit for neon fragments.) The two fitting parameters were the fission barrier and the level density of the parent nucleus at the saddle point (129). At present, it is not clear that a fission-like transition state model (69) provides a better description than an evaporative model (67) of compound nuclear intermediate mass fragment emission. More work is needed to resolve this issue.

### 4.3    Nonequilibrium Intermediate Mass Fragment Emission

4.3.1    REACTION DYNAMICS    Fields et al (125) observed that most coincident nonequilibrium intermediate mass fragments and nonequilibrium light particles produced by 720-MeV $^{32}$S ions incident on silver are emitted in a common plane containing the beam axis. Since previous measurements indicated that most nonequilibrium light particles are emitted in the reaction plane at these energies (130), it was concluded that the nonequilibrium intermediate mass fragments are also emitted predominantly in the reaction plane (125).

Measurements of the circular polarization of $\gamma$ rays emitted by the heavy reaction residue that accompanies an intermediate mass fragment provide some indication of the time scale for formation of nonequilibrium intermediate mass fragments (131). The positive polarizations ($\approx 0.4$) observed for 490-MeV $^{14}$N ions incident on $^{154}$Sm, shown in Figure 13, indicate that most ($\approx 70\%$) nonequilibrium intermediate mass fragments are preferentially emitted to the side of the target residue that lies across the beam axis from the point of initial contact between projectile and target (131). Polarizations for nonequilibrium $\alpha$ particles and protons, shown in the figure as dashed and solid lines, respectively, are also positive (132), in agreement with numerical solutions of the Boltzmann equation (132). Figure 13 suggests that most nonequilibrium intermediate mass fragments are emitted after significant numbers of nucleons are swept collectively by the nuclear mean field to negative angles. Such a collective motion suggests strongly the importance of a dynamical treatment of heavy fragment emission.

4.3.2    EMISSION TEMPERATURES    In the presence of collective motion, the internal excitation of the prefragment is difficult to extract from kinetic energy spectra. One may try to assess the internal excitation by extracting "emission temperatures" from the relative populations of ground and excited states of emitted fragments (133–136b). Although calculations (98, 135) indicate that the populations of many states are strongly influenced by sequential decay, the populations of particular particle-unstable states of $^5$Li, $^6$Li, and $^8$Be nuclei appear to be less strongly affected (98).

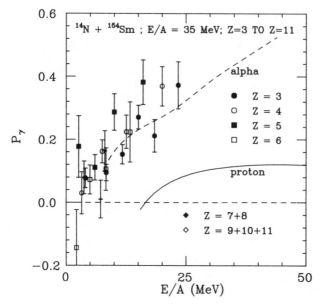

*Figure 13*  Gamma-ray polarizations associated with intermediate mass fragments produced by 490-MeV $^{14}$N ions incident on $^{154}$Sm (131).

Figure 14 shows the excitation energy spectrum corresponding to the decay $^6$Li → d + α for 2.4-GeV $^{40}$Ar ions incident on $^{197}$Au (134). Structures arising from the sequential decay of the known zero-isospin excited states of $^6$Li, given in Table 1, can be seen at the appropriate excitation energies. In the rest frame of $^6$Li, the excitation energy spectrum d$n$/d$E$ for $^6$Li emitted by a system in thermal equilibrium is given by (134)

$$\frac{dn(E^*)}{dE^*} = Ne^{-E^*/T} \sum_i \left| \frac{(2J_i+1)\Gamma_{\alpha,i}/2\pi}{(E^*-E_i)^2+\Gamma_i^2/4} \right|, \qquad 15.$$

where the summation includes the excited states of $^6$Li given in Table 1; $E^*$ and $E_i$ are the observed excitation energy and the excitation energy of the states in $^6$Li; $\Gamma_i$ and $\Gamma_{\alpha,i}$ are the total and alpha decay widths of the states, and $T$ is the emission temperature of the system. Calculated excitation energy spectra, shown as curves in Figure 14, include corrections for the efficiency of the detection array. The excitation energy spectrum is consistent with a mean (averaged over all experimentally observed $^6$Li nuclei) emission temperature of 5 MeV. Mean emission temperatures of 4–5 MeV were also obtained from excited states of $^5$Li and $^8$Be for this reaction (134). Surprisingly, virtually identical mean emission tempera-

**Table 1**  Particle-unstable excited states of $^6$Li

| $E^*$ (MeV) | $J^\pi$ | $\Gamma_{\alpha,i}/\Gamma_i$ | $\Gamma_i$ (MeV) |
|---|---|---|---|
| 2.186 | $3^+$ | 1.00 | 0.024 |
| 4.31 | $2^+$ | 0.97 | 1.3 |
| 5.65 | $1^+$ | 0.74 | 1.9 |

tures are obtained for 490-MeV $^{14}$N ions incident on $^{197}$Au (136), a reaction that is almost a factor of five lower in beam energy.

Low and constant emission temperatures could imply that local thermal equilibrium is not achieved (133). Boal suggests, however, that low temperatures could be caused by a quenching of the excited state populations, which is induced by collisions with thermally emitted neutrons while the fragments are leaving the system (137). Low temperatures could also be characteristic of the emission process if the prefragment expands and cools during breakup. In molecular dynamical calculations, Pandharipande and coworkers (76) find that all fragmentation events have similar final temperatures irrespective of the temperature and density assumed for the

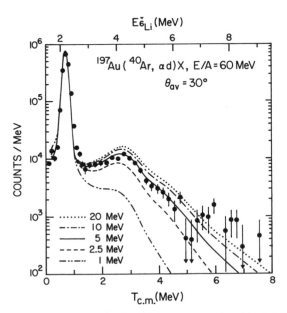

*Figure 14*  Excitation energy spectrum for $^6$Li fragments produced by 2.4-GeV $^{40}$Ar ions incident on $^{197}$Au (134). The curves are calculated from Equation 15 for different emission temperatures.

initial configuration. Further experimental and theoretical investigations are needed to clarify this issue.

## 5.  CONCLUSION

In the last several years, it has become apparent that fragmentation is a process universally present when nuclear systems are highly excited. Ample experimental evidence has accumulated to demonstrate the importance of fragmentation in collisions with energetic but slow-moving heavy projectiles.

Considerable theoretical and experimental progress has been made toward the eventual understanding of the fragmentation process. We now have a relatively clear global picture of the features of inclusive fragmentation data at relativistic energies. The approximate validity of the concepts of factorization and the limiting fragmentation implies that inclusive measurements are relatively insensitive to the entrance channel. These concepts also allow one to extrapolate from the measured to the unmeasured cross sections. We also have a variety of fragmentation models and we know a great deal about their predictive capabilities. Many of the models are very successful at reproducing simple observables such as the mass and charge multiplicities of intermediate mass fragments.

Concerning the confrontation between experiment and theory, we find mixed success. We are most sure of our conclusions when the excitation energy is low, and we can apply our experience with the excitation and decay of compound nuclei. Even there, interesting questions remain having to do with the differences between a fission-like and an evaporative description of intermediate mass fragment emission. The ambiguities are greatest for prefragments with high excitation energies. Here additional experimental information is clearly required, especially regarding the impact parameter, prefragment mass and excitation energy distribution of events leading to intermediate mass fragment emission. It is important to determine the energy deposition accurately; otherwise we cannot test the fragmentation models quantitatively. It is disturbing that the intranuclear cascade models, used extensively to estimate the energy deposition, overpredict significantly the linear momentum transfer to prefragments that produce spallation residues. Finally, we need more experimental information so that we can get a better notion of how and when the prefragment comes to thermal equilibrium. This would indicate where thermal equilibrium is a good theoretical assumption and where it is not.

There is always the question about the internal consistency of the theoretical assumptions that underlie the present theoretical calculations. To a large extent this occurs because of the imposition of static statistical concepts on a time-dependent phenomenon. It is encouraging to see so much recent work on rate equations and fully dynamical fragmentation calculations. Perhaps through these efforts we can delineate the limits under which we can apply our stationary models and, one hopes, draw some important distinctions between models that are presently equally successful at describing the experimental data.

## ACKNOWLEDGMENTS

The author gratefully acknowledges the helpful suggestions of Dr. M. B. Tsang and Dr. L. P. Csernai and helpful conversations with Dr. S. B. Kaufman. This work was supported in part by the National Science Foundation under Grant No. PHY 83-12245. In addition, the author acknowledges the receipt of a Presidential Young Investigator award.

*Literature Cited*

1. Bujak, A., Finn, J. E., Gutay, L. J., Hirsch, A. S., Minich, R. W., et al. *Phys. Rev.* C32: 620 (1985)
2. Gross, D. H. E., Satpathy, L., Meng, T. C., Satpathy, M. *Z. Phys.* A309: 41 (1982)
3. Dostrovsky, I., Fraenkel, Z., Friedlander, G. *Phys. Rev.* C116: 683 (1959)
4. Campi, X., Hüfner, J. *Phys. Rev.* C24: 2199 (1981)
5. Abul-Magd, A. Y., Friedman, W. A., Hüfner, J. *Phys. Rev.* C34: 113 (1986)
6. Kaufman, S. B., Henderson, M. S., Henderson, D. J., Steinberg, E. P., Wilkins, B. D., et al. *Phys. Rev.* C26: 2694 (1982)
7. Wilkins, B. D., Kaufman, S. B., Steinberg, E. P., Urbon, J. A., Henderson, D. J. *Phys. Rev. Lett.* 43: 1080 (1979)
8. Sobotka, L. G., Padgett, M. L., Wozniak, G. L., Guarino, G., Pacheco, A. J., et al. *Phys. Rev. Lett.* 51: 2187 (1983)
9. Green, R. E. L., Korteling, R. G. *Phys. Rev.* C22: 1594 (1980)
10. Fields, D. J., Lynch, W. G., Chitwood, C. B., Gelbke, C. K., Tsang, M. B., et al. *Phys. Rev.* C30: 1912 (1984)
11. Kwiatkowski, K., Bashkin, J., Karwowski, H., Fatyga, M., Viola, V. E. *Phys. Lett.* B171: 41 (1986)
12. Finn, J. E., et al. *Phys. Rev. Lett.* 49: 1321 (1982)
13. Kaufman, S. B., Steinberg, E. P. *Phys. Rev.* C22: 167 (1980)
14. Chitwood, C. B., Fields, D. J., Gelbke, C. K., Lynch, W. G., Panagiotou, A. D., et al. *Phys. Lett.* B131: 289 (1983);

B152: 443 (1985)
15. Hüfner, J. *Phys. Rep.* 125: 129 (1985)
16. Friedlander, E. M., Heckman, H. H. In *Treatise on Heavy Ion Science*, Vol. 4, ed. D. A. Bromley. New York: Plenum (1985)
17. Csernai, L. P., Kapusta, J. I. *Phys. Rep.* 131: 223 (1986)
18. Saint-Laurent, F., Conjeaud, M., Dayras, R., Harar, S., Oeschler, H., Volant, C. *Nucl. Phys.* A422: 307 (1984)
19. Fatyga, M., Kwiatkowski, K., Viola, V. E., Chitwood, C. B., Fields, D. J., et al. *Phys. Rev. Lett.* 55: 1376 (1985)
20. Serber, R. *Phys. Rev.* 72: 1114 (1947)
21. Metropolis, N., Bivins, R., Storm, M., Turkevich, A., Miller, J. M., et al. *Phys. Rev.* 110: 185 (1958); Metropolis, N., Bivins, R., Storm, M., Miller, J. M., Friedlander, G., et al. *Phys. Rev.* 110: 204 (1958)
22. Chen, K., Fraenkel, Z., Friedlander, G., Grover, J. R., Miller, J. M., et al. *Phys. Rev.* C166: 949 (1968)
23. Bertini, H. W. *Phys. Rev.* C6: 631 (1972)
24. Yariv, Y., Fraenkel, Z. *Phys. Rev.* C20: 2227 (1979)
25. Gudima, K. K., Iwe, H., Toneev, V. D. *J. Phys.* G5: 229 (1979)
26. Cugnon, J., Mizutani, T., Vandermeulen, J. *Nucl. Phys.* A352: 505 (1981)
27. Hüfner, J., Schäfer, H., Schürmann, B. *Phys. Rev.* C12: 1888 (1975)
28. Cugnon, J. *Nucl. Phys.* A462: 751 (1987)
29. Gosset, J., Gutbrod, H. H., Meyer, W.

G., Poskanzer, A. M., Sandoval, A., et al. *Phys. Rev.* C16: 629 (1977)
30. Frankel, K. A., Stevenson, J. D. *Phys. Rev.* C23: 1511 (1981)
31. Jacak, B. V., Westfall, G. D., Gelbke, D. K., Harwood, L. H., Lynch, W. G., et al. *Phys. Rev. Lett.* 51: 1846 (1983)
32. Baumgardt, H. G., Schott, J. U., Sakamoto, Y., Schopper, E., Stöcker, H., et al. *Z. Phys.* A273: 359 (1975)
33. Amsden, A. A., Bertsch, G. F., Harlow, F. H., Nix, J. R. *Phys. Rev. Lett.* 35: 905 (1975)
34. Stöcker, H., Griener, W. *Phys. Rep.* 137: 279 (1986)
35. Wilets, L., Yariv, Y., Chestnut, R. *Nucl. Phys.* A301: 359 (1978)
36. Bodmer, A. R., Panos, C. N., Mackellar, A. D. *Phys. Rev.* C22: 1025 (1980)
37. Bertsch, G., Kruse, H., Das Gupta, S. *Phys. Rev.* C29: 673 (1984)
38. Kruse, H., Jacak, B. V., Molitoris, J. J., Westfall, G. D., Stöcker, H. *Phys. Rev.* C31: 1770 (1985)
39. Grégoire, C., Remaud, B., Scheuter, F., Sébille, F. *Nucl. Phys.* A436: 365 (1985)
40. Bertsch, G., Cugnon, J. *Phys. Rev.* C24: 2514 (1981)
41. Csernai, L. P., Barz, H. W. *Z. Phys.* A296: 173 (1980)
42. Fài, G., Randrup, J. *Nucl. Phys.* A381: 557 (1982)
43. Jacak, B. V., Stöcker, H., Westfall, G. D. *Phys. Rev.* C29: 1744 (1984)
44. Sauer, G., Chandra, H., Mosel, U. *Nucl. Phys.* A264: 221 (1976)
45. Jaqaman, H. R., Mekjian, A. Z., Zamick, L. *Phys. Rev.* C27: 2782 (1983); C29: 2067 (1984)
46. Curtin, M. W., Toki, H., Scott, D. K. *Phys. Lett.* B123: 289 (1983)
47. Rose, J. H., Vary, J. P., Smith, J. R. *Phys. Rev. Lett.* 53: 344 (1984)
48. Lopez, J. A., Siemens, P. J. *Nucl. Phys.* A431: 728 (1984)
49. Siemens, P. J. *Nature* 305: 410 (1983)
50. Fischer, M. E. *Physics (NY)* 3: 255 (1967); Fischer, M. E. In *Proc. Int. Sch. Phys., Enrico Fermi, Course LI, Critical Phenomena*, ed. M. S. Green. New York: Academic (1971)
51. Minich, R. W., Agarwal, S., Bujak, A., Chuang, J., Finn, J. E., et al. *Phys. Lett.* B118: 458 (1982)
52. Hirsch, A. S., Bujak, A., Finn, J. E., Gutay, L. J., Minich, R. W., et al. *Phys. Rev.* C29: 508 (1984)
53. Panagiotou, A. D., Curtin, M. W., Toki, H., Scott, D. K., Siemens, P. J. *Phys. Rev. Lett.* 52: 496 (1984); Panagiotou, A. D., Curtin, M. W., Scott, D. K. *Phys. Rev.* C31: 55 (1985)

54. Bertsch, G., Siemens, P. J. *Phys. Lett.* B126: 9 (1983); Bertsch, G., Mundinger, D. *Phys. Rev.* C17: 1646 (1978)
55. Csernai, L. P. *Phys. Rev. Lett.* 54: 639 (1985)
56. Schulz, H., Kämpfer, B., Barz, H. W., Röpke, G., Bondorf, J. *Phys. Lett.* B147: 17 (1984)
57. Grant, C. R. *Phys. Rev.* C34: 1950 (1986)
58. Boal, D. H. *Phys. Rev.* C28: 2568 (1983)
59. Randrup, J., Koonin, S. E. *Nucl. Phys.* A356: 223 (1981)
60. Stöcker, H., Buchwald, G., Graebner, G., Subramanian, P., Maruhn, J. A., et al. *Nucl. Phys.* A400: 63c (1983)
61. Sa, B. H., Gross, D. H. E. *Nucl. Phys.* A437: 643 (1985); Gross, D. H. E., Zhang, X. Z., Xu, S. Y. *Phys. Rev. Lett.* 56: 1544 (1986)
61a. Siemens, P. J., Kapusta, J. I. *Phys. Rev. Lett.* 43: 1486 (1979)
62. Bohrmann, S., Hüfner, J., Nemes, M. C. *Phys. Lett.* B120: 59 (1983); Hüfner, J., Sommerman, H. M. *Phys. Rev.* C27: 2090 (1983)
63. Aichelin, J., Hüfner, J. *Phys. Lett.* B136: 15 (1984); Aichelin, J., Hüfner, J., Ibarra, R. *Phys. Rev.* C30: 107 (1984)
64. Mekjian, A. Z. *Phys. Rev. Lett.* 38: 640 (1977)
65. Morrissey, D. J., Marsh, W. R., Otto, R. J., Loveland, W., Seaborg, G. T. *Phys. Rev.* C18: 1267 (1978)
66. Dostrovsky, I., Fraenkel, Z., Rabinowitz, P. *Phys. Rev.* 118: 791 (1960)
67. Friedman, W. A., Lynch, W. G. *Phys. Rev.* C28: 950 (1983); C28: 16 (1983)
68. Weisskopf, V. F. *Phys. Rev.* 52: 295 (1937)
69. Moretto, L. G. *Phys. Lett.* B40: 185 (1972); *Nucl. Phys.* A247: 211 (1975)
70. Csernai, L. P., Stöcker, H. *Phys. Rev.* C28: 2001 (1983)
71. Sagawa, H., Bertsch, G. F. *Phys. Lett.* B155: 11 (1985)
72. Strack, B., Knoll, J. *Z. Phys.* A315: 249 (1984); Knoll, J., Strack, B. *Phys. Lett.* B149: 45 (1984)
73. Gale, C., Das Gupta, S. *Phys. Lett.* B162: 35 (1985)
74. Aichelin, J. *Phys. Lett.* B175: 120 (1986)
74a. Bauer, W., Bertsch, G. F., Das Gupta, S. *Phys. Rev. Lett.* 58: 863 (1987)
75. Aichelin, J., Stöcker, H. *Phys. Lett.* B176: 14 (1986)
76. Lenk, R. J., Pandharipande, V. R. *Phys. Rev.* C34: 177 (1986); Vicentini, A., Jacucci, G., Pandharipande, V. R. *Phys. Rev.* C31: 1783 (1985)

77. Bauer, W., Dean, D. R., Mosel, U., Post, U. *Phys. Lett.* B150: 53 (1985)
78. Biro, T. S., Knoll, J., Richert, J. *Nucl. Phys.* A459: 692 (1986)
78a. Campi, X. *J. Phys.* A19: L917 (1986)
79. Bondorf, J. P., Donangelo, R., Mishustin, I. N., Pethick, C. J., Schulz, H., et al. *Nucl. Phys.* A443: 321 (1985); Bondorf, J. P., Donangelo, R., Mishustin, I. N., Schulz, H. *Nucl. Phys.* A444: 460 (1985); Bondorf, J. P., Donangelo, R., Mishustin, I. N., Pethick, C. J., Sneppen, K. *Phys. Lett.* B150: 57 (1985)
80. Warwick, A. I., Wieman, H. H., Gutbrod, H. H., Maier, M. R., Peter, J., et al. *Phys. Rev.* C27: 1083 (1983); Warwick, A. I., Baden, A., Gutbrod, H. H., Maier, M. R., Peter, J., et al. *Phys. Rev. Lett.* 48: 1719 (1982); Gutbrod, H. H., Warwick, A. I., Wieman, H. H. *Nucl. Phys.* A387: 177c (1982)
81. Alexander, J. M., Baltzinger, C., Gazdik, M. F. *Phys. Rev.* 129: 1826 (1963)
82. Friedlander, G., Friedman, L., Gordon, B., Yaffe, L. *Phys. Rev.* 129: 1809 (1963)
83. de Saint Simon, M., Haan, S., Audi, G., Coc, A., Epherre, M., et al. *Phys. Rev.* C26: 2447 (1982)
83a. Biswas, S., Porile, N. T. *Phys. Rev.* C20: 1467 (1979)
84. Feynman, R. P. *Phys. Rev. Lett.* 23: 1415 (1969)
85. Benecke, J., Chou, T. T., Yang, C. N., Yen, E. *Phys. Rev.* 188: 2159 (1969)
86. Cumming, J. B., Haustein, P. E., Ruth, T. J., Virtes, G. J. *Phys. Rev.* C17: 1632 (1978)
87. Kaufman, S. B., Steinberg, E. P., Wilkins, B. D., Henderson, D. J. *Phys. Rev.* C22: 1897 (1980)
88. Porile, N. T., Cole, G. D., Rudy, C. R. *Phys. Rev.* C19: 2288 (1979)
89. Cumming, J. B., Haustein, P. E., Stoenner, R. W., Mausher, L., Naumann, R. A. *Phys. Rev.* C10: 739 (1974)
90. Rudstam, G. *Z. Naturforsch.* 21a: 1027 (1966)
91. Campi, X., Debois, J., Lipparini, E. *Phys. Lett.* B138: 353 (1984)
92. Morrissey, D. J., Loveland, W., de Saint Simon, M., Seaborg, G. T. *Phys. Rev.* C21: 1783 (1980)
93. Kaufman, S. B., Steinberg, E. P., Weisfield, M. W. *Phys. Rev.* C18: 1349 (1978)
94. Morrissey, D. J., Oliverira, L. F., Rasmussen, J. O., Seaborg, G. T., Yariv, Y., et al. *Phys. Rev. Lett.* 43: 1139 (1979).
95. Crespo, V. P., Cumming, J. B., Alexander, J. M. *Phys. Rev.* C2: 1777 (1970)
96. Sugarman, N., Münzel, H., Panontin, J. A., Wielgoz, K., Ramanih, M. V., et al. *Phys. Rev.* 143: 952 (1966)
97. Ajzenberg-Selove, F. *Nucl. Phys.* A449: 1 (1986)
98. Fields, D. J., Gelbke, C. K., Lynch, W. G., Pochodzalla, J. *Phys. Lett.* B187: 257 (1987)
98a. Cole, G. D., Porile, N. T. *Phys. Rev.* C24: 2038 (1981)
99. Aichelin, J., Campi, X. *Phys. Rev.* C34: 1643 (1986)
100. Yu, Y.-W., Porile, N. T. *Phys. Rev.* C1: 1597 (1973)
101. Sauvageon, H., Regnier, S., Simonoff, G. N. *Phys. Rev.* C25: 466 (1982); Z. *Phys.* A314: 181 (1983); Sauvageon, H. *Phys. Rev.* C24: 2667 (1981)
102. Hudis, J., Kirsten, T., Stoenner, R. W., Shaeffer, O. A. *Phys. Rev.* C1: 2019 (1970)
103. Westfall, G. D., Symons, T. J. M., Greiner, D. E., Heckman, H. H., Lindstrom, P. J., et al. *Phys. Rev. Lett.* 43: 1859 (1979)
104. Ku, T. H., Karol, P. J. *Phys. Rev.* C16: 1984 (1977)
105. Winsberg, L. *Nucl. Instrum. Methods* 150: 465 (1978); Pierson, W. R., Sugarman, N. *Phys. Rev.* 130: 2417 (1963)
106. Porile, N. T., *Phys. Rev.* 120: 572 (1960)
107. Cumming, J. B. *Phys. Rev. Lett.* 44: 17 (1980)
108. Cumming, J. B., Haustein, P. E., Stoenner, R. W. *Phys. Rev.* C33: 926 (1986)
109. Fortney, D. R., Porile, N. T. *Phys. Rev.* C21: 2511 (1980); C22: 670 (1980)
110. Remsberg, L. P., Perry, D. G. *Phys. Rev. Lett.* 35: 361
111. Urbon, J. A., Kaufman, S. B., Henderson, D. J., Steinberg, E. P. *Phys. Rev.* C21: 1048 (1980)
112. Poskanzer, A. M., Butler, G. W., Hyde, E. K. *Phys. Rev.* C3: 882 (1971); Hyde, E. K., Butler, G. M., Poskanzer, A. M. *Phys. Rev.* C4: 1159 (1971)
113. Cumming, J. B., Cross, R. J. Jr., Hudis, J., Poskanzer, A. M. *Phys. Rev.* 134: B167 (1964)
114. Price, P. B., Stevenson, J. B. *Phys. Lett.* B78: 197 (1978)
115. Winsberg, L. *Phys. Rev.* C22: 2116, 2123 (1980)
116. Feshbach, H., Huang, K. *Phys. Lett.* B47: 300 (1973)
117. Goldhaber, A. S. *Phys. Lett.* B53: 306 (1974)
118. Greiner, D. E., Lindstrom, P. J., Heckman, H. H., Cork, B., Bieser, F. S. *Phys. Rev. Lett.* 35: 152 (1975)

119. Stockstad, R. G. *Comments Nucl. Part. Phys.* 13: 231 (1984)
120. Westfall, G. D., Sextro, R. G., Poskanzer, A. M., Zebelman, A. M., Butler, G. W., et al. *Phys. Rev.* C17: 1368 (1978)
121. Gaidos, J. A., Gutay, L. J., Hirsch, A. S., Mitchell, R., Ragland, T. V., et al. *Phys. Rev. Lett.* 42: 82 (1979)
122. Siemens, P. J., Rasmussen, J. O. *Phys. Rev. Lett.* 42: 880 (1979)
123. Stöcker, H., Ogloblin, A. A., Griener, W. Z. *Phys.* A303: 259 (1981)
124. Trockel, R., Hildenbrand, K. D., Lynen, U., Müller, W. F. J., Rabe, H., et al. *Prog. Part. Nucl. Phys.* 15: 225 (1985)
125. Fields, D. J., Lynch, W. G., Nayak, T. K., Tsang, M. B., Chitwood, C. B., et al. *Phys. Rev.* C34: 536 (1986)
126. Charity, R. J., McMahan, M. A., Bowman, D. R., Liu, Z. H., McDonald, R. J., et al. *Phys. Rev. Lett.* 56: 1354 (1986)
127. Bowman, D. R., et al. *Phys. Lett.* B189: 282 (1987)
128. Pelte, D., Winkler, U., Bühler, M., Weissmann, B., Gobbi, A., et al. *Phys. Rev.* C34: 1673 (1986)
129. McMahan, M. A., Moretto, L. G., Padgett, M. L., Wozniak, G. J., Sobotka, L. G., et al. *Phys. Rev. Lett.* 54: 1995 (1985)
130. Tsang, M. B., Chitwood, C. B., Fields, D. J., Gelbke, C. K., Klesch, D. R., et al. *Phys. Rev. Lett.* 52: 1967 (1984)
131. Tsang, M. B., Trautmann, W., Ronningen, R., Gelbke, C. K., Lynch, W. G., et al. To be published.
132. Tsang, M. B., Ronningen, R. M., Bertsch, G., Chen, Z., Chitwood, C. B., et al. *Phys. Rev. Lett.* 57: 559 (1986)
133. Morrissey, D. J., Benenson, W., Kashy, E., Sherrill, B., Panagiotou, A. D., et al. *Phys. Lett.* B148: 423 (1984); Morrissey, D. J., Benenson, W., Kashy, E., Bloch, C., Lowe, M., et al. *Phys. Rev.* C32: 877 (1985)
134. Pochodzalla, J., Friedman, W. A., Gelbke, C. K., Lynch, W. G., Maier, M., et al. *Phys. Rev. Lett.* 55: 177 (1985); Pochodzalla, J., Friedman, W. A., Gelbke, C. K., Lynch, W. G., Maier, M., et al. *Phys. Lett.* B161: 275 (1985)
135. Xu, H. M., Fields, D. J., Lynch, W. G., Tsang, M. B., Gelbke, C. K., et al. *Phys. Lett.* B182: 155 (1986)
136. Chitwood, C. B., Gelbke, C. K., Pochodzalla, J., Chen, Z., Fields, D. J., et al. *Phys. Lett.* B172: 27 (1986)
136a. Morrissey, D. J., Bloch, C., Benenson, W., Kashy, E., Blue, R. A. *Phys. Rev.* C34: 761 (1986)
136b. Sobotka, L. G., Sarantites, D. G., Pachta, H., Dilmanian, F. A., Jääskeläinen, M. *Phys. Rev.* C34: 917 (1986)
137. Boal, D. H. *Phys. Rev.* C30: 749 (1984)

*Ann. Rev. Nucl. Part. Sci. 1987. 37: 537–65*

# RECOIL MASS SPECTROMETERS IN LOW-ENERGY NUCLEAR PHYSICS

## Thomas M. Cormier

Department of Physics and Astronomy, and Nuclear Structure Research Laboratory, University of Rochester, Rochester, New York 14627

CONTENTS

## 1. INTRODUCTION

Heavy-ion fusion reactions have become an important tool in nuclear spectroscopy in recent years. With laboratory projectile energies ranging from 5 MeV to typically $< 10$ MeV per nucleon, such reactions provide the principal means of producing high-spin reaction products. To date, they have allowed the study of the evolution of nuclear structure to spins as high as $60\hbar$ (1). Similarly, heavy-ion fusion reactions provide the most direct means for producing exotic, neutron-deficient nuclei near or beyond $N = Z$. While reactions produced by heavy nuclei are central to new

537

0163–8998/87/1201–0537$02.00

developments in many areas of nuclear spectroscopy, for the purposes of this introduction we limit our attention to these two areas of research.

Progress in these fields has been significantly keyed to progress in experimental techniques, perhaps more so than in many other areas of nuclear science. The reason for this lies in the frequently overwhelming complexity of the final states produced in nucleus-nucleus collisions. This general feature has spurred the development of ever more sensitive experimental methods. The best example of this is the recent development of Compton-suppressed germanium gamma-ray detector arrays coupled with high-resolution multiplicity arrays and the corresponding breakthroughs in nuclear physics at very high spin (2). These instruments perform two essential functions on an event-by-event basis. The first function is the obvious ability to select, by using gamma-ray multiplicity, those events that originate at the highest angular momentum. The second all-important function is the identification of the final product nucleus via the observation of known low-lying transitions.

In the study of nuclei far from stability the techniques described above find immediate application. As ever more neutron-deficient nuclei are probed, however, limitations set in that derive from the limited sensitivity of gamma-ray detector arrays to identify weakly produced final product nuclei in ever more complex backgrounds. In this area the solution has been to impose additional indirect channel identifiers. Examples of these include arrays of charged-particle or neutron detectors. Such arrays are used to create a subset of events that are then further scrutinized in the gamma detector arrays. In an early work of this type, Lister (3), for example, studied the evolution of the structure of the light strontium isotopes. Using the $^{40}Ca + ^{40}Ca$ reaction, he identified the isotope $^{77}Sr$ and studied its band structure using a comparison of gamma spectra in coincidence with one and two protons or one and two neutrons. In this instance the ratio of gamma intensities for these various gating conditions leads to an unambiguous identification of the ($^{40}Ca$, 2p1n) channel, which leads to $^{77}Sr$. A parallel study of the beta decay of $^{77}Sr$ supplied additional information.

While $^{77}Sr$ is rather nicely accessible using the techniques described above, the yield of the $N = Z$ product $^{76}Sr$ is already too small to permit a similar study. This is a general feature of research in this area. A useful rule of thumb is that production cross sections fall roughly a factor of ten per neutron for products near $N = Z$. These facts suggest the need for a significant revision in the approach to the identification problem. Rather than implementing additional filtering conditions, a method allowing direct detection and identification of the recoiling product nucleus has the potential of furnishing the significant improvement in sensitivity that is required.

The examples cited above, as well as numerous others that could be drawn from diverse areas of nuclear spectroscopy and reaction studies, point to the need for a general scheme for the direct detection and identification of low-energy heavy ions. The need to apply this device to heavy-ion-induced fusion reactions while preserving a time correlation between the detected heavy ion and the associated radiations from the event severely restricts its design. In particular these restrictions preclude the application of such conventional approaches as time-of-flight telescopes or on-line isotope separators.

First, while on-line isotope separators readily provide the requisite identification sensitivity, the essential time correlation is lost during the transport, thermalization, and ionization processes that occur in the ion source of such a device. It is unlikely that the total holding time in such spectrometers can be made significantly less than 10 ms, a time greatly exceeding the lifetimes of all but isomeric transitions.

Time-of-flight telescopes, on the other hand, preserve the time correlation between heavy-ion detection and associated prompt radiations, but they lack the necessary sensitivity. The kinematics of heavy-ion fusion reactions result in a strong forward focusing of the heavy recoiling nuclei. In the most interesting cases, the relevant reaction products will have angular distributions that strongly peak at zero degrees. In the extreme situation where the projectile is heavier than the target (inverse kinematics), the angular distribution in the laboratory may be dominated by multiple scattering in the target with a full width of only a few degrees. Thus in the angular range of interest the heavy recoiling reaction products are necessarily embedded in the intense elastic scattering flux or, indeed, the primary beam itself. Thus for any reasonable mass-resolving power that might be achieved with time-of-flight techniques, the ultimate sensitivity will be severely limited by beam-related background.

## 2.  GENERAL PERFORMANCE CRITERIA

The problems discussed above (and others omitted for space considerations) have been largely solved with the advent of a class of instruments known as recoil mass spectrometers. This name encompasses a very broad range of devices constructed since 1971, starting with the initial work by Enge and coworkers (4). A thorough discussion of the development of the present generation of recoil mass spectrometers from their origins in earlier devices is beyond the scope of the present review. Rather, the reader is referred to References (4–10) for a perspective on the development of these instruments. In this review we limit consideration to the present generation of spectrometers, typified by those in operation at the University of Rochester (5, 6), California Institute of Technology (7), Michigan State University (8), and Daresbury Laboratory (9), under construction at Legnaro National Laboratory (10), and proposed at both Argonne (ANL) and Oak Ridge (ORNL) National Laboratories. Theses spectrometers share all of their essential features in common and thus are not all discussed in equal detail.

We define the recoil mass spectrometers considered here as on-line isotope separators in which the usual ion source is replaced by the flux of reaction products directly as they recoil from the target. Clearly such a device overcomes the problems associated with the ion source in a conventional on-line isotope separator insofar as it impacts the class of experiments discussed above. Furthermore, if designed with sufficient mass resolution and careful primary beam separation, it does not suffer from the limitations discussed in connection with time-of-flight techniques.

Serious ion-optical challenges are presented by recoil mass spectrometers as defined above. The phase space occupied by the mono-energetic isotope beam from an ion source is extremely small compared to that of typical nuclear reaction products recoiling from a target. Angular divergences of a few degrees and energy spreads of $\pm 10\%$ are reasonable for typical fusion products. Thus a useful recoil mass spectrometer will have a large geometric solid angle and have a broad energy acceptance. A nominal mass-resolving power of 300 is adequate for the type of experiments discussed above. In several cases discussed below, however, we see that a mass-resolving power in the range of 1000 is required. To achieve mass resolution in this range obviously requires simultaneous control of geometric and chromatic aberration. With energy spreads as large as $\pm 10\%$, the need to control energy-dependent or chromatic aberration is perhaps the most obvious.

The technique used here is an extension of the velocity dispersion matching first used in the Aston mass spectrometer (11). The principal is readily illustrated in Figure 1. This figure shows a beam of particles of energy $E$, mass $m$, and atomic charge state $q$ impinging on an electrostatic

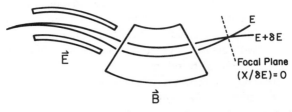

*Figure 1*   A simple combination of an electrostatic deflector and a magnetic dipole allows energy dispersion matching and forms one possible basis of a recoil mass spectrometer.

deflector followed by a magnetic deflector. The electric and magnetic fields are chosen such that the particles follow some desired path. This path is called the optic axis of the device and $E$, $m$, and $q$ are called the central energy, mass, and charge respectively. Consider now particles of energy $E + \delta E$ and mass $m + \delta m$. Such particles necessarily follow a different path through the device, which we characterize by the coordinate $x$, defined as the perpendicular displacement from the optic axis. The independent variable can be taken as $z$, defined as distance measured along the optic axis. [For a more rigorous treatment of these coordinates, as well as a complete discussion of ion-optical techniques, see (12).] Thus the trajectory of any particle is $x = x(z)$, and $x(z) = 0$ defines the optic axis. The functions $x(z)$ depend implicitly on initial conditions specified at the input to the device. The specification of $\delta E$ and $\delta m$ constitute such initial conditions. For $\delta m/m \ll 1$ and $\delta E/E \ll 1$, the following Taylor expansion to lowest order in $\delta E$ and $\delta m$ of $x(z)$ is obtained:

$$x(z) = (x/\delta E)\delta E + (x/\delta m)\delta m + \ldots, \qquad 1.$$

where the symbols $(x/\delta E)$ and $(x/\delta m)$ are $dx/dE$ and $dx/dm$ evaluated at $z$ and are called the first-order energy and mass dispersion respectively. In an Aston mass spectrometer, the geometric properties of the electric and magnetic deflectors are chosen to yield zero first-order energy dispersion for some value of $z$ on the output side of the device. At this value of $z$, which defines one point on the focal plane, the $x$ coordinates of particles are linearly related to mass.

It is easy to show to an order consistent with Equation 1 that the choice $(x/\delta E) = 0$ is possible. For particles of a given charge state, the displacement $x$ of particles from the optic axis is proportional to $\delta E/E$ in the electric deflector and proportional to $\delta p/p$ in the magnetic deflector, where $p$ is the momentum. Thus, neglecting the drift space between the elements, we obtain $x$ at the exit of the magnet

$$x = a\delta E/E + b\delta p/p$$

or, for

$$\delta p/p = \delta E/2E + \delta m/2m,$$

$$x = (a + b/2)\delta E/E + (b/2)\delta m/m. \qquad 2.$$

The constants $a$ and $b$ are functions of the radius of curvature of the central trajectory and angle of deflection of the electric and magnetic dipoles, respectively,[1] and the design selection $a+b/2 = 0$ can be made in an infinite number of ways to yield $x = -a\delta m/m$. The simple device shown in Figure 1 is thus a mass spectrometer with a mass dispersion $D_m = x/(\delta m/m) = -a$. Because the device is nondispersive in energy, it is also a candidate recoil mass spectrometer.

## 3.  ION OPTICS OF RECOIL MASS SPECTROMETERS

In a practical spectrometer, the full set of initial conditions will contain the point of origin and initial direction of the particle at the target as well as $\delta E$ and $\delta m$. The $y$ direction is taken perpendicular to $x$ and $z$. The $xz$ plane is the median plane and $yz$ is the transverse plane. Thus particles originate at a point $x_0$, $y_0$ at the target ($z = 0$), making angle $\theta$ and $\phi$ with respect to the optic axis in the median and transverse planes respectively. At the output of the spectrometer, the coordinates of a particle are $x$ and $y$ and it makes angles $x'$ and $y'$ in the transverse and median planes, respectively. These coordinates are measured at the $z$ position of the focal plane as defined below. In analogy with Equation 1, we write

$$x = (x/x_0)x_0+(x/\theta)\theta+(x/\delta E)\delta E+(x/\delta m)\delta m+(x/x_0^2)x_0^2$$
$$+(x/\theta^2)\theta^2+(x/\delta E^2)\delta E^2+(x/\delta m^2)\delta m^2+(x/x_0\theta)x_0\theta$$
$$+(x/x_0\delta E)x_0\delta E+(x/x_0\delta m)x_0\delta m+(x/\theta\delta E)\theta\delta E+(x/\theta\delta m)\theta\delta m$$
$$+(x/\delta E\delta m)\delta E\delta m+(x/x_0^3)x_0^3+(x/\theta^3)\theta^3+\dots. \qquad\qquad 3.$$

In this equation the parenthetic symbols again represent derivatives in the Taylor expansion of various order. Similar equations can clearly be given relating $y$, $x'$, and $y'$ to the initial conditions as well.

The first-order focal plane of a spectrometer is conventionally defined at the $z$ coordinate determined by the condition $(x/\theta) = 0$. In a recoil mass spectrometer this geometric focus must be arranged to coincide with the locus of the mass focus defined by the constraint $(x/\delta E) = 0$. Additionally, it is useful if possible to arrange coincidence with the geometric focus in the transverse plane as well $(x/\phi) = 0$.

The series presented in Equation 3 converges very slowly for recoil mass spectrometers, where one may find that significant contributions can be made even through fifth order. In practical calculations, therefore, trans-

---

[1] For a cylindrical electric dipole or magnetic dipole of radius of curvature $R$ and deflection angle $\theta$, the constants $a$ and $b$ in Equation 2 are $R[1-\cos(\sqrt{2}\theta)]$ and $R(1-\cos\theta)$, respectively (see 12, 14).

port matrix methods (12, 14) based on Equation 3 are used only for initial orientation, and final designs are based on complete numerical integration of the equations of motion as incorporated, for example, in the computer code RAYTRACE (15). Using the output of such a code then permits the numerical evaluation of the coefficients in Equation 3, which provide a convenient summary of the important optical properties of a spectrometer.

Although the arrangement of dispersive components within a recoil mass spectrometer is not strictly fixed, certain design options are commonly regarded as optimal. At least three dispersive components are required to provide adequate primary beam rejection. Furthermore, the first quantity dispersed after products leave the target should be either velocity or energy but not momentum. Figure 2 shows the arrangement of components in the Rochester Recoil Mass Spectrometer. A mirror-symmetric configuration of two electric dipoles (ED) and a magnetic dipole (MD) is used to cancel the first-order energy dispersion $(x/\delta E) = 0$. Energy is the first dispersed quantity. This basic arrangement is also used in the Legnaro spectrometer (10) and in both spectrometers currently in the proposal stage at ANL and ORNL. A significant variation from this basic geometry appears in the Daresbury spectrometer (9), where crossed field velocity filters are used in place of electric dipoles, but the basic function is unchanged.

The need for a minimum of three dispersive elements is dictated by the high degree of primary beam rejection required for operation at zero degrees. Frequently, the large difference in energy or velocity (see below) between the primary beam and the recoiling fusion products results in primary beam scatter within the first dispersive component. For the layout shown in Figure 2, for example, the primary beam normally stops within the first electrostatic deflector somewhere on the surface of the positive plate. While this creates absolutely no electrical difficulties in the operation of the deflector it does produce a strong flux of scattered beam particles

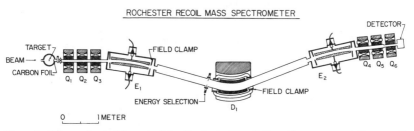

*Figure 2*  The mechanical layout of the Rochester Recoil Mass Spectrometer. The overall length is 10 meters. The letters Q, E, and D label magnetic quadrupoles, parallel plate electrostatic dipole deflectors, and a magnetic dipole, respectively.

with a continuous energy and atomic charge state spectrum that proceeds deeper into the spectrometer. The dipole magnet in Figure 2 then selects a number of momentum slices out of this continuous energy spectrum, one for each charge state. A third dispersive device is then required to suppress the momentum-selected, scattered beam particles. The Rochester geometry of Figure 2 is very effective overall. A total primary beam suppression factor of $10^{13}$ has been achieved under realistic conditions (7).

The mirror-symmetric ED-MD-ED configuration of Figure 2 is chosen to provide efficient separation of the beam from the reaction products as early in the spectrometer as possible. For a primary beam of mass $M_b$ and energy $E_b$ and target of mass $M_t$ the energy of fusion products $E_f$ is approximately that of the compound nucleus

$$E_f = M_b E_b/(M_t + M_b). \qquad\qquad 4.$$

In most cases of interest this results in a large difference between $E_f$ and $E_b$ and hence to very effective separation in electric fields. The mass factor in Equation 4 also relates reaction product and beam velocities; thus, Wien filters, which provide velocity dispersion, also serve as effective first components in recoil mass spectrometers (9). The momentum of the primary beam, however, is equal to the average momentum of recoiling fusion products, and so in most cases a magnetic dipole is not suitable as the first dispersive element.

The first-order optics of the Rochester Recoil Mass Spectrometer are shown in Figure 3. Note that, following convention, the optic axis of the system is drawn as a straight line. Figure 3A shows six rays traced in the median plane. The rays are chosen with $x_0 = \pm 0.5$ mm, $\theta = 0, \pm 20$ mr and all have $\delta E/E = 0.1$. This bundle of rays returns to the optic axis on the output side of the instrument at the focal plane defined by $(x/\theta) = 0$. The spectrometer is thus energy dispersionless. In Figure 3B we show the same set of rays with $\delta E = 0$ and $\delta m/m = 0.05$ illustrating the mass dispersion of the instrument for a 5% mass change. Finally, in Figure 3C the transverse plane is illustrated for rays with $y_0 = \pm 0.5$ mm and $\phi_0 = 0$, $\pm 50$ mr. The first-order focus in the transverse plane, $(y/\phi) = 0$, is thus coincident with the median plane focus as well as the mass focus. This triple focus has a number of important advantages for detector size and the compactness of collected reaction products. This last feature is very significant in experiments that study the subsequent decays of mass-separated reaction products after collection on the focal plane. While the triple focus is very advantageous in many applications, it does create some difficulties in correcting the spectrometer to higher order, as discussed below.

The first difficulty is already evident in Figure 3A. The focal plane is

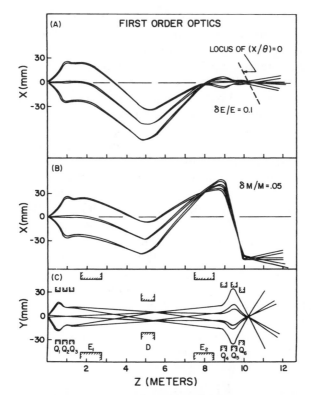

*Figure 3*  First-order optics of the Rochester Recoil Mass Spectrometer for a solid angle of 4 msr. (*A*) Six rays in the median plane with $x_0 = \pm 0.5$ mm, $\theta = \pm 20$ mr, $\delta E/E = 0.1$, $\delta m/m = 0$ illustrating energy dispersion matching. (*B*) Six rays in the median plane with $x_0 = \pm 0.5$ mm, $\theta = \pm 20$ mr, $\delta E/E = 0$, $\delta m/m = 0.05$ illustrating mass dispersion. (*C*) Six rays in the transverse plane with $y_0 = \pm 0.5$ mm, $\phi = \pm 50$ mr, $\delta E/E = 0$, $\delta m/m = 0$ illustrating vertical focusing.

inclined at a rather steep angle with respect to the optic axis. This is a simple consequence of the varying magnetic rigidity of particles of varying mass. The effect occurs almost exclusively in the focusing quadrupoles. Since detectors placed at the end of the spectrometer are of necessity mounted nearly perpendicular to the optic axis, the effect of the steeply inclined focal plane is to limit the useful mass range of the spectrometer. For masses too removed from the central mass, the reaction products are necessarily out of focus.

Of greater importance than the mass aberration discussed above are certain second-order aberrations that dominate the deviation from the ideal first-order behavior shown in Figure 3. These include $(x/\delta E^2)$,

$(x/\theta\delta E)$, and $(x/\theta^2)$. In the Rochester Recoil Mass Spectrometer, for reasons of cost, no systematic attempt was made to correct these aberrations. Rather, the magnetic sextupole field created in the fringe field region of the magnetic dipole by incorporating a curved boundary was used to minimize $(x/\delta E^2)$. The philosophy here is that $\theta$-dependent aberration can be controlled by limiting the spectrometer solid angle in cases where the best resolution is required. There is, however, no way to limit the energy acceptance of the spectrometer effectively since there is no internal focus within the spectrometer where the energy dispersion is nonzero. In practice, a solid angle of 1 msr is used when the best mass resolution is required ($< 1/1000$), while up to 5 msr are available for cases where poorer mass resolution can be tolerated ($< 1/100$).

Figure 4 shows an experimental mass spectrum from the $^{35}Cl + {}^{58}Ni$ fusion reaction. The detector was a simple position-sensitive silicon detector, and events are plotted here as position on the focal plane versus the measured energy of the detected particle. Reaction products of mass 89 and 90 are visible. The residual $(x/\delta E^2)$ aberration can be seen in the slight quadratic dependence of focal plane position on energy. Because both energy and focal plane position are determined on an event-by-event basis, a software correction of this energy-dependent aberration is possible. Applying such a correction to this data results in a mass resolution of 1/640 for this measurement made at 1.5 msr solid angle. Figure 5 shows a

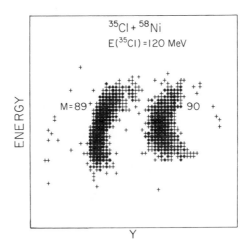

*Figure 4*  A focal plane spectrum for the $^{35}Cl + {}^{58}Ni$ reaction at 120 MeV measured at zero degrees. Reaction product masses 89 and 90 are identified. No software corrections have been applied, and the visible curvature of the mass lines with energy reflects a nonzero $(x/\delta E^2)$.

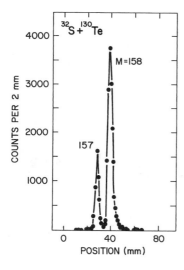

*Figure 5*  A mass spectrum with software corrections applied for $(x/\delta E^2)$ of the $^{32}\text{S} + ^{130}\text{Te}$ reaction at $E(^{32}\text{S}) = 160$ MeV.

mass spectrum made with the same focal plane detector for heavier reaction products at a solid angle of 2.0 msr.

All of the discussion to this point has implicitly been limited to reaction products of a given atomic charge state. In practice, reaction products are typically produced with a distribution of charge states that is four or five charge units wide. It is important to remember, therefore, that the quantity dispersed along the focal plane is the mass-to-charge ratio $m/Q$. For heavy nuclei in which average atomic charge states are approximately 15, the mass spacing on the focal plane may be only one tenth of the spacing between neighboring charge states. In a typical case where the evaporation mass spectrum is four units wide, there is no overlap of different charge states on the focal plane. Below mass 80, however, the reaction mechanism tends to produce wider mass distributions as alpha decay becomes significant. At the same time these higher energy reaction products achieve higher average atomic charge states. Thus, in this mass region, mass peaks from two or three charge states may be present on the focal plane at one time. Distinguishing the near degeneracies in $m/Q$ that can result requires high mass-resolving power. The resolving power of 640 shown in Figure 4 is adequate in most cases to resolve ambiguities, particularly when combined with measurements made at neighboring charge states.

## 4.  SECOND-ORDER FOCUSING

The resolution of the Rochester Recoil Mass Spectrometer is limited at large solid angle principally as a result of a few aberrations, for example

$(x/\theta\delta E)$, $(x/\theta^2)$, and $(x/\delta E^2)$. Additionally, the useful mass range is limited mainly by the focal plane tilt illustrated in Figure 3, which is a result of $(x/\theta\delta m)$. While these are the most important second-order aberrations, a great many others make very significant contributions. Among these $(x/\phi^2)$ is usually the largest. The construction of new spectrometers of significantly larger solid angle and useful mass range requires at least some correction of these strongest aberrations. Using multipole fields placed at locations of optimal dispersion makes it possible to couple to these second-order aberrations. In at least one instance, a spectrometer has been designed with complete second-order focusing (H. Enge, private communication, 1985). While this was a significant undertaking, it amounted to little more than a mathematical triumph because third and higher order aberrations appeared that had their origin in the multipole fields used to remove the second-order aberrations. These induced, higher order aberrations can easily be worse than the second-order aberrations that have been cancelled.

A practical approach is thus required in which substantial second-order aberrations are allowed to remain while suppressing induced higher order effects. The spectrometer under construction at Legnaro (10) and those proposed for ORNL and ANL were designed with this philosophy in mind. The mechanical layout of the proposed ORNL spectrometer is shown in Figure 6. The layouts of the Legnaro and proposed ANL spectrometers are essentially identical except for the magnetic quadrupole triplet near the detector, which is omitted in these latter two spectrometers. This difference results in substantially different ion optics in these instruments.

The median plane optics of the ORNL design is shown in Figure 7 and for the Legnaro instrument in Figure 8. In Figure 7, rays have been calculated by numerical integration of the equations of motion and thus include aberrations through all orders. In each case, the second sextupole

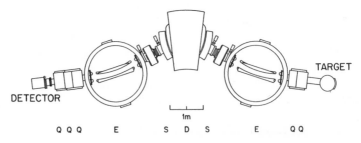

*Figure 6*   Mechanical layout of the proposed ORNL Recoil Mass Spectrometer. The overall geometry is very similar to that in Figure 2, with the addition of two sextupole magnets labelled S.

## MASS AND ENERGY ACCEPTANCE

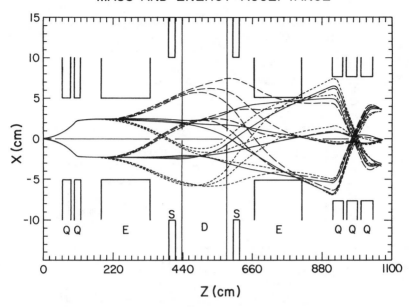

$$Z \, (cm)$$

*Figure 7*   Median plane optics through all orders of the proposed ORNL spectrometer. A variety of rays are traced illustrating mass dispersion with $\delta m/m = 0$, $\pm 0.05$ and energy dispersion with $\delta E/E = 0$ (*solid lines*), $\pm 0.1$ (*long-* and *short-dashed lines* respectively).

field, which is situated in a region of high mass dispersion, is adjusted to minimize the focal plane tilt $(x/\theta \delta m) \approx 0$; this results in a broad useful mass range that exceeds $\pm 5\%$. The remaining dominant aberrations $(x/\delta E^2)$ $(x/\theta^2)$, and $(x/\theta \delta E)$ are handled similarly in the two cases. In both cases the first sextupole, which is situated in the region of maximum energy dispersion and nearly zero mass dispersion, is used to couple to the two chromatic aberrations. The largest contribution to $(x/\theta^2)$ actually comes from the sextupole fields themselves. There is sufficient differential variation introduced, however, if an additional sextupole component is added to the boundary of the magnetic dipole to control this aberration.

In both spectrometer designs, the correction of median plane chromatic aberrations creates large chromatic aberrations in the transverse plane. These are apparent in Figure 8c where they result in a vertical spot size on the detector of nearly 6 cm. The vertical focus in this instrument is produced with a long image distance by a quadrupole field component added to the magnetic dipole boundary. In the ORNL design, the final quadrupole triplet allows significant refocusing in the vertical plane with a short image distance; this results in a vertical height at the detector of

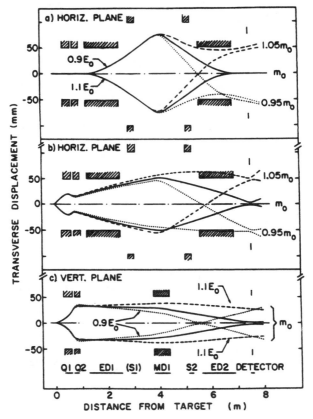

*Figure 8* First-order optics of the proposed ANL spectrometer. (*a*) Energy and mass acceptance at zero solid angle in the median plane. (*b*) Mass dispersion at full solid angle in the median plane. (*c*) Transverse plane at full solid angle illustrating the severe chromatic aberrations in the vertical focus.

1.5 cm under the same conditions. This reduced vertical extent is very important in experiments studying the decays of collected, mass-separated reaction products where the source size is important.

The improved vertical focusing in the ORNL design is not achieved without consequences in the median plane. There is no effective way, short of superposing multipole fields, to correct chromatic aberrations created in the last quadrupole triplet. These effects are mitigated by the focusing technique used in the median plane in which a crossover occurs in the center of the triplet. This crossover allows the momentum dispersion created in the first element to be largely removed by the last element. An additional $(x/\delta E^2)$ aberration remains that is not completely corrected with the two sextupole magnets.

In both spectrometer designs, priority is assigned to correcting the $(x/\theta\delta E)$ aberration. This is because residual $(x/\delta E^2)$ and $(x/\theta^2)$, which cannot be completely removed for reasonable values of the sextupole fields, are easily corrected in software by detectors that measure energy and angle of incidence (e.g. 18).

It is interesting to note that neither design addresses the out-of-plane aberration $(x/\phi^2)$. This is true despite the large $\phi$ dispersion found inside the entrance quadrupole doublets of each design at a location where energy and mass dispersion are still zero in first order. While the superposition of sextupole fields on the entrance quadrupoles does indeed permit complete correction of $(x/\phi^2)$, there results an explosive increase in geometric and chromatic aberrations associated with the sextupole field itself. The net result is thus a significant loss in spectrometer performance.

The final theoretical performance of the two proposed spectrometers is very similar. In each case a compromise is struck to minimize the most significant second-order aberrations without making heroic, and probably futile, efforts at a complete solution. In each case a mass-resolving power near 300 is predicted at 5 msr solid angle, rising to 500–600 at 1 msr. In the ORNL proposed design, the limiting mass-resolving power at very small solid angle for a focal plane detector with 1% energy resolution is 1300.

# 5.  APPLICATIONS

In this section some recent applications of recoil mass spectrometers are considered. The purpose here is to be illustrative rather than exhaustive, and a number of important results are passed over in the interest of space. Many of the important applications of recoil mass spectrometers are yet to be realized; some mention is made of those areas that seem promising. As with any device that provides new access to experimental observables, the best applications of this class of spectrometer are yet to be devised.

## 5.1  *Nuclei Far From Stability*

In the first generation of such experiments, gamma-ray spectra were detected in an array of germanium detectors located at the target in coincidence with mass-identified reaction products at the focal plane. The experimental delay between gamma ray and particle detection caused by the 10-meter flight path through the spectrometer can be as large as three microseconds, although a time of less than one microsecond is more typical. The recoil mass spectrometer is highly isochronous with respect to $\theta$ and $\phi$ but, to maintain reasonable coincidence-resolving times, the focal plane detector must supply the energy of each detected ion to permit a correction for time of flight through the spectrometer. For a detector with 1% energy resolution, the limiting coincidence-resolving time is less than 5 ns as a result of flight time and solid angle effects in typical cases. Frequently, the contribution from the intrinsic time resolution of the detectors themselves is of comparable magnitude.

In one of the first series of experiments of this type, several reactions were used to search for $^{73}$Kr and the $N = Z$ $^{72}$Kr. Many previously unobserved gamma transitions were identified in coincidence with virtually every mass. In most cases these were simply new high-spin transitions or previously unobserved branches. The mass-72 and mass-73 spectra were dominated by known transitions from strongly produced isobars, but several candidate transitions could be identified as potential low-lying lines in $^{72}$Kr and $^{73}$Kr. In these experiments conducted on the Rochester spectrometer in an effort led by the Vanderbilt University group, it was originally hoped that mass-identified candidate transitions could then be located in conventional $\gamma$-$\gamma$ coincidence experiments. This turned out not to be the case, however, even when neutron coincidence filtering was included. The two problems to be addressed then are the unambiguous association of candidate gamma transitions to specific nuclei and the construction of the nuclear level scheme.

The first of these two problems is easily solved by including charged-particle and neutron multiplicity filtering of mass-gated gamma-ray spectra. Figure 9 is a schematic view of such an experimental setup at the target position. An annular, five-segment, neutron multiplicity detector array occupies a large solid angle in the forward hemisphere. The backward hemisphere contains an array of intrinsic germanium gamma detectors, and charged-particle detectors fill as much remaining space as possible 90° to the beam. The neutron detector, which is the most effective filter for products close to $N = Z$, is placed at forward angles to take advantage of the considerable kinematic focusing of neutrons resulting from the high product recoil velocity. The gamma detectors, on the other hand, are placed as close to 180° as possible to minimize the doppler spread due to the large gamma detector solid angle and high product recoil velocities. The large Doppler shift that this placement entails presents no significant difficulty since it is easily incorporated in the gamma calibration, and the individual reaction product energies are measured at the focal plane.

Figure 10 shows the results of one experiment using a setup similar to that shown in Figure 9. This figure shows the low-energy region of mass-gated gamma spectra produced in the $^{54}$Fe($^{28}$Si, $xn\,yp$) reaction at 130 MeV. The top panel, labeled Single, is actually the gamma spectrum

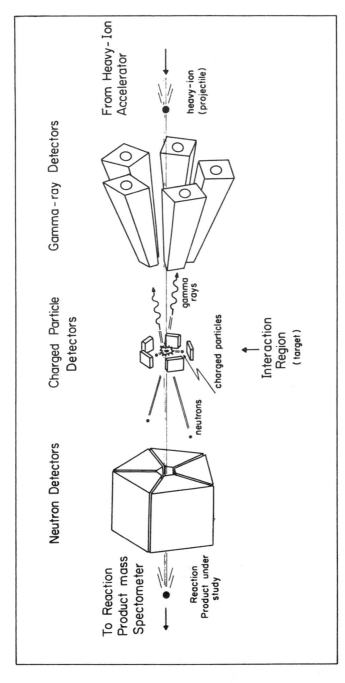

*Figure 9*  Schematic arrangement at the target in an experiment to study gamma spectroscopy of nuclei near $N = Z$. Gamma-ray coincidences with mass-identified reaction products are tagged by the associated neutron, proton, and alpha particle multiplicity on an event-by-event basis.

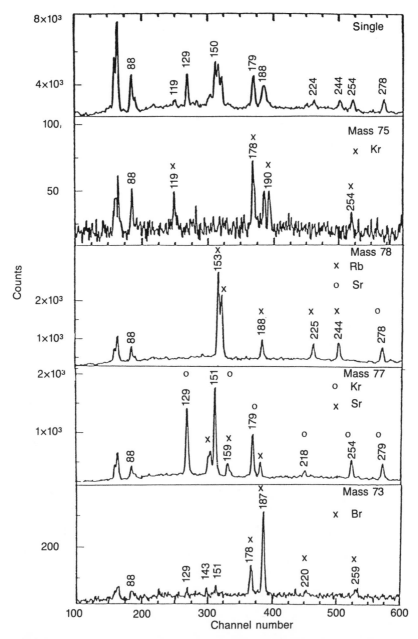

*Figure 10* Low-energy portion of gamma-ray spectra recorded with an apparatus similar to that of Figure 9 on the Rochester Recoil Mass Spectrometer. The top panel labeled Single contains those gamma events in coincidence with any focal plane event. Subsequent panels are labeled by mass, and individual gamma transitions are identified using associated neutron and charged-particle multiplicities. Note the complexity of the region between 178 and 190 keV in the single spectrum and its decomposition in the mass- and atomic-number-sorted spectra.

in coincidence with anything that hits the spectrometer focal plane. This spectrum is already vastly simpler than a true singles spectrum would be. Subsequent panels of Figure 10 are labeled by the coincident mass; individual gamma transitions are associated with specific elements using charged-particle and neutron multiplicities. A number of new high-spin states were observed in several nuclei and direct corroboration of the identification of $^{77}$Sr was obtained. The bottom panel shows the first in-beam spectrum for $^{73}$Br (19), which is of particular interest in view of a predicted (20) change from prolate to oblate ground-state shapes in this region. No significant evidence for $^{73}$Kr was found with this reaction, which suggests that the best estimates of production cross sections for these very neutron-deficient products are in error. In an alternative approach using a cold compound nucleus, we have tentatively identified some transitions in $^{72}$Kr produced in the $^{58}$Ni($^{16}$O, 2n) reaction (A. V. Rammaya, private communication).

The experiments described above are triple coincidence measurements. The construction of nuclear level schemes for products near $N = Z$ will require $\gamma$-$\gamma$ coincidence measurements as well. It is already clear that these measurements must be made in coincidence with recoil mass spectrometers. With larger arrays of germanium detectors, such experiments are not expected to pose much additional difficulty—particularly as higher solid angle spectrometers come into operation.

An elegant recent development in this area has been the direct determination of reaction product atomic numbers at the focal plane of the Daresbury spectrometer. Using the inverted reaction $^{54}$Fe($^{74}$Se, $xn\,y$p) at 300 MeV, one can detect reaction products with kinetic energies near 1.4 MeV/u at the focal plane. Products in this energy region are amenable[2] to atomic number determination using differential energy loss measurements in a high-resolution ion chamber. Figure 11 shows gamma spectra in coincidence with mass and atomic number (23). This first study of $^{124}$Ce has revealed a surprisingly rapid increase in deformation with decreasing neutron number.

Clearly the technique developed on the Daresbury spectrometer promises to provide the most direct approach in this field. The ability to transport the relatively energetic products produced in inverted reactions through the spectrometer is an essential feature. This is not entirely trivial, however, in spectrometers that utilize electrostatic deflectors. The very high operating voltages combined with the large physical size of these devices result in very high stored energy with significant potential for

---

[2] Below about 1 MeV/u, differential energy loss measurements are not useful for atomic number determination in the $Z$ range of interest in nuclear spectroscopy.

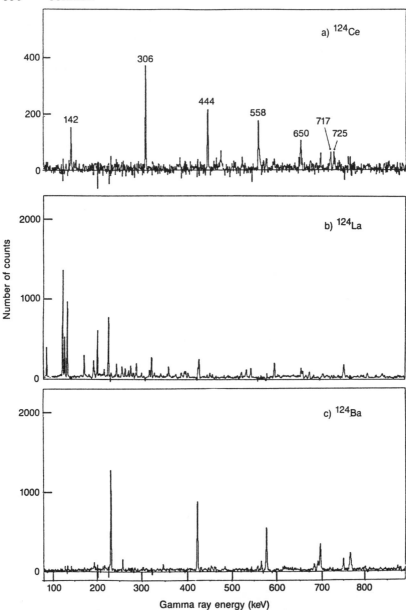

*Figure 11*  Gamma-ray spectra in prompt coincidence with mass-124 reaction products from the Daresbury Recoil Mass Spectrometer. Events associated with Ba ($Z = 56$), La ($Z = 57$), and Ce ($Z = 58$) have been isolated using differential energy loss measurements at the focal plane. This identification is facilitated by the use of inverse reaction kinematics, which produces relatively energetic reactions products.

damage during electrical breakdown. Minimizing the stored energy in an electrostatic deflector has certain ion-optical implications, but these are beyond the scope of this review.

## 5.2    Nuclei at High Spin and Excitation Energy

Large Compton-suppressed germanium detector arrays with high-resolution multiplicity filters are well suited for exploring very high spins in nuclei, and impressive progress has been made recently (2). The ultimate spin that is visible in such experiments is limited by angular momentum fractionation in the decay of the compound nucleus and the need to study exit channels that contain a significant fraction of the flux. Thus in the study of a particular $xn$ evaporation product, a compromise is struck between the need to maximize the input angular momentum and hence run at high beam energy and the need to maintain a significant cross section in the $xn$ channel as opposed to the $(x+1)n$ channel. This compromise can be largely circumvented if a recoil mass coincidence is included. A specific example will illustrate the point.

Nuclei with $N = 82$–$86$, for example $^{154}Er$, are spherical in their ground states and gradually become oblate at higher spins through the alignment of successive high-$j$ nucleons along the symmetry axis (24). This pattern of building angular momentum continues to the highest spins studied along the yrast line. Continuum gamma-ray experiments, however, clearly show the onset of collectivity at higher excitation energies and spins (25). The nature of this collectivity, as well as the structure of yrast states above spin 40, is as yet unknown. In order to produce $^{154}Er$ at higher excitation energies and spins than are possible using the dominant $xn$ channel, the reaction $^{92}Zr(^{64}Ni, 2n)^{154}Er$ is proposed (26) at a high beam energy, where the 3n and 4n channels dominate with cross sections on the order of 150 mb each and the 2n channel cross section has fallen to 10 mb. With a reasonable beam intensity of 10 pnA, an array of eight Compton-suppressed gamma detectors, and a recoil mass spectrometer solid angle of 5 msr, the data rate of $^{154}Er$-$\gamma$-$\gamma$ triple coincidences will be 350/s for a total data set of $> 10^7$ events in a $^{154}Er$-$\gamma$-$\gamma$ matrix in a four-day run. The fact that this study of $^{154}Er$ would write a factor of 30 less magnetic tape than a similar study without the recoil mass spectrometer is itself noteworthy.

Recoil mass–gamma coincidences also have a large impact in studies of the gamma-ray continuum. The most demanding of such experiments are those that examine the high-temperature regime, where the presence of isovector giant modes dominate the gamma spectrum. Figure 12 shows the results of one such experiment (27). Here the continuum gamma-ray spectrum following $^{35}Cl + ^{54}Fe$ fusion is studied out to $E_\gamma = 20$ MeV using a large NaI detector in coincidence with the Rochester spectrometer. The

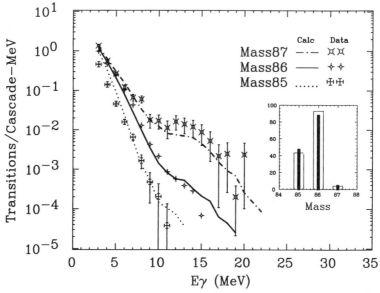

*Figure 12*  Gamma-ray continuum spectra for mass-identified reaction products in the range of $E_\gamma = 3$ to 20 MeV. The ordinate gives the absolute probability that a gamma ray of the given energy is produced per reaction. The lines through the data are the result of a statistical model calculation including 80% of the energy-weighted dipole sum rule of each excited state. The preponderance of giant dipole transitions in nuclei close to the compound nucleus is observed. The inset compares the experimental (*solid bars*) and theoretical (*open bars*) mass spectra calculated in the same model.

observed mass dependence of the gamma spectrum clearly shows the influence of the giant dipole resonance near 15 MeV and supports the theoretical prediction that these highest energy gamma rays are emitted directly from the compound nucleus. The lines through the data are statistical model calculations incorporating 80% of the energy-weighted dipole sum rule and give an excellent account of the data. The mass-87 product, which is two particles removed from the compound nucleus and accounts for 2% of the fusion cross section, is easily observed here in coincidence with gamma decays, which occur at a probability approaching $10^{-4}$ per MeV.

## 5.3    *Subbarrier Reactions*

There is long-standing interest in the fusion of heavy nuclei at subbarrier energies (28). More recently there is revived interest in near Coulomb transfer reactions between massive heavy ions (29). In both cases the interest stems from the clear involvement of internal nuclear degrees of freedom of the reacting partners even at large separation distances. These dynamics result in the participation of high entrance channel $L$-waves in the reaction at energies well below the classical $L = 0$ barrier. Very direct evidence for this is seen in Figure 13. Here we show a mass spectrum from the $^{58}$Ni + $^{64}$Ni fusion reaction at a center-of-mass energy of 90 MeV, some 10 MeV below the S-wave barrier, a point where the total fusion cross section is approximately 100 $\mu$b. In this experiment a 25 × 25 cm NaI sum spectrometer surrounded the target and subtended 90% of $4\pi$. Figure 13 shows gamma-ray sum energy spectra coincident with the dominant product masses of 119 and 120. The large difference in the mean gamma-ray sum energy of these two reaction channels results from angular momentum fractionation. In a fusion reaction where $L = 0$ dominates, the total energy removed by gamma rays is independent of the number of evaporated particles and is simply related to the lowest particle binding energy. The results in Figure 13, on the other hand are consistent with the participation of partial waves through $L = 4\hbar$.

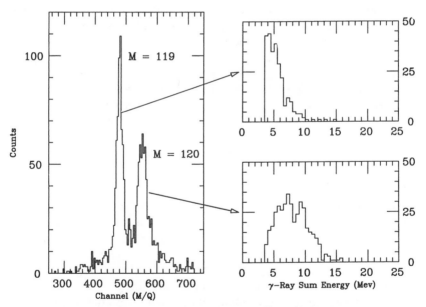

*Figure 13*    Mass spectrum from the subbarrier fusion of $^{58}$Ni + $^{64}$Ni and associated coincident gamma-ray sum energy spectra. The sum spectrometer threshold was 3.5 MeV. Evidence for angular momentum fractionation is apparent in the large difference in mean gamma-ray sum energy for products of mass 119 and 120, which implies the participation of partial waves through $L = 4$ in the fusion of these nuclei some 10 MeV below the nominal S-wave barrier.

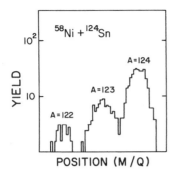

*Figure 14*   Mass spectrum at 180° center-of-mass scattering angle for the subbarrier nucleon transfer reactions of $^{58}$Ni + $^{124}$Sn. Reaction products were detected at a laboratory angle of 0° with the Daresbury spectrometer.

The coupling of entrance channel to quasi-elastic nucleon transfer channels is one mechanism contributing to enhanced subbarrier fusion of heavy nuclei. The standard techniques for studying reactions of this type near the interaction barrier involve the detection and identification of heavy reaction products at scattering angles that approach 180° at the barrier. In interesting cases, say with projectiles of mass 50 or greater, the kinematics of the reaction and particle identification difficulties make standard techniques impossible. An alternative approach is to observe the recoiling target-like nucleus that is emitted near zero degrees at the barrier. The problem is thus similar to the detection of heavy-ion fusion products. The first experiment of this type (30) has recently been performed on the Daresbury spectrometer. A sample mass spectrum from the $^{58}$Ni + $^{124}$Sn reaction at 213 MeV is shown in Figure 14. Elastic scattering and the one- and two-nucleon transfer channels are observed.

## 5.4   Nuclear Astrophysics

The $^{12}$C$(\alpha, \gamma)^{16}$O reaction is an important rate-determining step in the nucleosynthesis of elements from carbon to iron. At the present time, extension of existing measurements to cross sections $<0.1$ nb are needed to resolve large variations in theoretical extrapolations down to astrophysically interesting energies (31). Limitations of the existing experiments result mainly from the irreducible background in gamma-ray detection. In a California Institute of Technology experiment (31), a novel approach is being taken to enhance the detection sensitivity. A $^4$He gas target is employed with a high-intensity $^{12}$C beam resulting in energetic $^{16}$O products emitted close to zero degrees. Detecting the $^{16}$O recoil in coincidence with the gamma ray provides a unique signature of the reaction with a dramatic reduction in background.

## 5.5    *Hyperfine Interactions*

In all of the applications considered above, the recoil mass spectrometer is employed as a trigger condition. A broad range of applications also exist in which the spectrometer functions as a very fast isotope separator. The placement of a mass-defining aperture at the focal plane provides a mass-separated source located a nominal distance of 10 meters from the high background of the target with delay times in the range of $>300$ ns to $<3$ $\mu$s. In experiments studying the subsequent decay of isomeric states in these products, a large additional background reduction is achieved with the use of an active mass-defining aperture such as a thin parallel plate avalanche counter that signals the arrival of each new source particle. The full range of collection techniques normally used in conventional on-line isotope separators are applicable at the focal plane of a recoil mass spectrometer as well.

In the area of hyperfine interactions, a new option is available as consequence of the relatively high energy of the reaction products. An array of tilted foils placed after the mass-defining aperture produces a net nuclear polarization. Implanting these products in an appropriate crystal lattice produces a well-localized, polarized source. The first application of this technique on a recoil mass spectrometer for nuclear magnetic resonance measurement (32) is illustrated in Figure 15. Mass-selected reaction products are polarized on passing through a collection of tilted thin carbon foils and then implanted in a crystalline host located between the poles of

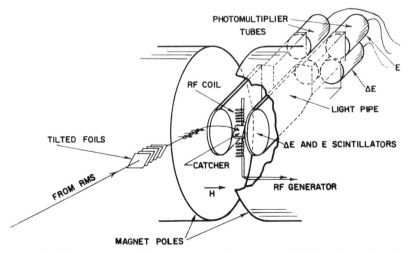

*Figure 15*   Schematic view of the nuclear magnetic resonance (NMR) device used on the focal plane of the Rochester spectrometer.

a uniform magnetic field. The polarization of the resultant source is detected via a pair of scintillation telescopes that detect the beta-decay asymmetry along the polarization direction. The array of polarizing foils is periodically rotated through 180° about the beam axis to flip the polarization direction. This allows for a complete cancellation of instrumental asymmetries. As in a conventional NMR instrument, a pair of radio-frequency coils is used to modulate the precession of the nuclear moment. On resonance the polarization of the ensemble is destroyed and the resultant loss of beta-decay asymmetry is detected. Figure 16 shows the results of a calibration experiment to check the method against the known magnetic moment of $^{12}$B. The reaction was $^2$H($^{11}$B, $^{12}$B) for which the spectrometer efficiency is close to 100%. A similar experiment shown in Figure 17 used a $^{32}$S projectile to produce the first measurement of the magnetic moment of $^{33}$Cl.

## 5.6   Atomic Charge State Distributions

Reaction products recoil from the target into vacuum in a time on the order of $5 \times 10^{-15}$ s. Such times are long enough to permit particle decay of the compound nucleus and complete stabilization of the inner atomic shells of the product ion. The gamma deexcitation of the reaction product takes considerably longer, however, and the nuclear ground state in, say, a rare earth reaction product where the total gamma-ray multiplicity may

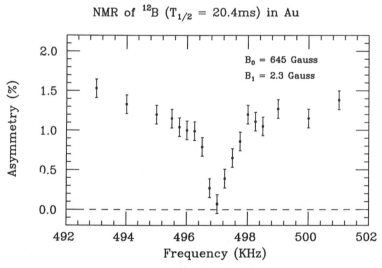

Figure 16   Nuclear magnetic resonance of $^{12}$B obtained with the apparatus shown in Figure 15. The maximum beta-decay asymmetry of 1.5% is extinguished on resonance.

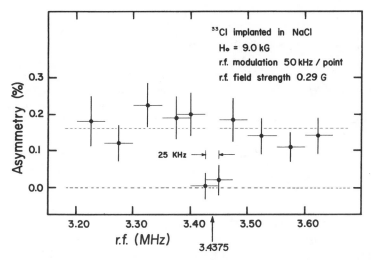

*Figure 17*  Nuclear magnetic resonance of $^{33}$Cl obtained with the apparatus shown in Figure 15. This case is considerably more difficult than that of Figure 16 as a result of the smaller nuclear spin and long beta-decay half-life. The maximum beta-decay asymmetry observed here is only 0.2%.

be 30, is not reached until the ion has recoiled several millimeters from the target. In most cases, then, the entire nuclear gamma decay takes place in vacuum. This fact has significant impact on the atomic charge state of the recoiling ion when the nuclear gamma cascade contains an internally converted transition. Consider a K-converted transition that creates an atomic K-shell vacancy, thus raising the mean atomic charge state by one unit. These inner shell vacancies decay in a time that on the average is much shorter than the time interval between gamma transitions. The decay of the inner shell vacancy may be viewed as an outward propagation of a hole with which a significant ionic excitation energy is associated. Auger processes are responsible for dissipating much of this excitation energy, and as many as six additional electrons may be emitted per initial K-shell vacancy. Thus a single converted transition in the gamma cascade can raise the ionic charge state of a reaction product by six or seven units. Since the atomic processes are somewhat faster than the nuclear gamma decay, it is observed that the shifts in atomic charge state are approximately additive for multiple converted transitions.

Shifts of this type are routinely observed. Even when the probablility of a converted transition is significantly less than one, the net effect on the charge state distribution is significant. Figure 18 shows atomic charge state distributions for $^{158}$Er in a reaction where the probability that at least one converted transition occurs in the gamma cascade is approximately 35%.

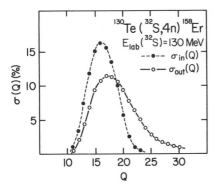

*Figure 18*  Charge state distributions with and without a second carbon foil at the entrance to the spectrometer. Anomalously high charge states due to internal conversion events in vacuum are evident in $\sigma_{out}(Q)$.

To fix the atomic charge states at time zero, a carbon backing is deposited on the downstream side of the target. To evelate those changes in the charge state distributions that occur in vacuum as a result of the gamma cascade, a second carbon foil is mounted several centimeters downstream from the target (see Figure 2) at a location encountered well after the gamma cascade is concluded. Figure 18 compares the observed charge states with, $\sigma_{in}(Q)$, and without, $\sigma_{out}(Q)$, the second carbon foil. The second carbon foil thus reestablishes an equilibrium atomic charge state distribution and a comparison of $\sigma_{in}(Q)$ with $\sigma_{out}(Q)$ provides a measure of the integral probability that a converted transition occurred in the cascade.

As a function of angular momentum, selected using either gamma-ray multiplicity or sum energy, the atomic charge state distributions of evaporation residues thus provide a sensitive window on changes in the average gamma-ray multipolarity in the high-spin continuum. By varying the distance between the target and the second carbon foil, the mean lifetime of the converting transitions can be investigated. A number of important developments remain to be made in this area.

## 6.   CONCLUSIONS

By the ion-optical standards of modern spectrometers in nuclear physics, the recoil mass spectrometer is a rather inelegant device. Satisfactory correction of the dominant second-order aberrations have expanded the useful mass range and solid angle of the new spectrometers but not without real loss of mass-resolving power. The new spectrometers will involve compromises between resolution and efficiency. Despite their ion-optical inelegance, existing recoil mass spectrometers have opened important new approaches in low-energy nuclear physics, and the new spectrometers promise similar developments. We have considered a very limited sample of the potential applications of these instruments. Many new applications will follow the development of the new, higher efficiency spectrometers.

*Literature Cited*

1. Twin, P. J., et al. *Daresbury-Liverpool-NBI preprint* (1986)
2. Garrett, J. D., Hagemann, G. B., Herskind, B. *Ann. Rev. Nucl. Part. Sci.* 36: 419–73 (1986)
3. Lister, C. J., et al. *Phys. Rev. Lett.* 49: 308–11 (1982)
4. Enge, H., et al. *Nucl. Instrum. Methods* 97: 449 (1971)
5. Cormier, T. M., Stwertka, P. M. *Nucl. Instrum. Methods* 184: 423–30 (1981)
6. Cormier, T. M., et al. *Nucl. Instrum. Methods* 212: 185–93 (1983)
7. Filipone, B. W. In *Accelerated Radioactive Beams Workshop*, ed. L. Buchman, J. M. d'Auria, TRI-85-1. TRIUMF (1985)
8. Harwood, L. H., Nolen, J. A. Jr. *Nucl. Instrum. Methods* 186: 435 (1981)
9. Price, H. G. *Daresbury Lab. Rep.* DL/NUC.R19 (1979)
10. Spolaore, P., et al. *Nucl. Instrum. Methods* A238: 381–92 (1985)
11. Aston, F. W. *Philos. Mag.* 38: 707 (1919)
12. Enge, H. In *Focusing of Charged Particles*, Volume II, ed. R. Septier, 4.2: 203–64. New York/London: Academic. 471 pp. (1967)
13. Deleted in proof
14. Wollnik, H. See Ref. 12, 4.1: 163–202
15. Kowalski, S., Enge, H. *MIT Lab. for Nucl. Sci. Rep.* (1985)
16. Deleted in proof
17. Deleted in proof
18. Shapira, D., et al. *Nucl. Instrum. Methods* 129: 123 (1975)
19. Wen, S., et al. *J. Phys. G* 11: L173–78 (1985)
20. Leander, G. *Nucl. Phys.* In press (1987)
21. Deleted in proof
22. Deleted in proof
23. Ying, K. L., et al. *J. Phys. G* In press (1987)
24. Sletten, G., et al. *Phys. Lett.* 135B: 33 (1984)
25. Radford, D., et al. *Phys. Rev. Lett.* 55: 1727 (1985)
26. A Proposal for a Fragment Mass Analyzer at ATLAS. Argonne Natl. Lab. Rep. (unpublished)
27. Herman, M. PhD Thesis. Univ. Rochester (1986)
28. Steadman, S. G., Rhoades-Brown, M. J. *Ann. Rev. Nucl. Part. Sci.* 36: 649–81 (1986)
29. Cline, D. *Ann. Rev. Nucl. Part. Sci.* 36: 683–716 (1986)
30. Betts, R., et al. Proc. ANL Symp. on Heavy Ion Fusion, March 1986 (unpublished)
31. Filipone, B. W. *Ann. Rev. Nucl. Part. Sci.* 36. 717–43 (1986)
32. Rogers, W. PhD Thesis. Univ. Rochester (1986)

# CUMULATIVE INDEXES

## CONTRIBUTING AUTHORS, VOLUMES 28–37

# 568    CONTRIBUTING AUTHORS

# CHAPTER TITLES, VOLUMES 28–37

# Annual Reviews Inc.

## A NONPROFIT SCIENTIFIC PUBLISHER

4139 El Camino Way
P.O. Box 10139
Palo Alto, CA 94303-0897 • USA

Annual Reviews Inc. publications may be ordered directly from our office by mail or use our Toll Free Telephone line (for orders paid by credit card or purchase order, and customer service calls only); through booksellers and subscription agents, worldwide; and through participating professional societies. Prices subject to change without notice. ARI Federal I.D. #94-1156476

- **Individuals:** Prepayment required on new accounts by check or money order (in U.S. dollars, check drawn on U.S. bank) or charge to credit card — American Express, VISA, MasterCard.
- **Institutional buyers:** Please include purchase order number.
- **Students:** $10.00 discount from retail price, per volume. Prepayment required. Proof of student status must be provided (photocopy of student I.D. or signature of department secretary is acceptable). Students must send orders direct to Annual Reviews. Orders received through bookstores and institutions requesting student rates will be returned. You may order at the Student Rate for a maximum of 3 years.
- **Professional Society Members:** Members of professional societies that have a contractual arrangement with Annual Reviews may order books through their society at a reduced rate. Check with your society for information.
- **Toll Free Telephone orders:** Call 1-800-523-8635 (except from California) for orders paid by credit card or purchase order and customer service calls only. California customers and all other business calls use 415-493-4400 (not toll free). Hours: 8:00 AM to 4:00 PM, Monday-Friday, Pacific Time.

**Regular orders:** Please list the volumes you wish to order by volume number.
**Standing orders:** New volume in the series will be sent to you automatically each year upon publication. Cancellation may be made at any time. Please indicate volume number to begin standing order.
**Prepublication orders:** Volumes not yet published will be shipped in month and year indicated.
**California orders:** Add applicable sales tax.
**Postage paid** (4th class bookrate/surface mail) by **Annual Reviews Inc.** Airmail postage or UPS, extra.

| ANNUAL REVIEWS SERIES | | Prices Postpaid per volume USA & Canada/elsewhere | Regular Order Please send: | Standing Order Begin with: |
|---|---|---|---|---|
| | | | Vol. number | Vol. number |
| **Annual Review of ANTHROPOLOGY** | | | | |
| Vols. 1-14 | (1972-1985) | $27.00/$30.00 | | |
| Vols. 15-16 | (1986-1987) | $31.00/$34.00 | | |
| Vol. 17 | (avail. Oct. 1988) | $35.00/$39.00 | Vol(s). _____ | Vol. _____ |
| **Annual Review of ASTRONOMY AND ASTROPHYSICS** | | | | |
| Vols. 1-2, 4-20 | (1963-1964; 1966-1982) | $27.00/$30.00 | | |
| Vols. 21-25 | (1983-1987) | $44.00/$47.00 | | |
| Vol. 26 | (avail. Sept. 1988) | $47.00/$51.00 | Vol(s). _____ | Vol. _____ |
| **Annual Review of BIOCHEMISTRY** | | | | |
| Vols. 30-34, 36-54 | (1961-1965; 1967-1985) | $29.00/$32.00 | | |
| Vols. 55-56 | (1986-1987) | $33.00/$36.00 | | |
| Vol. 57 | (avail. July 1988) | $35.00/$39.00 | Vol(s). _____ | Vol. _____ |
| **Annual Review of BIOPHYSICS AND BIOPHYSICAL CHEMISTRY** | | | | |
| Vols. 1-11 | (1972-1982) | $27.00/$30.00 | | |
| Vols. 12-16 | (1983-1987) | $47.00/$50.00 | | |
| Vol. 17 | (avail. June 1988) | $49.00/$53.00 | Vol(s). _____ | Vol. _____ |
| **Annual Review of CELL BIOLOGY** | | | | |
| Vol. 1 | (1985) | $27.00/$30.00 | | |
| Vols. 2-3 | (1986-1987) | $31.00/$34.00 | | |
| Vol. 4 | (avail. Nov. 1988) | $35.00/$39.00 | Vol(s). _____ | Vol. _____ |

| ANNUAL REVIEWS SERIES | Prices Postpaid per volume USA & Canada/elsewhere | Regular Order Please send: | Standing Order Begin with: |
|---|---|---|---|
| | | Vol. number | Vol. number |

### Annual Review of COMPUTER SCIENCE
| | | | |
|---|---|---|---|
| Vols. 1-2 | (1986-1987)..................$39.00/$42.00 | | |
| Vol. 3 | (avail. Nov. 1988)..............$45.00/$49.00 | Vol(s). _____ | Vol. _____ |

### Annual Review of EARTH AND PLANETARY SCIENCES
| | | | |
|---|---|---|---|
| Vols. 1-10 | (1973-1982)..................$27.00/$30.00 | | |
| Vols. 11-15 | (1983-1987)..................$44.00/$47.00 | | |
| Vol. 16 | (avail. May 1988)..............$49.00/$53.00 | Vol(s). _____ | Vol. _____ |

### Annual Review of ECOLOGY AND SYSTEMATICS
| | | | |
|---|---|---|---|
| Vols. 2-16 | (1971-1985)..................$27.00/$30.00 | | |
| Vols. 17-18 | (1986-1987)..................$31.00/$34.00 | | |
| Vol. 19 | (avail. Nov. 1988)..............$34.00/$38.00 | Vol(s). _____ | Vol. _____ |

### Annual Review of ENERGY
| | | | |
|---|---|---|---|
| Vols. 1-7 | (1976-1982)..................$27.00/$30.00 | | |
| Vols. 8-12 | (1983-1987)..................$56.00/$59.00 | | |
| Vol. 13 | (avail. Oct. 1988)..............$58.00/$62.00 | Vol(s). _____ | Vol. _____ |

### Annual Review of ENTOMOLOGY
| | | | |
|---|---|---|---|
| Vols. 10-16, 18-30 | (1965-1971; 1973-1985)........$27.00/$30.00 | | |
| Vols. 31-32 | (1986-1987)..................$31.00/$34.00 | | |
| Vol. 33 | (avail. Jan. 1988)..............$34.00/$38.00 | Vol(s). _____ | Vol. _____ |

### Annual Review of FLUID MECHANICS
| | | | |
|---|---|---|---|
| Vols. 1-4, 7-17 | (1969-1972, 1975-1985)........$28.00/$31.00 | | |
| Vols. 18-19 | (1986-1987)..................$32.00/$35.00 | | |
| Vol. 20 | (avail. Jan. 1988)..............$34.00/$38.00 | Vol(s). _____ | Vol. _____ |

### Annual Review of GENETICS
| | | | |
|---|---|---|---|
| Vols. 1-19 | (1967-1985)..................$27.00/$30.00 | | |
| Vols. 20-21 | (1986-1987)..................$31.00/$34.00 | | |
| Vol. 22 | (avail. Dec. 1988)..............$34.00/$38.00 | Vol(s). _____ | Vol. _____ |

### Annual Review of IMMUNOLOGY
| | | | |
|---|---|---|---|
| Vols. 1-3 | (1983-1985)..................$27.00/$30.00 | | |
| Vols. 4-5 | (1986-1987)..................$31.00/$34.00 | | |
| Vol. 6 | (avail. April 1988)..............$34.00/$38.00 | Vol(s). _____ | Vol. _____ |

### Annual Review of MATERIALS SCIENCE
| | | | |
|---|---|---|---|
| Vols. 1, 3-12 | (1971, 1973-1982)............$27.00/$30.00 | | |
| Vols. 13-17 | (1983-1987)..................$64.00/$67.00 | | |
| Vol. 18 | (avail. August 1988)...........$66.00/$70.00 | Vol(s). _____ | Vol. _____ |

### Annual Review of MEDICINE
| | | | |
|---|---|---|---|
| Vols. 1-3, 6, 8-9 11-15, 17-36 | (1950-1952, 1955, 1957-1958) (1960-1964, 1966-1985)........$27.00/$30.00 | | |
| Vols. 37-38 | (1986-1987)..................$31.00/$34.00 | | |
| Vol. 39 | (avail. April 1988)..............$34.00/$38.00 | Vol(s). _____ | Vol. _____ |

### Annual Review of MICROBIOLOGY
| | | | |
|---|---|---|---|
| Vols. 18-39 | (1964-1985)..................$27.00/$30.00 | | |
| Vols. 40-41 | (1986-1987)..................$31.00/$34.00 | | |
| Vol. 42 | (avail. Oct. 1988)..............$34.00/$38.00 | Vol(s). _____ | Vol. _____ |